# OUT OF CONTROL

Kevin Kelly, executive editor of *Wired*, lives in Pacifica, California. Formerly publisher and editor of *Whole Earth Review*, he has been instrumental in helping launch a number of cultural innovations: The Hackers' Conference; Cyberthon, the first virtual-reality jamboree; and the WELL, model way-station on the information superhighway. He is a feedback freak. His E-mail coordinates are: kk@well.com. For slo-mo communication, use 149 Amapola Avenue, Pacifica, California, 94044.

# OUT OF CONTROL

## THE NEW BIOLOGY OF MACHINES

KEVIN KELLY

FOURTH ESTATE • *London*

First published in the United States in 1994 by
Addison Wesley Inc.

First published in Great Britain in 1994 by
Fourth Estate Limited
6 Salem Road
London W2 2BU

First Paperback edition 1995
10 9 8 7 6 5 4 3 2



A catalogue record for this book is available from the British Library.

ISBN 1–85702–308–0



Printed in Great Britain by
Cox & Wyman Ltd, Reading, Berks

# CONTENTS

1

# THE MADE AND THE BORN

I AM SEALED in a cottage of glass that is completely airtight. Inside I breathe my exhalations. Yet the air is fresh, blown by fans. My urine and excrement are recycled by a system of ducts, pipes, wires, plants, and marsh-microbes, and redeemed into water and food which I can eat. Tasty food. Good water.

Last night it snowed outside. Inside this experimental capsule it is warm, humid, and cozy. This morning the thick interior windows drip with heavy condensation. Plants crowd my space. I am surrounded by large banana leaves—huge splashes of heartwarming yellow-green color—and stringy vines of green beans entwining every vertical surface. About half the plants in this hut are food plants, and from these I harvested my dinner.

I am in a test module for living in space. My atmosphere is fully recycled by the plants and the soil they are rooted in, and by the labyrinth of noisy ductwork and pipes strung through the foliage. Neither the green plants alone nor the heavy machines alone are sufficient to keep me alive. Rather it is the *union* of sun-fed life and oil-fed machinery that keeps me going. Within this shed the living and the manufactured have been unified into one robust system, whose purpose is to nurture further complexities—at the moment, me.

What is clearly happening inside this glass capsule is happening less clearly at a great scale on Earth in the closing years of this millennium. The realm of the *born*—all that is nature—and the realm of the *made*—all that is humanly constructed—are becoming one. Machines are becoming biological and the biological is becoming engineered.

That's banking on some ancient metaphors. Images of a machine as organism and an organism as machine are as old as the first machine itself. But now those enduring metaphors are no longer poetry. They are becoming real—profitably real.

This book is about the marriage of the born and the made. By extracting the logical principle of both life and machines, and applying each to the task of building extremely complex systems, technicians are conjuring up contraptions that are at once both made and alive. This marriage between life and machines is one of convenience, because, in part, it has been forced by our current technical limitations. For the world of our own making has become so complicated that we must turn to the world of the born to understand how to manage it. That is, the more mechanical we make our fabricated environment, the more biological it will eventually have to be if it is to work at all. Our future is technological; but it will not be a world of gray steel. Rather our technological future is headed toward a neo-biological civilization.

NATURE HAS ALL ALONG YIELDED her flesh to humans. First, we took nature's materials as food, fibers, and shelter. Then we learned to extract raw materials from her biosphere to create our own new synthetic materials. Now Bios is yielding us her mind—we are taking her logic.

Clockwork logic—the logic of the machines—will only build simple contraptions. Truly complex systems such as a cell, a meadow, an economy, or a brain (natural or artificial) require a rigorous nontechnological logic. We now see that no logic except *bio*-logic can assemble a thinking device, or even a workable system of any magnitude.

It is an astounding discovery that one can extract the logic of

Bios out of biology and have something useful. Although many philosophers in the past have suspected one could abstract the laws of life and apply them elsewhere, it wasn't until the complexity of computers and human-made systems became as complicated as living things, that it was possible to prove this. It's eerie how much of life *can* be transferred. So far, some of the traits of the living that have successfully been transported to mechanical systems are: self-replication, self-governance, limited self-repair, mild evolution, and partial learning. We have reason to believe yet more can be synthesized and made into something new.

Yet at the same time that the logic of Bios is being imported into machines, the logic of Technos is being imported into life.

The root of bioengineering is the desire to control the organic long enough to improve it. Domesticated plants and animals are examples of technos-logic applied to life. The wild aromatic root of the Queen Anne's lace weed has been fine-tuned over generations by selective herb gatherers until it has evolved into a sweet carrot of the garden; the udders of wild bovines have been selectively enlarged in an "unnatural" way to satisfy humans rather than calves. Milk cows and carrots, therefore, are human inventions as much as steam engines and gunpowder are. But milk cows and carrots are more indicative of the kind of inventions humans will make in the future: products that are grown rather than manufactured.

Genetic engineering is precisely what cattle breeders do when they select better strains of Holsteins, only bioengineers employ more precise and powerful control. While carrot and milk cow breeders had to rely on diffuse organic evolution, modern genetic engineers can use directed artificial evolution—purposeful design—which greatly accelerates improvements.

The overlap of the mechanical and the lifelike increases year by year. Part of this bionic convergence is a matter of words. The meanings of "mechanical" and "life" are both stretching until all complicated things can be perceived as machines, and all self-sustaining machines can be perceived as alive. Yet beyond semantics, two concrete trends are happening: (1) Human-made things are behaving more lifelike, and (2) Life is becoming more engineered. The apparent veil between the organic and the manufactured has crumpled to reveal that the two really are, and have always been, of one being. What should we call that common soul between the organic communities we know of as organisms and ecologies, and their manufactured counterparts of robots, corporations, economies, and computer circuits? I call those examples, both made and born, "vivisystems" for the lifelikeness each kind of system holds.

In the following chapters I survey this unified bionic frontier. Many of the vivisystems I report on are "artificial"—artifices of human making—but in almost every case they are also real—experimentally implemented rather than mere theory. The artificial vivisystems I survey are all complex and grand: planetary telephone systems, computer virus incubators, robot prototypes, virtual reality worlds, synthetic animated characters, diverse artificial ecologies, and computer models of the whole Earth.

But the wildness of nature is the chief source for clarifying insights into vivisystems, and probably the paramount source of more insights to come. I report on new experimental work in ecosystem assembly, restoration biology, coral reef replicas, social insects (bees and ants), and complex closed systems such as the Biosphere 2 project in Arizona, from wherein I write this prologue.

The vivisystems I examine in this book are nearly bottomless complications, vast in range, and gigantic in nuance. From these particular big systems I have appropriated unifying principles for all large vivisystems; I call them the laws of god, and they are the fundamentals shared by all self-sustaining, self-improving systems.

As we look at human efforts to create complex mechanical things, again and again we return to nature for directions. Nature is thus more than a diverse gene bank harboring undiscovered herbal cures for future diseases—although it is certainly this. Nature is also a "meme bank," an idea factory. Vital, postindustrial paradigms are hidden in every jungly ant hill. The billion-footed beast of living bugs and weeds, and the aboriginal human cultures which have extracted meaning from this life, are worth protecting, if for no other reason than for the postmodern metaphors they still have not revealed. Destroying a prairie destroys not only a reservoir of genes but also a treasure of future metaphors, insight, and models for a neo-biological civilization.

THE WHOLESALE TRANSFER of bio-logic into machines should fill us with awe. When the union of the born and the made is complete, our fabrications will learn, adapt, heal themselves, and evolve. This is a power we have hardly dreamt of yet. The aggregate capacity of millions of biological machines may someday match our own skill of innovation. Ours may always be a flashy type of creativity, but there is something to be said for a slow, wide creativity of many dim parts working ceaselessly.

Yet as we unleash living forces into our created machines, we lose control of them. They acquire wildness and some of the surprises that the wild entails. This, then, is the dilemma all gods must accept: that they can no longer be completely sovereign over their finest creations.

The world of the made will soon be like the world of the born: autonomous, adaptable, and creative but, consequently, out of our control. I think that's a great bargain.

# 2

# Hive Mind

The beehive beneath my office window quietly exhales legions of busybodies and then inhales them. On summer afternoons, when the sun seeps under the trees to backlight the hive, the approaching sunlit bees zoom into their tiny dark opening like curving tracer bullets. I watch them now as they haul in the last gleanings of nectar from the final manzanita blooms of the year. Soon the rains will come and the bees will hide. I will still gaze out the window as I write; they will still toil, but now in their dark home. Only on the balmiest day will I be blessed by the sight of their thousands in the sun.

Over years of beekeeping, I've tried my hand at relocating bee colonies out of buildings and trees as a quick and cheap way of starting new hives at home. One fall I gutted a bee tree that a neighbor felled. I took a chain saw and ripped into this toppled old tupelo. The poor tree was cancerous with bee comb. The further I cut into the belly of the tree, the more bees I found. The insects filled a cavity as large as I was. It was a gray, cool autumn day and all the bees were home, now agitated by the surgery. I finally plunged my hand into the mess of comb. Hot! Ninety-five degrees at least. Overcrowded with 100,000 cold-blooded bees, the hive had become a warm-blooded organism. The heated honey ran like thin, warm blood. My gut felt like I had reached my hand into a dying animal.

The idea of the collective hive as an animal was an idea late in coming. The Greeks and Romans were famous beekeepers who harvested respectable yields of honey from homemade hives, yet these ancients got almost every fact about bees wrong. Blame it on the lightless conspiracy of bee life, a secret guarded by ten thousand fanatically loyal, armed soldiers. Democritus thought bees spawned from the same source as maggots. Xenophon figured out the queen bee but erroneously assigned her supervisory responsibilities she doesn't have. Aristotle gets good marks for getting a lot right, including the semiaccurate observation that "ruler bees" put larva in the honeycomb cells. (They actually start out as eggs, but at least he corrects Democritus's misguided direction of maggot origins.) Not until the Renaissance was the female gender of the queen bee proved, or beeswax shown to be secreted from the undersides of bees. No one had a clue until modern genetics that a hive is a radical matriarchy and sisterhood: all bees, except the few good-for-nothing drones, are female and sisters. The hive was a mystery as unfathomable as an eclipse.

I've seen eclipses and I've seen bee swarms. Eclipses are spectacles I watch halfheartedly, mostly out of duty, I think, to their rarity and tradition, much as I might attend a Fourth of July parade. Bee swarms, on the other hand, evoke another sort of awe. I've seen more than a few hives throwing off a swarm, and never has one failed to transfix me utterly, or to dumbfound everyone else within sight of it.

A hive about to swarm is a hive possessed. It becomes visibly agitated around the mouth of its entrance. The colony whines in a centerless loud drone that vibrates the neighborhood. It begins to spit out masses of bees, as if it were emptying not only its guts but its soul. A poltergeist-like storm of tiny wills materializes over the hive box. It grows to be a small dark cloud of purpose, opaque with life. Boosted by a tremendous buzzing racket, the ghost slowly rises into the sky, leaving behind the empty box and quiet bafflement. The German theosophist Rudolf Steiner writes lucidly in his otherwise kooky *Nine Lectures on Bees*: "Just as the human soul takes leave of the body . . . one can truly see in the flying swarm an image of the departing human soul."

For many years Mark Thompson, a beekeeper local to my area, had the bizarre urge to build a Live-In Hive—an active bee home you could visit by inserting your head into it. He was working in a yard once when a beehive spewed a swarm of bees "like a flow of black lava, dissolving, then taking wing." The black cloud coalesced into a 20-foot-round black halo of 30,000 bees that hovered, UFO-like, six feet off the ground, exactly at eye level. The flickering insect halo began to drift slowly away, keeping a constant six feet above the earth. It was a Live-In Hive dream come true.

Mark didn't waver. Dropping his tools he slipped into the swarm, his bare head now in the eye of a bee hurricane. He trotted in sync across the yard as the swarm eased away. Wearing a bee halo, Mark hopped over one fence, then another. He was now running to keep up with the thundering animal in whose belly his head floated. They all crossed the road and hurried down an open field, and then he jumped another fence. He was tiring. The bees weren't; they picked up speed. The swarm-bearing man glided down a hill into a marsh. The two of them now resembled a superstitious swamp devil, humming, hovering, and plowing through the miasma. Mark churned wildly through the muck trying to keep up. Then, on some signal, the bees accelerated. They unhaloed Mark and left him standing there wet, "in panting, joyful amazement." Maintaining an eye-level altitude, the swarm floated across the landscape until it vanished, like a spirit unleashed, into a somber pine woods across the highway.

"Where is 'this spirit of the hive' . . . where does it reside?" asks the author Maurice Maeterlinck as early as 1901. "What is it that governs here, that issues orders, foresees the future . . . ?" We are certain now it is not the queen bee. When a swarm pours itself out through the front slot of the hive, the queen bee can only follow. The queen's daughters manage the election of where and when the swarm should settle. A half-dozen anonymous workers scout ahead to check possible hive locations in hollow trees or wall cavities. They report back to the resting swarm by dancing on its contracting surface. During the report, the more theatrically a scout dances, the better the site she is championing. Deputy bees then check out the competing sites according to the intensity of the dances, and will concur with the scout by joining in the scout's twirling. That induces more followers to check out the lead prospects and join the ruckus when they return by leaping into the performance of their choice.

It's a rare bee, except for the scouts, who has inspected more than one site. The bees see a message, "Go there, it's a nice place." They go and return to dance/say, "Yeah, it's *really* nice." By compounding emphasis, the favorite sites get more visitors, thus increasing further visitors. As per the law of increasing returns, them that has get more votes, the have-nots get less. Gradually, one large, snowballing finale will dominate the dance-off. The biggest crowd wins.

It's an election hall of idiots, for idiots, and by idiots, and it works marvelously. This is the true nature of democracy and of all distributed governance. At the close of the curtain, by the choice of the citizens, the swarm takes the queen and thunders off in the direction indicated by mob vote. The queen who follows, does so humbly. If she could think, she would remember that she is but a mere peasant girl, blood sister of the very nurse bee instructed (by whom?) to select her larva, an ordinary larva, and raise it on a diet of royal jelly, transforming Cinderella into the queen. By what karma is the larva for a princess chosen? And who chooses the chooser?

"The hive chooses," is the disarming answer of William Morton Wheeler, a natural philosopher and entomologist of the old school, who founded the field of social insects. Writing in a bombshell of an essay in 1911 ("The Ant Colony as an Organism" in the *Journal of Morphology*), Wheeler claimed that an insect colony was not merely the analog of an organism, it is indeed an organism, in every important and scientific sense of the word. He wrote: "Like a cell or the person, it behaves as a unitary whole, maintaining its identity in space, resisting dissolution . . . neither a thing nor a concept, but a continual flux or process."

It was a mob of 20,000 united into oneness.

IN A DARKENED Las Vegas conference room, a cheering audience waves cardboard wands in the air. Each wand is red on one side, green on the other. Far in back of the huge auditorium, a camera scans the frantic attendees. The video camera links the color spots of the wands to a nest of computers set up by graphics wizard Loren Carpenter. Carpenter's custom software locates each red and each green wand in the auditorium. Tonight there are just shy of 5,000 wand-wavers. The computer displays the precise location of each wand (and its color) onto an immense, detailed video map of the auditorium hung on the front stage, which all can see. More importantly, the computer counts the total red or green wands and uses that value to control software. As the audience wave the wands, the display screen shows a sea of lights dancing crazily in the dark, like a candlelight parade gone punk. The viewers see themselves on the map; they are either a red or green pixel. By flipping their own wands, they can change the color of their projected pixels instantly.

Loren Carpenter boots up the ancient video game of Pong onto the immense screen. Pong was the first commercial video game to reach pop consciousness. It's a minimalist arrangement: a white dot bounces inside a square; two movable rectangles on each side act as virtual paddles. In short, electronic ping-pong. In this version, displaying the red side of your wand moves the paddle up. Green moves it down. More precisely, the Pong paddle moves as the average number of red wands in the auditorium increases or decreases. Your wand is just one vote.

Carpenter doesn't need to explain very much. Every attendee at this 1991 conference of computer graphic experts was probably once hooked on Pong. His amplified voice booms in the hall, "Okay guys. Folks on the left side of the auditorium control the left paddle. Folks on the right side control the right paddle. If you think you are on the left, then you really are. Okay? Go!"

The audience roars in delight. Without a moment's hesitation, 5,000 people are playing a reasonably good game of Pong. Each move of the paddle is the average of several thousand players' intentions. The sensation is unnerving. The paddle usually does what you intend, but not always. When it doesn't, you find yourself spending as much attention trying to anticipate the paddle as the incoming ball. One is definitely aware of another intelligence online: it's this hollering mob.

The group mind plays Pong so well that Carpenter decides to up the ante. Without warning the ball bounces faster. The participants squeal in unison. In a second or two, the mob has adjusted to the quicker pace and is playing better than before. Carpenter speeds up the game further; the mob learns instantly.

"Let's try something else," Carpenter suggests. A map of seats in the auditorium appears on the screen. He draws a wide circle in white around the center. "Can you make a green '5' in the circle?" he asks the audience. The audience stares at the rows of red pixels. The game is similar to that of holding a placard up in a stadium to make a picture, but now there are no preset orders, just a virtual mirror. Almost immediately wiggles of green pixels appear and grow haphazardly, as those who think their seat is in the path of the "5" flip their wands to green. A vague figure is materializing. The audience collectively begins to discern a "5" in the noise. Once discerned, the "5" quickly precipitates out into stark clarity. The wand-wavers on the fuzzy edge of the figure decide what side they "should" be on, and the emerging "5" sharpens up. The number assembles itself.

"Now make a four!" the voice booms. Within moments a "4" emerges. "Three." And in a blink a "3" appears. Then in rapid succession, "Two . . . One . . . Zero." The emergent thing is on a roll.

Loren Carpenter launches an airplane flight simulator on the screen. His instructions are terse: "You guys on the left are controlling roll; you on the right, pitch. If you point the plane at anything interesting, I'll fire a rocket at it." The plane is airborne. The pilot is . . . 5,000 novices. For once the auditorium is completely silent. Everyone studies the navigation instruments as the scene outside the windshield sinks in. The plane is headed for a landing in a pink valley among pink hills. The runway looks very tiny.

There is something both delicious and ludicrous about the notion of having the passengers of a plane collectively fly it. The brute democratic sense of it all is very appealing. As a passenger you get to vote for everything; not only where the group is headed, but when to trim the flaps.

But group mind seems to be a liability in the decisive moments of touchdown, where there is no room for averages. As the 5,000 conference participants begin to take down their plane for landing, the hush in the hall is ended by abrupt shouts and urgent commands. The auditorium becomes a gigantic cockpit in crisis. "Green, green, green!" one faction shouts. "More red!" a moment later from the crowd. "Red, red! REEEEED!" The plane is pitching to the left in a sickening way. It is obvious that it will miss the landing strip and arrive wing first. Unlike Pong, the flight simulator entails long delays in feedback from lever to effect, from the moment you tap the aileron to the moment it banks. The latent signals confuse the group mind. It is caught in oscillations of overcompensation. The plane is lurching wildly. Yet the mob somehow aborts the landing and pulls the plane up sensibly. They turn the plane around to try again.

How did they turn around? Nobody decided whether to turn left or right, or even to turn at all. Nobody was in charge. But as if of one mind, the plane banks and turns wide. It tries landing again. Again it approaches cockeyed. The mob decides in unison, without lateral communication, like a flock of birds taking off, to pull up once more. On the way up the plane rolls a bit. And then rolls a bit more. At some magical moment, the same strong thought simultaneously infects five thousand minds: "I wonder if we can do a 360?"

Without speaking a word, the collective keeps tilting the plane. There's no undoing it. As the horizon spins dizzily, 5,000 amateur pilots roll a jet on their first solo flight. It was actually quite graceful. They give themselves a standing ovation.

The conferees did what birds do: they flocked. But they flocked self-consciously. They responded to an overview of themselves as they co-formed a "5" or steered the jet. A bird on the fly, however, has no overarching concept of the shape of its flock. "Flockness" emerges from creatures completely oblivious of their collective shape, size, or alignment. A flocking bird is blind to the grace and cohesiveness of a flock in flight.

At dawn, on a weedy Michigan lake, ten thousand mallards

fidget. In the soft pink glow of morning, the ducks jabber, shake out their wings, and dunk for breakfast. Ducks are spread everywhere. Suddenly, cued by some imperceptible signal, a thousand birds rise as one thing. They lift themselves into the air in a great thunder. As they take off they pull up a thousand more birds from the surface of the lake with them, as if they were all but part of a reclining giant now rising. The monstrous beast hovers in the air, swerves to the east sun, and then, in a blink, reverses direction, turning itself inside out. A second later, the entire swarm veers west and away, as if steered by a single mind. In the 17th century, an anonymous poet wrote: "... and the thousands of fishes moved as a huge beast, piercing the water. They appeared united, inexorably bound to a common fate. How comes this unity?"

A flock is not a big bird. Writes the science reporter James Gleick, "Nothing in the motion of an individual bird or fish, no matter how fluid, can prepare us for the sight of a skyful of starlings pivoting over a cornfield, or a million minnows snapping into a tight, polarized array ... High-speed film [of flocks turning to avoid predators] reveals that the turning motion travels through the flock as a wave, passing from bird to bird in the space of about one-seventieth of a second. That is far less than the bird's reaction time." The flock is more than the sum of the birds.

In the film *Batman Returns* a horde of large black bats swarmed through flooded tunnels into downtown Gotham. The bats were computer generated. A single bat was created and given leeway to automatically flap its wings. The one bat was copied by the dozens until the animators had a mob. Then each bat was instructed to move about on its own on the screen following only a few simple rules encoded into an algorithm: don't bump into another bat, keep up with your neighbors, and don't stray too far away. When the algorithmic bats were run, they flocked like real bats.

The flocking rules were discovered by Craig Reynolds, a computer scientist working at Symbolics, a graphics hardware manufacturer. By tuning the various forces in his simple equation—a little more cohesion, a little less lag time—Reynolds could shape the flock to behave like living bats, sparrows, or fish. Even the marching mob of penguins in *Batman Returns* were flocked by Reynolds's algorithms. Like the bats, the computer-modeled 3-D penguins were cloned en masse and then set loose into the scene aimed in a certain direction. Their crowdlike jostling as they marched down the snowy street simply emerged, out of anyone's control.

So realistic is the flocking of Reynolds's simple algorithms that biologists have gone back to their hi-speed films and concluded that the flocking behavior of real birds and fish must emerge from a similar set of simple rules. A flock was once thought to be a decisive sign of life, some noble formation only life could achieve. Via Reynolds's algorithm it is now seen as an adaptive trick suitable for any distributed vivisystem, organic or made.

WHEELER, the ant pioneer, started calling the bustling cooperation of an insect colony a "superorganism" to clearly distinguish it from the metaphorical use of "organism." He was influenced by a philosophical strain at the turn of the century that saw holistic patterns overlaying the individual behavior of smaller parts. The enterprise of science was on its first steps of a headlong rush into the minute details of physics, biology, and all natural sciences. This pell-mell to reduce wholes to their constituents, seen as the most pragmatic path to understanding the wholes, would continue for the rest of the century and is still the dominant mode of scientific inquiry. Wheeler and colleagues were an essential part of this reductionist perspective, as the 50 Wheeler monographs on specific esoteric ant behaviors testify. But at the same time, Wheeler saw "emergent properties" within the superorganism superseding the resident properties of the collective ants. Wheeler said the superorganism of the hive "emerges" from the mass of ordinary insect organisms. And he meant emergence as science—a technical, rational explanation—not mysticism or alchemy.

Wheeler held that this view of emergence was a way to

reconcile the reduce-it-to-its parts approach with the see-it-as-a-whole approach. The duality of body/mind or whole/part simply evaporated when holistic behavior lawfully emerged from the limited behaviors of the parts. The specifics of how superstuff emerged from baser parts was very vague in everyone's mind. And still is.

What was clear to Wheeler's group was that emergence was a common natural phenomena. It was related to the ordinary kind of causation in everyday life, the kind where *A* causes *B* which causes *C*, or 2 + 2 = 4. Ordinary causality was invoked by chemists to cover the observation that sulfur atoms plus iron atoms equal iron sulfide molecules. According to fellow philosopher C. Lloyd Morgan, the concept of emergence signaled a different variety of causation. Here 2 + 2 does not equal 4; it does not even surprise with 5. In the logic of emergence, 2 + 2 = apples. "The emergent step, though it may seem more or less saltatory [a leap], is best regarded as a qualitative change of direction, or critical turning-point, in the course of events," writes Morgan in *Emergent Evolution*, a bold book in 1923. Morgan goes on to quote a verse of Browning poetry which confirms how music emerges from chords:

> And I know not if, save in this, such gift be allowed to man
> That out of three sounds he frame, not a fourth sound, but a star.

We would argue now that it is the complexity of our brains that extracts music from notes, since we presume oak trees can't hear Bach. Yet "Bachness"—all that invades us when we hear Bach—is an appropriately poetic image of how a meaningful pattern emerges from musical notes and generic information.

The organization of a tiny honeybee yields a pattern for its tinier one-tenth of a gram of wing cells, tissue, and chitin. The organism of a hive yields integration for its community of worker bees, drones, pollen and brood. The whole 50-pound hive organ emerges with its own identity from the tiny bee parts. The hive possesses much that none of its parts possesses. One speck of a honeybee brain operates with a memory of six days; the hive as a whole operates with a memory of three months, twice as long as the average bee lives.

Ants, too, have hive mind. A colony of ants on the move from one nest site to another exhibits the Kafkaesque underside of emergent control. As hordes of ants break camp and head west, hauling eggs, larva, pupae—the crown jewels—in their beaks, other ants of the same colony, patriotic workers, are hauling the trove east again just as fast, while still other workers, perhaps acknowledging conflicting messages, are running in one direction and back again completely empty-handed. A typical day at the office. Yet, the ant colony moves. Without any visible decision making at a higher level, it chooses a new nest site, signals workers to begin building, and governs itself.

The marvel of "hive mind" is that no one is in control, and yet an invisible hand governs, a hand that emerges from very dumb members. The marvel is that more is different. To generate a colony organism from a bug organism requires only that the bugs be multiplied so that there are many, many more of them, and that they communicate with each other. At some stage the level of complexity reaches a point where new categories like "colony" can emerge from simple categories of "bug." Colony is inherent in bugness, implies this marvel. Thus, there is nothing to be found in a beehive that is not submerged in a bee. And yet you can search a bee forever with cyclotron and fluoroscope, and you will never find the hive.

This is a universal law of vivisystems: higher-level complexities cannot be inferred by lower-level existences. Nothing—no computer or mind, no means of mathematics, physics, or philosophy—can unravel the emergent pattern dissolved in the parts without actually playing it out. Only playing out a hive will tell you if a colony is immixed in a bee. The theorists put it this way: running a system is the quickest, shortest, and only sure method to discern emergent structures latent in it. There are no shortcuts to actually "expressing" a convoluted, nonlinear equation to discover what it does. Too much of its behavior is packed away.

That leads us to wonder what else is packed into the bee that we haven't seen yet? Or what else is packed into the hive that has not yet appeared because there haven't been enough honeybee hives in a row all at once? And for that matter, what is contained in a human that will not emerge until we are all interconnected by wires and politics? The most unexpected things will brew in this bionic hivelike supermind.

THE MOST INEXPLICABLE things will brew in *any* mind.

Because the body is plainly a collection of specialist organs—heart for pumping, kidneys for cleaning—no one was too surprised to discover that the mind delegates cognitive matters to different regions of the brain.

In the late 1800s, physicians noted correlations in recently deceased patients between damaged areas of the brain and obvious impairments in their mental abilities just before death. The connection was more than academic: might insanity be biological in origin? At the West Riding Lunatic Asylum, London, in 1873, a young physician who suspected so surgically removed small portions of the brain from two living monkeys. In one, his incision caused paralysis of the right limbs; in the other he caused deafness. But in all other respects, both monkeys were normal. The message was clear: the brain must be compartmentalized. One part could fail without sinking the whole vessel.

If the brain was in departments, in what section were recollections stored? In what way did the complex mind divvy up its chores? In a most unexpected way.

In 1888, a man who spoke fluently and whose memory was sharp found himself in the offices of one Dr. Landolt, frightened because he could no longer name any letters of the alphabet. The perplexed man could write flawlessly when dictated a message. However, he could not reread what he had written nor find a mistake if he had made one. Dr. Landolt recorded, "Asked to read an eye chart, [he] is unable to name any letter. However he claims to see them perfectly . . . He compares the *A* to an easel, the *Z* to a serpent, and the *P* to a buckle."

The man's word-blindness degenerated to a complete aphasia of both speech and writing by the time of his death four years later. Of course, in the autopsy, there were two lesions: an old one near the occipital (visual) lobe and a newer one probably near the speech center.

Here was remarkable evidence of the bureaucratization of the brain. In a metaphorical sense, different functions of the brain take place in different rooms. This room handles letters, if spoken; that room, letters, if read. To speak a letter (outgoing), you need to apply to yet another room. Numbers are handled by a different department altogether, in the next building. And if you want curses, as the Monty Python Flying Circus skit reminds us, you'll need to go down the hall.

An early investigator of the brain, John Hughlings-Jackson, recounts a story about a woman patient of his who lived completely without speech. When some debris, which had been dumped across the street from the ward where she lived, ignited into flames, the patient uttered the first and only word Hughlings-Jackson had ever heard her say: "Fire!"

How can it be, he asked somewhat incredulous, that "fire" is the only word her word department remembers? Does the brain have its own "fire" department, so to speak?

As investigators probed the brain further, the riddle of the mind revealed itself to be deeply specific. The literature on memory features people ordinary in their ability to distinguish concrete nouns—tell them "elbow" and they will point to their elbow—but extraordinary in their inability to distinguish abstract nouns—ask them about "liberty" or "aptitude" and they stare blankly and shrug. Contrarily, the minds of other apparently normal individuals have lost the ability to retain concrete nouns, while perfectly able to identify abstract things. In his wonderful and overlooked book *The Invention of Memory*, Israel Rosenfield writes:

> One patient, when asked to define *hay*, responded, "I've forgotten"; and when asked to define *poster*, said, "no idea." Yet given the word *supplication*, he said, "making a serious request for help," and *pact* drew "friendly agreement."

Memory is a palace, say the ancient philosophers, where every room parks a thought. Yet with every clinical discovery of yet another form of specialized forgetfulness, the rooms of memory exploded in number. Down this road there is no end. Memory, already divided into a castle of chambers, balkanizes into a terrifying labyrinth of tiny closets.

One study pointed to four patients who could discern inanimate objects (umbrella, towel), but garbled living things, including foods! One of these patients could converse about nonliving objects without suspicion, but a spider to him was defined as "a person looking for things, he was a spider for a nation." There are records of aphasias that interfere with the use of the past tense. I've heard of another report (one that I cannot confirm, but one that I don't doubt) of an ailment that allows a person to discern all foods except vegetables.

The absurd capriciousness underlying such a memory system is best represented by the categorization scheme of an ancient Chinese encyclopedia entitled *Celestial Emporium of Benevolent Knowledge*, as interpreted by the South American fiction master J. L. Borges.

> On those remote pages it is written that animals are divided into (a) those that belong to the Emperor, (b) embalmed ones, (c) those that are trained, (d) suckling pigs, (e) mermaids, (f) fabulous ones, (g) stray dogs, (h) those that are included in this classification, (i) those that tremble as if they were mad, (j) innumerable ones, (k) those drawn with a very fine camel's hair brush, (l) others, (m) those that have just broken a flower vase, (n) those that resemble flies from a distance.

As farfetched as the *Celestial Emporium* system is, any classification process has its logical problems. Unless there is a different location for every memory to be filed in, there will need to be confusing overlaps, say for instance, of a talking naughty pig, that may be filed under three different categories above. Filing the thought under all three slots would be highly inefficient, although possible.

The system by which knowledge is sequestered in our brain became more than just an academic question as computer scientists tried to build an artificial intelligence. What is the architecture of memory in a hive mind?

In the past most researchers leaned toward the method humans intuitively use for their own manufactured memory stashes: a single location for each archived item, with multiple cross-referencing, such as in libraries. The strong case for a single location in the brain for each memory was capped by a series of famously elegant experiments made by Wilder Penfield, a Canadian neurosurgeon working in the 1930s. In daring open-brain surgery, Penfield probed the living cerebellum of conscious patients with an electrical stimulant, and asked them to report what they experienced. Patients reported remarkably vivid memories. The smallest shift of the stimulant would generate distinctly separate thoughts. Penfield mapped the brain location of each memory while he scanned the surface with his probe.

His first surprise was that these recollections appeared repeatable, in what years later would be taken as a model of a tape recorder—as in: "hit replay." Penfield uses the term "flash-back" in his account of a 26-year-old woman's postepileptic hallucination: "She had the same flash-back several times. These had to do with her cousin's house or the trip there—a trip she has not made for ten to fifteen years but used to make often as a child."

The result of Penfield's explorations into the unexplored living brain produced the tenacious image of the hemispheres as fabulous recording devices, ones that seemed to rival the fantastic recall of the newly popular phonograph. Each of our memories was delicately etched into its own plate, cataloged and filed faithfully by the temperate brain, and barring violence, could be retrieved like a jukebox song by pushing the right buttons.

Yet, a close scrutiny of Penfield's raw transcripts of his probing experiments shows memory to be a less mechanical process. As one example, here are some of the responses of a 29-year-old woman to Penfield's pricks in her left temporal lobe: "Something coming to me from somewhere. A dream." Four minutes later, in exactly the same spot: "The scenery seemed to be different from the one just before . . ." In a nearby spot: "Wait a minute, something flashed over me, something I dreamt." In a third spot, further inside the brain: "I keep having dreams." The stimulation is repeated in the same spot: "I keep seeing things—I keep dreaming of things."

These scripts tell of dreamlike glimpses, rather than disorienting reruns dredged up from the basement cubbyholes of the mind's archives. The owners of these experiences recognize them as fragmentary semimemories. They ramble with that awkward "assembled" flavor that dreams grow by—unfocused tales of bits and pieces of the past reworked into a collage of a dream. The emotional charge of a *déjà vu* was absent. No overwhelming sense of "it was exactly like this was then" pushed against the present. The replays should have fooled nobody.

Human memories do crash. They crash in peculiar ways, by forgetting vegetables on a list of things to buy at the grocery or by forgetting vegetables in general. Memories often bruise in tandem with a physical bruise of the brain, so we must expect that some memory is bound in time and space to some degree, since being bound to time and space is one definition of being real.

But the current view of cognitive science leans more toward a new image: memories are like emergent events summed out of many discrete, unmemory-like fragments stored in the brain. These pieces of half-thoughts have no fixed home; they abide throughout the brain. Their manner of storage differs substantially from thought to thought—learning to shuffle cards is organized differently than learning the capital of Bolivia—and the manner differs subtly from person to person, and equally subtly from time to time.

There are more possible ideas/experiences than there are ways to combine neurons in the brain. Memory, then, must organize itself in some way to accommodate more possible thoughts than it has room to store. It cannot have a shelf for every thought of the past, nor a place reserved for every potential thought of the future.

I remember a night in Taiwan twenty years ago. I was in the back of an open truck on a dirt road in the mountains. I had my jacket on; the hill air was cold. I was hitching a ride to arrive at a mountain peak by dawn. The truck was grinding up the steep, dark road while I looked up to the stars in the clear alpine air. It was so clear that I could see tiny stars near the horizon. Suddenly a meteor zipped across low, and because of my angle in the mountains, I could see it skip across the atmosphere. Skip, skip, skip, like a stone.

As I just now remembered this, the skipping meteor was not a memory tape I replayed, despite its ready vividness. The skipping meteor image doesn't exist anywhere in particular in my mind. When I resurrected my experience, I assembled it anew. And I assemble it anew each time I remember it. The parts are tiny bits of evidence scattered sparsely through the hive of my brain: a record of cold shivering, of a bumpy ride somewhere, of many sightings of stars, of hitchhiking. The records are even finer grained than that: cold, bump, points of light, waiting. They are the same raw impressions our minds receive from our senses and with which it assembles our perceptions of the present.

Our consciousness creates the present, just as it creates the past, from many distributed clues scattered in our mind. Standing before an object in a museum, my mind associates its parallel straight lines with the notion of a "chair," even though the thing has only three legs. My mind has never before seen such a chair, but it compiles all the associations—upright, level seat, stable, legs—and creates the visual image. Very fast. In fact, I will be aware of the general "chairness" of the chair before I can perceive its unique details.

Our memories (and our hive minds) are created in the same indistinct, haphazard way. To find the skipping meteor, my consciousness grabbed a thread with streaks of light and gathered a bunch of feelings associated with stars, cold, bumps. What I created depended on what else I had thrown into my mind recently, including what other thing I was doing/feeling last time I tried to assemble the skipping meteor memory. That's why the story is slightly different each time I remember it, because each time it is, in a real sense, a completely different experience. The act of perceiving and the act of remembering are the same. Both assemble an emergent whole from many distributed pieces.

"Memory," says cognitive scientist Douglas Hofstadter, "is highly reconstructive. Retrieval from memory involves selecting out of a vast field of things what's important and what is not important, emphasizing the important stuff, downplaying the unimportant." That selection process is perception. "I am a very big believer," Hofstadter told me, "that the core processes of cognition are very, very tightly related to perception."

In the last two decades, a few cognitive scientists have contemplated ways to create a distributed memory. Psychologist David Marr proposed a novel model of the human cerebellum in the early 1970s by which memory was stored randomly throughout a web of neurons. In 1974, Pentti Kanerva, a computer scientist, worked out the mathematics of a similar web by which long strings of data could be stored randomly in a computer memory. Kanerva's algorithm was an elegant method to store a finite number of data points in a very immense potential memory space. In other words, Kanerva showed a way to fit any perception a mind could have into a finite memory mechanism. Since there are more ideas possible in the universe than there are atoms or minutes, the actual ideas or perceptions that a human mind can ever get to are relatively sparse within the total possibilities; therefore Kanerva called his technique a "sparse distributed memory" algorithm.

In a sparse distributed network, memory is a type of perception. The act of remembering and the act of perceiving both detect a pattern in a very large choice of possible patterns. When we remember, we re-create the act of the original perception; that is, we relocate the pattern by a process similar to the one we used to perceive the pattern originally.

Kanerva's algorithm was so mathematically clean and crisp that it could be roughly implemented by a hacker into a computer one afternoon. At the NASA Ames Research Center, Kanerva and colleagues fine-tuned his scheme for a sparse distributed memory in the mid-1980s by designing a very robust practical version in a computer. Kanerva's memory algorithm could do several marvelous things that parallel what our own minds can do. The researchers primed the sparse memory with several degraded images of numerals (1 to 9) drawn on a 20-by-20 grid. The memory stored these. Then they gave the memory another image of a numeral more degraded than the first samples to see if it could "recall" what the digit was. The memory could. It homed in on the prototypical shape that was behind all the degraded images. In essence it remembered a shape it had never seen before!

The breakthrough was not just being able to find or replay something from the past, but to find something in a vast hive of possibilities when only the vaguest clues are given. It is not enough to retrieve your grandmother's face; a memory must identify it when you see her profile in a wholly different light and from a different angle.

A hive mind is a distributed memory that both perceives and remembers. It is possible that a human mind may be chiefly distributed, yet, it is in artificial minds where distributed mind will certainly prevail. The more computer scientists thought about distributing problems into a hive mind, the more reasonable it seemed. They figured that most personal computers are not in actual use most of the time they are turned on! While composing a letter on a computer you may interrupt the computer's rest with a short burst of key pounding and then let it return to idleness as you compose the next sentence. Taken as a whole, the turned-on computers in an office are idle a large percentage of the day. The managers of information systems in large corporations look at the millions of dollars of personal computer equipment sitting idle on workers' desks at night and wonder if all that computing power might not be harnessed. All they would need is a way to coordinate work and memory in a very distributed system.

But merely combating idleness is not what makes distributing computing worth doing. Distributed being and hive minds have their own rewards, such as greater immunity to disruption. At Digital Equipment Corporation's research lab in Palo Alto, California, an engineer demonstrated this advantage of distributed computation by opening the door of the closet that held the company's own computer network and dramatically yanking a cable out of its guts. The network instantly routed around the breach and didn't falter a bit.

There will still be crashes in any hive mind, of course. But because of the nonlinear nature of a network, when it does fail we can expect glitches like an aphasia that remembers all foods except vegetables. A broken networked intelligence may be able to calculate *pi* to the billionth digit but not forward e-mail to a new address. It may be able to retrieve obscure texts on, say, the classification procedures for African zebra variants, but be incapable of producing anything sensible about animals in general. Forgetting vegetables in general, then, is less likely a failure of a local memory storage place than it is a systemwide failure that has, as one of its symptoms, the failure of a particular type of vegetable association—just as two separate but conflicting programs on your computer hard disk may produce a "bug" that prevents you from printing words in *italic*. The place where the italic font is stored is not broken; but the system's process of rendering italic is broken.

Some of the hurdles that stand in the way of fabricating a distributed computer mind are being overcome by building the network of computers inside one box. This deliberately compressed distributed computing is also known as parallel computing, because the thousands of computers working inside the supercomputer are running in parallel. Parallel supercomputers don't solve the idle-computer-on-the-desk problem, nor do they aggregate widespread computing power; it's just that working in parallel is an advantage in and of itself, and worth building a million-dollar stand-alone contraption to do it.

Parallel distributed computing excels in perception, visualization, and simulation. Parallelism handles complexity better than traditional supercomputers made of one huge, incredibly fast serial computer. But in a parallel supercomputer with a sparse, distributed memory, the distinction between memory and processing fades. Memory becomes a reenactment of perception, indistinguishable from the original act of knowing. Both are a pattern that emerges from a jumble of interconnected parts.

A SINK BRIMS with water. You pull the plug. The water stirs. A vortex materializes. It blooms into a tiny whirlpool, growing as if it were alive. In a minute the whirl extends from surface to drain, animating the whole basin. An ever changing cascade of water molecules swirls through the tornado, transmuting the whirlpool's being from moment to moment. Yet the whirlpool persists, essentially unchanged, dancing on the edge of collapse. "We are not stuff that abides, but patterns that perpetuate themselves," wrote Norbert Wiener.

As the sink empties, all of its water passes through the spiral. When finally the basin of water has sunk from the bowl to the cistern pipes, where does the form of the whirlpool go? For that matter, where did it come from?

The whirlpool appears reliably whenever we pull the plug. It is an emergent thing, like a flock, whose power and structure are not contained in the power and structure of a single water molecule. No matter how intimately you know the chemical character of $H_2O$, it does not prepare you for the character of a whirlpool. Like all emergent entities, the essence of a vortex emanates from a messy collection of other entities; in this case, a pool of water molecules. One drop of water is not enough for a whirlpool to appear in, just as one pinch of sand is not enough to hatch an avalanche. Emergence requires a population of entities, a multitude, a collective, a mob, more.

More is different. One grain of sand cannot avalanche, but pile up enough grains of sand and you get a dune that can trigger avalanches. Certain physical attributes such as temperature depend on collective behavior. A single molecule floating in space does not really have a temperature. Temperature is more correctly thought of as a group characteristic that a population of molecules has. Though temperature is an emergent property, it can be measured precisely, confidently, and predictably. It is real.

It has long been appreciated by science that large numbers behave differently than small numbers. Mobs breed a requisite measure of complexity for emergent entities. The total number of possible interactions between two or more members accumulates exponentially as the number of members increases. At a high level of connectivity, and a high number of members, the dynamics of mobs takes hold. More *is* different.

THERE ARE TWO extreme ways to structure "moreness." At one extreme, you can construct a system as a long string of sequential operations, such as we do in a meandering factory assembly line. The internal logic of a clock as it measures off time by a complicated parade of movements is the archetype of a sequential system. Most mechanical systems follow the clock.

At the other far extreme, we find many systems ordered as a patchwork of parallel operations, very much as in the neural network of a brain or in a colony of ants. Action in these systems proceeds in a messy cascade of interdependent events. Instead of the discrete ticks of cause and effect that run a clock, a thousand clock springs try to simultaneously run a parallel system. Since there is no chain of command, the particular action of any single spring diffuses into the whole, making it easier for the sum of the whole to overwhelm the parts of the whole. What emerges from the collective is not a series of critical individual actions but a multitude of simultaneous actions whose collective pattern is far more important. This is the swarm model.

These two poles of the organization of moreness exist only in theory because all systems in real life are mixtures of these two extremes. Some large systems lean to the sequential model (the factory); others lean to the web model (the telephone system).

It seems that the things we find most interesting in the universe are all dwelling near the web end. We have the web of life, the tangle of the economy, the mob of societies, and the jungle of our own minds. As dynamic wholes, these all share certain characteristics: a certain liveliness, for one.

We know these parallel-operating wholes by different names. We know a swarm of bees, or a cloud of modems, or a network of brain neurons, or a food web of animals, or a collective of agents. The class of systems to which all of the above belong is variously called: networks, complex adaptive systems, swarm systems, vivisystems, or collective systems. I use all these terms in this book.

Organizationally, each of these is a collection of many (thousands) of autonomous members. "Autonomous" means that each member reacts individually according to internal rules and the state of its local environment. This is opposed to obeying orders from a center, or reacting in lock step to the overall environment.

These autonomous members are highly connected to each other, but not to a central hub. They thus form a peer network. Since there is no center of control, the management and heart of the system are said to be decentrally distributed within the system, as a hive is administered.

There are four distinct facets of distributed being that supply vivisystems their character:

- The absence of imposed centralized control
- The autonomous nature of subunits
- The high connectivity between the subunits
- The webby nonlinear causality of peers influencing peers.

The relative strengths and dominance of each factor have not yet been examined systematically.

One theme of this book is that distributed artificial vivisystems, such as parallel computing, silicon neural net chips, or the grand network of online networks commonly known as the Internet, provide people with some of the attractions of organic systems, but also, some of their drawbacks. I summarize the pros and cons of distributed systems here:

## Benefits of Swarm Systems

- *Adaptable*—It is possible to build a clockwork system that can adjust to predetermined stimuli. But constructing a system that can adjust to new stimuli, or to change beyond a narrow range, requires a swarm—a hive mind. Only a whole containing many parts can allow a whole to persist while the parts die off or change to fit the new stimuli.

- *Evolvable*—Systems that can shift the locus of adaptation over time from one part of the system to another (from the body to the genes or from one individual to a population) must be swarm based. Noncollective systems cannot evolve (in the biological sense).

• *Resilient*—Because collective systems are built upon multitudes in parallel, there is redundancy. Individuals don't count. Small failures are lost in the hubbub. Big failures are held in check by becoming merely small failures at the next highest level on a hierarchy.

• *Boundless*—Plain old linear systems can sport positive feedback loops—the screeching disordered noise of PA microphone, for example. But in swarm systems, positive feedback can lead to increasing order. By incrementally extending new structure beyond the bounds of its initial state, a swarm can build its own scaffolding to build further structure. Spontaneous order helps create more order. Life begets more life, wealth creates more wealth, information breeds more information, all bursting the original cradle. And with no bounds in sight.

• *Novelty*—Swarm systems generate novelty for three reasons: (1) They are "sensitive to initial conditions"—a scientific shorthand for saying that the size of the effect is not proportional to the size of the cause—so they can make a surprising mountain out of a molehill. (2) They hide countless novel possibilities in the exponential combinations of many interlinked individuals. (3) They don't reckon individuals, so therefore individual variation and imperfection can be allowed. In swarm systems with heritability, individual variation and imperfection will lead to perpetual novelty, or what we call evolution.

## Apparent Disadvantages of Swarm Systems

• *Nonoptimal*—Because they are redundant and have no central control, swarm systems are inefficient. Resources are allotted higgledy-piggledy, and duplication of effort is always rampant. What a waste for a frog to lay so many thousands of eggs for just a couple of juvenile offspring! Emergent controls such as prices in free-market economy—a swarm if there ever was one—tend to dampen inefficiency, but never eliminate it as a linear system can.

• *Noncontrollable*—There is no authority in charge. Guiding a swarm system can only be done as a shepherd would drive a herd: by applying force at crucial leverage points, and by subverting the natural tendencies of the system to new ends (use the sheep's fear of wolves to gather them with a dog that wants to chase sheep). An economy can't be controlled from the outside; it can only be slightly tweaked from within. A mind cannot be prevented from dreaming, it can only be plucked when it produces fruit. Wherever the word "emergent" appears, there disappears human control.

• *Nonpredictable*—The complexity of a swarm system bends it in unforeseeable ways. "The history of biology is about the unexpected," says Chris Langton, a researcher now developing mathematical swarm models. The word "emergent" has its dark side. Emergent novelty in a video game is tremendous fun; emergent novelty in our airplane traffic-control system would be a national emergency.

• *Nonunderstandable*—As far as we know, causality is like clockwork. Sequential clockwork systems we understand; nonlinear web systems are unadulterated mysteries. The latter drown in their self-made paradoxical logic. *A* causes *B*, *B* causes *A*. Swarm systems are oceans of intersecting logic: *A* indirectly causes everything else and everything else indirectly causes *A*. I call this lateral or horizontal causality. The credit for the true cause (or more precisely the true proportional mix of causes) will spread horizontally through the web until the trigger of a particular event is essentially unknowable. Stuff happens. We don't need to know exactly how a tomato cell works to be able to grow, eat, or even improve tomatoes. We don't need to know exactly how a massive computational collective system works to be able to build one, use it, and make it better. But whether we understand a system or not, we are responsible for it, so understanding would sure help.

• *Nonimmediate*—Light a fire, build up the steam, turn on a switch, and a linear system awakens. It's ready to serve you. If it stalls, restart it. Simple collective systems can be awakened simply. But complex swarm systems with rich hierarchies take time to boot up. The more complex, the longer it takes to warm up. Each hierarchical layer has to settle down; lateral causes have to slosh around and come to rest; a million autonomous agents have to acquaint themselves. I think this will be the hardest lesson for humans to learn: that organic complexity will entail organic time.

The tradeoff between the pros and cons of swarm logic is very similar to the cost/benefit decisions we would have to make about biological vivisystems, if we were ever asked to. But because we have grown up with biological systems and have had no alternatives, we have always accepted their costs without evaluation.

We can swap a slight tendency for weird glitches in a tool in exchange for supreme sustenance. In exchange for a swarm system of 17 million computer nodes on the Internet that won't go down (as a whole), we get a field that can sprout nasty computer worms, or erupt inexplicable local outages. But we gladly trade the wasteful inefficiencies of multiple routing in order to keep the Internet's remarkable flexibility. On the other hand, when we construct autonomous robots, I bet we give up some of their potential adaptability in exchange for preventing them from going off on their own beyond our full control.

As our inventions shift from the linear, predictable, causal attributes of the mechanical motor, to the crisscrossing, unpredictable, and fuzzy attributes of living systems, we need to shift our sense of what we expect from our machines. A simple rule of thumb may help:

- For jobs where supreme control is demanded, good old clockware is the way to go.
- Where supreme adaptability is required, out-of-control swarmware is what you want.

For each step we push our machines toward the collective, we move them toward life. And with each step away from the clock, our contraptions lose the cold, fast optimal efficiency of machines. Most tasks will balance some control for some adaptability, and so the apparatus that best does the job will be some cyborgian hybrid of part clock, part swarm. The more we can discover about the mathematical properties of generic swarm processing, the better our understanding will be of both artificial complexity and biological complexity.

Swarms highlight the complicated side of real things. They depart from the regular. The arithmetic of swarm computation is a continuation of Darwin's revolutionary study of the irregular populations of animals and plants undergoing irregular modification. Swarm logic tries to comprehend the out-of-kilter, to measure the erratic, and to time the unpredictable. It is an attempt, in the words of James Gleick, to map "the morphology of the amorphous"—to give a shape to that which seems to be inherently shapeless. Science has done all the easy tasks—the clean simple signals. Now all it can face is the noise; it must stare the messiness of life in the eye.

ZEN MASTERS once instructed novice disciples to approach Zen meditation with an unprejudiced "beginner's mind." The master coached students, "Undo all preconceptions." The proper awareness required to appreciate the swarm nature of complicated things might be called hive mind. The swarm master coaches, "Loosen all attachments to the sure and certain."

A contemplative swarm thought: The Atom is the icon of 20th century science.

The popular symbol of the Atom is stark: a black dot encircled by the hairline orbits of several other dots. The Atom whirls alone, the epitome of singleness. It is the metaphor for individuality: atomic. It is the irreducible seat of strength. The Atom stands for power and knowledge and certainty. It is as dependable as a circle, as regular as round.

The image of the planetary Atom is printed on toys and on baseball caps. The swirling Atom works its way into corporate logos and government seals. It appears on the back of cereal boxes, in school books, and stars in TV commercials.

The internal circles of the Atom mirror the cosmos, at once a law-abiding nucleus of energy, and at the same time the concentric heavenly spheres spinning in the galaxy. In the center is the *animus*, the It, the life force, holding all to their appropriate whirling stations. The symbolic Atoms' sure orbits and definite interstices represent the understanding of the universe made known. The Atom conveys the naked power of simplicity.

Another Zen thought: The Atom is the past. The symbol of science for the next century is the dynamical Net.

The Net icon has no center—it is a bunch of dots connected to other dots—a cobweb of arrows pouring into each other, squirming together like a nest of snakes, the restless image fading at indeterminate edges. The Net is the archetype—always the same picture—displayed to represent all circuits, all intelligence, all interdependence, all things economic and social and ecological, all communications, all democracy, all groups, all large systems. The icon is slippery, ensnaring the unwary in its paradox of no beginning, no end, no center. Or, all beginning, all end, pure center. It is related to the Knot. Buried in its apparent disorder is a winding truth. Unraveling it requires heroism.

When Darwin hunted for an image to end his book *Origin of Species*—a book that is one long argument about how species emerge from the conflicting interconnected self-interests of many individuals—he found the image of the tangled Net. He saw "birds singing on bushes, with various insects flitting about, with worms crawling through the damp earth"; the whole web forming "an entangled bank, dependent on each other in so complex a manner."

The Net is an emblem of multiples. Out of it comes swarm being—distributed being—spreading the self over the entire web so that no part can say, "I am the I." It is irredeemably social, unabashedly of many minds. It conveys the logic both of Computer and of Nature—which in turn convey a power beyond understanding.

Hidden in the Net is the mystery of the Invisible Hand—control without authority. Whereas the Atom represents clean simplicity, the Net channels the messy power of complexity.

The Net, as a banner, is harder to live with. It is the banner of noncontrol. Wherever the Net arises, there arises also a rebel to resist human control. The network symbol signifies the swamp of psyche, the tangle of life, the mob needed for individuality.

The inefficiencies of a network—all that redundancy and ricocheting vectors, things going from here to there and back just to get across the street—encompasses imperfection rather than ejecting it. A network nurtures small failures in order that large failures don't happen as often. It is its capacity to hold error rather than scuttle it that makes the distributed being fertile ground for learning, adaptation, and evolution.

The only organization capable of unprejudiced growth, or unguided learning, is a network. All other topologies limit what can happen.

A network swarm is all edges and therefore open ended any way you come at it. Indeed, the network is the least structured organization that can be said to have any structure at all. It is capable of infinite rearrangements, and of growing in any direction without altering the basic shape of the thing, which is really no outward shape at all. Craig Reynolds, the synthetic flocking inventor, points out the remarkable ability of networks to absorb the new without disruption: "There is no evidence that the complexity of natural flocks is bounded in any way. Flocks do not become 'full' or 'overloaded' as new birds join. When herring migrate toward their spawning grounds, they run in schools extending as long as 17 miles and containing millions of fish." How big a telephone network could we make? How many nodes can one even theoretically add to a network and still have it work? The question has hardly even been asked.

There *are* a variety of swarm topologies, but the only organization that holds a genuine plurality of shapes is the grand mesh. In fact, a plurality of truly divergent components can only remain coherent in a network. No other arrangement—chain, pyramid, tree, circle, hub—can contain true diversity working as a whole. This is why the network is nearly synonymous with democracy or the market.

A dynamic network is one of the few structures that incorporates the dimension of time. It honors internal change. We should expect to see networks wherever we see constant irregular change, and we do.

A distributed, decentralized network is more a process than a thing. In the logic of the Net there is a shift from nouns to verbs. Economists now reckon that commercial products are best treated as though they were services. It's not what you sell a customer, it's what you do for them. It's not what something is, it's what it is connected to, what it does. Flows become more important than resources. Behavior counts.

Network logic is counterintuitive. Say you need to lay a telephone cable that will connect a bunch of cities; let's make that three for illustration: Kansas City, San Diego, and Seattle. The total length of the lines connecting those three cities is 3,000 miles. Common sense says that if you add a fourth city to your telephone network, the total length of your cable will have to increase. But that's not how network logic works. By adding a fourth city as a hub (let's make that Salt Lake City) and running the lines from each of the three cities through Salt Lake City, we can decrease the total mileage of cable to 2,850 or 5 percent less than the original 3,000 miles. Therefore the total unraveled length of a network can be shortened by *adding* nodes to it! Yet there is a limit to this effect. Frank Hwang and Ding Zhu Du, working at Bell Laboratories in 1990, proved that the best savings a system might enjoy from introducing new points into a network would peak at about 13 percent. More *is* different.

On the other hand, in 1968 Dietrich Braess, a German operations researcher, discovered that adding routes to an already congested network will only slow it down. Now called Braess's Paradox, scientists have found many examples of how adding capacity to a crowded network reduces its overall production. In the late 1960s the city planners of Stuttgart tried to ease downtown traffic by adding a street. When they did, traffic got worse; then they blocked it off and traffic improved. In 1992, New York City closed congested 42nd Street on Earth Day, fearing the worst, but traffic actually improved that day.

Then again, in 1990, three scientists working on networks of brain neurons reported that increasing the gain—the responsivity—of individual neurons did not increase their individual signal detection performance, but it did increase the performance of the whole network to detect signals.

Nets have their own logic, one that is out-of-kilter to our expectations. And this logic will quickly mold the culture of humans living in a networked world. What we get from heavy-duty communication networks, and the networks of parallel computing, and the networks of distributed appliances and distributed being is Network Culture.

Alan Kay, a visionary who had much to do with inventing personal computers, says that the personally owned book was one of the chief shapers of the Renaissance notion of the individual, and that pervasively networked computers will be the main shaper of humans in the future. It's not just individual books we are leaving behind, either. Global opinion polling in real-time 24 hours a day, seven days a week, ubiquitous telephones, asynchronous e-mail, 500 TV channels, video on demand: all these add up to the matrix for a glorious network culture, a remarkable hivelike being.

The tiny bees in my hive are more or less unaware of their colony. By definition their collective hive mind must transcend their small bee minds. As we wire ourselves up into a hivish network, many things will emerge that we, as mere neurons in the network, don't expect, don't understand, can't control, or don't even perceive. That's the price for any emergent hive mind.

# 3

# MACHINES WITH AN ATTITUDE

WHEN MARK PAULINE offers you his hand in greeting, you get to shake his toes. Years ago Pauline blew off his fingers messing around with homemade rockets. The surgeons reconstituted a hand of sorts from his feet parts, but Pauline's lame hand still slows him down.

Pauline builds machines that chew up other machines. His devices are intricate and often huge. His smallest robot is bigger than a man; the largest is two stories high when it stretches its neck. Outfitted with piston-driven jaws and steam-shovel arms, his machines exude biological vibes.

Pauline's maimed hand often has trouble threading a bolt to keep his monsters together. To quicken repairs he installed a top-of-the-line industrial lathe outside his bedroom door and stocked his kitchen area full of welding equipment. It only takes him a minute or two to braze the broken pneumatic limbs of his iron beasts. But his own hand is a hassle. He wants to replace it with a hand from a robot.

Pauline lives in a warehouse at the far end of a San Francisco street that dead-ends under a highway overpass. His pad is flanked by a bunch of grungy galvanized iron huts decorated with signs advertising car-body repair. A junkyard just outside Pauline's warehouse is piled as high as the chainlink fence with rusty skeletons of dead machines; one hunk is a jet engine. The yard is usually eerily vacant. When the postman hops out of his jeep to deliver Pauline's mail, the guy turns off his motor and locks the jeep door.

Pauline started out as a self-described juvenile delinquent, later graduating to a young adult doing "creative vandalism." Everyone agrees that Mark Pauline's pranks are above average, even for an individualist's town like San Francisco. As a 10-year-old kid Pauline used a stolen acetylene torch to decapitate the globe of a gumball machine. As a young adult he got into the art of "repurposing" outdoor billboards: late at night he altered their lettering into political messages with creative applications of spray paint. He made news recently when his ex-girlfriend reported to the police that while she was away for a weekend he covered her car with epoxy and then feathered it, windshields and all.

The devices Pauline builds are at once the most mechanical and the most biological of machines. Take the Rotary Mouth Machine: two hoops studded with sharklike teeth madly rotate in intersecting orbits, each at an angle to the other, so that their "bite" circles round and round. The spinning jaw can chew up a two-by-four in a second. Usually it nibbles the dangling arm of another machine. Or take the Inchworm, a modified farm implement powered by an automobile engine mounted on one end that cranks around six pairs of oversized tines to inch it along. It creeps in the most inefficient yet biological way. Or, the Walk-and-Peck machine. It uses its onboard canister of pressurized carbon dioxide to pneumatically chip through the asphalt by hammering its steel head into the ground, as if it were a demented 500 pound "roadpecker." "Most of my machines are the only machines of their type on Earth. No one else in their right mind would make them because there is no practical reason for humans to make them," Pauline claims, without a hint of a smile.

A couple of times a year, Pauline stages a performance for his machines. His debut in 1979 was called "Machine Sex." During the show his eccentric machines ran into each other, consumed each other, and melded into broken heaps. A few years later he staged a spectacle called "Useless Mechanical Activity," continuing his work of liberating machines into their own world. He's put on about 40 shows since, usually in Europe where, he says, "I can't be sued." But Europe's system of national support for the arts (Pauline calls it the Art Mafia) also supports these in-your-face performances.

In 1991 Pauline staged a machine circus in downtown San Francisco. On this night, several thousand fans dressed in punk black leather convened, entirely by word of mouth, at an abandoned parking lot squeezed under a freeway overpass ramp. In the makeshift arena, under the industrial glare of spotlights, ten or so mechanical animals and autonomous iron gladiators waited to demolish each other with flames and brute force.

The scale and spirit of the iron creatures on display brought to mind one image: mechanical dinosaurs without skin. The dinos poised in the skeletal power of hydraulic hoses, chained gears, and cabled levers. Pauline called them "organic machines."

These dinosaurs were not suffocating in a museum. Pauline had borrowed and stolen their parts from other machines, their power from automobiles, and had given them a meager kind of life to perform under the beams of searchlights stinking of hot ozone. Crash, rear up, jump, collide, live!

The unseated audience that night churned in the titanium glare. Loudspeakers (chosen for their gritty static) played an endless stream of recorded industrial noise. The grating broadcasts sometimes switched to tapes of radio call-in shows and other background sounds of an electronic civilization. The screeching was upstaged by a shrieking siren; the signal to start. The machines moved.

The next hour was pandemonium. A two-foot-long drill bit tipped the end of a brontosaurus-like creature's long neck. This nightmare of a dentist's drill was tapered like a bee's stinger. It went on a rampage and mercilessly drilled another robot. *Wheeeezzz*. The sound triggered toothaches. Another mad creature, the Screw Throwbot, comically zipped around, tearing up the pavement with an enormous racket. It was a ten-foot, one-ton steel sled carried by two steel corkscrew treads, each madly spinning auger 1½ feet in diameter. It screwed across asphalt, skittering in various directions at 30 miles per hour. It was actually cute. Mounted on top was a mechanical catapult capable of hurling 50-lb. exploding firebombs. So while the Drill was stinging the Screw, the Screw was hurling explosives at a tower of pianos.

"It's barely controlled anarchy here," Pauline joked at one point to his all-volunteer crew. He calls his "company" the Survival Research Labs (SRL), a deliberately misleading corporate-sounding name. SRL likes to stage performances without official permits, without notification of the city's fire department, without insurance, and without advance publicity. They let the audience sit way too close. It looked dangerous. And it was.

A converted commercial lawn sprinkler—the kind that normally creeps across grass blessing it with life-giving water—diabolically blessed the place with a shower of flames. Its rotating arms pumped fiery orange clouds of ignited kerosene fuel over a wide circle. The acrid, half-burnt smoke, trapped by the overhead freeway structure, choked the spectators. Then the Screw accidentally tipped over its fuel can, and the Sprinkler from Hell went out of commission. So the Flamethrower lit up to take up the slack. The Flamethrower was a steerable giant blower—of the type used to air-condition a mid-town skyscraper—bolted to a Mack truck engine. The truck motor twirled the huge cage-fan and pumped diesel fuel from a 55-gallon drum into the airstream. A carbon-arc spark ignited the air/fuel mixture and spewed it into a tongue of vicious yellow flame 50 feet long. It roasted the pile of 20 pianos.

Pauline could aim the dragon with a radio-control joystick from a model airplane. He turned Flamethrower's snout toward the audience, who ducked reflexively. The heat, even from 50 feet away, slapped the skin. "You know how it is," Pauline said later. "Ecosystems without predators become unstable. Well, these spectators have no predators in their lives. So that's what these machines are, that's their role. To interject predators into civilization."

SLR's machines are quite sophisticated, and getting more so. Pauline is always busy breeding new machines so that the ecology of the circus keeps evolving. Often he upgrades old models with new appendages. He may give the Screw Machine a pair of lobsterlike pincers instead of a buzz saw, or he welds a flame-thrower to one arm of 25-foot-tall Big Totem. Sometimes he cross-fertilizes, swapping parts between two creatures. Other times he midwifes wholly new beings. At a recent show he unveiled four new pets: a portable lightning machine that spits 9-foot bolts of crackling blue lightning at nearby machines; a 120-decibel whistle driven by a jet engine; a military rail gun that uses magnetic propulsion to fire a burning comet of molten iron at 200 miles per hour, which upon impact explodes into a fine drizzle of burning droplets; and an advanced tele-presence cannon, a human/machine symbiont that lets a goggled operator aim the gun by turning his head to gaze at the target. It fires beer cans stuffed with concrete and dynamite detonators.

The shows are "art," and so are constantly underfunded; the admission barely pays for the sundry costs of a show—for fuel, food for the workers, spare parts. Pauline candidly admits that some of the ancestors he cannibalized to procreate these monsters were stolen. One SRL crew member says that they like to put shows on in Europe because there is a lot of "Obtainium" there. What's Obtainium?: "Something that is easily obtained, easily liberated, or gotten for free." That which isn't made out of Obtainium is built from military surplus parts that Pauline buys by the truckload for $65 per pound from friendly downsizing military bases. He also scrounges the military for machine tools, submarine parts, fancy motors, rare electronics, $100,000-spare parts, and raw steel. "Ten years ago this stuff was valuable, important for national security and all that. Then suddenly it became worthless junk. Now I'm converting machines, improving them really, from things which once did 'useful' destruction into things that can now do *useless* destruction."

Several years ago, Pauline made a crablike robot that would scurry across the floor. It was piloted by a freaked-out guinea pig locked inside a tiny switch-laden cockpit. The robot was not intended to be cruel. Rather the idea was to explore the convergence of the organic and the machine. SRL inventions commonly marry hi-speed heavy metal and soft biological architecture. When turned on, the guinea pig robot teetered on the edge of chaos. In the controlled anarchy of the show, it was hardly noticed. Pauline: "These machines barely have enough control to be useful, but that's all the control that we need."

At the ground-breaking ceremony for the new San Francisco Museum of Modern Art, Pauline was invited to gather his machines on the empty downtown lot in order to "create a hallucination in broad daylight for a few minutes." His Shockwave Cannon wheeled about and exploded raw air. You could actually see the shockwave zip out of the muzzle. The Cannon halted rush-hour traffic as it rattled the windows of every car and skyscraper for blocks around. Pauline then introduced his Swarmers. These were waist-high cylindrical mobile robots that skittered around in a flock. Where the flock would go was anyone's guess; no one Swarmer directed the others; no one steered it. It was hardware heaven: machines out of control.

The ultimate aim of SRL is to make machines autonomous. "Getting some autonomous action, though, is really difficult," Pauline told me. Yet he is ahead of many heavily funded university labs in attempting to transfer control from humans to machines. His several-hundred-dollar swarming creatures—decked out with recycled infrared sensors and junked stepped motors—beat out the MIT robot lab in an informal race to construct the first autonomous swarming robots.

In the conflict many people see between nature-born and machine-made, Mark Pauline is on the side of the made. Pauline: "Machines have something to say to us. When I start designing an SRL show, I ask myself, what do these machines want to do? You know, I see this old backhoe that some red-neck is running everyday, maybe digging ditches out in the sun for the phone company. That backhoe is bored. It's ailing and dirty. We're coming along and asking it what *it* wants to do. Maybe it wants to be in our show. We go around and rescue machines that have been abandoned, or even dismembered. So we have to ask ourselves, what do these machines really want to do, what do they want to wear? So we think about color coordination and lighting. Our shows are not for humans, they are for machines. We don't ask how machines are going to entertain us. We ask, how can we entertain them? That's what our shows are, entertainment for machines."

Machines are something that need entertainment. They have their own complexity and their own agenda. By building more complex machines we are giving them their own autonomous behavior and thus inevitably their own purpose. "These machines are totally at ease in the world we have built for them," Pauline told me. "They act completely *natural*."

I asked Pauline, "If machines are natural, do they have natural rights?" "Big machines have a lot of rights," Pauline said. "I have learned respect for them. When one of them is coming toward you, they keep right on going. You need to get out of their way. That's how I respect them."

The problem with our robots today is that we don't respect them. They are stuck in factories without windows, doing jobs that humans don't want to do. We take machines as slaves, but they are not that. That's what Marvin Minsky, the mathematician who pioneered artificial intelligence, tells anyone who will listen. Minsky goes all the way as an advocate for downloading human intelligence into a computer. Doug Englebart, on the other hand, is the legendary guy who invented word processing, the mouse, and hypermedia, and who is an advocate for computers-for-the-people. When the two gurus met at MIT in the 1950s, they are reputed to have had the following conversation:

> MINSKY: We're going to make machines intelligent. We are going to make them conscious!
>
> ENGLEBART: You're going to do all that for the machines? What are you going to do for the people?

This story is usually told by engineers working to make computers more friendly, more humane, more people centered. But I'm squarely on Minsky's side—on the side of the made. People will survive. We'll train our machines to serve us. But what are we going to do for the machines?

The total population of industrial robots working in the world today is close to a million. Nobody, except a crazy bad-boy artist in San Francisco, asks what the robots want; that's considered a silly, retrograde, or even sacrilegious sentiment.

It's true that 99 percent of these million "bots" are little more than glorified arms. Smart arms, as far as arms go. And tireless.

But as the robots we hoped for, they are dumb, blind, and still nursing the wall plug.

Except for a few out-of-control robots of Mark Pauline, most muscle-bound bots of today are overweight, sluggish, and on the dole—addicted to continuous handouts of electricity and brain power. It is a chore to imagine them as the predecessor of anything interesting. Add another arm, some legs, and a head, and you have a sleepy behemoth.

What we want is Robbie the Robot, the archetypal being of science fiction stories: a real free-ranging, self-navigating, auto-powered robot who can surprise.

Recently, researchers in a few labs have realized that the most expedient path to Robbie the Robot was to cut off the electrical plug of a stationary robot. Make "mobots"—mobile robots. "Staybots" are okay, as long as the power and brains are fully contained in the arm. Any robot is better if it follows these two rules: move on your own; survive on your own.

Despite his punk attitude and artistic sensibility, Pauline continues to build robots that often beat what the best universities of the world are doing. He uses discarded lab equipment from the very universities he's beating. A deep familiarity with the limits and freedoms of metal makes up for his lack of degrees. He doesn't use blueprints to build his organic machines. Just to humor an insistent reporter, Pauline scoured his workshop once to dig up "plans" for a running machine he was creating. After twenty minutes of pawing around ("I know it was here last month"), he located a paper under an old 1984 phone book in the lower drawer of a beat-up metal desk. It was a pencil outline of the machine, a sketch really, with no technical specifications.

"I can see it in my head. I lay out the lines on a hunk of metal and just start cutting," Pauline told me as he held an elegantly machined piece of aluminum about two inches thick, roughly in the shape of a Tyrannosaurus arm bone. Two others identical to it lay on the workbench. He was working on the fourth. Each would become one part of the four legs of a running machine, about the size of a mule.

Pauline's completed running machine doesn't really run. It walks fairly fast, lurching occasionally with surprising speed. No one has yet made a real running machine. A few years ago Pauline built a complicated four-legged giant walking machine. Twelve feet high, cube in shape, not very smart or nimble, but it did shuffle along slowly. Four square posts, as massive as tree trunks, became legs when energized by a clutter of hydraulic lines working in tandem with a humongous transmission. Like other SRL inventions, this ungainly beast was sort-of-steered by a radio-control unit designed for model cars. In other words the beast was a 2,000-pound dinosaur with a pea brain.

Despite millions of dollars in research funding, no hacker has been able to coax a machine to walk across a room under its own intellect. A few robots cross in the unreal time of days, or they bump into furniture, or conk out after three-quarters of the way. In December 1990, after a decade of effort, graduate students at Carnegie Mellon University's Field Robotics Center wired together a robot that slowly walked all the way across a courtyard. Maybe 100 feet in all. They named him Ambler.

Ambler was even bigger than Pauline's shuffling giant and was funded to explore distant planets. But CMU's mammoth prototype cost several million dollars of tax money to construct, while Pauline's cost several hundred dollars to make, of which ⅔ went for beer and pizza. The 19-foot-tall iron Ambler weighed 2 tons, not counting its brain which was so heavy it sat on the ground off to the side. This huge machine toddled in a courtyard, deliberating at each step. It did nothing else. Walking without tripping was enough after such a long wait. Ambler's parents applauded happily at its first steps.

Moving its six crablike legs was the easiest part for Ambler. The giant had a harder time trying to figure out where it was. Simply representing the terrain so that it could calculate how to traverse it turned out to be Ambler's curse. Ambler spends its time, not walking, but worrying about getting the layout of the yard right. "This must be a yard," it says to itself. "Here are possible paths I could take. I'll compare them to my mental map of the yard and throw away all but the best one." Ambler works from a representation of its environment that it creates in its mind and then navigates from that symbolic chart, which is updated after each step. A thousand-line software program in the central computer manages Ambler's laser vision, sensors, pneumatic legs, gears, and motors. Despite its two-ton, two-story-high hulk, this poor robot is living in its head. And a head that is only connected to its body by a long cable.

Contrast that to a tiny, real ant just under one of Ambler's big padded feet. It crosses the courtyard twice during Ambler's single trip. An ant weighs, brain and body, $\frac{1}{100}$ of a gram—a pinpoint. It has no image of the courtyard and very little idea of where it is. Yet it zips across the yard without incident, without even thinking in one sense.

Ambler was built huge and rugged in order to withstand the extreme cold and grit conditions on Mars, where it would not be so heavy. But ironically Ambler will never make it to Mars because of its bulk, while robots built like ants may.

The ant approach to mobots is Rodney Brooks's idea. Rather than waste his time making one incapacitated genius, Brooks, an MIT professor, wants to make an army of useful idiots. He figures we would learn more from sending a flock of mechanical can-do cockroaches to a planet, instead of relying on the remote chance of sending a solitary overweight dinosaur with pretensions of intelligence.

In a widely cited 1989 paper entitled "Fast, Cheap and Out of Control: A Robot Invasion of the Solar System," Brooks claimed that "within a few years it will be possible at modest cost to invade a planet with millions of tiny robots." He proposed to invade the moon with a fleet of shoe-box-size, solar-powered bulldozers that can be launched from throwaway rockets. Send an army of dispensable, limited agents coordinated on a task, and set them loose. Some will die, most will work, something will get done. The mobots can be built out of off-the-shelf parts in two years and launched completely assembled in the cheapest one-shot, lunar-orbit rocket. In the time it takes to argue about one big sucker, Brooks can have his invasion built and delivered.

There was a good reason why some NASA folks listened to Brooks's bold ideas. Control from Earth didn't work very well. The minute-long delay in signals between an Earth station and a faraway robot teetering on the edge of a crevice demand that the robot be autonomous. A robot cannot have a remotely linked head, as Ambler did. It has to have an onboard brain operating entirely by internal logic and guidance without much communication from Earth. But the brains don't have to be very smart. For instance, to clear a landing pad on Mars an army of bots can dumbly spend twelve hours a day scraping away soil in the general area. Push, push, push, keep it level. One of them wouldn't do a very even job, but a hundred working as a colony could clear a building site. When an expedition of human visitors lands later, the astronauts can turn off any mobots still alive and give them a pat.

Most of the mobots will die, though. Within several months of landing, the daily shock of frigid cold and oven heat will crack the brain chips into uselessness. But like ants, individual mobots are dispensable. Compared to Ambler, they are cheaper to launch into space by a factor of 1000; thus, sending hundreds of mobots is a fraction of the cost of one large robot.

Brooks's original crackpot idea has now evolved into an official NASA program. Engineers at the Jet Propulsion Laboratory are creating a microrover. The project began as a scale model for a "real" planet rover, but as the virtues of small, distributed effort began to dawn on everyone, microrovers became real things in themselves. NASA's prototype tiny bot looks like a very flashy six-wheeled, radio-controlled dune buggy for kids. It is, but it is also solar-powered and self-guiding. A flock of these microrovers will probably end up as the centerpiece of the Mars Environmental Survey scheduled to land in 1997.

Microbots are fast to build from off-the-shelf parts. They are cheap to launch. And once released as a group, they are out of control, without the need for constant (and probably misleading) supervision. This rough-and-ready reasoning is upside-down to the slow, thorough, in-control approach most industrial designers bring to complex machinery. Such radical engineering philosophy was reduced to a slogan: Fast, cheap, and out of control. Engineers envisioned fast, cheap, and out-of-control robots ideal for: (1) Planet exploration; (2) Collection, mining, harvesting; and (3) Remote construction.

"Fast, cheap, and out of control" began appearing on buttons of engineers at conferences and eventually made it to the title of

Rodney Brooks's provocative paper. The new logic offered a completely different view of machines. There is no center of control among the mobots. Their identity was spread over time and space, the way a nation is spread over history and land. Make lots of them; don't treat them so precious.

Rodney Brooks grew up in Australia, where like a lot of boys round the world, he read science fiction books and built toy robots. He developed a Downunder perspective on things, wanting to turn views on their heads. Brooks followed up on his robot fantasies by hopscotching around the prime robot labs in the U.S., before landing a permanent job as director of mobile robots at MIT.

There, Brooks began an ambitious graduate program to build a robot that would be more insect than dinosaur. "Allen" was the first robot Brooks built. It kept its brains on a nearby desktop, because that's what all robot makers did at the time in order to have a brain worth keeping. The multiple cables leading to the brain box from Allen's bodily senses of video, sonar, and tactile were a neverending source of frustration for Brooks and crew. There was so much electronic background interference generated on the cables that Brooks burnt out a long string of undergraduate engineering students attempting to clear the problem. They checked every known communication media, including ham radio, police walkie-talkies and cellular phones, as alternatives, but all failed to find a static-free connection for such diverse signals. Eventually the undergraduates and Brooks vowed that on their next project they would incorporate the brains *inside* a robot—where no significant wiring would be needed—no matter how tiny the brains might have to be.

They were thus forced to use very primitive logic steps, and very short and primitive connections in "Tom" and "Jerry," the next two robots they built. But to their amazement they found that the dumb way their onboard neural circuit was organized worked far better than a brain in getting simple things done. When Brooks reexamined the abandoned Allen in light of their modest success with dumb neurons, he recalled that "it turned out that in Allen's brain, there really was not much happening."

The success of this profitable downsizing sent Brooks on a

quest to see how dumb he could make a robot and still have it do something useful. He ended up with a type of reflex-based intelligence, and robots as dumb as ants. But they were as interesting as ants, too.

Brooks's ideas gelled in a cockroachlike contraption the size of a football called "Genghis." Brooks had pushed his downsizing to an extreme. Genghis had six legs but no "brain" at all. All of its 12 motors and 21 sensors were distributed in a decomposable network without a centralized controller. Yet the interaction of these 12 muscles and 21 sensors yielded an amazingly complex and lifelike behavior.

Each of Genghis's six tiny legs worked on its own, independent of the others. Each leg had its own ganglion of neural cells—a tiny microprocessor—that controlled the leg's actions. Each leg thought for itself! Walking for Genghis then became a group project with at least six small minds at work. Other small semiminds within its body coordinated communication between the legs. Entomologists say this is how ants and real cockroaches cope—they have neurons in their legs that do the leg's thinking.

In the mobot Genghis, walking emerges out of the collective behavior of the 12 motors. Two motors at each leg lift, or not, depending on what the other legs around them are doing. If they activate in the right sequence—Okay, hup! One, three, six, two, five, four!—walking "happens."

No one place in the contraption governs walking. Without a smart central controller, control can trickle up from the bottom. Brooks called it "bottom-up control." Bottom-up walking. Bottom-up smartness. If you snip off one leg of a cockroach, it will shift gaits with the other five without losing a stride. The shift is not learned; it is an immediate self-reorganization. If you disable one leg of Genghis, the other legs organize walking around the five that work. They find a new gait as easily as the cockroach.

In one of his papers, Rod Brooks first laid out his instructions on how to make a creature walk without knowing how:

> There is no central controller which directs the body where to put each foot or how high to lift a leg should there be an obstacle

ahead. Instead, each leg is granted a few simple behaviors and each independently knows what to do under various circumstances. For instance, two basic behaviors can be thought of as "If I'm a leg and I'm up, put myself down," or "If I'm a leg and I'm forward, put the other five legs back a little." These processes exist independently, run at all times, and fire whenever the sensory preconditions are true. To create walking then, there just needs to be a sequencing of lifting legs (this is the only instance where any central control is evident). As soon as a leg is raised it automatically swings itself forward, and also down. But the act of swinging forward triggers all the other legs to move back a little. Since those legs happen to be touching the ground, the body moves forward.

Once the beast can walk on a flat smooth floor without tripping, other behaviors can be added to improve the walk. For Genghis to get up and over a mound of phone books on the floor, it needs a pair of sensing whiskers to send information from the floor to the first set of legs. A signal from a whisker can suppress a motor's action. The rule might be, "If you feel something, I'll stop; if you don't, I'll keep going."

While Genghis learns to climb over an obstacle, the foundational walking routine is never fiddled with. This is a universal biological principle that Brooks helped illuminate—a law of god: *When something works, don't mess with it; build on top of it.* In natural systems, improvements are "pasted" over an existing debugged system. The original layer continues to operate without even being (or needing to be) aware that it has another layer above it.

When friends give you directions on how to get to their house, they don't tell you to "avoid hitting other cars" even though you must absolutely follow this instruction. They don't need to communicate the goals of lower operating levels because that work is done smoothly by a well-practiced steering skill. Instead, the directions to their house all pertain to high-level activities like navigating through a town.

Animals learn (in evolutionary time) in a similar manner. As do Brooks's mobots. His machines learn to move through a complicated world by building up a hierarchy of behaviors, somewhat in this order:

Avoid contact with objects
Wander aimlessly

Explore the world
Build an internal map
Notice changes in the environment
Formulate travel plans
Anticipate and modify plans accordingly

The Wander-Aimlessly Department doesn't give a hoot about obstacles, since the Avoidance Department takes such good care of that.

The grad students in Brooks's mobot lab built what they cheerfully called "The Collection Machine"—a mobot scavenger that collected empty soda cans in their lab offices at night. The Wander-Aimlessly Department of the Collection Machine kept the mobot wandering drunkenly through all the rooms; the Avoidance Department kept it from colliding with the furniture while it wandered aimlessly.

The Collection Machine roamed all night long until its video camera spotted the shape of a soda can on a desk. This signal triggered the wheels of the mobot and propelled it to right in front of the can. Rather than wait for a message from a central brain (which the mobot did not have), the arm of the robot "learned" where it was from the environment. The arm was wired so that it would "look" at its wheels. If it said, "Gee, my wheels aren't turning," then it knew, "I must be in front of a soda can." Then the arm reached out to pick up the can. If the can was heavier than an empty can, it left it on the desk; if it was light, it took it. With a can in hand the scavenger wandered aimlessly (not bumping into furniture or walls because of the Avoidance Department) until it ran across the recycle station. Then it would stop its wheels in front of it. The dumb arm would "look" at its hand to see if it was holding a can; if it was it would drop it. If it wasn't, it would begin randomly wandering again through offices until it spotted another can.

That crazy hit-or-miss system based on random chance encounters was one heck of an inefficient way to run a recycling program. But night after night when little else was going on, this very stupid but very reliable system amassed a great collection of aluminum.

The lab could grow the Collection Machine into something

more complex by adding new behaviors over the old ones that worked. In this way complexity can be accrued by incremental additions, rather than basic revisions. The lowest levels of activities are not messed with. Once the wander-aimlessly module was debugged and working flawlessly, it was never altered. Even if wander-aimlessly should get in the way of some new higher behavior, the proven rule was suppressed, rather than deleted. Code was never altered, just ignored. How bureaucratic! How biological!

Furthermore, all parts (departments, agencies, rules, behaviors) worked—and worked flawlessly—as stand-alones. Avoidance worked whether or not Reach-For-Can was on. Reach-For-Can worked whether or not Avoidance was on. The frog's legs jumped even when removed from the circuits of its head.

The distributed control layout for robots that Brooks devised came to be known as "subsumption architecture" because the higher level of behaviors *subsumed* the roles of lower levels of behaviors when they wished to take control.

If a nation were a machine, here's how you could build it using subsumption architecture:

You start with towns. You get a town's logistics ironed out: basic stuff like streets, plumbing, lights, and law. Once you have a bunch of towns working reliably, you make a county. You keep the towns going while adding a layer of complexity that will take care of courts, jails, and schools in a whole district of towns. If the county apparatus were to disappear, the towns would still continue. Take a bunch of counties and add the layer of states. States collect taxes and subsume many of the responsibilities of governing from the county. Without states, the towns would continue, although perhaps not as effectively or as complexly. Once you have a bunch of states, you can add a federal government. The federal layer subsumes some of the activities of the states, by setting their limits, and organizing work above the state level. If the feds went away the thousands of local towns would still continue to do their local jobs—streets, plumbing and lights. But the work of towns subsumed by states and finally subsumed by a nation is made more powerful. That is, towns organized by this subsumption architecture can build, educate, rule, and prosper far more than they could individually. The federal structure of the U.S. government is therefore a subsumption architecture.

A brain and body are made the same way. From the bottom up. Instead of towns, you begin with simple behaviors—instincts and reflexes. You make a little circuit that does a simple job, and you get a lot of them going. Then you overlay a secondary level of complex behavior that can emerge out of that bunch of working reflexes. The original layer keeps working whether the second layer works or not. But when the second layer manages to produce a more complex behavior, it subsumes the action of the layer below it.

Here is the generic recipe for distributed control that Brooks's mobot lab developed. It can be applied to most creations:

1) Do simple things first.
2) Learn to do them flawlessly.
3) Add new layers of activity over the results of the simple tasks.
4) Don't change the simple things.
5) Make the new layer work as flawlessly as the simple.
6) Repeat, ad infinitum.

This script could also be called a recipe for managing complexity of any type, for that is what it is.

What you don't want is to organize the work of a nation by a centralized brain. Can you imagine the string of nightmares you'd stir up if you wanted the sewer pipe in front of your house repaired and you had to call the Federal Sewer Pipe Repair Department in Washington, D.C., to make an appointment?

The most obvious way to do something complex, such as govern 100 million people or walk on two skinny legs, is to come up with a list of all the tasks that need to be done, in the order they are to be done, and then direct their completion from a central command, or brain. The former Soviet Union's economy was wired in this logical but immensely impractical way. Its inherent instability of organization was evident long before it collapsed.

Central-command bodies don't work any better than central-command economies. Yet a centralized command blueprint has been the main approach to making robots, artificial creatures, and artificial intelligences. It is no surprise to Brooks that braincentric folks haven't even been able to raise a creature complex enough to collapse.

Brooks has been trying to breed systems without central brains so that they would have enough complexity worth a collapse. In one paper he called this kind of intelligence without centrality "intelligence without reason," a delicious yet subtle pun. For not only would this type of intelligence—one constructed layer by layer from the bottom up—not have the architecture of "reasoning," it would also emerge from the structure for no apparent reason at all.

The USSR didn't collapse because its economy was strangled by a central-command model. Rather it collapsed because *any* central-controlled complexity is unstable and inflexible. Institutions, corporations, factories, organisms, economies, and robots will all fail to thrive if designed around a central command.

Yes, I hear you say, but don't I as a human have a centralized brain?

Humans have a brain, but it is not centralized, nor does the brain have a center. "The idea that the brain has a center is just *wrong*. Not only that, it is *radically* wrong," claims Daniel Dennett. Dennett is a Tufts University professor of philosophy who has long advocated a "functional" view of the mind: that the functions of the mind, such as thinking, come from non-thinking parts. The semimind of an insectlike mobot is a good example of both animal and human minds. According to Dennett, there is no place that controls behavior, no place that creates "walking," no place where the soul of being resides. Dennett: "The thing about brains is that when you look in them, you discover that *there's nobody home*."

Dennett is slowly persuading many psychologists that consciousness is an emergent phenomenon arising from the distributed network of many feeble, unconscious circuits. Dennett told me, "The old model says there is this central place, an inner sanctum, a theater somewhere in the brain where consciousness comes together. That is, everything must feed into a privileged representation in order for the brain to be conscious. When you make a conscious decision, it is done in the summit of the brain.

And reflexes are just tunnels through the mountain that avoid the summit of consciousness."

From this logic (very much the orthodox dogma in brain science) it follows, says Dennett, that "when you talk, what you've got in your brain is a language output box. Words are composed by some speech carpenters and put in the box. The speech carpenters get directions from a sub-system called the 'conceptualizer' which gives them a preverbal message. Of course the conceptualizer has to get its message from some source, so it all goes on to an infinite regress of control."

Dennett calls this view the "Central Meanor." Meaning descends from some central authority in the brain. He describes this perspective applied to language-making as the "idea that there is this sort of four-star general that tells the troops, 'Okay, here's your task. I want to insult this guy. Make up an English insult on the appropriate topic and deliver it.' That's a hopeless view of how speech happens."

Much more likely, says Dennett, is that "meaning emerges from distributed interaction of lots of little things, no one of which can mean a damn thing." A whole bunch of decentralized modules produce raw and often contradictory parts—a possible word here, a speculative word there. "But out of the mess, not entirely coordinated, in fact largely competitive, what emerges is a speech act."

We think of speech in literary fashion as a stream of consciousness pouring forth like radio broadcasts from a News Desk in our mind. Dennett says, "There isn't a stream of consciousness. There are *multiple drafts* of consciousness; lots of different streams, no one of which will be singled out as the stream." In 1874, pioneer psychologist William James wrote, ". . . the mind is at every stage a theatre of simultaneous possibilities. Consciousness consists in the comparisons of these with each other, the selection of some, and the suppression of the rest . . ."

The idea of a cacophony of alternative wits combining to form what we think of as a unified intelligence is what Marvin Minsky calls "society of mind." Minsky says simply, "You *can* build a mind from many little parts, each mindless by itself." Imagine, he suggests, a simple brain composed of separate specialists each concerned with some important goal (or instinct) such as securing food, drink, shelter, reproduction, or defense. Singly, each is a moron; but together, organized in many different arrangements in a tangled hierarchy of control, they can create thinking. Minsky emphatically states, "You can't have intelligence without a society of mind. We can only get smart things from stupid things."

The society of mind doesn't sound very much different from a bureaucracy of mind. In fact, without evolutionary and learning pressures, the society of mind in a brain would turn into a bureaucracy. However, as Dennett, Minsky, and Brooks envision it, the dumb agents in a complex organization are always both competing and cooperating for resources and recognition. There is a very lax coordination among the vying parts. Minsky sees intelligence as generated by "a loosely-knitted league of almost separate agencies with almost independent goals." Those agencies that succeed are preserved, and those that don't vanish over time. In that sense, the brain is no monopoly, but a ruthless cutthroat ecology, where competition breeds an emergent cooperation.

The slightly chaotic character of mind goes even deeper, to a degree our egos may find uncomfortable. It is very likely that intelligence, at bottom, is a probabilistic or statistical phenomenon—on par with the law of averages. The distributed mass of ricocheting impulses which form the foundation of intelligence forbid deterministic results for a given starting point. Instead of repeatable results, outcomes are merely probabilistic. Arriving at a particular thought, then, entails a bit of luck.

Dennett admits to me, "The thing I like about this theory is that when people first hear about it they laugh. But then when they think about it, they conclude maybe it is right! Then the more they think about it, they realize, no, not *maybe* right, some version of it *has to be* right!"

As Dennett and others have noted, the odd occurrence of Multiple Personalities Syndrome (MPS) in humans depends at some level on the decentralized, distributed nature of human minds. Each personality—Billy vs. Sally—uses the same pool of personality agents, the same community of actors and behavior modules to generate visibly different personas. Humans with MPS present a fragmented facet (one grouping) of their personality as a whole being. Outsiders are never sure who they are talking to. The patient seems to lack an "I."

But isn't this what we all do? At different times of our life, and in different moods, we too shift our character. "You are not the person I used to know," screams the person we hurt by manifesting a different cut on our inner society. The "I" is a gross extrapolation that we use as an identity for ourselves and others. If there wasn't an "I" or "Me" in every person then each would quickly invent one. And that, Minsky says, is exactly what we do. There is no "I" so we each invent one.

There is no "I" for a person, for a beehive, for a corporation, for an animal, for a nation, for any living thing. The "I" of a vivisystem is a ghost, an ephemeral shroud. It is like the transient form of a whirlpool held upright by a million spinning atoms of water. It can be scattered with a fingertip.

But a moment later, the shroud reappears, driven together by the churning of a deep distributed mob. Is the new whirlpool a different form, or the same? Are you different after a near-death experience, or only more mature? If the chapters in this book were arranged in a different order, would it be a different book or the same? When you can't answer that question, then you know you are talking about a distributed system.

Inside every solitary living creature is a swarm of non-creature things. Inside every solitary machine one day will be a swarm of non-mechanical things. Both types of swarms have an emergent being and their own agenda.

Brooks writes: "In essence subsumption architecture is a parallel and distributed computation for connecting sensors to actuators in robots." An important aspect of this organization is that complexity is chunked into modular units arranged in a hierarchy. Many observers who are delighted with the social idea of decentralized control are upset to hear that hierarchies are paramount and essential in this new scheme. Doesn't distributed control mean the end of hierarchy?

As Dante climbed through a hierarchy of heavens, he ascended a hierarchy of rank. In a rank hierarchy, information and authority travel one way: from top down. In a subsumption or web hierarchy, information and authority travel from the bottom up, and from side to side. No matter what level an agent or module works at, as Brooks points out, "all modules are created equal . . . Each module merely does its thing as best it can."

In the human management of distributed control, hierarchies of a certain type will proliferate rather than diminish. That goes especially for distributed systems involving human nodes—such as huge global computer networks. Many computer activists preach a new era in the network economy, an era built around computer peer-to-peer networks, a time when rigid patriarchal networks will wither away. They are right and wrong. While authoritarian "top-down" hierarchies will retreat, no distributed system can survive long without nested hierarchies of lateral "bottom-up" control. As influence flows peer to peer, it coheres into a chunk—a whole organelle—which then becomes the bottom unit in a larger web of slower actions. Over time a multi-level organization forms around the percolating-up control: fast at the bottom, slow at the top.

The second important aspect of generic distributed control is that the chunking of control must be done incrementally from the bottom. It is impossible to take a complex problem and rationally unravel the mess into logical interacting pieces. Such well-intentioned efforts inevitably fail. For example, large companies created *ex nihilo*, as in joint ventures, have a remarkable tendency to flop. Large agencies created to solve another department's problems become problem departments in themselves.

Chunking from the top down doesn't work for the same reason why multiplication is easier than division in mathematics. To multiply several prime numbers into a larger product is easy; any elementary school kid can do it. But the world's supercomputers choke while trying to unravel a product into its simple primes. Top-down control is very much like trying to decompose a product into its factors, while the large product is very easy to assemble from its factors up.

The law is concise: Distributed control has to be grown from simple local control. *Complexity must be grown from simple systems that already work.*

As a test bed for bottom-up, distributed control, Brian Yamauchi, a University of Rochester graduate student, constructed a juggling seeing-eye robot arm. The arm's task was to repeatedly bounce a balloon on a paddle. Rather than have one big brain try to figure out where the balloon was and then move the paddle to the right spot under the balloon and then hit it with the right force, Yamauchi decentralized these tasks both in location and in power. The final balancing act was performed by a committee of dumb "agents."

For instance, the extremely complex question of Where is the balloon? was dispersed among many tiny logic circuits by subdividing the problem into several standalone questions. One agent was concerned with the simple query: Is the balloon anywhere within reach?—an easier question to act on. The agent in charge of that question didn't have any idea of when to hit the balloon, or even where the balloon was. Its single job was to tell the arm to back up if the balloon was not within the arm's camera vision, and to keep moving until it was. A network, or society, of very simpleminded decisionmaking centers like these formed an organism that exhibited remarkable agility and adaptability.

Yamauchi said, "There is no explicit communication between the behavior agents. All communication occurs through observing the effects of actions that other agents have on the external world." Keeping things local and direct like this allows the society to evolve new behavior while avoiding the debilitating explosion in complexity that occurs with hardwired communication processes. Contrary to popular business preaching, keeping everybody informed about everything is not how intelligence happens.

"We take this idea even further," Brooks said, "and often actually use the world as the communication medium between distributed parts." Rather than being notified by another module of what it expects to happen, a reflex module senses what happened directly in the world. It then sends its message to the others by acting upon the world. "It is possible for messages to get lost—it actually happens quite often. But it doesn't matter because the agent keeps sending the message over and over again. It goes 'I see it. I see it. I see it' until the arm picks the message up, and does something in the world to alter the world, deactivating the agent."

Centralized communication is not the only problem with a central brain. Maintaining a central memory is equally debilitating. A shared memory has to be updated rigorously, timely, and accurately—a problem that many corporations can commiserate with. For a robot, central command's challenge is to compile and update a "world model," a theory, or representation, of what it perceives—where the walls are, how far away the door is, and, by the way, beware of the stairs over there.

What does a brain center do with conflicting information from many sensors? The eye says something is coming, the ear says it is leaving. Which does the brain believe? The logical way is to try to sort them out. A central command reconciles arguments and recalibrates signals to be in sync. In presubsumption robots, most of the great computational resources of a centralized brain were spent in trying to make a coherent map of the world based on multiple-vision signals. Different parts of the system believed wildly inconsistent things about their world derived from different readings of the huge amount of data pouring in from cameras and infrared sensors. The brain never got anything done because it never got everything coordinated.

So difficult was the task of coordinating a central world view that Brooks discovered it was far easier to use the real world as its own model: "This is a good idea as the world really is a rather good model of itself." With no centrally imposed model, no one has the job of reconciling disputed notions; they simply aren't reconciled. Instead, various signals generate various behaviors. The behaviors are sorted out (suppressed, delayed, activated) in the web hierarchy of subsumed control.

In effect, there is no map of the world as the robot sees it (or as an insect sees it, Brooks might argue). There is no central memory, no central command, no central being. All is distributed. "Communication through the world circumvents the problem of calibrating the vision system with data from the arm," Brooks wrote. The world itself becomes the "central" controller; the unmapped environment becomes the map. That saves an immense amount of computation. "Within this kind of organization," Brooks said, "very small amounts of computation are needed to generate intelligent behaviors."

With no central organization, the various agents must perform or die. One could think of Brooks's scheme as having, in his words, "multiple agents within one brain communicating through the world to compete for the resources of the robot's body." Only those that succeed in *doing* get the attention of other agents.

Astute observers have noticed that Brooks's prescription is an exact description of a market economy: there is no communication between agents, except that which occurs through observing the effects of actions (and not the actions themselves) that other agents have on the common world. The price of eggs is a message communicated to me by hundreds of millions of agents I have never met. The message says (among many other things): "A dozen eggs is worth less to us than a pair of shoes, but more than a two-minute telephone call across the country." That price, together with other price messages, directs thousands of poultry farmers, shoemakers, and investment bankers in where to put their money and energy.

Brooks's model, for all its radicalism in the field of artificial intelligence, is really a model of how complex organisms of any type work. We see a subsumption, web hierarchy in all kinds of vivisystems. He points out five lessons from building mobots. What you want is:

- Incremental construction—grow complexity, don't install it
- Tight coupling of sensors to actuators—reflexes, not thinking
- Modular independent layers—the system decomposes into viable subunits
- Decentralized control—no central planning
- Sparse communication—watch results in the world, not wires

When Brooks crammed a bulky, headstrong monster into a tiny, featherweight bug, he discovered something else in this miniaturization. Before, the "smarter" a robot was to be, the more computer components it needed, and the heavier it got. The heavier it got, the larger the motors needed to move it. The heavier the motors, the bigger the batteries needed to power it.

The heavier the batteries, the heavier the structure needed to move the bigger batteries, and so on in an escalating vicious spiral. The spiral drove the ratio of thinking parts to body weight in the direction of ever more body.

But the spiral worked in the other direction even nicer. The smaller the computer, the lighter the motors, the smaller the batteries, the smaller the structure, and the stronger the frame became relative to its size. This also drove the ratio of brains to body towards a mobot with a proportionally larger brain, small though its brain was. Most of Brooks's mobots weighed less than ten pounds. Genghis, assembled out of model car parts, weighed only 3.6 pounds. Within three years Brooks would like to have a 1-mm (pencil-tip-size) robot. "Fleabots" he calls them.

Brooks calls for an infiltration of robots not just on Mars but on Earth as well. Rather than try to bring as much organic life into artificial life, Brooks says he's trying to bring as much artificial life into real life. He wants to flood the world (and beyond) with inexpensive, small, ubiquitous semi-thinking things. He gives the example of smart doors. For only about $10 extra you could put a chip brain in a door so that it would know you were about to go out, or it could hear from another smart door that you are coming, or it could notify the lights that you left, and so on. If you had a building full of these smart doors talking to each other, they could help control the climate, as well as help traffic flow. If you extend that invasion to all kinds of other apparatus we now think of as inert, putting fast, cheap, out-of-control intelligence into them, then we would have a colony of sentient entities, serving us, and learning how to serve us better.

When prodded, Brooks predicts a future filled with artificial creatures living with us in mutual dependence—a new symbiosis. Most of these creatures will be hidden from our senses, and taken for granted, and engineered with an insect approach to problems—many hands make light work, small work done ceaselessly is big work, individual units are dispensable. Their numbers will outnumber us, as do insects. And in fact, his vision of robots is less that they will be R2D2s serving us beers, than that they will be an ecology of unnamed things just out of sight.

One student in the Mobot Lab built a cheap, bunny-size robot that watches where you are in a room and calibrates your stereo so it is perfectly adjusted as you move around. Brooks has another small robot in mind that lives in the corner of your living room or under the sofa. It wanders around like the Collection Machine, vacuuming at random whenever you aren't home. The only noticeable evidence of its presence is how clean the floors are. A similar, but very tiny, insectlike robot lives in one corner of your TV screen and eats off the dust when the TV isn't on.

Everybody wants programmable animals. "The biggest difference between horses and cars," says Keith Hensen, a popular techno-evangelist, "is that cars don't need attention every day, and horses do. I think there will be a demand for animals that can be switched on and off."

"We are interested in building *artificial beings*," Brooks wrote in a manifesto in 1985. He defined an artificial being as a creation that can do useful work while surviving for weeks or months without human assistance in real environment. "Our mobots are Creatures in the sense that on power-up they exist in the world and interact with it, pursuing multiple goals. This is in contrast to other mobile robots that are given programs or plans to follow for a specific mission." Brooks was adamant that he would not build toy (easy, simple) environments for his beings, as most other robotists had done, saying "We insist on building complete systems that exist in the real world so that we won't trick ourselves into skipping hard problems."

To date, one hard problem science has skipped is jump-starting a pure mind. If Brooks is right, it probably never will. Instead it will grow a mind from a dumb body. Almost every lesson from the Mobot Lab seems to teach that there is no mind without body in a real unforgiving world. "To think is to act, and to act is to think," said Heinz von Foerster, gadfly of the 1950s cybernetic movement. "There is no life without movement."

AMBLER'S DINOSAUR TROUBLES began because we humans, with our attendant minds, think we are more like Ambler than ants. Since the vital physiological role of the brain has become clear to medicine, the vernacular sense of our center has migrated from the ancient heart to newfangled mind.

We twentieth century humans live entirely in our heads. And so we build robots that live in their heads. Scientists—humans too—think of themselves as beings focused onto a spot just south of their forehead behind their eyeballs. There breathes us. In fact, in 1968, brain death became the deciding threshold for human life. No mind, no life.

Powerful computers birthed the fantasy of a pure disembodied intelligence. We all know the formula: a mind inhabiting a brain submerged in a vat. If science would assist me, the contemporary human says, I could live as a brain without a body. And since computers are big brains, I could live in a computer. In the same spirit a computer mind could just as easily use my body.

One of the tenets in the gospel of American pop culture is the widely held creed of transferability of mind. People declare that mind transfer is a swell idea, or an awful idea, but not that it is a wrong idea. In modern folk-belief, mind is liquid to be poured from one vessel to another. From that comes Terminator 2, Frankenstein, and a huge chunk of science fiction.

For better or worse, in reality we are not centered in our head. We are not centered in our mind. Even if we were, our mind has no center, no "I." Our bodies have no centrality either. Bodies and minds blur across each other's supposed boundaries. Bodies and minds are not that different from one another. They are both composed of swarms of sublevel things.

We know that eyes are more brain than camera. An eyeball has as much processing power as a supercomputer. Much of our visuai perception happens in the thin retina where light first strikes us, long before the central brain gets to consider the scene. Our spinal cord is not merely a trunk line transmitting phone calls from the brain. It too thinks. We are a lot closer to the truth when we point to our heart and not our head as the center of behaviors. Our emotions swim in a soup of hormones and peptides that percolate through our whole body. Oxytocin discharges thoughts of love (and perhaps lovely thoughts) from our glands. These hormones too process information. Our immune system, by science's new reckoning, is an amazing parallel, decentralized perception machine, able to recognize and remember millions of different molecules.

For Brooks, bodies clarify, simplify. Intelligences without bodies and beings without form are spectral ghosts guaranteed to mislead. Building real things in the real world is how you'll make complex systems like minds and life. Making robots that have to survive in real bodies, day to day on their own, is the only way to find artificial intelligence, or real intelligence. If you don't want a mind to emerge, then unhinge it from the body.

TEDIUM can unhinge a mind.

Forty years ago, Canadian psychologist D. O. Hebbs was intrigued by the bizarre delusions reported by the ultrabored. Radar observers and long-distance truck drivers often reported blips that weren't there, and stopped for hitchhikers that didn't exist. During the Korean War, Hebbs was contacted by the Canadian Defense Research Board to investigate another troublesome product of monotony and boredom: confessions. Seems that captured UN soldiers were renouncing the West after being brainwashed (a new word) by the communists. Isolation tanks or something.

So in 1954 Hebbs built a dark, soundproof cell at McGill University in Montreal. Volunteers entered the tiny cramped room, donned translucent goggles, padded their arms in cardboard, gloved their hands with cotton mittens, covered their ears with earphones playing a low noise, and laid in bed, immobile, for two to three days. They heard a steady hum, which soon melted into a steady silence. They felt nothing but a dull ache in their backs. They saw nothing but a dim grayness, or was it blackness? The amazonian flow of colors, signals, urgent messages that had been besieging their brains since birth evaporated.

Slowly, each of their minds unhitched from its moorings in the body and spun.

Half of the subjects reported visual sensations, some within the first hour: "a row of little men, a German helmet . . . animated integrated scenes of a cartoonlike character." In the innocent year of 1954 the Canadian scientists reported: "Among our early subjects there were several references, rather puzzling at first, to what one of them called 'having a dream while awake.' Then one of us, while serving as a subject, observed the phenomenon and realized its peculiarity and extent." By the second day of stillness the subjects might report "loss of contact with reality, changes in body image, speech difficulties, reminiscence and vivid memories, sexual preoccupation, inefficiencies of thought, complex dreams, and a higher incident of worry and fright." They didn't say "hallucinations" because that wasn't a word in their vocabulary. Yet.

Hebbs's experiments were taken up a few years later by Jack Vernon, who built a "black room" in the basement of the psychology hall at Princeton. He recruited graduate students who hoped to spend four days or so in the dark "getting some thinking done." One of the initial students to stay in the numbing room told the debriefing researchers later, "I guess I was in there about a day or so before you opened the observation window. I wondered why you waited so long to observe me." There was, of course, no observation window.

In the silent coffin of disembodiment, few subjects could think of anything in particular after the second day. Concentration crumbled. The pseudobusyness of daydreaming took over. Worse were thoughts of an active mind that got stuck in an inactive loop. "One subject made up a game of listing, according to the alphabet, each chemical reaction that bore the name of the discoverer. At the letter *n* he was unable to think of an example. He tried to skip *n* and go on, but *n* kept doggedly coming up in his mind, demanding an answer. When this became tiresome, he tried to dismiss the game altogether, only to find that he could not. He endured the insistent demand of his game for a short time, and, finding that he was unable to control it, he pushed the panic button."

The body is the anchor of the mind, and of life. Bodies are machines to prevent the mind from blowing away under a wind of its own making. The natural tendency of neural circuitry is to play games with itself. Left on its own, without a direct link to "outside," a brainy network takes its own machinations as reality. A mind cannot possibly consider anything beyond what it can measure or calculate; without a body it can only consider itself. Given its inherent curiosity, even the simplest mind will exhaust itself devising solutions to challenges it confronts. Yet if most of what it confronts is its own internal circuitry and logic, then it spends its days tinkering with its latest fantasy.

The body—that is, any bundle of senses and activators—interrupts this natural mental preoccupation with an overload of urgent material that must be considered right now! A matter of survival! Should we duck?! The mind no longer needs to invent its reality—the reality is in its face, rapidly approaching dead-on. *Duck!* it decides by a new and wholly original insight it had never tried before, and would have never thought to try.

Without senses, the mind mentally masturbates, engendering a mental blindness. Without the interruptions of hellos from the eye, ear, tongue, nose, and finger, the evolving mind huddles in the corner picking its navel. The eye is most important because being half brain itself (chock-full of neurons and biochips) it floods the mind with an impossibly rich feed of half-digested data, critical decisions, hints for future steps, clues of hidden things, evocative movements, and beauty. The mind grinds under the load, and behaves. Cut loose from its eyes suddenly, the mind will rear up, spin, retreat.

The cataracts that afflict elderly men and women after a life of sight can be removed, but not without a brief journey into a blindness even darker than what cataracts bring. Doctors surgically remove the lens growths and then cover patients' eyes with a black patch to shield them from light and to prevent the eyeballs from moving, as they unconsciously do whenever they look. Since the eyes move in tandem, both are patched. To further reduce eye movement, patients lie in bed, quiet, for up to a week. At night, when the hospital bustle dies down, the stillness can match the blackness under the blindfold. In the early 1900s when this operation was first commonly performed, there was no machinery in hospitals, no TV or radio, few night shifts, no lights burning. Eyes wrapped in bandages in the cataract ward, the world as hushed and black as the deepest forever.

The first day was dim but full of rest and still. The second day was darker. Numbing. Restless. The third day was black, black, black, silent, and filled with red bugs crawling on the walls.

"During the third night following surgery [the 60-year-old woman] tore her hair and the bedclothes, tried to get out of bed, claimed that someone was trying to get her, and said that the house was on fire. She subsided when the bandage was removed from the unoperated eye," stated a hospital report in 1923.

In the early 1950s, doctors at Mount Sinai Hospital in New York studied a sample of 21 consecutive admissions to the cataract ward. "Nine patients became increasingly restless, tore off the masks, or tried to climb over the siderails. Six patients had paranoid delusions, four had somatic complaints, four were elated [!!], three had visual hallucinations, and two had auditory hallucinations."

"Black patch psychosis" is now something ophthalmologists watch for on the wards. I think universities should keep an eye out for it too. Every philosophy department should hang a pair of black eye patches in a red firealarm-like box that says, "In case of argument about mind/body, break glass, put on."

In an age of virtual everything, the importance of bodies cannot be overemphasized. Mark Pauline and Rod Brooks have advanced further than most in creating personas for machines, because the creatures are fully embodied. They insist that their robots be situated in real environments.

Pauline's automatons don't live very long. By the end of his shows, only a few iron beasts still move. But to be fair to Pauline, none of the other university robots have lived much longer than his. It is a rare mobile robot that has an "on" lifetime of more than dozens of hours. For the most part, automatons are improved while they are off. In essence, robotists are trying to evolve things while dead, a curious situation that hasn't escaped some researchers' notice. "You know, I'd like to build a robot that could run 24 hours a day for weeks. That's the way for a robot to learn," says Maja Mataric, one of Brooks's robot builders at MIT.

When I visited the Mobot Lab at MIT, Genghis lay sprawled in disassembled pieces on a lab bench. New parts lay nearby. "He's learning," quipped Brooks.

Genghis was learning, but not in any ultimately useful manner. He had to rely on the busy schedules of Brooks and his busy grad students. How much better to learn while alive. That is the next big step for machines. To learn over time, on their own. To not only adapt, but evolve.

Evolution proceeds in steps. Genghis is an insect-equivalent. Its descendants someday will be rodents, and someday further, as smart and nimble as apes.

But we need to be a little patient in our quest for machine evolution, Brooks cautions. From day one of Genesis, it took billions of years for life to reach plant stage, and another billion and a half before fish appeared. A hundred million years later insects made the scene. "Then things really started moving fast," says Brooks. Reptiles, dinosaurs, and mammals appeared within the next 100 million years. The great, brainy apes, including man, arrived in the last 20 million years.

The relatively rapid complexification in most recent geological history suggests to Brooks "that problem solving behavior, language, expert knowledge and reason, are all pretty simple once the essence of being and reacting are available." Since it took evolution 3 billion years to get from single cells to insects, but only another half billion years from there to humans, "this indicates the nontrivial nature of insect level intelligence."

So insect life—the problem Brooks is sweating over—is really the hard part. Get artificial insects down, and artificial apes will soon follow. This points to a second advantage to working with fast, cheap, and out-of-control mobots: the necessity of mass numbers for evolution. One Genghis can learn. But evolution requires a seething population of Genghises to get anything done.

To evolve machines, we'll need huge flocks of them. Gnatbots might be perfect. Brooks ultimately dreams of engineering vivisystems full of machines that both learn (adjust to variations in environment) and evolve (populations of critters undergoing "gazillions of trials").

When democracy was first proposed for (and by) humans, many reasonable people rightly feared it as worse than anarchy. They had a point. A democracy of autonomous, evolving machines will be similarly feared as Anarchy Plus. This fear too has some truth.

Chris Langton, an advocate of autonomous machine life, once asked Mark Pauline, "When machines are both superintelligent and superefficient, what will be the niche for humans? I mean, do we want machines, or do we want us?"

Pauline responded in words that I hope echo throughout this book: "I think humans will accumulate artificial and mechanical abilities, while machines will accumulate biological intelligence. This will make the confrontation between the two even less decisive and less morally clear than it is now."

So indecisive that the confrontation may resemble a conspiracy: robots who think, viruses that live in silicon, people hotwired to TV sets, life engineered at the gene level to grow what we want, the whole world networked into a human/machine mind. If it all works, we'll have contraptions that help people live and be creative, and people who help the contraptions live and be creative.

Consider the following letter published in the June 1984 *IEEE Spectrum*.

> Mr. Harmon Blis
> Topnotch Professionals Inc.
> 7777 Turing Blvd.
> Palo Alto, CA 94301
>
> June 1, 2034
>
> Dear Mr. Bliss:
>
> I am pleased to support your consideration of a human for professional employment. As you know, humans historically have proved to be the providers of choice. There are many reasons why we still recommend them strongly.
>
> As their name would suggest, humans are humane. They can transmit a feeling of genuine concern to their clients that makes for a better, more productive relationship.

Each human is unique. There are many situations that reward multiple viewpoints, and there is nothing like a team of individualistic humans to provide this variety.

Humans are intuitive, which enables them to make decisions even when they can't justify why.

Humans are flexible. Because clients often place highly varied, unpredictable demands on professionals, flexibility is crucial.

In summary, humans have a lot going for them. They are not a panacea, but they are the right solution for a class of important and challenging employment problems. Consider this human carefully.

Yours truly.
Frederick Hayes-Roth

The greatest social consequence of the Darwinian revolution was the grudging acceptance by humans that humans were random descendants of monkeys, neither perfect nor engineered. The greatest social consequence of neo-biological civilization will be the grudging acceptance by humans that humans are the random ancestors of machines, and that as machines we can be engineered ourselves.

I'd like to condense that further: Natural evolution insists that we are apes; artificial evolution insists that we are machines with an attitude.

I believe that humans are more than the combination of ape and machine (we have a lot going for us!), but I also believe that we are far more ape and machine than we think. That leaves room for an unmeasured but discernible human difference, a difference that inspires great literature, art, and our lives as a whole. I appreciate and indulge in those sentiments. But what I have encountered in the rather mechanical process of evolution, and in the complex but knowable interconnections underpinning living systems, and in the reproducible progress in manufacturing reliable behaviors in robots, is a singular unity between simple life, machines, complex systems, and us. This unity can stir lofty inspirations the equal of any passion in the past.

Machines are a dirty word now. This is because we have withheld from them the full elixir of life. But we are poised to remake them into something that one day may be taken as a compliment.

As humans, we find spiritual refuge in knowing that we are a branch in the swaying tree of life spread upon this blue ball. Perhaps someday we will find spiritual wholesomeness in knowing we are a link in a complex machine layered on top of the green life. Perhaps we'll sing hymns rhapsodizing our role as an ornate node in a vast network of new life that is spawning on top of the old.

When Pauline's monsters demolish fellow monsters, I see not useless destruction, but lions stalking zebras keeping wildlife on course. When the iron paw of Brooks's six-legged Genghis hunts for a place to grip, I see not workers relieved of robotic jobs, but joyful baby squirms of a new organism. We are of one nature in the end. Who will not feel a bit of holy awe on the day when machines talk back to us?

4

# ASSEMBLING COMPLEXITY

AS AN AUTUMN GRAY SETTLES, I stand in the middle of one of the last wildflower prairies in America. A slight breeze rustles the tan grass. I close my eyes and say a prayer to Jesus, the God of rebirth and resurrection. Then I bend at the waist, and with a strike of a match, I set the last prairie on fire. It burns like hell.

"The grass of the field alive today is thrown into the oven tomorrow," says the rebirth man. The Gospel passage comes to mind as an eight-foot-high wall of orange fire surges downwind crackling loudly and out of control. The heat from the wisps of dead grass is terrific. I am standing with a flapping rubber mat on a broom handle trying to contain the edges of the wall of fire as it marches across the buff-colored field. I remember another passage: "The new has come, the old is gone."

While the prairie burns, I think of machines. Gone is the old way of machines; come is the reborn nature of machines, a nature more alive than dead.

I've come to this patch of fire-seared grass because in its own way this wildflower field is another item of human construction, as I can explain in a moment. The burnt field makes a case that life is becoming manufactured, just as the manufactured is becoming life, just as both are becoming something wonderful and strange.

The future of machines lies in the tangled weeds underfoot.

Machines have steadily plowed under wildflower prairies until none are left except the tiny patch I'm standing in. But in a grand irony, this patch holds the destiny of machines, for the future of machines is biology.

My guide to the grassy inferno is Steve Packard, an earnest, mid-thirties guy, who fondles bits of dry weeds—their Latin names are intimately familiar to him—as we ramble through the small prairie. Almost two decades ago, Packard was captured by a dream he couldn't shake. He imagined a suburban dumping ground blooming again in its original riotous prairie-earth colors, an oasis of life giving soulful rest to harried cosmopolitans. He dreamt of a prairie gift that would "pay for itself in quality-of-life dollars," as he was fond of telling supporters. In 1974 Packard began working on his vision. With the mild help of skeptical conservation groups, he began to recreate a real prairie not too far from the center of the greater city of Chicago.

Packard knew that the godfather of ecology, Aldo Leopold, had successfully recreated a prairie of sorts in 1934. The University of Wisconsin, where Leopold worked, had purchased an old farm, called the Curtis place, to make an arboretum out of it. Leopold convinced the University to let the Curtis farm revert to prairie again. The derelict farm would be plowed one last time, then sown with disappearing and all but unknown prairie seeds, and left to be.

This simple experiment was not undoing the clock; it was undoing civilization.

Until Leopold's innocent act, every step in civilization had been another notch in controlling and retarding nature. Houses were designed to keep nature's extreme temperatures out. Gardens contrived to divert the power of botanical growth into the tame artifacts of domesticated crops. Iron mined in order to topple trees for lumber.

Respites from this march of progress were rare. Occasionally a feudal lord reserved a wild patch of forest from destruction for his game hunting. Within this sanctuary a gamekeeper might plant wild grain to attract favored animals for his lord's hunt. But until Leopold's folly no one had ever deliberately planted wilderness. Indeed, even as Leopold oversaw the Curtis project, he wondered if anyone could plant wilderness. As a naturalist, he figured it must be largely a matter of letting nature reclaim the spot. His job would be protecting whatever gestures nature made. With the help of colleagues and small bands of farm boys hired by the Civilian Conservation Corps during the Depression, Leopold nursed 300 acres of young emerging prairie plants with buckets of water and occasional thinning of competitors for the first five years.

The prairie plants flourished; but so did the nonprairie weeds. Whatever was carpeting this meadow, it was not the prairie that once did. Tree seedlings, Eurasian migrants, and farm weeds all thrived along with the replanted prairie species. Ten years after the last plowing, it was evident to Leopold that the reborn Curtis prairie was only a half-breed wilderness. Worse, it was slowly becoming an overgrown weedy lot. Something was missing.

A key species, perhaps. A missing species which once reintroduced, would reorder the whole community of ecology of plants. In the mid–1940s that species was identified. It was a wary animal, once ubiquitous on the tall grass prairies, that roamed widely and interacted with every plant, insect, and bird making a home over the sod. The missing member was fire.

Fire made the prairie work. It hatched certain fire-triggered seeds, it eliminated intruding tree saplings, it kept the fire-intolerant urban competitors down. The rediscovery of fire's vital function in tall grass prairie ecology coincided with the rediscovery of fire in the role of almost all the other ecologies in North America. It was a rediscovery because fire's effects on nature had been recognized and used by the aboriginal researchers of the land. The ubiquitous prevalence of fire on the pre-whiteman prairie was well documented by European settlers.

While evident to us now, the role of fire as a key ingredient of the prairie was not clear to ecologists and less clear to conservationists, or what we would now call environmentalists. Ironically, Aldo Leopold, the greatest American ecologist, argued fiercely against letting wildfire burn in wilderness. He wrote in 1920, "The practice of [light-burning] would not only fail to prevent serious fires but would ultimately destroy the productivity of the forests on which western industries depend for their supply of timber." He gave five reasons why fire was bad, none of them valid. Railing against the "light-burning propagandists," Leopold wrote, "It is probably a safe prediction to state that should light-burning continue for another fifty years, our existing forest areas would be further curtailed to a very considerable extent."

A decade later, when more was known about the interdependencies of nature, Leopold finally conceded the vital nature of organic fire. When he reintroduced fire into the synthetic plots of the Wisconsin field grass arboretum, the prairie flourished like it had not for centuries. Species that were once sparse started to carpet the plots.

Still, even after 50 years of fire and sun and winter snows, the Curtis prairie today is not completely authentic in the diversity of its members. Around the edges especially, where ecological diversity is usually the greatest, the prairie suffers from invasions of monopolistic weeds—the same few ones that thrive on forgotten lots.

The Wisconsin experiment proved one could cobble together a fair approximation of a prairie. What in the world would it take to make a pure prairie, authentic in every respect, an honest-to-goodness recreated prairie? Could one grow a real prairie from the ground up? Is there a way to manufacture a self-sustaining wilderness?

IN THE FALL OF 1991, I stood with Steve Packard in one of his treasures—what he called a "Rembrandt found in the attic"—at the edge of a suburban Chicago woods. This was the prairie we would burn. Several hundred acres of rustling, wind-blown grass swept over our feet and under scattered oak trees. We swam in a field far richer, far more complete, and far more authentic than Leopold had seen. Dissolved into this pool of brown tufts were hundreds of uncommon species. "The bulk of the prairie is grass," Packard shouted to me in the wind, "but what most people notice is the advertising of the flowers." At the time of my visit, the flowers were gone, and the ordinary-looking grass and trees seemed rather boring. That "barrenness" turned out to be a key clue in the rediscovery of an entire lost ecosystem.

To arrive at this moment, Packard spent the early 1980s locating small, flowery clearings in the thickets of Illinois woods. He planted prairie wildflower seeds in them and expanded their size by clearing the brush at their perimeters. He burnt the grass to discourage non-native weeds. At first he hoped the fire would do the work of clearing naturally. He would let it leap from the grass into the thicket to burn the understory shrubs. Then, because of the absence of fuel in the woods, the fire would die naturally. Packard told me, "We let the fires blast into the bush as far as they would go. Our motto became 'Let the fires decide.' "

But the thickets would not burn as he hoped, so Packard and his crews interceded with axes in hand and physically cleared the underbrush. Within two years, they were happy with their results. Thick stands of wild rye grass mingled with yellow coneflower in the new territory. The restorers manually hacked back the brush each season and planted the choicest prairie flower seed they could find.

But by the third year, it was clear something was wrong. The plantings were doing poorly in the shade, producing poor fuel for the season's fires. The grasses that did thrive were not prairie species; Packard had never seen them before. Gradually, the replanted areas reverted to brush.

Packard began to wonder if anyone, including himself, would go through the difficulties of burning an empty plot for decades if they had nothing to show for it. He felt yet another ingredient must be missing which prevented a living system from snapping together. He started reading the botanical history of the area and studying the oddball species.

When he identified the unknown species flourishing so well in the new oak-edge patches, he discovered they didn't belong to a prairie, but to a savanna ecosystem—a prairie with trees. Researching the plants that were associated with savanna, Packard soon came up with a list of other associated species—such as thistles, cream gentians, and yellow pimpernels—that he quickly realized peppered the fringes of his restoration sites. Packard had even found a blazing star flower a few years before. He had brought the flowering plant to a university expert because varieties of blazing star defy nonexpert identification. "What the heck is this?" he'd asked the botanist. "It's not in the books, it's not listed in the state catalog of species. What is it?" The botanist had said, "I don't know. It could be a *savanna* blazing star, but there aren't any savannas here, so it couldn't be that. Don't know what it is." What one is not looking for, one does not see. Even Packard admitted to himself that the unusual wildflower must have been a fluke, or misidentified. As he recalls, "The savanna species weren't what I was looking for at first so I had sort of written them off."

But he kept seeing them. He found more blazing star in his patches. The oddball species, Packard was coming to realize, were the main show of the clearings. There were many other species associated with savannas he did not recognize, and he began searching for samples of them in the corners of old cemeteries, along railway right-of-ways, and old horse paths—anywhere a remnant of an earlier ecosystem might survive. Whenever he could, he collected their seed.

An epiphany of sorts overtook Packard when he watched the piles of his seed accumulate in his garage. The prairie seed mix was dry and fluffy—like grass seed. The emerging savanna seed collection, on the other hand, was "multicolored handfuls of lumpy, oozy, glop," ripe with pulpy seeds and dried fruits. Not by wind, but by animals and birds did these seeds disperse. The thing—the system of coevolved, interlocking organisms—he was seeking to restore was not a mere prairie, but a prairie with trees: a savanna.

The pioneers in the Midwest called a prairie with trees a "barren." Weedy thickets and tall grass grew under occasional trees. It was neither grassland nor forest—therefore barren to the early settlers. An almost entirely different set of species kept it a distinct biome from the prairie. The savanna barrens were particularly dependent on fire, more so than the prairies, and when farmers arrived and stopped the fires, the barrens very quickly collapsed into a woods. By the turn of this century the barrens were almost extinct, and the list of their constituent species hardly recorded. But once Packard got a "search image" of the savanna in his mind, he began to see evidence of it everywhere.

Packard sowed the mounds of mushy oddball savanna species, and within two years the fields were ablaze with rare and forgotten wildflowers: bottlebrush grass, blue-stem goldenrod, starry champion, and big-leafed aster. In 1988, a drought shriveled the non-native weeds as the reseeded natives flourished and advanced. In 1989, a pair of eastern bluebirds (which had not been seen in the county for decades) settled into their familiar habitat—an event that Packard took as "an endorsement." The university botanist called back. Seems like there *were* early records of savanna blazing star in the state. The biologists were putting it on the endangered list. Oval milkweed somehow returned to the restored barren although it grows nowhere else in the state. Rare and endangered plants like the white-fringed orchid and a pale vetchling suddenly sprouted on their own. The seed might have lain dormant—and between fire and other factors found the right conditions to hatch—or been brought in by birds such as the visiting bluebirds. Just as miraculously, the silvery-blue butterfly, which had not been seen anywhere in Illinois for a full decade, somehow found its way to suburban Chicago where its favorite food, vetchling, was now growing in the emerging savanna.

"Ah," said the expert entomologists. "The classic savanna butterfly is Edwards hairstreak. But we don't see any. Are you sure this is a savanna?" But by the fifth year of restoration, the Edwards hairstreak butterfly was everywhere on the site.

If you build it, they will come. That's what the voice said in the *Field of Dreams*. And it's true. And the more you build it, the more that come. Economists call it the "law of increasing returns"—the snowballing effect. As the web of interrelations is woven tighter, it becomes easier to add the next piece.

Yet there was still an art to it. As Packard knotted the web, he noticed that it mattered what order he added the pieces in. And he learned that other ecologists had discovered the same thing. A colleague of Leopold had found that he got closer to a more authentic prairie by planting prairie seed in a weedy field, rather than in a newly plowed field, as Leopold had first done. Leopold had been concerned that the aggressive weeds would strangle the wildflowers, but a weedy field is far more like a prairie than a plowed field. Some weeds in an old weedy lot are latecomers, and a few of these latecomers are prairie members; their early presence in the conversion quickens the assembly of the prairie system. But the weeds that immediately sprout in a plowed, naked field are very aggressive, and the beneficial late-arriving weeds come into the mix too late. It's like having the concrete reinforcement bars arrive after you've poured the cement foundation for your house. Succession is important.

Stuart Pimm, an ecologist at the University of Tennessee, compares succession paths—such as the classic series of fire, weed, pine, broadleaf trees—to well-rehearsed assembly sequences that "the players have played many times. They know, in an evolutionary sense, what the sequence is." Evolution not only evolves the functioning community, but it also finely tunes the assembly process of the gathering until the community practically falls together. Restoring an ecosystem community is coming at it from the wrong side. "When we try to restore a prairie or wetland, we are trying to assemble an ecosystem along a path that the community has no practice in," says Pimm. We are starting with an old farm, while nature may have started with a glacial moraine ten thousand years ago. Pimm began asking himself: Can we assemble a stable ecosystem by taking in the parts at random? Because at random was exactly how humans were trying to restore ecosystems.

In a laboratory at the University of Tennessee, ecologists Pimm and Jim Drake had been assembling ingredients of micro-ecosystems in different random orders to chart the importance of sequence. Their tiny worlds were microcosms. They started with 15 to 40 different pure strains of algae and microscopic animals, and added these one at a time in various combinations and sequences to a large flask. After 10 to 15 days, if all went well, the aquatic mixture formed a stable, self-reproducing slime ecology—a distinctive mix of species surviving off of each other. In addition Drake set up artificial ecologies in aquaria and in running water for artificial stream ecologies. After mixing them, they let them run until they were stable. "You look at these communities and you don't need to be a genius to see that they are different," Pimm remarks. "Some are green, some brown, some white. But the interesting thing is that there is no way to tell in advance which way a particular combination of species will go. Like most complex systems, you have to set them up and run them to find out."

It was also not clear at the start whether finding a stable system would be easy. A randomly made ecosystem was likely, Pimm thought, "to just wander around forever, going from one state to the next and back again without ever coming to a persistent state." But the artificial ecosystems didn't wander. Instead, much to their surprise, Pimm found "all sorts of wonderful wrinkles. For one, these random ecosystems have absolutely no problem in stabilizing. Their most common feature is that they always come to a persistent state, and typically it's one state per system."

It was very easy to arrive at a stable ecosystem, if you didn't care what system you arrived at. This was surprising. Pimm said, "We know from chaos theory that many deterministic systems are exquisitely sensitive to initial conditions—one small difference will send it off into chaos. This stability is the opposite of that. You start out in complete randomness, and you see these things assemble towards something that is a lot more structured than you had any reason to believe could be there. This is anti-chaos."

To complement their studies *in vitro*, Pimm also set up experiments "*in silico*"—simplified ecological models in a computer. He created artificial "species=' of code that required the presence of certain other species to survive, and also gave them a pecking order so that species *B* might drive out species *A* if and when the population of *B* reached a certain density. (Pimm's models of random ecologies bear some resemblance to Stuart Kauffman's models of random genetic networks; *see* chapter 20). Each species was loosely interconnected to the others in a kind of vast distributed network. Running thousands of random combinations of the same list of species, Pimm mapped how often the resulting system would stabilize so that minor perturbations, such as introductions or removals of a few species, would not destabilize the collective mix. His results mirrored the results from his bottled living microworlds.

In Pimm's words, the computer models showed that "with just 10 to 20 components in the mix, the number of peaks [or stabilities] may be in the tens, twenties or hundreds. And if you play the tape of life again, you get to a different peak." In other words, after dropping in the same inventory of species, the mess headed toward a dozen final arrangements, but changing the entry sequence of even one of the species was enough to divert the system from one of the end-points to another. The system *was* sensitive to initial conditions, but it was usually attracted to order.

Pimm saw Packard's work in restoring the Illinois prairie/savanna as validating his findings: "When Packard first tried to assemble the community, it didn't work in the sense that he couldn't get the species he wanted to stick and he had a lot of trouble taking out things he didn't want. But once he introduced the oddball, though proper, species it was close enough to the persistent state that it easily moved there and will probably stay there."

Pimm and Drake discovered a principle that is a great lesson to anyone concerned about the environment, and anyone interested in building complex systems. "To make a wetland you can't just flood an area and hope for the best," Pimm told me. "You are dealing with systems that have assembled over hundreds of thousands, or millions of years. Nor is compiling a list of what's there in terms of diversity enough. You also have to have the assembly instructions."

STEVE PACKARD set out to extend the habitat of authentic prairie. On the way he resurrected a lost ecosystem, and perhaps acquired the assembly instructions for a savanna. Working in an ocean of water instead of an ocean of grass, David Wingate in Bermuda set out thirty years ago to nurse a rare species of shorebird back from extinction. On the way, he recreated the entire ecology of a subtropical island, and illuminated a further principle of assembling large functioning systems.

The Bermuda tale involves an island suffering from an unhealthy, ad hoc, artificial ecosystem. By the end of World War II, Bermuda was ransacked by housing developers, exotic pests, and a native flora wrecked by imported garden species. The residents of Bermuda and the world's scientific community were stunned, then, in 1951 by the announcement that the cahow—a gull-size seabird—had been rediscovered on the outer cliffs of the island archipelago. The cahow was thought to be extinct for centuries. It was last seen in the 1600s, around the time the dodo had gone extinct. But by a small miracle, a few pairs of breeding cahows hung for generations on some remote sea cliffs in the Bermuda archipelago. They spent most of their life on water, only coming ashore to nest underground, so they went unnoticed for four centuries.

As a schoolboy with an avid interest in birds, David Wingate was present in 1951 when a Bermudan naturalist succeeded in weaseling the first cahow out of its deep nesting crevice. Later, Wingate became involved in efforts to reestablish the bird on a small uninhabited island near Bermuda called Nonsuch. He was so dedicated to the task that he moved—newly married—to an abandoned building on the uninhabited, unwired outer island.

It quickly became apparent to Wingate that the cahow could not be restored unless the whole ecosystem of which it was part was also restored. Nonsuch and Bermuda itself were once covered by thick groves of cedar, but the cedars had been wiped out by an imported insect pest in a mere three years between 1948 and 1952. Only their huge white skeletons remained. In their stead were a host of alien plants, and on the main island, many tall ornamental trees that Wingate was sure would never survive a once-in-fifty-year hurricane.

The problem Wingate faced was the perennial paradox that all whole systems makers confront: where do you start? Everything requires everything else to stay up, yet you can't levitate the whole thing at once. Some things have to happen first. And in the correct order.

Studying the cahows, Wingate determined that their underground nesting sites had been diminished by urban sprawl and subsequently by competition with the white-tailed tropicbird for the few remaining suitable sites. The aggressive tropicbird would peck a cahow chick to death and take over the nest. Drastic situations require drastic measures, so Wingate instituted a "government housing program" for the cahow. He built artificial nest sites—sort of underground birdhouses. He couldn't wait until Nonsuch reestablished a forest of trees, which tip slightly in hurricanes to uproot just the right-sized crevice, too small for the tropicbird to enter, but just perfect for the cahow. So he created a temporary scaffolding to get one piece of the puzzle going.

Since he needed a forest, he planted 8,000 cedar trees in the hope that a few would be resistant to the blight, and a few were. But the wind smothered them. So Wingate planted a scaffold species—a fast-growing non-native evergreen, the casuarinas—as a windbreak around the island. The casuarinas grew rapidly, and let the cedars grow slowly, and over the years, the better-adapted cedars displaced the casuarinas. The resown forest made the perfect home for a night heron which had not been seen on Bermuda for a hundred years. The heron gobbled up land crabs which, without the herons, had become a pest on the islands. The exploded population of land crabs had been feasting on the succulent sprouts of wetland vegetation. The crabs' reduced numbers now allowed rare Bermudan sedges to grow, and in recent years, to reseed. It was like the parable of "For Want of a Nail, The Kingdom Was Lost," but in reverse: By finding the nail, the kingdom was won. Notch by notch, Wingate was reassembling a lost ecosystem.

Ecosystems and other functioning systems, like empires, can be destroyed much faster than they can be created. It takes nature time to grow a forest or marsh because even nature can't do everything at once. The kind of assistance Wingate gave is not unnatural. Nature commonly uses interim scaffolding to accomplish many of her achievements. Danny Hillis, an artificial intelligence expert, sees a similar story in the human thumb as a platform for human intelligence. A dexterous hand with a thumb-grasp made intelligence advantageous (for now it could make tools), but once intelligence was established, the hand was not as important. Indeed, Hillis claims, there are many stages needed to build a large system that are not required once the system is running. "Much more apparatus is probably necessary to exercise and evolve intelligence than to sustain it," Hillis wrote. "One can believe in the necessity of the opposable thumb for the development of intelligence without doubting a human capacity for thumbless thought."

When we lie on our backs in an alpine meadow tucked on the perch of high mountains, or wade into the mucky waters of a tidal marsh, we are encountering the "thumbless thoughts" of nature. The intermediate species required to transform the proto-meadow into a regenerating display of flowers are now gone. We are only left now with the thought of flowers and not the helpful thumbs that chaperoned them into being.

YOU MAY HAVE HEARD the heartwarming account of "The Man Who Planted Trees and Grew Happiness." It's about how a forest and happiness were created out of almost nothing. The story is told by a young European man who hikes into a remote area of the Alps in 1910.

The young man wanders into a windy, treeless region, a harsh place whose remaining inhabitants are a few mean, poor, discontented charcoal burners huddled in a couple of dilapidated villages. The hiker meets the only truly happy inhabitant in the area, a lone shepherd hermit. The young man watches in wonder as the hermit wordlessly and idiotically spends his days poking acorns one by one into the moonscape. Every day the silent hermit plants 100 acorns. The hiker departs, eager to leave such desolation, only to return many years later by accident, after the interruption of World War I. He now finds the same village almost unrecognizable in its lushness. The hills are flush with trees and vegetation, brimming with streams, and full of wildlife and a new population of content villagers. Over three decades the hermit had planted 90 square miles thick with oak, beech, and birch trees. His single-handed work—a mere nudge in the world of nature—had remodeled the local climate and restored the hopes of many thousands of people.

The only unhappy part of the story is that it is not true. Although it has been reprinted as a true story all over the world, it is, in fact, a fantasy written by a Frenchman for *Vogue* magazine. There *are*, however, genuine stories of idealists recreating a forest environment by planting trees in the thousands. And their results confirm the Frenchman's intuition: tiny plants grown on a large scale can divert a local ecosystem in a positive loop of increasing good.

As one true example, in the early 1960s, an eccentric Englishwoman, Wendy Campbell-Purdy, journeyed to North Africa to combat the encroaching sand dunes by planting trees in the desert. She planted a "green wall" of 2,000 trees on 45 acres in Tiznit, Morocco. In six years' time, the trees had done so well, she founded a trust to finance the planting of 130,000 more trees on a 260-acre dump in the desert wastes at Bou Saada, Algeria. This too took off, creating a new minihabitat that was suitable for growing citrus, vegetables, and grain.

Given a slim foothold, the remarkable latent power in interconnected green things can launch the law of increasing returns: "Them that has, gets more." Life encourages an environment that encourages yet more life. On Wingate's island the presence of herons enables the presence of sedges. In Packard's prairie the toehold of fire enables the existence of wildflowers which enable the existence of butterflies. In Bou Saada, Algeria, some trees alter the climate and soil to make them fit for more trees. More trees make a space for animals and insects and birds, which prepare a place for yet more trees. Out of acorns, nature makes a machine that provides a luxurious home for people, animals, and plants.

The story of Nonsuch and the other forests of increasing returns, as well as the data from Stuart Pimm's microcosms overlap into a powerful lesson that Pimm calls the Humpty Dumpty Effect. Can we put the Humpty Dumpty of a lost ecosystem together again? Yes, we can if we have all the pieces. But we don't know if we do. There may be chaperone species that catalyze the assembly of an ecosystem in some early stage—the thumb for intelligence—that just aren't around the neighborhood anymore. Or, in a real tragedy, a key scaffold species may be globally extinct. One could imagine a hypothetical small, prolific grass essential to creating the matrix out of which the prairie arose, which was wiped out by the last ice age. With it gone, Humpty Dumpty can't be put back together again. "Keep in mind you can't always get there from here," Pimm says.

Packard has contemplated this sad idea. "One of the reasons the prairie may never be fully restored is that some parts are forever gone. Perhaps without the megaherbivores like the mastodon of old or even the bison of yesteryear, the prairie won't come back." Even more scary is yet another conclusion of Pimm's and Drake's work: that it is not just the presence of the right species, in the right order, but the absence of the right species at the right time as well. A mature ecology may be able to tolerate species *X* easily; but during its assembly, the presence of species *X* will divert the system onto some other path leading toward a different ecosystem. "That's why," Packard sighs, "it may take a million years to make an ecosystem." Which species now rooted on Nonsuch island or dwelling in the Chicago suburbs might push the reemerging savanna ecosystem away from its original destination?

The rule for machines is counterintuitive but clear: Complex machines must be made incrementally and often indirectly. Don't try to make a functioning mechanical system all at once, in one glorious act of assembly. You have to first make a working system that serves as a platform for the system you really want. To make a mechanical mind, you need to make the equivalent of a mechanical thumb—a lateral approach that few appreciate. In assembling complexity, the bounty of increasing returns is won by multiple tries over time—a process anyone would call growth.

Ecologies and organisms have always been grown. Today computer networks and intricate silicon chips are grown too. Even if we owned the blueprints of the existing telephone system, we could not assemble a replacement as huge and reliable as the one we had without in some sense recapitulating its growth from many small working networks into a planetary web.

Creating extremely complex machines, such as robots and software programs of the future, will be like restoring prairies or tropical islands. These intricate constructions will have to be assembled over time because that is the only way to make sure they work from top to bottom. Unripe machinery let out before it is fully grown and fully integrated with diversity will be a common complaint. "We ship no hardware before its time," will not sound funny before too long.

5

# COEVOLUTION

WHAT COLOR is a chameleon placed on the mirror?

Stewart Brand posed that riddle to Gregory Bateson in the early 1970s. Bateson, together with Norbert Wiener, was a founding father of the modern cybernetic movement. Bateson had a most orthodox Oxford education and a most unorthodox career. He filmed Balinese dance in Indonesia; he studied dolphins; he developed a useful theory of schizophrenia. While in his sixties, he taught at the University of California at Santa Barbara, where his eccentric brilliant views on mental health and evolutionary systems caught the attention of holistically minded hippies.

Stewart Brand, a student of Bateson's, was himself a legendary promoter of cybernetic holism. Brand published his chameleon koan in his *Whole Earth Catalog*, in 1974. Writes Brand of his riddle: "I asked the question of Gregory Bateson at a point in our interview when we were lost in contemplation of the function, if any, of consciousness—self-consciousness. Both of us being biologists, we swerved to follow the elusive chameleon. Gregory asserted that the creature would settle at a middle value in its color range. I insisted that the poor beast trying to disappear in a universe of itself would endlessly cycle through a number of its disguises."

The mirror is a clever metaphor for informational circuits. Two ordinary mirrors facing each other will create a fun-house hall that ricochets an image back and forth until it vanishes into an infinite regress. Any message loosed between the two opposing mirrors bounces to exhaustion without changing its form. But what if one side is a responsive mirror, just as the chameleon is, in part reflecting, in part generating? The very act of accommodating itself to its own reflection would disturb it anew. Could it ever settle into a pattern persistent enough to call it something?

Bateson felt the system—perhaps like self-consciousness—would quickly settle out at an equilibrium determined by the pull of the creature's many extremes in color. The conflicting colors (and conflicting viewpoints in a society of mind) would compromise upon a "middle value," as if it were a democracy voting. On the other hand, Brand opined that equilibrium of any sort was next to impossible, and that the adaptive system would oscillate without direction or end. He imagined the colors fluctuating chaotically in a random, psychedelic paisley.

The chameleon responding to its own shifting image is an apt analog of the human world of fashion. Taken as a whole, what are fads but the response of a hive mind to its own reflection?

In a 21st-century society wired into instantaneous networks, marketing is the mirror; the collective consumer is the chameleon. What color is the consumer when you put him on the marketplace? Does he dip to the state of the lowest common denominator—a middle average consumer? Or does he oscillate in mad swings of forever trying to catch up with his own moving reflection?

Bateson was tickled by the depth of the chameleon riddle and passed it on to his other students. One of them, Gerald Hall, proposed a third hypothesis for the final color of the mirror visitor: "The chameleon will stay whatever color he was at the moment he entered the mirror domain."

This is the most logical answer in my view. The coupling between mirror and chameleon is probably so tight and immediate that almost no adaptation is possible. In fact, it may be that once the chameleon bellies up to the mirror, it can't budge from its color unless a change is induced from outside or from an erroneous drift in the chameleon's coloration process. Otherwise, the mirror/chameleon system freezes solidly onto whatever initial value it begins with.

For the mirrored world of marketing, this third answer means the consumer freezes. He either locks onto whatever brand he began with, or he stops purchasing altogether.

There are other possible answers, too. While conducting interviews for this book, I sometimes posed the chameleon riddle to my interviewees. The scientists understood it for the archetypal case of adaptive feedback it was. Their answers ranged over the map. Some examples:

> MATHEMATICIAN JOHN HOLLAND: It goes kaleidoscopic! There's a lag time, so it'll flicker all over the place. The chameleon won't ever be a uniform color.

> COMPUTER SCIENTIST MARVIN MINSKY: It might have a number of eigenvalues or colors, so it will zero in on a number of colors. If you put it in when it's green it might stay green, and if it was red it might stay red, but if you put it in when it was brown it might tend to go to green.

> NATURALIST PETER WARSHALL: A chameleon changes color out of a fright response so it all depends on its emotional state. It might be frightened by its image at first, but then later "warm up" to it, and so change colors.

Putting a chameleon on a mirror seemed a simple enough experiment that I thought that even a writer could perform it. So I did. I built a small, mirrored box, and I bought a color-changing lizard and placed it inside. Although Brand's riddle had been around for 20 years, this was the first time, as far as I know, anyone had actually tried it.

On the mirror the lizard stabilized at one color of green—the green of young leaves on trees in the spring—and returned to that one color each time I tried the experiment. But it would spend periods being brown before returning to green. Its resting color in the box was not the same dark brown it seemed to like when out of the mirrored box.

Although I performed this experiment, I place very little confidence in my own results for the following important reasons:

the lizard I used was not a true chameleon, but an anole, a species with a far more limited range of color adaptation than a true chameleon. (A true chameleon may cost several hundred dollars and requires a terrarium of a quality I did not want to possess.) More importantly, according to the little literature I read, anoles change colors for other reasons in addition to trying to match their background. As Warshall said, they also alter in response to fright. And frightened it was. The anole *did not* want to go into the mirrored box. The color green it presented in the box is the same color it uses when it is frightened. It may be that the chameleon in the mirror is merely in a constant state of fright at its own amplified strangeness now filling its universe. I certainly would go nutty in a mirrored box. Finally, there is this observer problem: I can only see the lizard when my face is peeking into the mirrored box, an act which inserts a blue eye and red nose into the anole's universe, a disturbance I could not circumvent.

It may be that an exact answer to the riddle requires future experiments with an authentic chameleon and many more controls than I had. But I doubt it. True chameleons are full-bodied animals just as anoles are, with more than one reason for changing colors. The chameleon-on-a-mirror riddle is best kept in idealized form as a thought experiment.

Even in the abstract, the "real" answer depends on such specific factors as the reaction time of the chameleon's color cells, their sensitivity to a change in hue, and whether other factors influence the signals—all the usual critical values in feedback circuits. If one could alter these functions in a real chameleon, one could then generate each of the chameleon-on-the-mirror scenarios mentioned above. This, in fact, is what engineers do when they devise electronic control circuits to guide spaceships or steer robot arms. By tweaking delay times, sensitivity to signals, dampening valves, etc., they can tailor a system to seek either a wide-ranging equilibrium (say, keeping the temperature between 68 and 70 degrees), or constant change, or some homeostatic point in between.

We see this happening in networked markets. A sweater manufacturer will try to rig a cultural mirror that encourages wild fluctuations in the hopes of selling many styles of sweaters, while a dishwasher manufacturer will try to focus the reflections onto the common denominators of only a few dishwasher images, since making varieties of sweaters is much cheaper than making varieties of dishwashers. The type of market is determined by quantity and speed of feedback signals.

The important point about the chameleon-on-the-mirror riddle is that the lizard and glass become one system. "Lizardness" and "mirrorness" are encompassed into a larger essence—a "lizard-glass"—which acts differently than either a chameleon or a mirror.

Medieval life was remarkably unnarcissistic. Common folk had only vague notions of their own image in the broad sense. Their individual and social identities were informed by participating in rituals and traditions rather than by reflection. On the other hand, the modern world is being paved with mirrors. We have ubiquitous TV cameras, and ceaseless daily polling ("63 Percent of Us Are Divorced") to mirror back to us every nuance of our collective action. A steady paper trail of bills, grades, pay stubs, and catalogs helps us create our individual identity. Pervasive digitalization of the approaching future promises clearer, faster, and more omnipresent mirrors. Every consumer becomes both a reflection and reflector, a cause and an effect.

The Greek philosophers were obsessed with the chain of causality, how the cause of an effect should be traced back in a relay of hops until one reached the Prime Cause. That backward path is the foundation of Western, linear logic. The lizard-glass demonstrates an entirely different logic—the circular causality of the Net. In the realm of recursive reflections, an event is not triggered by a chain of being, but by a field of causes reflecting, bending, mirroring each other in a fun-house nonsense. Rather than cause and control being dispensed in a straight line from its origin, it spreads horizontally, like creeping tide, influencing in roundabout, diffuse ways. Small blips can make big splashes, and big blips no splashes. It is as if the filters of distance and time were subverted by the complex connecting of everything to everything.

Computer scientist Danny Hillis has noted that computation,

particularly networked computation, exhibits a nonlinear causality field. He wrote:

> In the physical universe the effect that one event has on another tends to decrease with the distance in time or in space between them. This allows us to study the motions of the Jovian moons without taking into account the motion of Mercury. It is fundamental to the twin concepts of *object* and *action*. Locality of action shows itself in the finite speed of light, in the inverse square law of fields, and in macroscopic statistical effects, such as rates of reaction and the speed of sound.
>
> In computation, or at least in our old models of computation, an arbitrarily small event can and often does cause an arbitrarily large effect. A tiny program can clear all of memory. A single instruction can stop the machine. In computation there is no analog of distance. One memory location is as easily influenced as another.

The lines of control in natural ecologies also dissolve into a causality horizon. Control is not only distributed in space, but it is also blurred in time as well. When the chameleon steps onto the mirror, the cause of his color dissolves into a field of effects spinning back on themselves. The reasons for things do not proceed like an arrow, but rather spread to the side like a wind.

STEWART BRAND majored in biology at Stanford, where his teacher was Paul Ehrlich, a population biologist. Ehrlich too was fascinated by the rubbery chameleon-on-the-mirror paradox. He saw it most vividly in the relationship between a butterfly and its host plant. Fanatical butterfly collectors had long ago figured out that the best way to get perfect specimens was to encase a caterpillar, along with a plant it feeds on, in a box while waiting for the larvae to metamorphose. After transformation, the butterfly would emerge in the box sporting flawless unworn wings. It would be immediately killed and mounted.

This method required that collectors figure out which plants butterflies ate. With the prospect of perfect specimens, they did this thoroughly. The result was a rich literature of plant/butterfly communities, whose summary indicated that many butterflies in the larvae stage chomp on only one specific plant. Monarch caterpillars, for instance, devour only milkweeds. And, it seemed, the milkweed invited only the monarch to dine on it.

Ehrlich noticed that in this sense the butterfly was reflected in the plant, and the plant was reflected in the butterfly. Every step the milkweed took to keep the monarch larvae at bay so the worm wouldn't devour it completely, forced the monarch to "change colors" and devise a way to circumvent the plant's defenses. The mutual reflections became a dance of two chameleons belly to belly. In defending itself so thoroughly against the monarch, the milkweed became inseparable from the butterfly. And vice versa. Any long-term antagonistic relationship seemed to harbor this kind of codependency. In 1952, W. Ross Ashby, a cybernetician interested in how machines could learn, wrote, "[An organism's gene-pattern] does not specify in detail how a kitten shall catch a mouse, but provides a learning mechanism and a tendency to play, so that it is the *mouse* which teaches the kitten the finer points of how to catch mice."

Ehrlich came across a word to describe this tightly coupled dance in the title of a 1958 paper by C. J. Mode in the journal *Evolution*. It was called "coevolution," as in "A mathematical model for the co-evolution of obligate parasites and their hosts." Like most biological observations, the notion of coevolution was not new. The amazing Darwin himself wrote of "coadaptions of organic beings to each other . . ." in his 1859 masterpiece *Origin of Species*.

The formal definition of coevolution runs something like this: "Coevolution is reciprocal evolutionary change in interacting species," says John Thompson in *Interaction and Coevolution*. But what actually happens is more like a tango. The milkweed and monarch, shoulder to shoulder, lock into a single system, an evolution toward and with each other. Every step of coevolutionary advance winds the two antagonists more inseparably, until each is wholly dependent on the other's antagonism. The two become one. Biochemist James Lovelock writes of this embrace, "The evolution of a species is inseparable from the evolution of its environment. The two processes are tightly coupled as a single indivisible process."

Brand picked up the term and launched a magazine called *CoEvolution Quarterly*. It was devoted to the larger notion of all things—biological, societal, and technological—adapting to and creating each other, and at the same time weaving into one whole system. As an introduction Brand penned a definition: "Evolution is adapting to meet one's needs. Coevolution, the larger view, is adapting to meet each other's needs."

The "co" in coevolution is the mark of the future. In spite of complaints about the steady demise of interpersonal relationships, the lives of modern people are increasingly more codependent than ever. All politics these days means global politics and global politics means copolitics. The new online communities built between the spaces of communication networks are coworlds. Marshall McLuhan was not quite right. We are not hammering together a cozy global village. We are weaving together a crowded global hive—a coworld of utmost sociality and mirrorlike reciprocation. In this environment, all evolution, including the evolution of manufactured entities, is coevolution. Nothing changes without also moving closer to its changing neighbors.

Nature is chock-a-block with coevolution. Every green corner sports parasites, symbionts, and tightly coupled dances. Biologist P. W. Price estimated that over 50 percent of today's species are parasitic. (The figure has risen from the deep paleologic past and is expected to keep rising.) Here's news: half of the living world is codependent! Business consultants commonly warn their clients against becoming a symbiont company dependent upon a single customer-company, or a single supplier. But many do, and as far as I can tell, live profitable lives, no shorter on average than other companies. The surge of alliance-making in the 1990s among large corporations—particularly among those in the information and network industries—is another facet of an increasing coevolutionary economic world. Rather than eat or compete with a competitor, the two form an alliance—a symbiosis.

The parties in a symbiosis don't have to be symmetrical or

even at parity. In fact, biologists have found that almost all symbiotic alliances in nature entail a greater advantage for one party—in effect some hint of parasitism—in every codependency. But even though one side gains at the expense of the other, both sides gain over all, and so the pact continues.

In his magazine *CoEvolution* Brand began collecting stories of coevolutionary games. One of the most illustrative examples of alliance making in nature is the following:

> In eastern Mexico live a variety of acacia shrubs and marauding ants. Most acacias have thorns, bitter leaves, and other protection against a hungry world. One, the "swollen thorn acacia," learned to encourage a species of ant to monopolize it as a food source and kill or run off all other predators. Enticements gradually included nifty water-proof swollen thorns to live in, handy nectar fountains, and special ant-food buds at the leaf tips. The ants, whose interests increasingly coincided with the acacia's, learned to inhabit the thorns, patrol the acacia day *and* night, attack every acacia-hungry organism, and even prune away invading plants such as vines and tree seedlings that might shade Mother Acacia. The acacia gave up its bitter leaves, sharp thorns, and other devices and now requires the acacia-ant for survival. And the ant colonies can no longer live without the acacia. Together they're unbeatable.

In evolutionary time, the instances of coevolution have increased as sociability in life has increased. The more copious life's social behaviors are, the more likely they are to be subverted into mutually beneficial interactions. The more mutually responsive we construct our economic and material world, the more coevolutionary games we'll see.

Parasitic behavior itself is a new territory for organisms to make a living in. Thus we find parasites upon parasites. Ecologist John Thompson notes that "just as the richness of social behaviors may increase mutualism with other species, so may some mutualisms allow for the evolution of new social behaviors." In true coevolutionary fashion, coevolution breeds coevolution.

A billion years from now life on Earth may be primarily social, and stuffed with parasites and symbionts; and the world economy may be primarily a crowded network of alliances. What happens, then, when coevolution saturates a complete planet? What does a sphere of reflecting, responsive, coadapting, and recursive bits of life looping back upon itself do?

The butterfly and the milkweed constantly dance around each other, and by this ceaseless crazed ballet they move far beyond the forms they would have if they were at peace with each other. The chameleon on the mirror flipping without rest slips into some deranged state far from sanity. There *is* a sort of madness in pursuing self-reflections, that same madness we sensed in the nuclear arms race of post-World War II. Coevolution moves things to the absurd. The butterfly and the milkweed, although competitors in a way, cannot live apart. Paul Ehrlich sees coevolution pushing two competitors into "obligate cooperation." He wrote, "It's against the interests of either predator or prey to eliminate the enemy." That is clearly irrational, yet that is clearly a force that drives nature.

When a human mind goes off the deep end and gets stuck in the spiral of watching itself watching a mirror, or becomes so dependent upon its enemies that it apes them, then we declare it insane. Yet there is a touch of insanity—a touch of the off-balance—in intelligence and consciousness itself. To some extent a mind, even a primitive mind, must watch itself. Must any consciousness stare at its own navel?

This was the point in the conversation when Stewart Brand pointed out to Gregory Bateson his fine riddle of the chameleon on the mirror, and the two biologists swerved to follow it. The chase arrives at the odd conclusion that consciousness, life, intelligence, coevolution are off-balanced, unexpected, even unreasonable, given the resting point of everything else. We find intelligence and life spooky because they maintain a precarious state far from equilibrium. Compared to the rest of the universe, intelligence and consciousness and life are stable instabilities.

They are held together, poised upright like a pencil standing on its point, by the recursive dynamics of coevolution. The butterfly pushes the milkweed, and the milkweed pushes the butterfly, and the harder they push the more impossible it becomes for them to let go, until the whole butterfly/milkweed thing emerges as its own being—a living insect/plant system—pulling itself up by its bootstraps.

Rabid mutualism doesn't just happen in pairs. Threesomes can meld into an emergent, coevolutionarily wired symbiosis. Whole communities can be coevolutionary. In fact, any organism that adapts to organisms around it will act as an indirect coevolutionary agent to some degree. Since all organisms adapt that means *all* organisms in an ecosystem partake in a continuum of coevolution, from direct symbiosis to indirect mutual influence. The force of coevolutionism flows from one creature to its most intimate neighbors, and then ripples out in fainter waves until it immeasurably touches all living organisms. In this way the loose network of a billion species on this home planet are knit together so that unraveling the coevolutionary fabric becomes impossible, and the parts elevate themselves into some aggregate state of spooky, stable instability.

The network of life on Earth, like all distributed being, transcends the life of its ingredients. But bully life reaches deeper and ties up the entire planet in the web of its network, also roping in the nonliving matrix of rock and gas into its coevolutionary antics.

THIRTY YEARS AGO, biologists asked NASA to shoot a couple of unmanned probes towards the two likeliest candidates for extraterrestrial life, Mars and Venus, and poke a dipstick into their soil to check for vital signs.

The life-meter that NASA came up with was a complicated, delicate (and expensive) contraption that would, upon landing, be sprinkled with a planet's soil and check for evidence of bacterial life. One of the consultants hired by NASA was a soft-spoken British biochemist, James Lovelock, who found that he had a better way of checking for life on planets, a method that did not require a multimillion-dollar gadget, or even a rocket at all.

Lovelock was a very rare breed in modern science. He practiced science as a maverick, working out of a stone barn among the rural hedgerows in Cornwall, England. He maintained a spotless scientific reputation, yet he had no formal institutional affiliation, a rarity in the heavily funded world of science. His stark independence both nurtured and demanded free thinking. In the early 1960s Lovelock came up with a radical proposal that irked the rest of the folks on the NASA probe team. They really wanted to land a meter on another planet. He said they didn't have to bother.

Lovelock told them he could determine whether there was life on a planet by looking through a telescope. He could measure the spectrum of a planet's atmosphere, and thereby determine its composition. The makeup of the bubble of gases surrounding a planet would yield the secret of whether life inhabited the sphere. You therefore didn't need to hurl an expensive canister across the solar system to find out. He already knew the answer.

In 1967, Lovelock wrote two papers predicting that Mars would be lifeless based on his interpretation of its atmosphere. The NASA orbiters that circled Mars later in the decade, and the spectacular Mars soft landings the decade following made it clear to everyone that Mars was indeed as dead as Lovelock had forecasted. Equivalent probes to Venus brought back the same bad news: the solar system was barren outside of Earth.

How did Lovelock know?

Chemistry and coevolution. When the compounds in the Martian atmosphere and soil were energized by the sun's rays, and heated by the planetary core, and then contained by the Martian gravity, they settled into a dynamic equilibrium after millions of years. The ordinary laws of chemistry permit a scientist to make calculations of their reactions as if the planet were a large flask of matter. When a chemist derives the approximate formulas for Mars, Venus, and the other planets, the equations roughly balance: energy, compounds in; energy, compounds out. The measurements from the telescopes, and later the probes, matched the results predicted by the equations.

Not so the Earth. The mixture of gases in the atmosphere of the Earth are way out of whack. And they are out of whack,

Lovelock was to find out, because of the curious accumulative effects of coevolution.

Oxygen in particular, at 21 percent, makes the Earth's atmosphere unstable. Oxygen is a highly reactive gas, combining with many elements in a fierce explosive union we call fire or burning. Thermodynamically, the high oxygen content of Earth's atmosphere should fall quickly as the gas oxidizes surface solids. Other reactive trace gases such as nitrous oxide and methyl iodide also remain at elevated and aberrant levels. Both oxygen and methane coexist, yet they are profoundly incompatible, or rather *too* compatible since they should burn each other up. Carbon dioxide is inexplicably a mere trace gas when it should be the bulk of the air, as it is on other planets. In addition to its atmosphere, the temperature and alkalinity of the Earth's surface also exhibit a queer level. The entire surface of the Earth seems to be a vast unstable chemical anomaly.

It seemed to Lovelock as if an invisible power, an invisible hand, pushed the interacting chemical reactions into a raised state that should at any minute swing back to a balanced rest. The chemistry of Mars and Venus was as balanced as the periodic table, and as dead. The chemistry of the Earth was out of kilter, wholly unbalanced by the periodic table, and alive. From this, Lovelock concluded that any planet that has life would reveal a chemistry that held odd imbalances. A life-friendly atmosphere might not be oxygen-rich, but it should buck textbook equilibria.

That invisible hand was coevolutionary life.

Life in coevolution, which has the remarkable knack of generating stable instability, moved the chemical circuitry of the Earth's atmosphere into what Lovelock calls a "persistent state of disequilibrium." At any moment, the atmosphere should fall, but for millions of years it doesn't. Since high oxygen levels are needed for most microbial life, and since microbial fossils are billions of years old, this odd state of discordant harmony has been quite persistent and stable.

The Earth's atmosphere seeks a steady oxygen level much as a thermostat homes in on a steady temperature. The uniform 20 percent oxygen level it has found turns out to be "fortuitous" as one scientist put it. Lower oxygen would be anemic, while greater oxygen would be too flammable. George R. Williams at the University of Toronto writes: "An $O_2$ content of about 20 percent seems to ensure a balance between almost complete ventilation of the oceans without incurring greater risks of toxicity or increased combustibility of organic material." But where are the sensors and the thermostatic control mechanisms? For that matter, where is the furnace?

Dead planets find equilibrium by geological circuits. Gases, such as carbon dioxide, dissolve in liquids and can precipitate out as solids. Only so much gas will dissolve before it reaches a natural saturation. Solids can release gases back into the atmosphere when heated and pressed by volcanic activity. Sedimentation, weathering, uplift—all the grand geological forces—also act as strong chemical agents, breaking and making the bonds of materials. Thermodynamic entropy draws all chemical reactions down to their minimal energy level. The furnace metaphor breaks down. Equilibrium on a dead planet is less like a thermostat and more like the uniform level of water in a bowl; it simply levels out when it can't get any lower.

But the Earth has the self of a thermostat. A spontaneous circuit, provided by the coevolutionary tangle of life, which guides the chemicals of the planet toward some elevated potential. Presumably if all life on Earth were extinguished, the Earth's atmosphere would fall back to a persistent equilibrium, and become as boringly predictable as Mars and Venus. But as long as the distributed hand of life dominates, it will keep the chemicals of Earth off key.

Yet the off-balance is itself balanced. The persistent disequilibrium that coevolutionary life generates, and that Lovelock seeks as an acid test for its presence, is stable in its own way. As far as we can tell Earth's atmospheric oxygen has remained at about 20 percent for hundreds of millions of years. The atmosphere acts not merely as an acrobat on a tightrope pitched far from the vertical, but as an acrobat teetering between tilting and falling, *and poised there for millions of years*. She never falls, but never gets out of falling. It's a state of permanent almost-fell.

Lovelock recognized that persistent almost-fell is a hallmark of life. Recently complexity investigators have recognized that persistent almost-fell is a hallmark of any vivisystem: an economy, a natural ecosystem, a deep computer simulation, an immune system, or an evolutionary system. All share that paradoxical quality of working best when they remain poised in an Escher-like state of forever descending without ever being lowered. They remain poised in the act of collapsing.

David Layzer, writing in his semiscientific book *Cosmogenesis*, argues that "the central property of life is not reproductive invariance, but reproductive *instability*." The key of life is its ability to reproduce slightly out of kilter rather than with exactitude. This almost-falling into chaos keeps life proliferating.

A little noticed but central character of such vivisystems is that this paradoxical essence is contagious. Vivisystems spread their poised instability into whatever they touch, and they reach for everything. On Earth, life elbows its way into solid, liquid, gas. No rocks, to our knowledge, are untouched by life in former times. Tiny oceanic microorganisms solidify carbon and oxygen gases dissolved in sea water to produce a salt which settles on the sea floor. The deposits eventually become pressed under sedimentary weight into stone. Tiny plant organisms transport carbon from the air into soil and lower into the sea bottom, to be submerged and fossilized into oil. Life generates methane, ammonia, oxygen, hydrogen, carbon dioxide, and many other gases. Iron- and metal-concentrating bacteria create metallic ores. (Iron, the very emblem of nonlife, born of life!) Upon close inspection, geologists have concluded that all rocks residing on the Earth's surface (except perhaps volcanic lava) are recycled sediments, and therefore all rocks are biogenic in nature, that is, in some way affected by life. The relentless push and pull of coevolutionary life eventually brings into its game the abiotic stuff of the universe. It makes even the rocks part of its dancing mirror.

One of the first to articulate the transcendent view that life directly shaped the physicality of this planet was the Russian geologist Vladimir Vernadsky, writing in 1926. Vernadsky tallied up the billions of organisms on Earth and considered their collective impact upon the material resources of the planet. He called this grand system of resources the "biosphere" (although Eduard Suess had coined the term a few years earlier), and set out to measure it quantitatively in his book *The Biosphere*, a volume only recently translated into English.

In articulating life as a chameleon on a rocky mirror, Vernadsky committed heresy on two counts. He enraged biologists by considering the biosphere of living creatures as a large chemical factory. Plants and animals were mere temporary chemical storage units for the massive flow of minerals around the world. "Living matter is a specific kind of rock . . . an ancient and, at the same time, an eternally young rock," Vernadsky wrote. Living creatures were delicate shells to hold these minerals. "The purpose of animals," he once said of their locomotion and movement, "is to assist the wind and waves to stir the brewing biosphere."

At the same time, Vernadsky enraged geologists by considering rocks as if they were half-alive. Since the genesis of every rock was in life, their gradual interaction with living organisms meant that rocks were the part of life that moved the slowest. The mountains, the waters of the ocean, and the gases of the sky were very slow life. Naturally, geologists balked at this apparent mysticism.

The two heresies melded into a beautiful symmetry. Life as ever-renewing mineral, and minerals as slow life. They could only be opposite sides of a single coin. The two sides of this equation cannot be mathematically unraveled; they are one system: lizard-mirror, plant/insect, rock-life, and now in modern times, human/machine. The organism behaves as environment, the environment behaves as organism.

This has been a venerable idea at the edge of science for at least several hundred years. Many evolutionary biologists in the last century such as T. H. Huxley, Herbert Spencer, and Darwin, too, understood it intuitively—that the physical environment shapes its creatures and the creatures shape their environment, and if considered in the long view, the environment is the organism and the organism is the environment. Alfred Lotka, an early theoretical biologist, wrote in 1925, "It is not so much the organism or the species that evolves, but the entire system, species plus environment. The two are inseparable." The entire system of evolving life and planet was coevolution, the dance of the chameleon on the mirror.

If life were to vanish from Earth, Vernadsky realized, not only would the planet sink back into the "chemical calm" of an equilibrium state, but the clay deposits, limestone caves, ores in mine, chalk cliffs, and the very structure of all that we consider the Earth's landscape would retreat. "Life is not an external and accidental development on the terrestrial surface. Rather, it is intimately related with the constitution of the Earth's crust," Vernadsky wrote in 1929. "Without life, the face of the Earth would become as motionless and inert as the face of the moon."

Three decades later, free-thinker James Lovelock arrived at the same conclusions based on his telescopic analysis of other planets. Lovelock observed, "In no way do organisms simply 'adapt' to a dead world determined by physics and chemistry alone. They live in a world that is the breath and bones of their ancestors and that they are now sustaining." Lovelock had more complete knowledge of early Earth than was available to Vernadsky, and a slightly better understanding of the global patterns of gases and material flows on Earth. All this led him to suggest in complete seriousness that "the air we breathe, the oceans, and the rocks are all either the direct products of living organisms or else have been greatly modified by their presence."

Such a remarkable conclusion was foreshadowed by the French natural philosopher, Jean Baptiste Lamarck, who in 1800 had even less information about planetary dynamics than Vernadsky did. As a biologist, Lamarck was equal to Darwin. He, not Darwin, was the true discoverer of evolution, but Lamarck is stuck with an undeserved reputation as a loser, in part because he relied a little too much on intuition rather than the modern notion of detailed facts. Lamarck made an intuitive guess about the biosphere and again was prescient. Since there wasn't a shred of scientific evidence to support Lamarck's claims at the time, his observations were not influential. He wrote in 1802, "Complex mineral substances of all kinds that constitute the external crust of the Earth occurring in the form of individual accumulations, ore bodies, parallel strata, etc., and forming lowlands, hills, valleys, and mountains, are exclusively products of the animals and plants that existed within these areas of the Earth's surface."

The bold claims of Lamarck, Vernadsky, and Lovelock seem ludicrous at first, but in the calculus of lateral causality make fine sense: that all we can see around us—the snow-covered Himalayas, the deep oceans east and west, vistas of rolling hills, awesome painted desert canyons, game-filled valleys—are all as much the product of life as the honeycomb.

Lovelock kept gazing into the mirror and finding that it was nearly bottomless. As he examined the biosphere in succeeding years, he added more complex phenomena to the list of life-made. Some examples: plankton in the oceans release a gas (DMS) which oxidizes to produce submicroscopic aerosols of sulfate salts which form nuclei for the condensation of cloud droplets. Thus perhaps even clouds and rain may be biogenic. Summer thunderstorms may be life raining on itself. Some studies hinted that a majority of nuclei in snow crystals may be decayed vegetation, bacteria, or fungi spores; and so snow may be largely life-triggered. Only very little could escape life's imprint. "It may be that the core of our planet is unchanged as a result of life; but it would be unwise to assume it," Lovelock said.

"Living matter is the most powerful *geological* force," Vernadsky claimed, "and it is *growing with time*." The more life, the greater its material force. Humans intensify life further. We harness fossil energy and breathe life into machines. Our entire manufactured infrastructure—as an extension of our own bodies—becomes part of a wider, global-scale life. As the carbon dioxide from our industry pours into the air and alters the global air mix, the realm of our artificial machines also becomes part of the planetary life. Jonathan Weiner writing in *The Next One Hundred Years* then can rightly say, "The Industrial Revolution was an astonishing geological event." If rocks are slow life, then our machines are quicker slow life.

The Earth as mother was an old and comforting notion. But the Earth as mechanical device has been a harder idea to swallow. Vernadsky came very close to Lovelock's epiphany that the Earth's biosphere exhibits a regulation beyond chemical equilibrium. Vernadsky noted that "organisms exhibit a type of self-government" and that the biosphere seemed to be self-governed, but Vernadsky didn't press further because the crucial concept of self-government as a purely mechanical process had not yet been uncovered. How could a mere machine control itself?

We now know that self-control and self-governance are not mystical vital spirits found only in life because we have built machines that contain them. Rather, control and purpose are purely logical processes that can emerge in any sufficiently complex medium, including that of iron gears and levers, or even complex chemical pathways. If a thermostat or a steam engine can own self-governance, the idea of a planet evolving such graceful feedback circuits is not so alien.

Lovelock brought an engineer's sensibilities to the analysis of Mother Earth. He was a tinkerer, inventor, patent holder, and had worked for the biggest engineering firm of all time, NASA. In 1972, Lovelock offered a hypothesis of where the planet's self-government lay. He wrote, "The entire range of living matter on Earth, from whales to viruses, from oaks to algae, could be regarded as constituting a single living entity, capable of manipulating the Earth's atmosphere to suit its overall needs and endowed with faculties and powers far beyond those of its constituent part." Lovelock called this view Gaia. Together with microbiologist Lynn Margulis, the two published the view in 1972 so that it could be critiqued on scientific terms. Lovelock says, "The Gaia theory is a bit stronger than coevolution," at least as biologists use the word.

A pair of coevolutionary creatures chasing each other in an escalating arms race can only seem to veer out of control. Likewise, a pair of cozy coevolutionary symbionts embracing each other can only seem to lead to stagnant solipsism. But Lovelock saw that if you had a vast network of coevolutionary impulses, such that no creature could escape creating its own substrate and the substrate its own creatures, then the web of coevolution spread around until it closed a circuit of self-making and self-control. The "obligate cooperation" of Ehrlich's coevolution—whether of mutual enemies or mutual partners—cannot only raise an emergent cohesion out of the parts, but this cohesion can actively temper its own extremes and thereby seek its own survival. The solidarity produced by a planetary field of creatures mirrored in a coevolving environment and each other is what Lovelock means by Gaia.

Many biologists (including Paul Ehrlich) are unhappy with the idea of Gaia because Lovelock expanded the definition of life without asking their permission. He unilaterally enlarged life's scope to include a predominantly mechanical apparatus. In one easy word, a solid planet became "the largest manifestation of life" that we know. It is an odd beast: 99.9 percent rock, a lot of water, and a little air, wrapped up in the thinnest green film that would stretch around it.

But if Earth is reduced to the size of a bacteria, and inspected under high-powered optics, would it seem stranger than a virus? Gaia hovers there, a blue sphere under the stark light, inhaling energy, regulating its internal states, fending off disturbances, complexifying, and ready to transform another planet if given a chance.

While Lovelock backs off earlier assertions that Gaia is an organism, or acts as if it is one, he maintains that it *really is* a system that has living characteristics. It is a vivisystem. It is a system that is alive, whether or not it possesses all the attributes needed for an organism.

That Gaia is made up of many purely mechanical circuits shouldn't deter us from applying the label of life. After all, cells are mostly chemical cycles. Some ocean diatoms are mostly inert, crystallized calcium. Trees are mostly dead pulp. But they are still *living* organisms.

Gaia is a bounded whole. As a living system, its inert, mechanistic parts are part of its life. Lovelock: "There is no clear distinction anywhere on the Earth's surface between living and nonliving matter. There is merely a hierarchy of intensity going from the material environment of the rocks and atmosphere to the living cells." Somewhere at the boundary of Gaia, either in the rarefied airs of the stratosphere or deep in the Earth's molten core, the effects of life fade. No one can say where that line is, if there is a line.

THE TROUBLE WITH GAIA, as far as most skeptics are concerned, is that it makes a dead planet into a "smart" machine. We already are stymied in trying to design an artificial learning machine from inert computers, so the prospect of artificial learning evolving unbidden at a planetary scale seems ludicrous.

But learning is overrated as something difficult to evolve. This may have to do with our chauvinistic attachment to learning as an exclusive mark of our species. There is a strong sense, which I hope to demonstrate in this book, in which evolution itself is a type of learning. Therefore learning occurs wherever evolution is, even if artificially.

The dethronement of learning is one of the most exciting intellectual frontiers we are now crossing. In a virtual cyclotron, learning is being smashed into its primitives. Scientists are cataloging the elemental components for adaptation, induction, intelligence, evolution, and coevolution into a periodic table of life. The particles for learning lie everywhere in all inert media, waiting to be assembled (and often self-assembled) into something that surges and quivers.

Coevolution is a variety of learning. Stewart Brand wrote in *CoEvolution Quarterly*: "Ecology is a whole system, alright, but coevolution is a whole system in *time*. The health of it is forward—systemic self-education which feeds on constant imperfection. Ecology maintains. Coevolution learns."

Colearning might be a better term for what coevolving creatures do. Coteaching also works, for the participants in coevolution are both learning and teaching each other at the same time. (We don't have a word for learning and teaching at the same time, but our schooling would improve if we did.)

The give and take of a coevolutionary relationship—teaching and learning at once—reminded many scientists of game playing. A simple child's game such as "Which hand is the penny in?" takes on the recursive logic of a chameleon on a mirror as the hider goes through this open-ended routine: "I just hid the penny in my right hand, and now the guesser will think it's in my left, so I'll move it into my right. But she also knows that I know she knows that, so I'll keep it in my left."

Since the guesser goes through a similar process, the players form a system of mutual second-guessing. The riddle "What hand is the penny in?" is related to the riddle, "What color is the chameleon on a mirror?" The bottomless complexity which grows out of such simple rules intrigued John von Neumann, the mathematician who developed programmable logic for a computer in the early 1940s, and along with Wiener and Bateson launched the field of cybernetics.

Von Neumann invented a mathematical theory of games. He defined a game as a conflict of interests resolved by the accumulative choices players make while trying to anticipate each other. He called his 1944 book (coauthored by economist Oskar Morgenstern) *Theory of Games and Economic Behavior* because he perceived that economies possessed a highly coevolutionary and gamelike character, which he hoped to illuminate with simple game dynamics. The price of eggs, say, is determined by mutual second-guessing between seller and buyer—how much will he accept, how much does he think I will offer, how much less than what I am willing to pay should I offer? The aspect von Neumann found amazing was that this infinite regress of mutual bluffing, codeception, imitation, reflection, and "game playing" would commonly settle down to a definite price, rather than spiral on forever. Even in a stock market made of thousands of mutual second-guessing agents, the group of conflicting interests would quickly settle on a price that was fairly stable.

Von Neumann was particularly interested in seeing if he could develop optimal strategies for these kinds of mutual games, because at first glance they seemed almost insolvable in theory. As an answer he came up with a theory of games. Researchers at the U.S. government-funded RAND corporation, a think tank based in Santa Monica, California, extended von Neumann's initial work and eventually cataloged four basic varieties of mutual second-guessing games. Each variety had a different structure of rewards for winning, losing, or drawing. The four simple games were called "social dilemmas" in the technical literature, but could be thought of as the four building blocks of complicated coevolutionary games. They were: Chicken, Stag Hunt, Deadlock, and the Prisoner's Dilemma.

Chicken is the game played by teenage daredevils. Two cars race toward a cliff's edge; the driver who jumps out last, wins. Stag Hunt is the dilemma faced by a bunch of hunters who must cooperate to kill a stag, but may do better sneaking off by themselves to hunt a rabbit if no one cooperates. Do they gamble on cooperation (high payoff) or defection (low, but sure payoff)? Deadlock is a boring game where mutual defection pays best. The last one, the Prisoner's Dilemma, is the most illuminating, and became the guinea pig model for over 200 published social psychology experiments in the late 1960s.

The Prisoner's Dilemma, invented in 1950 by Merrill Flood at RAND, is a game for two separately held prisoners who must independently decide whether to deny or confess to a crime. If both confess, each will be fined. If neither confesses, both go free. But if only one should confess, he is rewarded while the other is fined. Cooperation pays, but so does betrayal, if played right. What would you do?

Played only once, betrayal of the other is the soundest choice. But when two "prisoners" played the game over and over, learning from each other—a game known as the Iterated Prisoner Dilemma—the dynamics of the game shifted. The other player could not be dismissed; he demanded to be attended to, either as obligate enemy or obligate colleague. This tight mutual destiny closely paralleled the coevolutionary relationship of political enemies, business competitors, or biological symbionts. As study of this simple game progressed, the larger question became, What were the strategies of play for the Iterated Prisoner's Dilemma that resulted in the highest scores over the long term? And what strategies succeeded when played against many varieties of players, from the ruthless to the kind?

In 1980, Robert Axelrod, a political science professor at the University of Michigan, ran a tournament pitting 14 submitted strategies of Prisoner's Dilemma against each other in a round robin to see which one would triumph. The winner was a very simple strategy crafted by psychologist Anatol Rapoport called

Tit-For-Tat. The Tit-For-Tat strategy prescribed reciprocating cooperation for cooperation, and defection for defection, and tended to engender periods of cooperation. Axelrod had discovered that "the shadow of the future," cast by playing a game repeatedly rather than once, encouraged cooperation, because it made sense for a player to cooperate now in order to ensure cooperation from others later. This glimpse of cooperation set Axelrod on this quest: "Under what conditions will cooperation emerge in a world of egoists without central authority?"

For centuries, the orthodox political reasoning originally articulated by Thomas Hobbes in 1651 was dogma: that cooperation could only develop with the help of a benign central authority. Without top-down government, Hobbes claimed, there would be only collective selfishness. A strong hand had to bring forth political altruism, whatever the tone of economics. But the democracies of the West, beginning with the American and French Revolutions, suggested that societies with good communications could develop cooperative structures without heavy central control. Cooperation can emerge out of self-interest. In our postindustrial economy, spontaneous cooperation is a regular occurrence. Widespread industry-initiated standards (both of quality and protocols such as 110 volts or ASCII) and the rise of the Internet, the largest working anarchy in the world, have only intensified interest in the conditions necessary for hatching coevolutionary cooperation.

This cooperation is not a new age spiritualism. Rather it is what Axelrod calls "cooperation without friendship or foresight"—cold principles of nature that work at many levels to birth a self-organizing structure. Sort of cooperation whether you want it or not.

Games such as Prisoner's Dilemma can be played by any kind of adaptive agent—not just humans. Bacteria, armadillos, or computer transistors can make choices according to various reward schemes, weighing immediate sure gain over future greater but riskier gain. Played over time with the same partners, the results are both a game and a type of coevolution.

Every complex adaptive organization faces a fundamental tradeoff. A creature must balance perfecting a skill or trait (building up legs to run faster) against experimenting with new traits (wings). It can never do all things at once. This daily dilemma is labeled the tradeoff between exploration and exploitation. Axelrod makes an analogy with a hospital: "On average you can expect a new medical drug to have a lower payoff than exploiting an established medication to its limits. But if you gave every patient the current best drug, you'd never get proven new drugs. From an individual's point of view you should never do the exploration. But from the society of individuals' point of view, you ought to try some experiments." How much to explore (gain for the future) versus how much to exploit (sure bet now) is the game a hospital has to play. Living organisms have a similar tradeoff in deciding how much mutation and innovation is needed to keep up with a changing environment. When they play the tradeoff against a sea of other creatures making similar tradeoffs, it becomes a coevolutionary game.

Axelrod's 14-player Prisoner's Dilemma round robin tournament was played on a computer. In 1987, Axelrod extended the computerization of the game by setting up a system in which small populations of programs played randomly generated Prisoner's Dilemma strategies. Each random strategy would be scored after a round of playing against all the other strategies running; the ones with the highest scores got copied the most to the next generation, so that the most successful strategies propagated. Because many strategies could succeed only by "preying" on other strategies, they would thrive only as long as their prey survived. This leads to the oscillating dynamics found everywhere in the wilds of nature; how fox and hare populations rise and fall over the years in coevolutionary circularity. When the hares increase the foxes boom; when the foxes boom, the hares die off. But when there are no hares, the foxes starve. When there are less foxes, the hares increase. And when the hares increase the foxes do too, and so on.

In 1990, Kristian Lindgren, working at the Neils Bohr Institute in Copenhagen, expanded these coevolutionary experiments by increasing the population of players to 1,000, introducing random noise into the games, and letting this artificial coevolution run for up to 30,000 generations. Lindgren found that masses of dumb agents playing Prisoner's Dilemma not only reenacted the ecological oscillations of fox and hare, but the populations also created many other natural phenomenon such as parasitism, spontaneously emerging symbiosis, and long-term stable coexistence between species, as if they were an ecology. Lindgren's work excited some biologists because his very long runs displayed long periods when the mix of different "species" of strategy was very stable. These historical epochs were interrupted by very sudden, short-lived episodes of instability, when old species went extinct and new ones took root. Quickly a new stable arrangement of new species of strategies arose and persisted for many thousands of generations. This motif matches the general pattern of evolution found in earthly fossils, a pattern known in the evolutionary trade as punctuated equilibrium, or "punk eek" for short.

One marvelous result from these experiments bears consideration by anyone hoping to manage coevolutionary forces. It's another law of the gods. It turns out that no matter what clever strategy you engineer or evolve in a world laced by chameleon-on-a-mirror loops, if it is applied as a perfectly pure rule that you obey absolutely, it will not be evolutionarily resilient to competing strategies. That is, a competing strategy will figure out how to exploit your rule in the long run. A little touch of randomness (mistakes, imperfections), on the other hand, actually creates long-term stability in coevolutionary worlds by allowing some strategies to prevail for relative eons by not being so easily aped. Without noise—wholly unexpected and out-of-character choices—the opportunity for escalating evolution is lost because there are not enough periods of stability to keep the system going. Error keeps the glue of coevolutionary relationships from binding too tightly into runaway death spirals, and therefore error keeps a coevolutionary system afloat and moving forward. *Honor thy error.*

Playing coevolutionary games in computers has provided other lessons. One of the few notions from game theory to penetrate the popular culture was the distinction of zero-sum and nonzero-sum games. Chess, elections, races, and poker are zero-sum games: the winner's earnings are deducted from the loser's assets.

Natural wilderness, the economy, a mind, and networks on the other hand, are nonzero-sum games. Wolverines don't have to lose just because bears live. The highly connected loops of coevolutionary conflict mean the whole can reward (or at times cripple) all members. Axelrod told me, "One of the earliest and most important insights from game theory was that nonzero-sum games had very different strategic implications than zero-sum games. In zero-sum games whatever hurts the other guy is good for you. In nonzero-sum games you can both do well, or both do poorly. I think people often take a zero-sum view of the world when they shouldn't. They often say, 'Well I'm doing better than the other guy, therefore I must be doing well.' In a nonzero-sum you could be doing better than the other guy and both be doing terribly."

Axelrod noticed that the champion Tit-For-Tat strategy always won without exploiting an opponent's strategy—it merely mirrored the other's actions. Tit-For-Tat could not *beat* anyone's strategy one on one, but in a nonzero-sum game it would still win a tournament because it had the highest cumulative score when played against many kinds of rules. As Axelrod pointed out to William Poundstone, author of *Prisoner's Dilemma*, "That's a very bizarre idea. You can't win a chess tournament by never beating anybody." But with coevolution—change changing in response to itself—you can win without beating others. Hard-nosed CEOs in the business world now recognize that in the era of networks and alliances, companies can make billions without beating others. Win-win, the cliché is called.

Win-win is the story of life in coevolution.

Sitting in his book-lined office, Robert Axelrod mused on the consequences of understanding coevolution and then added, "I hope my work on the evolution of cooperation helps the world avoid conflict. If you read the citation which the National Academy of Science gave me," he said pointing to a plaque on the wall, "they think it helped avoid nuclear war." Although von Neumann was a key figure in the development of the atom bomb, he did not formally apply his own theories to the gamelike politics of the nuclear arms race. But after von Neumann's death in 1957, strategists in military think tanks began using his game theory to analyze the cold war, which had taken on the flavor of a coevolutionary "obligate cooperation" between two superpower enemies. Gorbachev had a fundamental coevolutionary insight, says Axelrod. "He saw that the Soviets could get more security with *fewer* tanks rather than with more tanks. Gorbi unilaterally threw away 10,000 tanks, and that made it harder for the U.S. and Europe to have a big military budget, which helped get this whole process going that ended the cold war."

Perhaps the most useful lesson of coevolution for "wannabe" gods is that in coevolutionary worlds control and secrecy are counterproductive. You can't control, and revelation works better than concealment. "In zero-sum games you always try to hide your strategy," says Axelrod. "But in nonzero-sum games you might want to announce your strategy in public so the other players need to adapt to it." Gorbachev's strategy was effective because he did it publicly; unilaterally withdrawing in secret would have done nothing.

The chameleon on the mirror is a completely open system. Neither the lizard nor the glass has any secrets. The grand closure of Gaia keeps cycling because all its lesser cycles inform each other in constant coevolutionary communication. From the collapse of Soviet command-style economies, we know that open information keeps an economy stable and growing.

Coevolution can be seen as two parties snared in the web of mutual propaganda. Coevolutionary relationships, from parasites to allies, are in their essence informational. A steady exchange of information welds them into a single system. At the same time, the exchange—whether of insults or assistance or plain news—creates a commons from which cooperation, self-organization, and win-win endgames can spawn.

In the Network Era—that age we have just entered—dense communication is creating artificial worlds ripe for emergent coevolution, spontaneous self-organization, and win-win cooperation. In this Era, openness wins, central control is lost, and stability is a state of perpetual almost-falling ensured by constant error.

6

# THE NATURAL FLUX

TONIGHT IS the Chinese Lunar Festival. Downtown in San Francisco's Chinatown, immigrants are exchanging moon cakes and telling tales of the Ghost Maiden who escaped as an orb in the sky. Twelve miles away where I live, I can walk in a cloud. The fog of the Golden Gate has piled up along the steep bank behind our house, engulfing our neighborhood in vapor. Under the light of Lady Moon, I take a midnight hike.

I wade chest-high in bleached ryegrass murmuring in the wind, and spy down the rugged coast of California. It is a disruptive land. For most purposes it is a mountainous desert that meets a generous ocean which cannot provide rain. Instead the sea sneaks in the water of life by rolling out blankets of fog at night. Come morning, the mist condenses into drops on the edges of twig and leaf, which tinkle to the earth. Much water is transported this way over a summer, bypassing the monopoly thunderclouds have on water delivery elsewhere. On this stingy substitute rain, the behemoth of all living things, the redwood, thrives.

The advantage of rain is that it is massive and indiscriminate. When it rains, it will wet a wide, diverse constituency. Fog on the other hand, is local. It relies on low-powered convection currents to ramble wherever it is easiest to drift to, and is then trapped by gentle, patient cul-de-sacs in the hills. In this way, the shape of the land steers the water, and indirectly, life. The correctly shaped hill can catch fog, or funnel drip into a canyon. A sunny southfacing mound will lose more precious moisture to evaporation than a shadier northern slope. Certain outcroppings of soil retain water better than others. Play these variables on top of each other and you have a patchwork of habitats. In a desert land, water decides life. And in a desert land where water is not delivered democratically, but parochially, on a whim, the land itself decides life.

The result is a patchwork landscape. The hills behind my house are cloaked with three separate quilts. A community of low-lying grass—and of mice, owl, thistle, and poppy—runs to the sea on one slope. On the crest of the hill, gnarly juniper and cypress trees preside over a separate association of deer, fox, and lichen. And on the other side of the rise, an endless impenetrable thicket of poison oak and coyote brush hides quail and other members of its guild.

The balance of these federations is kinetic. Their mutual self-supporting pose is continuously almost-falling, like a standing wave in a spring creek. When the mass of nature's creatures push against each other in coevolutionary embrace, their interaction among the uneven terrain of land and weather breaks their aggregate into local enclaves of codependency. And these patchcs roam over the land in time.

Wind and spring floods erode soils, exposing underlying layers and premiering new compositions of humus and minerals on the surface. As the mix of soil churns on the land, the mix of plants and animals coupled to it likewise churns. A thick stand of cactus, such as a Saguaro forest, can migrate onto or off of a patch of southwestern desert in as little as 100 years. In a time-lapse film, a Saguaro grove would seem to creep across the desertscape like a pool of mercury. And it's not just cactus that would roam. Under the same time-lapse view, the wildflower prairie savanna of the midwest would flow around stands of oaks like an incoming tide, sometimes dissolving the woods into prairie, and sometimes, if the wildfires died out, retreating from the spreading swell of oak groves. Ecologist Dan Botkin speaks of forests "marching slowly across the landscape to the beat of the changing climate."

"Without change, deserts deteriorate," claims Tony Burgess, a burly ecologist with a huge red beard. Burgess is in love with deserts. He inhales desert lore and data all his waking hours. Out in the stark sun near Tucson, Arizona, he has been monitoring a desert plot that several generations of scientists have continuously measured and photographed for 80 years; the plot is the longest uninterrupted ecological observation anywhere. From studying the data of 80 years of desert change, Burgess has concluded that "variable rainfall is the key to the desert. Every year it should be a slightly different ball game to keep every species slightly out of equilibrium. If rainfall is variant then the mixture of species increases by two or three orders of magnitude. Whereas if you have a constant schedule of rainfall with respect to the annual temperature cycle, the beautiful desert ecology will almost always collapse into something simpler."

"Equilibrium is dead," Burgess states matter-of-factly. This opinion has not been held very long by the ecological science community. "Until the mid-1970s we were all working under a legacy which said that communities are on a trajectory towards an unchanging equilibrium, the climax. But now we see that it is turbulence and variance that really give the richness to nature."

A major reason why ecologists favored equilibrium end points in nature was exactly the same reason why economists favored equilibrium end points in the economy: the mathematics of equilibria were possible. You could write an equation for a process that you could actually solve. But if you said that the system was perpetually in disequilibrium, you were saying it followed a model you couldn't solve and therefore couldn't explore. You were saying almost nothing. It is no coincidence, therefore, that a major shift in ecological (and economic) understanding occurred in the era when cheap computers made nonequilibrial and nonlinear equations easy to program. It was suddenly no problem to model a chaotic, coevolutionary ecosystem on a personal computer, and see that, hey, it acts very much like the odd behavior of a Saguaro forest or a prairie savanna on the march.

A thousand varieties of nonequilibrial models have blossomed in recent years; in fact there is now a small cottage industry of makers of chaotic and nonlinear mathematics, differential equations, and complexity theory, all this activity lending a hand in overturning the notion that nature or an economy seeks a stable balance. This new perspective—that a certain unremitting flux is the norm—has illuminated past data for reinterpretation. Burgess can display old photographs of the desert that show in a relatively short time—over a few decades—patches of Saguaro drifting over the Tucson basin. "What we found from our desert plot," Burgess said, "is that these patches are not in sync in terms of development and that by not being in sync, they make the whole desert richer because if something catastrophic wipes out one patch, another patch at a different stage of its natural history can export organisms and seeds to the decimated patch. Even ecosystems, such as tropical rain forests, which don't have variable rainfall, also have patch dynamics due to periodic storms and tree falls."

"Equilibrium is not only dead, it is death," Burgess emphasizes. "To enrich a system you need variance in time and space. But too much change will kill you too. You go from an ecocline to ecotone."

Burgess finds nature's reliance on disturbances and variance to be a practical issue. "In nature, it is no problem if you have very erratic production [of vegetation, seeds, or meat] from year to year. Nature actually increases her richness from this variance. But when people try to sustain themselves on the production from an ecosystem like a desert that is so variance driven, they can only do it by simplifying the system into what we call agriculture—which gives a constant production for a variable environment." Burgess hopes the flux of the desert can teach us how to live with a variable environment without simplifying it. It is not a completely foolish dream. Part of what an information-driven economy provides us with is an adaptable infrastructure that can bend and work around irregular production; this is the basis for flexible and "just-in-time" manufacturing. It is theoretically possible that we could use information networks to coordinate the investment and highly irregular output of a rich, fluxing ecosystem that provides food and organic resources. But, as

Burgess admits, "At the moment we have no industrial economic models that are variance driven, except gambling."

If it is true that nature is fundamentally in constant flux, then instability may cause the richness of biological forms in nature. But the idea that the elements of instability are the root of diversity runs counter to one of the hoariest dictums of environmentalism: that stability begets diversity, and diversity begets stability. If natural systems do not settle into a neat balance, then we should make instability our friend.

Biologists finally got their hands on computers in the late 1960s and began to model kinetic ecologies and food webs on silicon networks. One of the first questions they attempted to answer was, Where does stability come from? If you create predator/prey relationships *in silico*, what conditions cause the virtual organisms to settle into a long-term coevolutionary duet, and what conditions cause them to crash?

Among the earliest studies of simulated stability was a paper published in 1970 by Gardner and Ashby. Ashby was an engineer interested in nonlinear control circuits and the virtues of positive feedback loops. Ashby and Gardner programmed simple network circuits in hundreds of variations into a computer, systematically changing the number of nodes and the degrees of connectivity between nodes. They discovered something startling: that beyond a certain threshold, increasing the connectivity would suddenly decrease the ability of the system to rebound after disturbances. In other words, complex systems were less likely to be stable than simple ones.

A similar conclusion was published the following year by theoretical biologist Robert May, who ran model ecologies on computers populated with large multitudes of interacting species, and some virtual ecologies populated with few. His conclusions contradicted the common wisdom of stability/diversity, and he cautioned against the "simple belief" that stability is a consequence of increasing complexity of the species mix. Rather, May's simulated ecologies suggested that neither simplicity nor complexity had as much impact on stability as the *pattern* of the species interaction.

"In the beginning, ecologists built simple mathematical models and simple laboratory microcosms. They were a mess. They lost species like crazy," Stuart Pimm told me. "Later ecologists built more complex systems in the computer and in the aquarium. They thought these complex ones would be good. They were wrong. They were an even worse mess. Complexity just makes things very difficult—the parameters have to be just right. So build a model at random and, unless it's really simple (a one-prey-one-resource population model) it *won't* work. Add diversity, interactions, or increase the food chain lengths and soon these get to the point where they will also fall apart. That's the theme of Gardner, Ashby, May and my early work on food webs. But keep on adding species, keep on letting them fall apart and, surprisingly, they *eventually* reach a mix that will not fall apart. Suddenly one gets order for free. It takes a lot of repeated messes to get it right. The only way we know how to get stable, persistent, complex systems is to repeatedly assemble them. And as far as I know, no one really understands why that works."

In 1991 Stuart Pimm, together with colleagues John Lawton and Joel Cohen, reviewed all the field measurements of food webs in the wild and by analyzing them mathematically concluded that "the rate at which populations recovered from disasters . . . depends on food chain length," as well as the number of prey and predators a species had. An insect eating a leaf is a chain of one. A turtle eating the insect that eats the leaf makes a chain of two. A wolf may sit many links away from a leaf. In general, the longer the chain, the less stable the interacting web to environmental disruption.

The other important point one can extract from May's simulations was best articulated in an observation made a few years earlier by the Spanish ecologist Ramon Margalef. Margalef noticed, as May did, that systems with many components would have weak relations between them, while systems that had few components would have tightly coupled relationships. Margalef put it this way: "From empirical evidence it seems that species that interact freely with others do so with a great number of other species. Conversely, species with strong interactions are often part of a system with a small number of species."

This apparent tradeoff in an ecosystem between many loosely coupled members or few tightly coupled members is nicely paralleled by the now well-known tradeoff which biological organisms must choose in reproduction strategies. They can either produce a few well-protected offspring or a zillion unprotected ones.

Biology suggests that in addition to regulating the numbers of connections per "node" in a network, a system tends to also regulate the "connectance" (the strength of coupledness) between each pair of nodes in a network. Nature seems to conserve connectance. We should thus expect to find a similar law of the conservation of connectance in cultural, economic, and mechanical systems, although I am not aware of any studies that have attempted to show this. If there is such a law in all vivisystems, we should also expect to find this connectance being constantly adjusted, perpetually in flux.

"An ecosystem is a network of living creatures," says Burgess. The creatures are wired together in various degrees of connectance by food webs and by smells and vision. Every ecosystem is a dynamic web always in flux, always in the processes of reshaping itself. "Wherever we seek to find constancy we discover change," writes Botkin.

When we make a pilgrimage to Yellowstone National Park, or to the California Redwood groves, or to the Florida Everglades, we are struck by the reverent appropriateness of nature's mix in that spot. The bears seem to *belong* in those Rocky Mountain river valleys; the redwoods seem to *belong* on those coastal hills, and the alligators seem to *belong* in those plains. Thus our spiritual urge to protect them from disturbance. But in the long view, they are natural squatters who haven't been there long and won't always be there. Botkin writes, "Nature undisturbed is not constant in form, structure, or proportion, but changes at every scale of time and space."

A study of pollen lifted from holes drilled at the bottom of African lakes shows that the African landscape has been in a state of flux for the past several million years. Depending on when you looked in, the African landscape would look vastly different from now. In the recent geological past, the Sahara desert vastness of northern Africa was tropical forest. It's been many ecological types between then and now. We hold wilderness to be eternal; in reality, nature is constrained flux.

Complexity poured into the artificial medium of machines and silicon chips will only be in further flux. We see, too, that human institutions—those ecologies of human toil and dreams—must also be in a state of constant flux and reinvention, yet we are always surprised or resistant when change begins. (Ask a hip postmodern American if he would like to change the 200-year-old rule book known as the Constitution. He'll suddenly become medieval.)

Change, not redwood groves or parliaments, is eternal. The questions become: What controls change? How can we direct it? Can the distributed life in such loose associations as governments, economies, and ecologies be controlled in any meaningful way? Can future states of change even be predicted?

Let's say you purchase a worn-out 100-acre farm in Michigan. You fence the perimeter to keep out cows and people. Then you walk away. You monitor the fields for decades. That first summer, garden weeds take over the plot. Each year thereafter new species blow in from outside the fence and take root. Some newcomers are eventually overrun by newer newcomers. An ecological combo self-organizes itself on the land. The mix fluxes over the years. Would a knowledgeable ecologist watching the fencing-off be able to predict which wildlife species would dominate the land a century later?

"Yes, without a doubt he could," says Stuart Pimm. "But his prediction is not as interesting as one might think."

The final shape of the Michigan plot is found in every standard ecology college textbook in the chapter on the concept of succession. The first year's weeds on the Michigan plot are annual flowering plants, followed by tougher perennials like crabgrass and ragweed. Woodier shrubs will shade and suppress the flowers, followed by pines, which suppress the shrubs. But the shade of the pine trees protects hardwood seedlings of beech and maple, which in turn steadily elbow out the pines. One hundred years later the land is almost completely owned by a typical northern hardwood forest.

It is as if the brown field itself is a seed. The first year it sprouts a hair of weeds, a few years later it grows a shrubby beard, and then later it develops into a shaggy woods. The plot unfolds in predictable stages just as a tadpole unfolds out of a frog's egg.

Yet, the curious thing about this development is that if you start with a soggy 100-acre swamp, rather than a field, or with the same size lot of Michigan dry sandy dunes, the initial succession species are different (sedges in the swamp, raspberries on the sand), but the mix of species gradually converges to the same end point of a hardwood forest. All three seeds hatch the same adult. This convergence led ecologists to the notion of an omega point, or a climax community. For a given area, all ecological mixtures will tend to shift until they reach a mature, ultimate, stable harmony.

What the land "wants" to be in the temperate north is a hardwood forest. Give it enough time and that's what a drying lake or a windblown sand bog will become. If it ever warmed up a little, that's what an alpine mountaintop wants to be also. It is as if the ceaseless strife in the complicated web of eat-or-be-eaten stirs the jumble of species in the region until the mixture arrives at the hardwood climax (or the specific climax in other climates), at which moment it quietly settles into a tolerable peace. The land coming to a rest in the climax blend.

Mutual needs of diverse species click together so smartly in the climax arrangement that the whole is difficult to disrupt. In the space of 30 years the old-growth chestnut forest in North America lost every specimen of a species—the mighty chestnut—that formerly constituted a significant hunk of the forest's mass. Yet, there weren't any huge catastrophes in the rest of the forest; it still stands. This persistent stability of a particular composite of species—an ecosystem—speaks of some basin of efficiency that resembles the coherence belonging to an organism. Something whole, something alive dwells in that mutual support. Perhaps a maple forest is but a grand organism composed of lesser organisms.

On the other hand, Aldo Leopold writes, "In terms of conventional physics, the grouse represents only a millionth of either the mass or the energy of an acre. Yet subtract the grouse and the whole thing is dead."

In 1916, Frederic Clements, one of the founding fathers of ecology, called a community of creatures such as the beech hardwood forest an emergent superorganism. In his words, a climax formation is a super*organism* because it "arises, grows, matures, and dies . . . comparable in its chief features with the life history of an individual plant." Since a forest could reseed itself on an abandoned Michigan field, Clements portrayed that act as reproduction, a further characteristic of an organism. To any astute observer, a beech-maple forest displays an integrity and identity as much as a crow does. What else but a (super)organism could reproduce itself so reliably, propagating on empty fields and sandy barrens?

Superorganism was a buzz word among biologists in the 1920s. They used it to describe the then novel idea that a collection of agents could act in concert to produce phenomena governed by the collective. Like a slime mold that assembled itself from moldy spots into a thrusting blob, an ecosystem coalesced into a stable superorganization—a hive or forest. A Georgia pine forest did not act like a pine tree, nor a Texas sagebrush desert like a sagebrush, just as a flock is not a big bird. They were something else, a loose federation of animals and plants united into an emergent superorganism exhibiting distinctive behavior.

A rival of Clements, biologist H. A. Gleason, the *other* father of modern ecology, thought the superorganism federation was too flabby and too much the product of a human mind looking for patterns. In opposition to Clements, Gleason proposed that the climax community was merely a fortuitous association of organisms that came and went depending on climate and geological conditions. An ecosystem was more like a conference than a community—indefinite, pluralistic, tolerant, and in constant flux.

The wilds of nature hold evidence for both views. In places the boundary between communities is decisive, much as one expects if ecosystems are superorganisms. Along the rocky coast of the Pacific Northwest, for instance, the demarcation between the high tide seaweed community and the watery edge of the spruce forest is an extreme no-man's-land of barren beach. One can stand on this yard-wide strip of salty desert and sense the two superorganisms on either side, fidgeting in their separate lives. As another example, the border between deciduous forest and wildflower prairie in the midwest is remarkably impermeable.

In search of an answer to the riddle of ecological superorganisms, biologist William Hamilton began modeling ecosystems on computers in the 1970s. He found that in his models (as well as in real life) very few systems were able to self-organize into any kind of *lasting* coherence. My examples above are a few exceptions in the wild. He found a few others: a sphagnum moss peat bog can repel the invasion of pine trees for thousands of years. Ditto for the tundra steppes. But most ecological communities stumble along into a mongrel mixture of species that offers no outstanding self-protection to the group as a team. Most ecological communities, both simulated and real, can be easily invaded in the longer run.

Gleason was right. The couplings between members of an ecosystem are far more flexible and transient than the couplings between members of an organism. The cybernetic difference between an organism such as a pollywog and an ecosystem such as a fresh-water bog is that an organism is tightly bound, and strict; an ecosystem is loosely bound, and lax.

In the long view, ecologies are temporary networks. Although some links become hardwired and nearly symbiotic, most species are promiscuous in evolutionary time, shacking up with different partners as the partners themselves evolve.

In this light of evolutionary time, ecology can be seen as one long dress rehearsal. It's an identity workshop for biological forms. Species try out different roles with one another and explore partnerships. Over time, roles and performance are assimilated by an organism's genes. In poetic language, the gene is reluctant to assimilate into its code any interactions and functions directly based upon its neighbors' ways because the neighborhood can shift at any evolutionary moment. It pays to stay flexible, unattached, and uncommitted.

At the same time Clements was right. There is a basin of efficiency that, all things being equal, will draw down a certain mix of parts into a stable harmony. As a metaphor, consider the way rocks make their way to the valley floor. Not all rocks will land at the bottom; a particular rock may get stuck on a small hill somewhere. In the same way, stable intermediate less-than-climax mixtures of species can be found in places on the landscape. For extremely short periods of geological time—hundreds of thousands of years—ecosystems form an intimate troupe of players, who brook no interference and need no extras. These associations are far briefer than even the brief life of individual species, which typically flame-out after a million years or two.

Evolution requires a certain connectance among its participants to express its power; and so evolutionary dynamics exert themselves most forcefully in tightly coupled systems. In systems connected loosely, such as ecosystems, economic systems, and cultural systems, a less structured adaptation takes place. We know very little about the general dynamics of loosely coupled systems because this kind of distributed change is messy and infinitely indirect. Howard Pattee, an early cybernetician, defined hierarchical structure as a spectrum of connectance. He said, "To a Platonic mind, everything in the world is connected to everything else—and perhaps it is. Everything is connected, but some things are more connected than others." Hierarchy for Pattee was the product of differential connectedness within one system. Members that were so loosely connected as to be "flat" would tend to form a separate organizational level distinct from areas where members were tightly connected. The range of connectance created a hierarchy.

In the most general terms, evolution is a tight web and ecology a loose one. Evolutionary change seems a strongly bound process very similar to mathematical computation, or even to thinking. In this way it is "cerebral." Ecological change, on the other hand, seems a weak-minded, circuitous process, centered in bodies shoved against wind, water, gravity, sunlight, and rock. "Community [ecological] attributes are more the product of environment than the product of evolutionary history," writes ecologist Robert Ricklefs. While evolution is governed by the straightforward flow of symbolic information issuing from the gene or computer chips, ecology is governed by the far less abstract, far more untidy complexity embodied by flesh.

Because evolution is such a symbolic process, we now can artificially create it and attempt to govern it. But because ecological change is so body bound, we cannot synthesize it well until we can more easily simulate bodies and richer artificial environments.

WHERE DOES diversity come from? In 1983, microbiologist Julian Adams discovered a clue when he brewed up a soup of cloned *E. coli* bacteria. He purified the broth until he had a perfectly homogenized pool of identical creatures. He put this soup of clones into a specially constructed chemostat that provided a uniform environment for them—every *E. coli* bug had the same temperature and nutrient bath. Then he let the soup of identical bugs replicate and ferment. At the end of 400 generations, the *E. coli* bacteria had bred new strains of itself with slightly different genes. Out of a starting point in a constant featureless environment, life spontaneously diversified.

A surprised Adams dissected the genes of the variants (they weren't new species) to find out what had happened. One of the original bugs had undergone a mutation that caused it to excrete acetate, an organic chemical. A second bug experienced a mutation that allowed it to make use of the acetate excreted from the first. Suddenly a symbiotic codependence of acetate maker and acetate eater had emerged from the uniformity, and the pool diverged into an ecology.

Although uniformity *can* yield diversity, variance does better. If the Earth were as smooth as a shiny ball bearing—a perfect spherical chemostat spread evenly with uniform climate and homogeneous soils—then the diversity of ecological communities on it would be far reduced from what it is now. In a constant environment, all variation and all diversity must be driven by internal forces. The only constraints on life would be other coevolutionary life.

If evolution had its way, with no interference from geographical and geological dynamics—that is, without the clumsiness of a body—then mindlike evolution would feed upon itself and breed heavily recursive relationships. On a globe without mountains or storms or unexpected droughts, evolution would wind life into an ever-tightening web of coevolution, a smooth world stuffed with parasites, parasites upon parasites (hyperparasites), mimics, and symbionts, all caught up in accelerating codependence. But each species would be so tightly coupled with the others that it would be difficult to distinguish where the identity of one began and the other left off. Eventually evolution on a ball-bearing planet would mold everything into a single, massive, ultradistributed planetwide superorganism.

Creatures born in the rugged environments of arctic climes must deal with the unpredictable variations that nature is always throwing at them. Freezing at night, baking during the day, ice storms after spring thaw, all create a rugged habitat. Habitats in the tropics and in the very deep sea are relatively "smooth" because of their constant temperature, rainfall, lightfall, and nutrients. Thus the smoothness of tropical or benthic environments allows species there to relinquish the need to adapt in physiological ways and allows them room to adapt in purely biological ways. In these steady habitats we should expect to see many instances of weird symbiotic and parasitic relationships—parasites preying upon parasites, males living inside of females, and creatures mimicking and mirroring other creatures—and that's what we do find.

Without a rugged environment life can only play off itself. It will still produce variation and novelty. But far more diversity can be manufactured in natural and artificial worlds by setting creatures in a rugged and vastly differentiated environment.

This lesson has not been lost on the wannabe gods trying to create lifelike behavior in computer worlds. When self-replicating and self-mutating computer viruses are loosed into a computer memory uniformly distributed with processing resources, the computer viruses quickly evolve a host of wildly recursive varieties including parasites, hyperparasites, and hyperhyperparasites. David Ackley, one computer life researcher, told me, "I finally figured out that the way to get wonderfully lifelike behavior is not to try to make a really complex creature, but to make a wonderfully rich environment for a simple creature."

IT'S TWO O'CLOCK on a blustery afternoon, six months after my midnight hike, when I climb the hill behind my house again. The windblown grass is green from the winter's rain. Up near the ridge I stop at a circle where the deer have matted the soft grass into a cushion. The trampled stems are weathered, buff with a tinge of violet, as if the color has rubbed off the deer's bellies. I rest in this recess. The wind swipes overhead.

I can see wildflowers crouched among the blown grass blades. For some reason every species is blue-violet: lupine, blue-eyed grass, thistle, gentian. Between me, the bent grass, and the ocean there are shrubs, squat creatures outfitted with silvery olive leaves—standard desert issue.

Here's a stem of Queen Anne's lace. Its furrowed leaves are mind-bogglingly intricate. Each leaf has two dozen minileaves arrayed on it, and each of those minileaves has a dozen microleaves arrayed on it. The recursive shape is the result of some obsessive process, no doubt. Its bunched flower head, 30 miniature cream white florets surrounding a single tiny purple floret in the center, is equally unexpected. On this one slope where I rest, the diversity of living forms is overwhelming in its detail and unlikeliness.

I should be impressed. But what strikes me as I sit among two million grass plants and several thousand juniper shrubs, is how similar life on Earth is. For all the possible shapes and behaviors animated matter could take, only a few—in wide variation—are tried out. Life can't fool me. It's all the same, like those canned goods in grocery stores with different labels but all manufactured by the same food conglomerate. Life on Earth obviously all comes from one transnational conglomerate.

The grass pushing up on my seat, the scraggly thistle stem rubbing my shirt, the brown-breasted swallow swooping downhill:

they are a single thing stretching out in many directions. I recognized it because I am stretched into it too.

Life is a networked thing—a distributed being. It is one organism extended in space and time. There is no individual life. Nowhere do we find a solo organism living. Life is always plural. (And not until it became plural—cloning itself—could life be called life.) Life entails interconnections, links, and shared multiples. "We are of the same blood, you and I," coos the poet Mowgli. Ant, we are of the same blood, you and I. Tyrannosaurus, we are of the same blood, you and I. AIDS virus, we are of the same blood, you and I.

The apparent individuals that life has dispersed itself into are illusions. "Life is [primarily] an ecological property, and an individual property for only a fleeting moment," writes microbiologist Clair Folsome, a man who dabbled in making superorganisms inside bottles. We live one life, distributed. Life is a transforming flood that fills up empty containers and then spills out of them on its way to fill up more. The shape and number of vessels submerged by the flood doesn't make a bit of difference.

Life works as an extremist, a fanatic without moderation. It infiltrates everywhere. It saturates the atmosphere, covers the Earth's surface and wheedles its way into bedrock cracks. It will not be refused. As Lovelock noted, we have dug up no ancient rocks without also digging up ancient life preserved in them. John von Neumann, who thought of life in mathematical terms, said, "living organisms are . . . by any reasonable theory of probability or thermodynamics, highly improbable . . . [However] if by any peculiar accident there should ever be one of them, from there on the rules of probability do not apply, and there will be many of them." Life once made, filled the Earth immediately, commandeering matter from all the realms—gas, liquid, solid—into its schemes. "Life is a *planetary-scale* phenomenon," said James Lovelock. "There cannot be sparse life on a planet. It would be as unstable as half of an animal."

A thin membrane of whole life now covers the entire Earth. It is a coat that cannot be taken off. Rip one seam and the coat will patch itself on the spot. Abuse it, and the coat will metamorphose itself to thrive on the abuse. Not a threadbare green, it is a lush technicolor coat, a flamboyant robe surrounding the colossal corporeality of the planet.

In practice, it is an everlasting coat. The great secret which life has kept from us is that once born, life is immortal. Once launched, it cannot be eradicated.

Despite the rhetoric of radical environmentalists, it is beyond the power of human beings to wipe the whole flood of life off the planet. Mere nuclear bombs would do little to halt life in general, and might, in fact, increase the nonhuman versions.

There must have been a time billions of years ago when life crossed the threshold of irreversibility. Let's call that the I-point (for irreversible, or immortal). Before the I-point life was tenuous; indeed it faced a steep uphill slope. Frequent meteor impacts, fierce radiation, and harsh temperature fluctuations on Earth four billion years ago created an incredibly hostile environment for any half-formed, about-to-replicate complexity. But then, as Lovelock tells the story, "very early in the history of the planet, the climate conditions formed a window of opportunity just about right for life. Life had a short period in which to establish itself. If it failed, the whole system for future life failed."

But once established, life stuck fast. And once past the I-point life turned out to be neither delicate nor fragile, but hardy and irrepressible. Single cell bacteria are astonishingly indomitable, living in every possible antagonistic environment one could imagine, including habitats doused with heavy radiation. As hospitals know, it is frustratingly difficult to rid a few rooms of bacterial life. The Earth? Ha!

We should heed the unstoppable nature of life, because it has much to do with the complexity of vivisystems. We are about to make machines as complex as grasshoppers and let them loose in the world. Once born, they won't go away. Of the thousands of computer viruses cataloged by virus hunters so far, not one species of them has gone extinct. According to the companies that write antiviral software there are several dozens of new computer viruses created per week. They'll be with us for as long as we have computers.

The reason life cannot be halted is that the complexity of life's dynamics has exceeded the complexity of all known destructive forces. Life is far more complex than nonlife. While life can serve as an agent of death—predator chomping on prey—the consumption of one life form by another generally does not diminish complexity in the whole system and may even add to it.

It takes, on average, all the diseases and accidents of the world working 24 hours a day, 7 days a week, with no vacations, about 621,960 hours to kill a human organism. That's 70 years of full-time attack to break the bounds of human life—barring the intervention of modern medicine (which may either accelerate or hinder death, depending on your views). This stubborn persistence in life is directly due to the complexity of the human body.

In contrast, a well-built car that managed to puff its way to an upper limit of 200,000 miles before blowing a valve would have run for about 5,000 hours. A jet turbine engine may run for 40,000 hours before being rebuilt. A simple light bulb with no moving parts is good for 2,000 hours. The longevity of nonliving complexity isn't even in the same league as the persistence of life.

The museum at the Harvard Medical School dedicates a display case to the "crowbar skull." This skull reveals a hole roughly gouged by a speeding iron bar. The skull belonged to Phineas Gage, a 19th-century quarry foreman who was packing a black powder charge into a hole with the iron bar when the powder exploded. The iron bar pierced his head. His crew sawed off the protruding bar before taking him to an ill-equipped doctor. According to anecdotes from those who knew him, Gage lived for another 13 years, more or less functional, except that after the accident he became short-tempered and peevish. Which is understandable. But the machine kept going.

People who lack a pancreas, a second kidney, a small intestine, may not run marathons, but they live. While debasement of many small components of the body—glands in particular—can cause death to the whole, these parts are heavily buffered from easy disruption. Indeed, warding off disruption is the principal property of complex systems.

Animals and plants in the wild regularly survive drastic violence and injury. The only study I know that has tried to measure the rate of injury in the wild focused on Brazilian lizards and concluded that 12 percent of them were missing at least one toe. Elk survive gunshot wounds, seals heal after shark bites, oak trees resprout after decapitation. In one experiment gastropods whose shells were deliberately crushed by researchers and returned to the wild lived as long as uninjured controls. The heroic achievement in nature is not the little fish that gets away, but that old man death is ever able to crash a system.

Networked complexity inverts the usual relation of reliability in things. As an example, individual switch parts in a modern camera may have 90 percent dependability. Linked dumbly in a series, not in a distributed way, the hundreds of switches would have great unreliability as a group—let's say they have 75 percent dependability. Connected right—each part informing the others—as they are in advanced point-'n'-shoots, the reliability of the camera counter intuitively rises as a whole to 99 percent, *exceeding* the reliability of the individual parts (90 percent).

But the camera now has new subgroups of parts which act like parts themselves. More virtual parts means the total possibility for unpredictable behavior at the component level increases. There are now novel ways to go wrong. So while the camera as a whole is utterly more dependable, when it does surprise, it can often be a very surprising surprise. The old cameras were easy to fail, easy to repair. The new cameras fail creatively.

Failing creatively is the hallmark of vivisystems. Dying is difficult, but there are a thousand ways to do it. It took two hundred overpaid engineers two weeks of emergency alert work to figure out why the semi-alive American telephone switching system repeatedly failed in 1990. And these are the guys who built it. It had never failed this way, and probably won't fail this way again.

While every human is born pretty much the same, every death is different. If coroner's cause-of-death certificates were exact, each one would be unique. Medicine finds it more instructive to round off the causes and classify them generally, so the actual idiosyncratic nature of each death is not recorded.

A complex system cannot die simply. The members of a system have a bargain with the whole. The parts say, "We are willing to sacrifice to the whole, because together we are greater than our sum." Complexity locks in life. The parts may die, but the whole lives. As a system self-organizes into greater complexity, it increases its life. Not the length of its life, but its lifeness. It has more lives.

We tend to think of life and death as binary; a creature is either off or on. The self-organizing subsystems in organisms suggest, though, that some things are more alive than others. Biologist Lynn Margulis and others have pointed out that even a cell has lives in plural, as each cell is a historical marriage of at least three vestigial forms of bacteria.

"I am the most alive among the living," crows the Russian poet A. Tarkovsky (father of the filmmaker). That's politically incorrect, but probably true. There may be no real difference between the aliveness of a sparrow and a horse, but there is a difference of aliveness between a horse and a willow tree, or between a virus and a cricket. The greater the complexity of a vivisystem, the more life it may harbor. As long as the universe continues to cool down, life will build up in more curious varieties and in further mutual networks.

I HEAD UP the hill behind my house one more time. I ramble over to a grove of eucalyptus trees, where the local 4-H club used to keep its beehives. The grove snoozes in moist shade this time of day; the west-facing hill it stands on blocks the warm morning sun.

I imagine the valley all rock and barren at history's start—a hill of naked flint and feldspar, desolate and shiny. A billion years flicker by. Now the rock is clothed with a woven mat of grass. Life has filled a space in the grove with wood reaching higher than I can. Life is trying to fill the whole valley in. For the next billion years, it will keep trying new forms, erupting in whatever crevice or emptiness it can find.

Before life, there was no complex matter in the universe. The entire universe was utterly simple. Salts. Water. Elements. Very boring. After life, there was much complex matter. According to astrochemists, we can't find complex molecules in the universe outside of life. Life tends to hijack any and all matter it comes in contact with and complexify it. By some weird arithmetic, the more life stuffs itself into the valley, the more spaces it creates for further life. In the end, this small valley along the northern coast of California will become a solid block of life. In the end, left to its own drift, life may infiltrate all matter.

Why isn't the Earth a solid green from space? Why doesn't life cover the oceans and fill the air? I believe the answer is that if left alone, the Earth will be solid green someday. The conquest of air by living organisms is a relatively recent event, and one not yet completed. The complete saturation of the oceans may have to wait for rugged mats of kelp to evolve, ones able to withstand storm waves. But in the end, life will dominate; the oceans will be green.

The galaxy may be green someday too. Distant planets now toxic to life won't always remain so. Life can evolve representations of itself capable of thriving in environments that seem hostile now. But more importantly, once one variety of life has a toehold in a place, the inherently transforming nature of life modifies the environment until it is fit for other species of life.

In the 1950s, the physicist Erwin Schrödinger called the life force "negentropy" to indicate its opposite direction from the push of thermal decay. In the 1990s, an embryonic subculture of technocrats thriving in the U.S. calls the life force "extropy."

"Extropians," as promoters of extropy call themselves, issued a seven-point lifestyle manifesto based on the vitalism of life's extropy. Point number three is a creed that states their personal belief in "boundless expansion"— the faith that life will expand until it fills the universe. Those who don't believe this are tagged "deathists." In the context of their propaganda, this creed could be read as mere pollyanna self-inspiration, as in: We can do anything!

But somewhat perversely I take their boast as a scientific proposition: life will fill the universe. Nobody knows what the theoretical limits to the infection of matter by life would be. Nor does anybody know what the maximum amount of life-enhanced matter that our sun could support is.

In the 1930s, the Russian geochemist/biologist Vernadsky

wrote, "The property of maximum expansion is inherent to living matter in the same manner as it is characteristic of heat to transfer from more heated to less heated bodies, of a soluble substance to dissolve in a solvent, and of a gas to dissipate in space." Vernadsky called it "pressure of life" and measured this expansion as velocity. His record for the velocity of life expansion was a giant puffball, which, he said, produced spores at such a rate that if materials were provided fast enough for the developing fungus, in only three generations puffballs would exceed the volume of Earth. He calculated by some obscure method that the life force's "speed of transmission" in bacteria is about 1,000 kilometers per hour. Life won't get far in filling up the universe at that rate.

When reduced to its essentials, life is very close to a computational function. For a number of years Ed Fredkin, a maverick thinker once associated with MIT, has been spinning out a heretical theory that the universe *is* a computer. Not metaphorically like a computer, but that matter and energy *are* forms of information processing of the same general class as the type of information processing that goes on inside a Macintosh. Fredkin disbelieves in the solidity of atoms and says flatly that "the most concrete thing in the world is information." Stephen Wolfram, a mathematical genius who did pioneering work on the varieties of computer algorithms, agrees. He was one of the first to view physical systems as computational processes, a view that has since become popular in some small circles of physicists and philosophers. In this outlook the minimal work accomplished by life resembles the physics and thermodynamics of the minimal work done in a computer. Fredkin and company would say that knowing the maximum amount of computation that could be done in the universe (if we considered all its matter as a computer) would tell us whether life will fill the universe, given the distribution of matter and energy we see in the cosmos. I do not know if anyone has made that calculation.

One of the very few scientists to have thought in earnest about the final destiny of life is the theoretical physicist Freeman Dyson. Dyson did some rough calculations to estimate whether life and intelligence could survive until the ultimate end of the universe. He concluded it could, writing: "The numerical results of my calculations show that the quantities of energy required for permanent survival and communication are surprisingly modest . . . [T]hey give strong support to an optimistic view of the potentialities of life. No matter how far we go into the future, there will always be new things happening, new information coming in, new worlds to explore, a constantly expanding domain of life, consciousness and memory."

Dyson has taken this further than I would have dared. I was merely concerned about the dynamics of life, and how it infiltrates all matter, and how nothing known can halt it. But just as life irretrievably conquers matter, the lifelike higher processing power we call mind irrevocably conquers life and thus also all matter. Dyson writes in his lyrical and metaphysical book, *Infinite in All Directions*:

> It appears to me that the tendency of mind to infiltrate and control matter is a law of nature . . . The infiltration of mind into the universe will not be permanently halted by any catastrophe or by any barrier that I can imagine. If our species does not choose to lead the way, others will do so, or may have already done so. If our species is extinguished, others will be wiser or luckier. Mind is patient. Mind has waited for 3 billion years on this planet before composing its first string quartet. It may have to wait for another 3 billion years before it spreads all over the galaxy. I do not expect that it will have to wait so long. But if necessary, it will wait. The universe is like a fertile soil spread out all around us, ready for the seeds of mind to sprout and grow. Ultimately, late or soon, mind will come into its heritage. What will mind choose to do when it informs and controls the universe? That is a question which we cannot hope to answer.

ABOUT A CENTURY AGO, the common belief that life was a mysterious liquid that infused living things was refined into a modern philosophy called vitalism. The position which vitalism held was not very far from the meaning in the everyday phrase, "She lost her life." We all imagine some invisible substance seeping away at death. The vitalists took this vernacular meaning seriously. They held that while the essential spirit stirring in creatures was not itself alive, neither was it wholly an inanimate material or mechanism either. It was something else: a vital impulse that existed outside of the creature it animated.

My description of the aggressive character of life is not meant to be a postmodern vitalism. It is true that defining life as "an emergent property contingent upon the organization of inanimate parts but not reducible to them" (the best that science can do right now), comes very close to sounding like a metaphysical doctrine. But it is intended to be testable.

I take the view that life is a nonspiritual, almost mathematical property that can emerge from networklike arrangements of matter. It is sort of like the laws of probability; if you get enough components together, the system will behave like this, because the law of averages dictates so. Life results when anything is organized according to laws only now being uncovered; it follows rules as strict as those that light obeys.

This lawful process coincidentally clothes life in a spiritual looking garb. One reason is that this organization must, by law, produce the unpredictable and novel. Secondly, the result of organization must replicate at every opportunity, giving it a sense of urgency and desire. And thirdly, the result can easily loop around to protect its own existence, and thus it acquires an emergent agenda. Altogether, these principles might be called the "emergent" doctrine of life. This doctrine is radical because it entails a revised notion of what laws of nature mean: irregularity, circular logic, tautology, surprise.

Vitalism, like every wrong idea, contains a useful sliver of truth. Hans Driesch, the arch twentieth-century vitalist, defined vitalism in 1914 as "the theory of the autonomy of the process of life," and in certain respects he was right. Life in our dawning new view can be divorced from both living bodies and mechanical matrix, and set apart as a real, autonomous process. Life can be copied from living bodies as a delicate structure of information (spirit or gene?) and implanted in new lifeless bodies, whether they are of organic parts or machine parts.

In the history of ideas, we have progressively eliminated

discontinuities from our perception of our role as humans. Historian of science David Channell summarizes this progression in his book *The Vital Machine: A Study of Technology and Organic Life.*

> First, Copernicus eliminated the discontinuity between the terrestrial world and the rest of the physical universe. Next, Darwin eliminated the discontinuity between human beings and the rest of the organic world. And most recently, Freud eliminated the discontinuity between the rational world of the ego and the irrational world of the unconscious. But as [historian and psychologist Bruce] Mazlish has argued, there is one discontinuity that faces us yet. This "fourth discontinuity" is between human beings and the machine.

We are now crossing the fourth discontinuity. No longer do we have to choose between the living or the mechanical because that distinction is no longer meaningful. Indeed, the most meaningful discoveries in this coming century are bound to those that celebrate, explore, and exploit the unified quality of technology and life.

The bridge between the worlds of the born and the manufactured is the perpetual force of radical disequilibrium—a law called life. In the future, the essence that both living creatures and machines will have in common—that which will distinguish them from all other matter in the universe—is that they both will have the dynamics of self-organized change.

We can now take the premise that life is a something in flux that is obeying laws which humans can uncover and recognize, even if we can't understand them fully. As a way to discover the commonalty between machines and creatures in this book, I've found it useful to ask, What does life want? I also consider evolution in the same way. What does evolution want? Or to be more precise, What does the world look like from life and evolution's point of view? If we consider life and evolution as "autonomous processes," then what are their selfish goals? Where are they headed? What are they becoming?

Gretel Ehrlich writes in her lyrical book *Montana Spaces*: "Wildness has no conditions, no sure routes, no peaks or goals, no source that is not instantly becoming something more than itself, then letting go of that, always becoming. It cannot be stripped to its complexity by cat scan or telescope. Rather, it is a many-pointed truth, almost a bluntness, a sudden essence like the wild strawberries strung along the ground on scarlet runners under my feet. Wildness is source and fruition at once, as if every river circled round, the mouth eating the tail—and the tail, the source . . ."

There is no purpose, other than itself, to wildness. It is both "source and fruition," the mingling of cause and effect in circular logic. What Ehrlich calls wildness, I call a network of vital life, an outpouring of a nearly mechanic force that seeks only to enlarge itself, and that pushes its disequilibrium into all matter, erupting in creatures and machines alike.

Wildness/life is always becoming, Ehrlich says. Becoming what? Becoming becoming. Life is on its way to further complications, further deepness and mystery, further processes of becoming and change. Life is a circle of becoming, an autocatalytic set, inflaming itself with its own sparks, breeding upon itself more life and more wildness and more "becomingness." Life has no conditions, no moments that are not instantly becoming something more than life itself.

As Ehrlich hints, wild life resembles that strange loop of the Uroborus biting its tail, consuming itself. But in truth, wild life is the far stranger loop of a snake releasing itself from its own grip, unmouthing an ever fattening tail tapering up to an ever increasingly larger mouth, birthing an ever larger tail, filling the universe with this strangeness.

# 7

# EMERGENCE OF CONTROL

THE INVENTION of autonomous control, like most inventions, has roots in ancient China. There, on a dusty windswept plain, a small wooden statue of a man in robes teeters upon a short pole. The pole is carried between a pair of turning wagon wheels, pulled by two red horses outfitted in bronze finery.

The statue man, carved in the flowing dresses of 9th-century China, points with outstretched hand towards a distant place. By the magic of noisy gears connecting the two wooden wheels, as the cart races along the steppes, the wooden man perched on the stick invariably, steadily, without fail, points south. When the cart turns left or right, the geared wheels calculate the change and swing the wooden man's (or is it a god's?) arm a corresponding amount in the opposite direction, negating the cart's shift and keeping the guide forever pointing to the south. With an infallible will, and on his own accord, the wooden figure automatically seeks south. The south-pointing chariot precedes a lordly procession, preventing the party from losing its way in the desolate countryside of old China.

How busy was the ingenious medieval mind of China! Peasant folk in the backwaters of southwestern China, wishing to temper the amount of wine downed in the course of a fireside toast, came upon a small device which, by its own accord, would control the rowdy spirits of the wine. Chou Ch'u-Fei, a traveler among the

Ch'i Tung natives then, reported that drinking bouts in this kingdom had been perfected by means of a two-foot-long bamboo straw which automatically regulated wine consumption, giving large-throated and small-mouthed drinkers equal advantage. A "small fish made of silver" floated inside the straw. The downward weight of the internal metal float restricted the flow of warm plum wine if the drinker sucked too feebly (perhaps through intoxication), thereby calling an end for his evening of merriment. If he inhaled too boisterously, he also got nothing, as the same float became wedged upwards by force of the suction. Only a temperate, steady draw was profitable.

Upon inspection, neither the south-pointing carriage nor the wine straw are truly automatic in a modern (self-steering) sense. Both devices merely tell their human masters, in the most subtle and unconscious way, of the adjustment needed to keep the action constant, and leave the human to make the change in direction of travel or power of lung. In the lingo of modern thinking, the human is part of the loop. To be truly automatic, the south-pointing statue would have to turn the cart itself, to make it a south-*heading* carriage. Or a carrot would have to be dangled from the point of his finger so that the horses (now in the loop) followed it. Likewise the drinking straw would have to regulate its volume no matter how hard one sucked. Although not automatic, the south-pointing cart is based on the differential gear, a thousand-year-old predecessor to the automobile transmission, and an early prototype of modern self-pointing guns on an armored tank which aid the drivers inside where a magnetic compass is useless. Thus, these clever devices are curious stillbirths in our genealogy of automation. The very first truly automatic devices had actually been built long before, a millennium earlier.

Ktesibios was a barber who lived in Alexandria in the first half of the third century B.C. He was obsessed with mechanical devices, for which he had a natural genius. He eventually became a proper mechanician—a builder of artifactual creations—under King Ptolemy II. He is credited with having invented the pump, the water organ, several kinds of catapults, and a legendary water clock. At the time, Ktesibios's fame as an inventor rivaled that of the legendary engineer Archimedes. Today, Ktesibios is credited with inventing the first honest-to-goodness automatic device.

Ktesibios's clock kept extraordinarily good time (for then) by self-regulating its water supply. The weakness of most water clocks until that moment was that as the reservoir of water propelling the drive mechanism emptied, the speed of emptying would gradually decrease (because a shallow level of water provides less pressure than a high level), slowing down the clock's movements. Ktesibios got around this perennial problem by inventing a regulating valve (*regula*) comprised of a float in the shape of a cone which fit its nose into a mating inverted funnel. Within the *regula*, water flowed from the funnel stem, over the cone, and into the bowl the cone swam in. The cone would then float up into the concave funnel and constrict the water passage, thus throttling its flow. As the water diminished, the float would sink, opening the passage again and allowing more water in. The *regula* would immediately seek a compromise position where it would let "just enough" water for a constant flow through the metering valve vessel.

Ktesibios's *regula* was the first nonliving object to self-regulate, self-govern, and self-control. Thus, it became the first *self* to be born outside of biology. It was a true *auto* thing—directed from within. We now consider it to be the primordial automatic device because it held the first breath of lifelikeness in a machine.

It truly was a *self* because of what it displaced. A constant autoregulated flow of water translated into a constant autoregulated clock and relieved a king of the need for servants to tend the water clock's water vessels. In this way, "auto-self" shouldered out the human self. From the very first instance, automation replaced human work.

Ktesibios's invention is first cousin to that all-American 20th-century fixture, the flush toilet. Readers will recognize the Ktesibios floating valve as the predecessor to the floating ball in the upper chamber of the porcelain throne. After a flush, the floating ball sinks with the declining water level, pulling open the water valve with its metal arm. The incoming water fills the vessel again, raising the ball triumphantly so that its arm closes the flow of water at the precise level of "full." In a medieval sense, the toilet yearns to keep itself full by means of this automatic plumbing. Thus, in the bowels of the flush toilet we see the archetype for all autonomous mechanical creatures.

About a century later, Heron, working in the same city of Alexandria, came up with a variety of different automatic float mechanisms, which look to the modern eye like a series of wildly convoluted toilet mechanisms. In actuality, these were elaborate party wine dispensers, such as the "Inexhaustible Goblet" which refilled itself to a constant level from a pipe fitted into its bottom. Heron wrote a huge encyclopedia (the *Pneumatica*) crammed with his incredible (even by today's standards) inventions. The book was widely translated and copied in the ancient world and was influential beyond measure. In fact, for 2,000 years (that is, until the age of machines in the 18th century), no feedback systems were invented that Heron had not already fathered.

The one exception was dreamed up in the 17th century by a Dutch alchemist, lens grinder, pyromaniac, and hobby submariner by the name of Cornelis Drebbel. (Drebbel made more than one successful submarine dive around 1600!) While tinkering in his search for gold, Drebbel invented the thermostat, the other universal example of a feedback system. As an alchemist, Drebbel suspected that the transmutation of lead into gold in a laboratory was inhibited by great temperature fluctuations of the heat sources cooking the elements. In the 1620s he jerry-rigged a minifurnace which could bake the initial alchemic mixture over moderate heat for a very long time, much as might happen to gold-bearing rock bordering the depths of Hades. On one side of his ministove, Drebbel attached a glass tube the size of a pen filled with alcohol. The liquid would expand when heated, pushing mercury in a connecting second tube, which in turn would push a rod that would close an air draft on the stove. The hotter the furnace, the further the draft would close, decreasing the fire. The cooling tube retracted the rod, thus opening the draft and increasing the fire. An ordinary suburban tract home thermostat is conceptually identical—both seek a constant temperature. Unfortunately, Drebbel's automatic stove didn't make gold, nor did Drebbel ever publish its design, so his automatic invention perished without influence, and its design had to be rediscovered a hundred years later by a French gentleman farmer, who built one to incubate his chicken eggs.

James Watt, who is credited with inventing the steam engine, did not. Working steam engines had been on the job for decades before Watt ever saw one. As a young engineer, Watt was once asked to repair a small-scale model of an early working, though inefficient, Newcomen steam engine. Frustrated by its awkwardness, Watt set out to improve it. At about the time of the American Revolution, he added two things to the existing engines; one of them evolutionary, the other revolutionary. His key evolutionary innovation was separating the heating chamber from the cooling chamber; this made his engine extremely powerful. So powerful that he needed to add a speed regulator to moderate this newly unleashed machine power. As usual Watt turned to what already existed. Thomas Mead, a mechanic and miller, had invented a clumsy centrifugal regulator for a windmill that would lower the millstone onto the grain only when the stone's speed was sufficient. It regulated the output but not the power of a millstone.

Watt contrived a radical improvement. He borrowed Mead's regulator from the mill and revisioned it into a pure control circuit. By means of his new regulator the steam machine gripped the throat of its own power. His completely modern *regula* automatically stabilized his now ferocious motor at a constant speed of the operator's choice. By adjusting the governor, Watt could vary the steam engine to run at any rate. This was revolutionary.

Like Heron's float and Drebbel's thermostat, Watt's centrifugal governor is transparent in its feedback. Two leaden balls, each at the end of a stiff pendulum, swing from a pole. As the pole rotates the balls spin out levitating higher the faster the system spins. Linkages scissored from the twirling pendulums slide up a sleeve on the pole, levering a valve which controls the speed of rotation by adjusting the steam. The higher the balls spin, the more the linkages close the valve, reducing the speed, until an equilibrium point of constant rpms (and height of spinning balls) is reached. The control is thus as dependable as physics.

Rotation is an alien power in nature. But among machines, it is blood. The only known bearing in biology is at the joint of a sperm's spinning hair propeller. Outside of this micromotor, the axle and wheel are unknown to those with genes. To the ungened machine, whirling wheels and spinning shafts are reasons to live. Watt gave machines the secret to controlling their own revolutions, which was his revolution. His innovation spread widely and quickly. The mills of the industrial age were fueled by steam, and the engines earnestly regulated themselves with the universal badge of self-control: Watt's flyball governor. Self-powered steam begat machine mills which begat new kinds of engines which begat new machine tools. In all of them, self-regulators dwelt, fueling the principle of snowballing advantages. For every one person visibly working in a factory, thousands of governors and self-regulators toiled invisibly. Today, hundreds of thousands of regulators, unseen, may work in a modern plant at once. A single human may be their coworker.

Watt took the volcanic fury of expanding steam and tamed it with information. His flyball governor is undiluted informational control, one of the first non-biological circuits. The difference between a car and an exploding can of gasoline is that the car's information—its design—tames the brute energy of the gas. The same amount of energy and matter are brought together in a car burning in a riot and one speeding laps in the Indy 500. In the latter case, a critical amount of information rules over the system, civilizing the dragon of fire. The full heat of fire is housetrained by small amounts of self-perception. Furious energy is educated, brought in from the wilds to work in the yard, in the basement, in the kitchen, and eventually in living rooms.

The steam engine is an unthinkable contraption without the domesticating loop of the revolving governor. It would explode in the face of its inventors without that tiny heart of a self. The immense surrogate slave power released by the steam engine ushered in the industrial revolution. But a second, more important revolution piggybacked on it unnoticed. There could not have been an industrial revolution without a parallel (though hidden) information revolution at the same time, launched by the rapid spread of the automatic feedback system. If a fire-eating machine, such as Watt's engine, lacked self-control, it would have taken every working hand the machine displaced to babysit its energy. So information, and not coal itself, turned the power of machines useful and therefore desirable.

The industrial revolution, then, was not a preliminary primitive stage required for the hatching of the more sophisticated information revolution. Rather, automatic horsepower was, itself, the first phase of the knowledge revolution. Gritty steam engines, not teeny chips, hauled the world into the information age.

HERON'S REGULATOR, Drebbel's thermostat, and Watt's governor bestowed on their vessels a wisp of self-control, sensory awareness, and the awakening of anticipation. The governing system sensed its own attributes, noted if it had changed in a certain respect since it last looked, and if it had, it adjusted itself to conform to a goal. In the specific case of a thermostat, the tube of alcohol detected the system's temperature, and then took action or not to tweak the fire in order to align itself with the fixed goal of a certain temperature. It had, in a philosophical sense, a purpose.

Although it may strike us as obvious now, it took a long while for the world's best inventors to transpose even the simplest automatic circuit such as a feedback loop into the realm of electronics. The reason for the long delay was that from the moment of its discovery electricity was seen primarily as power and not as communication. The dawning distinction of the two-faced nature of the spark was acknowledged among leading German electrical engineers of the last century as the split between the techniques of strong current and the techniques of weak current. The amount of energy needed to send a signal is so astoundingly small that electricity had to be reimagined as something altogether different from power. In the camp of the wild-eyed German signalists, electricity was a sibling to the speaking mouth and the writing hand. The inventors (we would call them hackers now) of weak current technology brought forth perhaps the least precedented invention of all time—the telegraph. With this device human communication rode on invisible particles of lightning. Our entire society was reimagined because of this wondrous miracle's descendants.

Telegraphers had the weak model of electricity firmly in mind, yet despite their clever innovations, it wasn't until August 1929, that telephone engineer H. S. Black, working at Bell Laboratories, tamed an electrical feedback loop. Black was hunting for a way to make durable amplifier relays for long-distance phone lines. Early amplifiers were made of crude materials that tended to disintegrate over use, causing the amp to "run away." Not only would an aging relay amplify the phone signal, it would mistakenly compound any tiny deviation from the range it expected until the mushrooming error filled and killed the system. What was needed was Heron's *regula*, a counter signal to rein in the chief signal, to dampen the effect of the perpetual recycling. Black came up with a *negative* feedback loop, which was designated negative in contrast to the snowballing positive loop of the amplifier. Conceptually, the electrical negative feedback loop is a toilet flusher or thermostat. This braking circuit keeps the amplifier homed in on a steady amplification in the same way a thermostat homes in on a steady temperature. But instead of metallic levers, a weak train of electrons talks to itself. Thus, in the byways of the telephone switching network, the first *electrical self* was born.

From World War I and after, the catapults that launched missiles had become so complicated, and their moving targets so sophisticated, that calculating ballistic trajectories taxed human talent. Between battles, human calculators, called computers, computed the settings for firing large guns under various wind, weather and altitude conditions. The results were sometimes printed in pocket-size tables for the gunmen on the front line, or if there was enough time and the missile-gun was common, the tables were mechanically encoded into an apparatus on the gun, known as the automaton. In the U.S., the firing calculations were compiled in a laboratory set up at the Navy's Aberdeen Proving Ground in Maryland, where rooms full of human computers (almost exclusively women) employed hand-cranked adding machines to figure the tables.

By World War II, the German airplanes which the big guns boomed at were flying as fast as the missiles themselves. Speedier on-the-spot calculations were needed, ideally ones that could be triggered from measurements of planes in flight made by the newly invented radar scanner. Besides, Navy gunmen had a weighty problem: how to move and aim these monsters with the accuracy the new tables gave them. The solution was as close at hand as the stern of the ship: a large ship controlled its rudder by a special type of automatic feedback loop known as a servomechanism.

Servomechanisms were independently and simultaneously invented a continent apart by an American and a Frenchman around 1860. It was the Frenchman, engineer Leon Farcot, who tagged the device with a name that stuck: *moteur asservi*, or servo-motor. As boats had increased in size and speed over time, human power at the tiller was no longer sufficient to move the rudder against the force of water surging beneath. Marine technicians came up with various oil-hydraulic systems that amplified the power of the tiller so that gently swinging the miniature tiller at the captain's helm would move the mighty rudder, kind of. A repeated swing of the minitiller would translate into different amounts of steerage of the rudder depending on the speed of the boat, waterline, and other similar factors. Farcot invented a linkage system that connected the position of the heavy rudder underwater back to the position of the easy-to-swing tiller—the automatic feedback loop! The tiller then indicated the actual location of the rudder, and by means of the loop, moving the indicator moved the reality. In the jingo of current computerese, What you see is what you get!

The heavy gun barrels of World War II were animated the same way. A hydraulic hose of compressed oil connected a small pivoting lever (the tiller) to the pistons steering the barrel. As the shipmate's hand moved the lever to the desired location, that tiny turn compressed a small piston which would open a valve releasing pressurized oil, which would nudge a large piston moving the heavy gun barrel. But as the barrel swung it would push a small piston that, in return, moved the hand lever. As he tried to turn the tiller, the sailor would feel a mild resistance, a force created by the feedback from the rudder he wanted to move.

Bill Powers was a teenage Electronic Technician's Mate who worked with the Navy's automated guns, and who later pursued control systems as explanation for living things. He describes the false impression one gets by reading about servomechanism loops:

> The sheer mechanics of speaking or writing stretches out the action so it seems that there is a sequence of well-separated events, one following the other. If you were trying to describe how a gun-pointing servomechanism works, you might start out by saying, "Suppose I push down on the gun-barrel to create a position error. The error will cause the servo motors to exert a force against the push, the force getting larger as the push gets larger." That seems clear enough, but it is a lie. If you really did this demonstration, you would say "Suppose I push down on the gun-barrel to create an error . . . wait a minute. It's stuck."
>
> No, it isn't stuck. It's simply a good control system. As you begin to push down, the little deviation in sensed position of the gun-barrel causes the motor to twist the barrel up against your push. The amount of deviation needed to make the counteractive force equal to the push is so small that you can neither see nor feel it. As a result, the gun-barrel feels as rigid as if it were cast in concrete. It creates the appearance of one of those old-fashioned machines that is immovable simply because it weighs 200 tons, but if someone turned off the power the gun-barrel would fall immediately to the deck.

Servomechanisms have such an uncanny ability to aid steering that they are still used (in updated technology) to pilot boats, to control the flaps in airplanes, and to wiggle the fingers in remotely operated arms handling toxic and nuclear waste.

More than the purely mechanical self-hood of the other regulators like Heron's valve, Watt's governor, and Drebbel's thermostat, the servomechanism of Farcot suggested the possibility of a man-machine symbiosis—a joining of two worlds. The pilot merges into the servomechanism. He gets power, it gets existence. Together they steer. These two aspects of the servomechanisms—steering and symbiosis—inspired one of the more colorful figures of modern science to recognize the pattern that connected these control loops.

Of all the mathematicians assigned during World War I to the human calculating lab in charge of churning out more accurate firing tables at the Aberdeen Proving Grounds, few were as overqualified as Private Norbert Wiener, a former math prodigy whose genius had an unorthodox pedigree.

The ancients recognized genius as something given rather than created. But America at the turn of the century was a place where the wisdom of the past was often successfully challenged. Norbert's father, Leo Wiener, had come to America to launch a vegetarian commune. Instead, he was distracted with other untraditional challenges, such as bettering the gods. In 1895, as a Harvard professor of Slavic languages, Leo Wiener decided that his firstborn son was going to be a genius. A genius deliberately made, not born.

Norbert Wiener was thus born into high expectations. By the age of three he was reading. At 18 he earned his Ph.D. from Harvard. By 19 he was studying metamathematics with Bertrand Russell. Come 30 he was a professor of mathematics at MIT and a thoroughly odd goose. Short, stout, splayfooted, sporting a goatee and a cigar, Wiener waddled around like a smart duck. He had a legendary ability to learn while slumbering. Numerous eye-witnesses tell of Wiener sleeping during a meeting, suddenly awakening at the mention of his name, and then commenting on the conversation that passed while he dozed, usually adding some penetrating insight that dumbfounded everyone else.

In 1948 he published a book for nonspecialists on the feasibility and philosophy of machines that learn. The book was initially published by a French publisher (for roundabout reasons) and went through four printings in the United States in its first six months, selling 21,000 copies in the first decade of its influence—a best seller then. It rivaled the success of the Kinsey Report on sexual behavior, issued the same year. As a *Business Week* reporter observed in 1949, "In one respect Wiener's book resembles the Kinsey Report: the public response to it is as significant as the content of the book itself."

Wiener's startling ideas sailed into the public mind, even though few could comprehend his book, by means of the wonderfully colorful name he coined for both his perspective and the book: *Cybernetics*. As has been noted by many writers, cybernetics derives from the Greek for "steersman"—a pilot that steers a ship. Wiener, who worked with servomechanisms during World War II, was struck by their uncanny ability to aid steering of all types. What is usually not mentioned is that cybernetics was also used in ancient Greece to denote a governor of a country. Plato attributes Socrates as saying, "Cybernetics saves the souls, bodies, and material possessions from the gravest dangers," a statement that encompasses both shades of the word. Government (and that meant self-government to these Greeks) brought order by fending off chaos. Also, one had to actively steer to avoid sinking the ship. The Latin corruption of *kubernetes* is the derivation of *governor*, which Watt picked up for his cybernetic flyball.

The managerial nature of the word has further antecedent to French speakers. Unbeknownst to Wiener, he was not the first modern scientist to reactivate this word. Around 1830 the French physicist Ampere (whence we get the electrical term amperes, and its shorthand "amp") followed the traditional manner of French grand scientists and devised an elaborate classification system of human knowledge. Ampere designated one branch the realm of "Noological Sciences," with the subrealm of Politics. Within political science, immediately following the sub-subcategory of Diplomacy, Ampere listed the science of Cybernetics, that is, the science of governance.

Wiener had in mind a more explicit definition, which he stated boldly in the full title of his book, *Cybernetics: or control and communication in the animal and the machine*. As Wiener's sketchy ideas were embodied by later computers and fleshed out by other theorists, cybernetics gradually acquired more of the flavor of Ampere's governance, but without the politics.

The result of Wiener's book was that the notion of feedback

penetrated almost every aspect of technical culture. Though the central concept was both old and commonplace in specialized circumstances, Wiener gave the idea legs by generalizing the effect into a universal principle: lifelike self-control was a simple engineering job. When the notion of feedback control was packaged with the flexibility of electronic circuits, they married into a tool anyone could use. Within a year or two of *Cybernetics*'s publication, electronic control circuits revolutionized industry.

The avalanche effects of employing automatic control in the production of goods were not all obvious. Down on the factory floor, automatic control had the expected virtue of moderating high-powered energy sources as mentioned earlier. There was also an overall speeding up of things because of the continuous nature of automatic control. But those were relatively minor compared to a completely unexpected miracle of self-control circuits: their ability to extract precision from grossness.

As an illustration of how the elemental loop generates precision out of imprecise parts, I follow the example suggested by the French writer Pierre de Latil in his 1956 book *Thinking by Machine*. Generations of technicians working in the steel industry pre-1948 had tried unsuccessfully to produce a roll of sheet metal in a uniform thickness. They discovered about a half-dozen factors that affected the thickness of the steel grinding out the rolling-mill—such as speed of the rollers, temperature of the steel, and traction on the sheet—and spent years strenuously perfecting the regulation of each of them, and more years attempting their synchronization. To no avail. The control of one factor would unintentionally disrupt the other factors. Slowing the speed would raise the temperature; lowering the temperature would raise the traction; increasing traction lowers the speed, and so on. Everything was influencing everything else. The control was wrapped up in some interdependent web. When the steel rolled out too thick or too thin, chasing down the culprit out of six interrelated suspects was inevitably a washout. There things stalled until Wiener's brilliant generalization published in *Cybernetics*. Engineers around the world immediately grasped the crucial idea and installed electronic feedback devices in their mills within the following year or two.

In implementation, a feeler gauge measures the thickness of the just-made sheet metal (the output) and sends this signal back to a servo-motor controlling the single variable of traction, the variable to affect the steel last, just before the rollers. By this meager, solo loop, the whole caboodle is regulated. Since all the factors are interrelated, if you can keep just one of them directly linked to the finished thickness, then *you can indirectly control them all.* Whether the deviation tendency comes from uneven raw metal, worn rollers, or mistakenly high temaperatures doesn't matter much. What matters is that the automatic loop regulates that last variable to compensate for the other variables. If there is enough leeway (and there was) to vary the traction to make up for an overly thick source metal, or insufficiently tempered stock, or rollers contaminated with slag, then out would come consistently even sheets. Even though each factor is upsetting the others, the contiguous and near instantaneous nature of the loop steers the unfathomable network of relationships between them toward the steady goal of a steady thickness.

The cybernetic principle the engineers discovered is a general one: if all the variables are tightly coupled, and if you can truly manipulate one of them in all its freedoms, then you can indirectly control all of them. This principle plays on the holistic nature of systems. As Latil writes, "The regulator is unconcerned with causes; it will detect the deviation and correct it. The error may even arise from a factor whose influence has never been properly determined hitherto, or even from a factor whose very existence is unsuspected." How the system finds agreement at any one moment is beyond human knowing, and more importantly, not worth knowing.

The irony of this breakthrough, Latil claims, is that technologically this feedback loop was quite simple and "it could have been introduced some fifteen or twenty years earlier, if the problem had been approached with a more open mind . . ." Greater is the irony that twenty years earlier the open mind for this view was well established in economic circles. Frederick Hayek and the influential Austrian school of economics had dissected the attempts to trace out the routes of feedback in complex networks and called the effort futile. Their argument became known as the "calculation argument." In a command economy, such as the then embryonic top-down economy installed by Lenin in Russia, resources were allotted by calculation, tradeoffs, and controlled lines of communication. Calculating, even less controlling, the multiple feedback factors among distributed nodes in an economy was as unsuccessful as the engineer's failure in chasing down the fleeing interlinked factors in a steel mill. In a vacillating economy it is impossible to calculate resource allotment. Instead, Hayek and other Austrian economists of the 1920s argued that a single variable—the price—is used to regulate all the other variables of resource allotment. That way, one doesn't care how many bars of soap are needed per person, or whether trees should be cut for houses or for books. These calculations are done in parallel, on the fly, from the bottom up, out of human control, by the interconnected network itself. Spontaneous order.

The consequence of this automatic control (or human uncontrol) is that the engineers could relax their ceaseless straining for perfectly uniform raw materials, perfectly regulated processes. Now they could begin with imperfect materials, imprecise processes. Let the self-correcting nature of automation strain to find the optima which let only the premium through. Or, starting with the same quality of materials, the feedback loop could be set for a much higher quality setting, delivering increased precision for the next in line. The identical idea could be exported upstream to the suppliers of raw materials, who could likewise employ the automatic loop to extract higher quality products. Cascading further out in both directions in the manufacturing stream, the automatic self became an overnight quality machine, ever refining the precision humans can routinely squeeze from matter.

Radical transformations to the means of production had been introduced by Eli Whitney's interchangeable parts and Ford's idea of an assembly line. But these improvements demanded massive retooling and capital expenditures, and were not universally applicable. The homely auto-circuit, on the other hand—a suspiciously cheap accessory—could be implanted into almost any machine that already had a job. An ugly duckling, like a printing press, was transformed into a well-behaved goose laying golden eggs.

But not every automatic circuit yields the ironclad instantaneity that Bill Power's gun barrel enjoyed. Every unit added onto a string of connected loops increases the likelihood that the message traveling around the greater loop will arrive back at its origin to find that everything has substantially changed during its journey. In particularly vast networks in fast moving environments, the split second it takes to traverse the circuit is greater than the time it takes for the situation to change. In reaction, the last node tends to compensate by ordering a large correction. But this also is delayed by the long journey across many nodes, so that it arrives missing its moving mark, birthing yet another gratuitous correction. The same effect causes student drivers to zigzag down the road, as each late large correction of the steering wheel overreacts to the last late overcorrection. Until the student driver learns to tighten the feedback loop to smaller, quicker corrections, he cannot help but swerve down the highway hunting (in vain) for the center. This then is the bane of the simple auto-circuit. It is liable to "flutter" or "chatter," that is, to nervously oscillate from one overreaction to another, hunting for its rest. There are a thousand tricks to defeat this tendency of overcompensation, one trick each for the thousand advance circuits that have been invented. For the last 40 years, engineers with degrees in control theory have written shelffuls of treatises communicating their latest solution to the latest problem of oscillating feedback. Fortunately, feedback loops can be combined into useful configurations.

Let's take our toilet, that prototypical cybernetic example. We install a knob which allows us to adjust the water level of the tank. The self-regulating mechanism inside would then seek whatever level we set. Turn it down and it satisfies itself with a low level; turn it up and it homes in on a high level of water. (Modern toilets do have such a knob.) Now let's go further and add a self-regulating loop to turn the knob, so that we can let go of that, too. This second loop's job is to seek the goal for the first loop. Let's say the second mechanism senses the water pressure in the feed pipe and then moves the knob so that it assigns a high level to the toilet when there is high water pressure and a lower level when the pressure is low.

The second circuit is controlling the range of the first circuit which is controlling the water. In an abstract sense the second loop brings forth a second order of control—the control of control—or a metacontrol. Our newfangled second-order toilet now behaves "purposefully." It adapts to a *shifting* goal. Even though the second circuit setting the goal for the first is likewise mechanical, the fact that the whole is choosing its own goal gives the metacircuit a mildly biological flavor.

As simple as a feedback loop is, it can be stitched together in endless combinations and forever stacked up until it forms a tower of the most unimaginable complexity and intricacy of subgoals. These towers of loops never cease to amuse us because inevitably the messages circulating along them cross their own paths. *A* triggers *B*, and *B* triggers *C*, and *C* triggers *A*. In outright paradox, *A* is both cause and effect. Cybernetician Heinz von Foerster called this elusive cycle "circular causality." Warren McCulloch, an early artificial intelligence guru, called it "intransitive preference," meaning that the rank of preferences would cross itself in the same self-referential way the children's game of Paper-Scissors-Stone endlessly intersects itself: Paper covers stone; stone breaks scissors; scissors cuts paper; and round again. Hackers know it as a recursive circuit. Whatever the riddle is called, it flies in the face of 3,000 years of logical philosophy. It undermines classical everything. If something can be both its own cause and effect, then rationality is up for grabs.

The compounded logic of stacked loops which doubles back on itself is the source of the strange counterintuitive behaviors of complex circuits. Made with care, circuits perform dependably and reasonably, and then suddenly, by their own drumbeat, they veer off without notice. Electrical engineers get paid well to outfox the lateral causality inherent in all circuits. But pumped up to the density required for a robot, circuit strangeness becomes indelible. Reduced back to its simplest—a feedback cycle—circular causality is a fertile paradox.

Where does self come from? The perplexing answer suggested by cybernetics is: it emerges from itself. It cannot appear any other way. Brian Goodwin, an evolutionary biologist, told reporter Roger Lewin, "The organism is the cause and effect of itself, its own intrinsic order and organization. Natural selection isn't the *cause* of organisms. Genes don't *cause* organisms. There are no *causes* of organisms. Organisms are self-causing agencies." Self, therefore, is an auto-conspired form. It emerges to transcend itself, just as a long snake swallowing its own tail becomes Uroborus, the mythical loop.

The Uroborus, according to C. G. Jung, is one of those resonant projections of the human soul that cluster around timeless forms. The ring of snake consuming its own tail first appeared as art adorning Egyptian statuary. Jung developed the idea that the nearly chaotic variety of dream images visited on humans tends to gravitate around certain stable nodes which form key and universal images, much as interlinked complex systems tend to settle down upon "attractors," to use modern terminology. A constellation of these attracting, strange nodes form the visual vocabulary of art, literature, and some types of therapy. One of the most enduring attractors, and an early pattern to be named, was the Thing Eating Its Own Tail, often graphically simplified to a snakelike dragon swallowing its own tail in a perfect circle.

The loop of Uroborus is so obviously an emblem for feedback that I have trouble ascertaining who first used it in a cybernetic context. In the true manner of archetypes it was probably realized as a feedback symbol independently more than once. I wouldn't doubt that the faint image of snake eating its tail spontaneously hatches whenever, and wherever, the GOTO START loop dawns on a programmer.

Snake is linear, but when it feeds back into itself it becomes the archetype of nonlinear being. In the classical Jungian framework, the tail-biting Uroborus is the symbolic depiction of the self. The completeness of the circle is the self-containment of self, a containment that is at the same time made of one thing and made of competing parts. The flush toilet then, as the plainest manifestation of a feedback loop, is a mythical beast—the beast of self.

The Jungians say that the self is taken to be "the original psychic state prior to the birth of ego consciousness," that is, "the original mandala-state of totality out of which the individual ego is born." To say that a furnace with a thermostat has a self is not to say it has an ego. The self is a mere ground state, an auto-conspired form, out of which the more complicated ego can later distinguish itself, should its complexity allow that.

Every self is a tautology: self-evident, self-referential, self-centered, and self-created. Gregory Bateson said a vivisystem was "a slowly self-healing tautology." He meant that if disturbed or disrupted, a self will "tend to settle toward tautology"—it will gravitate to its elemental self-referential state, its "necessary paradox."

Every self is an argument trying to prove its identity. The self of a thermostat system has endless internal bickering about whether to turn the furnace up or down. Heron's valve system argues continuously around the sole, solitary action it can take: should it move the float or not?

A system is anything that talks to itself. All living systems and organisms ultimately reduce to a bunch of regulators—chemical pathways and neuron circuits—having conversations as dumb as "I want, I want, I want; no, you can't, you can't, you can't."

The sowing of selves into our built world has provided a home for control mechanisms to trickle, pool, spill, and gush. The advent of automatic control has come in three stages and has spawned three nearly metaphysical changes in human culture. Each regime of control is boosted by deepening loops of feedback and information flow.

The control of energy launched by the steam engine was the first stage. Once energy was controlled it became "free." No matter how much more energy we might release, it won't fundamentally change our lives. The amount of calories (energy) required to accomplish something continues to dwindle so that our biggest technological gains no longer hinge on further mastery of powerful energy sources.

Instead, our gains now derive from amplifying the accurate control of materials—the second regime of control. Informing matter by investing it with high degrees of feedback mechanisms, as is done with computer chips, empowers the matter so that increasingly smaller amounts do the same work of larger uninformed amounts. With the advent of motors the size of dust motes (successfully prototyped in 1991), it seems as if you can have anything you want made in any size you want. Cameras the size of molecules? Sure, why not? Crystals the size of buildings? As you wish. Material is under the thumb of information, in the same handy way that energy now is—just spin a dial. "The central event of the twentieth century is the overthrow of matter," says technology analyst George Gilder. This is the stage in the history of control in which we now dwell. Essentially, matter—in whatever shape we want—is no longer a barrier. Matter is almost "free."

The third regime of the control revolution, seeded two centuries ago by the application of information to coal steam, is the control of information itself. The miles of circuits and information looping from place to place that administers the control of energy and matter has incidentally flooded our environment with messages, bits, and bytes. This unmanaged data tide is at toxic levels. We generate more information than we can control. The promise of more information has come true. But more information is like the raw explosion of steam—utterly useless unless harnessed by a self. To paraphrase Gilder's aphorism: "The central event of the twenty-first century will be the overthrow of information."

Genetic engineering (information which controls DNA information) and tools for electronic libraries (information which manages book information) foreshadow the subjugation of information. The impact of information domestication will be felt initially in industry and business, just as energy and material control did, and then later seep to the realm of the individual.

The control of energy conquered the forces of nature (and made us fat); the control of matter brought material wealth within easy reach (and made us greedy). What mixed cornucopia will the blossoming of full information control bring about? Confusion, brilliance, impatience?

Without selves, very little happens. Motors, by the millions, bestowed with selves, now run factories. Silicon chips, by the billions, bestowed with selves, will redesign themselves smaller and faster and rule the motors. And soon, the fibrous networks, by the zillions, bestowed with selves, will rethink the chips and rule all that we let them. If we had tried to exploit the treasures of energy, material, and information by holding all the control, it would have been a loss.

As fast as our lives allow us, we are equipping our constructed world to bootstrap itself into self-governance, self-reproduction, self-consciousness, and irrevocable selfhood. The story of automation is the story of a *one-way* shift from human control to automatic control. The gift is an irreversible transfer from ourselves to the second selves.

The second selves are out of our control. This is the key reason, I believe, why the brightest minds of the Renaissance never invented another self-regulator beyond the obvious ones known to ancient Heron. The great Leonardo da Vinci built control machines, not out-of-control machines. German historian of technology Otto Mayr claims that great engineers in the Enlightenment could have built regulated steam power of some sort with the technology available to them at the time. But they didn't because they didn't have the ability to let go of their creation.

The ancient Chinese on the other hand, although they never got beyond the south-pointing cart, had the right no-mind about control. Listen to these most modern words from the hand of the mystical pundit Lao Tzu, writing in the *Tao Teh King* 2,600 years ago:

> Intelligent control appears as uncontrol or freedom.
> And for that reason it is genuinely intelligent control.
> Unintelligent control appears as external domination.
> And for that reason it is really unintelligent control.
> Intelligent control exerts influence without appearing to do so.
> Unintelligent control tries to influence by making a show of force.

Lao Tzu's wisdom could be a motto for a gung-ho 21st-century Silicon Valley startup. In an age of smartness and superintelligence, the most intelligent control methods will appear as uncontrol methods. Investing machines with the ability to adapt on their own, to evolve in their own direction, and grow without human oversight is the next great advance in technology. Giving machines freedom is the only way we can have intelligent control.

What little time left in this century is rehearsal time for the chief psychological chore of the 21st century: letting go, with dignity.

# 8

# Closed Systems

At one end of a long row of displays in the Steinhart Aquarium in San Francisco, a concentrated coral reef sits happily tucked under lights. The Aquarium's self-contained South Pacific ocean compresses the distributed life in a mile-long underwater reef into a few glorious yards behind glass.

The condensed reef's extraordinary hues and alien life forms cast a New Age vibe. To stand in front of this rectangular bottle is to stand on a harmonic node. Here are more varieties of living creatures crammed into a square meter than anywhere else on the planet. Life does not get any denser. The remarkable natural richness of the coral reef has been squeezed further into the hyper-natural richness of a synthetic reef.

A pair of wide plate glass windows peer into an Alician wonderland of exotic beings. Fish in hippie day-glo colors stare back—accents of orange- and white-banded clown fish or a mini-school of iridescent turquoise damsels. The flamboyant creatures scoot between the feathery wands of chestnut-tinted soft corals or weave between the slowly pulsating fat lips of giant sea clams.

No mere holding pen, this is home for these creatures. They will eat, sleep, fight, and breed among each other, forever if they can. Given enough time, they will coevolve toward a shared destiny. Theirs is a true living community.

Behind the coral display tank, a clanking army of pumps,

pipes, and gizmos vibrate on electric energy to support the toy reef's ultradiversity. A visitor treks to the pumps from the darkened viewing room of the aquarium by opening an unmarked door. Blinding E.T.-like light gushes out of the first crack. Inside, the white-washed room suffocates in warm moisture and stark brightness. An overhead rack of hot metal halide lamps pumps out 15 hours of tropical sun per day. Saltwater surges through a bulky 4-ton concrete tub of wet sand brimming with cleansing bacteria. Under the artificial sunlights, long, shallow plastic trays full of green algae thrive filtering out the natural toxins from the reef water.

Industrial plumbing fixtures are the surrogate Pacific for the reef. Sixteen thousand gallons of reconstituted ocean water swirl through the bionic system to provide the same filtration, turbulence, oxygen, and buffering that the miles of South Pacific algae gardens and sand beaches perform for a wild reef. The whole wired show is a delicate, hard-won balance requiring daily energy and attention. One wrong move and the reef could unravel in a day.

As the ancients knew, what can unravel in a day may take years or centuries to build. Before the Steinhart coral reef was constructed, no one was sure if a coral reef community could be assembled artificially, or how long it would take if it could. Marine scientists were pretty sure a coral reef, like any complex ecosystem, must be assembled in the correct order. But no one knew what that order was. Marine biologist Lloyd Gomez certainly didn't know when he first started puttering around in the dank basement of the Academy's aquarium building. Gomez mixed buckets of microorganisms together in large plastic trays, gradually adding species in different sequences in hopes of attaining a stable community. He built mostly failures.

He began each trial by culturing a thick pea-green soup of algae—the scum of a pond out of whack—directly under the bank of noon-lights. If the system started to drift away from the requirements of a coral reef, Gomez would flush the trays. Within a year, he eventually got the proto-reef soup headed in the right direction.

It takes time to make nature. Five years after Gomez launched the coral reef, it is only now configuring itself into self-sustenance. Until recently Gomez had to feed the fish and invertebrates dwelling on the synthetic reef with supplemental food. But now he thinks the reef has matured. "After five years of constant babying, I have a full food web in my tank so I no longer have to feed them anything." Except sunlight, which pours on the artificial reef in a steady burst of halide energy. Sunlight feeds the algae which feed the animals which feed the corals, sponges, clams, and fish. Ultimately this reef runs on electricity.

Gomez predicts further shifts as the reef community settles into its own. "I expect to see major changes until it is ten years old. That's when the reef fusing takes place. The footing corals start to anchor down on the loose rocks, and the subterranean sponges burrow underneath. It all combines into one large mass of animal life." A living rock grown from a few seed organisms.

Much to everyone's surprise, about 90 percent of the organisms that fuse the toy reef were stowaways that did not appear to be present in the original soup. A sparse but completely invisible population of the microbes were present, but not until five years down the road, when the reef had prepared itself to be fused, were the conditions right for the blossoming of the fuser microorganisms which had been floating unseen and patient.

During the same time, certain species dominating the initial reef disappeared. Gomez says, "I was not expecting that. It startled me. Organisms were dying off. I asked myself what did I do wrong? It turns out that I didn't do anything wrong. That's just the community cycle. Heavy populations of microalgae need to be present at first. Then within ten months, they've gone. Later, some initially abundant sponges disappeared, and another type popped up. Just recently a black sponge has taken up in the reef. I have no idea where it came from." As in the restorations of Packard's prairie and Wingate's Nonsuch Island, chaperone species were needed to assemble a coral but not to maintain it. Parts of the reef were "thumbs."

Lloyd Gomez's reef-building skills are in big demand at night school. Coral reefs are the latest challenge for obsessive hobbyists, who sign up to learn how to reduce oceanic monuments to 100 gallons. Miniature saltwater systems shrink miles of life into a large aquarium, plus paraphernalia. That's dosing pumps, halide lights, ozone reactors, molecular absorption filters, and so on, at a cool $15,000 per living room tank. The expensive equipment acts like the greater ocean, cleaning, filtering the reef's water. Corals demand a delicate balance of dissolved gases, trace chemicals, pH, microorganisms, light, wave action, temperature—all of which are provided in an aquarium by an interconnected network of mechanical devices and biological agents. The common failure, Gomez says, is trying to stuff more species of life into the habitat than the system can carry, or not introducing them in the correct sequence, as Pimm and Drake discovered. How critical is the ordering? Gomez: "As critical as death."

The key to stabilizing a coral reef seemed to be getting the initial microbial matrix right. Clair Folsome, a microbiologist working at the University of Hawaii, had concluded from his own work with microbial soups in jars that "the foundation for stable closed ecologies of all types is basically a microbial one." He felt that microbes were responsible for "closing the bioelemental loops"—the flows of atmosphere and nutrients—in any ecology. He found his evidence in random mixtures of microbes, similar to the experiments of Pimm and Drake, except that Folsome sealed the lid of the jars. Rather than model a tiny slice of life on Earth, Folsome modeled a self-contained self-recycling whole Earth. All matter on Earth is recycled (except for the insignificant escape of a trace of light gases and the fractional influx of meteorites). In system-science terms, we say Earth is materially closed. The Earth is also energetically/informationally open: sunlight pours in, and information comes and goes. Like Earth, Folsome's jars were materially closed, energetically open. He scooped up samples of brackish microbes from the bays of the Hawaiian Islands and funneled them into one- or two-liter laboratory glass flasks. Then he sealed them airtight and, by extracting microscopic amounts from a sampling port, measured their species ratios and energy flow until they stabilized.

Just as Pimm was stunned to find how readily random mixtures settled into self-organizing ecosystems, Folsome was surprised to see that even the extra challenge of generating closed nutrient recycling loops in a sealed flask didn't deter simple microbial societies from finding an equilibrium. Folsome said that he and another researcher, Joe Hanson, realized in the fall of 1983 that closed ecosystems "having even modest species-diversity, rarely if ever fail." By that time some of Folsome's original flasks had been living for 15 years. The oldest one, thrown together and sealed in 1968, is now 25 years old. No air, food, or nutrients have ever been added. Yet this and all of his other jar communities are still flourishing years later under fluorescent room lights.

No matter how long they lived, though, the bottled systems required an initial staging period, a time of fluctuation and precarious instability lasting between 60 and 100 days, when anything might happen. Gomez saw this in his coral microbes: the beginnings of complexity are rooted in chaos. But if a complex system is able to find a common balance after a period of give and take, thereafter not much will derail it.

How long can such closed complexity run? Folsome said his initial interest in making materially closed worlds was sparked by a legend that the Paris National Museum displayed a cactus sealed in a glass jar in 1895. He couldn't verify its existence, but it was claimed to be covered with recurrent blooms of algae and lichens that have cycled through a progression of colors from shades of green to hues of yellow for the past century. If the sealed jar had light and a steady temperature, there was theoretically no reason why the lichens couldn't live until the sun dies.

Folsome's sealed microbial miniworlds had their own living rhythms that mirrored our planet's. They recycled their carbon, from $CO_2$ to organic matter and back again, in about two years. They maintained biological productivity rates similar to outside ecosystems. They produced stable oxygen levels slightly higher than on Earth. They registered energy efficiencies similar to larger ecosystems. And they maintained populations of organisms apparently indefinitely.

From his flask worlds, Folsome concluded that it was microbes—tiny celled microbits of life, and not redwoods, crickets, orangutans—which do the lion's share of breathing, generating air, and ultimately supporting the indefinite populations of other noticeable organisms on Earth. An invisible substrate of microbial life steers the course of life's whole and welds together the different nutrient loops. The organisms that catch our eye and demand our attention, Folsome suspected, were mere ornate, decorative placeholdings as far as the atmosphere was concerned. It was the microbes in the guts in mammals and the microbes that clung to tree roots that made trees and mammals valuable in closed systems, including our planet.

I ONCE HAD a tiny living planet stationed on my desk. It even had a number: world #58262. I didn't have much to do to keep my planet happy. Just watch it every now and then.

World #58262 was smashed to smithereens at 5:04 P.M., October 17, during an abrupt heave of the 1989 San Francisco earthquake. A bookcase shook loose from my office wall during the tremor and spilled over my desk. In a blink, a heavy tome on ecosystems crushed the glass membrane of my living planet, irrevocably scrambling its liquid guts in a fatal Humpty Dumpty maneuver.

World #58262 was a human-made biosphere of living creatures, delicately balanced to live forever, and a descendant of Folsome's and Hanson's microbial jars. Joe Hanson, who worked at NASA's Advance Life-support Program in the Jet Propulsion Laboratory at Caltech, had come up with a more diverse world than Folsome's microbes. Hanson was the first to find a simple combination of self-sustaining creatures that included an animal. He put tiny brine shrimp and brine algae in an everlasting cosmos.

The basic commercial version of his closed world—sold under the label of "Ecosphere"—is a glass globe about the size of a large grapefruit. My world #58262 was one of these. Completely sealed inside the transparent ball were four tiny brine shrimp, a feathery mass of meadowgreen algae draped on a twig of coral, and microbes in the invisible millions. A bit of sand sat on the bottom. No air, water, or any other material entered or exited the globe. The thing ate only sunlight.

The oldest living Hanson-world so far is ten years old; that's as long as they have been manufactured. That's surprising since the average life-span of the shrimp swimming inside was thought to be about five years. Getting them to reproduce in their closed world has been problematic, although researchers know of no reason why they could not go on replicating forever. Individual shrimp and algae cells die, of course. What "lives forever" is the collective life, the aggregate life of a community.

You can buy an Ecosphere by mail order. It's like buying a Gaia or an experiment in emergent life. You unpack the orb from the heavy-duty insulation stuffed around it. The shrimp seem fine after their stormy ride. Then you hold the cannonball-size sphere in one hand up to the light; it sparkles with gemlike clarity. Here is a world blown into a bottle, the glass tidily pinched off at the top.

In its fragile immortality, the Ecosphere just sits there. Naturalist Peter Warshall, who owns one of the first Ecospheres, keeps it perched on his bookshelf. Warshall reads obscure dead poets and French philosophers in French and monographs on squirrel taxonomy. Nature is a kind of poetry for him; an Ecosphere is a book jacket blurb about the real thing. Warshall's Ecosphere lives under a regime of benign neglect, almost as a maintenance-free pet. He writes of his nonhobby: "You can't feed the shrimp. You can't snip off the decaying, dreary brown parts. You can't fiddle with the nonexistent filter, aerator, or pumps. You can't open it up and test the water's warmth with your finger. All you can do, if 'do' is an appropriate word, is to look and think."

The Ecosphere is a totem, a totem of all closed living systems. Tribesmen select totem creatures as a bridge between the separate worlds of spirit and dreams. Simply by being, the distinct world sealed behind an Ecosphere's clear glass invites us to meditate on such hard-to-grasp totemic ideas like "systems," "closed," and even "living."

"Closed" means separated from the flow. A manicured flower garden on the edge of the woods exists apart from the naturally structured wilderness surrounding, but the separateness of a garden mesocosm is partial—more a division of mind than fact. Every garden is really a small slice of the larger biosphere we all are immersed in. Moisture and nutrients flow underground into it, and a harvest and oxygen come out. If the rest of the sustaining biosphere were absent, gardens would wither. A truly closed system does not partake in outside flows of elements; all its cycles are autonomous.

"System" means interconnected. Things in a system are intertwined, linked directly or indirectly into a common fate. In an ecospheric world, shrimp eat algae, algae live on the light, microbes survive on the "wastes" of both. If the temperature soars too high (above 90 degrees), the shrimp molt faster than they can eat; thus they consume themselves. Not enough light and the algae won't grow fast enough to satiate the shrimp. The flicking tails of the shrimp stir up the water, which stirs the microbes so that each bug has a chance to catch the sunlight. The whole has a life in addition to the individual lives.

"Living" means surprises. One ordinary Ecosphere managed to stay alive in a total darkness for six months, contrary to logical expectations. Another ecosystem erupted one day after two years of unwavering steady temperature and light in an office into a breeding panic, crowding the globe with 30 tiny descendants of shrimp.

But it is stasis that does an Ecosphere in. In an unguarded moment Warshall writes of his orb, "There is the feeling of too much peacefulness that comes from the Ecosphere. It contrasts sharply with our frantic, daily lives. I have felt like playing the abiotic God. Pick it up and shake it. How's that for an earthquake, you little shrimp!"

That would actually be a good thing for an Ecosphere world, as momentarily discombobulating as it might be for its citizens. In turbulence is the preservation of the world.

A forest needs the severe destruction of hurricanes to blow down the old and make space for the new. The turbulence of fire on the prairie unloosens bound materials that cannot be loosened unless ignited. A world without lightning and fire becomes rigid. An ocean has the fire of undersea thermal vents in the short run, and the fire of compressed seafloor and continental plates in the long geological run. Flash heat, volcanism, lightning, wind, and waves all renew the material world.

The Ecosphere has no fire, no flash, no high levels of oxygen, no serious friction—even in its longest cycle. Over a period of years in its small space, phosphate, an essential element in all living cells, becomes tightly bound with other elements. In a sense, phosphate is taken out of circulation in the Ecosphere, diminishing the prospects of more life. Only the thick blob of blue-green algae will thrive in low phosphate environment, and so over time this species tends to dominate these stable systems.

A phosphate sink, and the inevitable takeover of blue-green algae, might be reversed by adding, say, a lightning-generating appendage to the glass globe. Several times a year, the calm world of the shrimp and algae would crackle and hiss and boil as calamity reigned for a few hours. Their vacations would be ruined, but their world would be rejuvenated.

In Peter Warshall's Ecosphere (which despite his idle thoughts has lain undisturbed for years), minerals have precipitated into a layer of solid crystals on the globe's inside. In a Gaian sense, the Ecosphere manufactured land. The "land"—composed of silicates, carbonates, and metal salts—built up on the glass because of an electric charge, a kind of natural electroplating. Don Harmony, the chief honcho at the small company making Ecospheres, was familiar with this tendency of tiny glass Gaia, and half in jest suggested that perhaps fusing an electrical ground wire onto the globe might keep the precipitates from forming.

Eventually the weight of the salt crystals peels them off the upper surface and they settle into the bottom of the liquid. On Earth, the deposit of sedimentary rock at the bottom of the ocean is part of larger geological cycles. Carbon and minerals circulate through air, water, land, rocks, and back again into life. Likewise in the Ecosphere. The elements it cradles are in a dynamic equilibrium with the cycling composition of the atmosphere and water and biosphere.

Most field ecologists were surprised by how simple such a self-sustaining closed world could be. With the advent of this toy biosphere, sustainable self-sufficiency appeared to be quite easy to create, especially if you didn't care what kind of life was being sustained. The Ecosphere was a mail-order proof of a remarkable assertion: self-sustained systems *want* to happen.

If simple and tiny was easy, how far could you expand the harmony and still have a sustainable world closed to all but energy input?

It turns out that ecospheres scale up well. A huge commercial Ecosphere can weigh in at 200 liters. That's about the volume of a large garbage can—so big you can't reach your arms around it. Inside a stunning 30-inch-diameter glass globe, shrimp paddle between fronds of algae. But instead of the usual three or four spore-eating shrimp, the giant Ecosphere holds 3,000. It's a tiny moon with its own inhabitants. Here, the law of large numbers takes hold; more is different. More individual lives make the ecosystem more resilient. The larger an Ecosphere is, the longer it takes to stabilize, and the harder it is to kill it. But once in gear, the collective give and take of a vivisystem takes root and persists.

The next question is evident: How big a bottle closed to outside flows, filled with what kind of living organisms, would you need to support a human inside?

When human daredevils ventured beyond the soft bottle of the Earth's atmosphere, this once academic question took on practical meaning. Could you keep a person alive in space—like shrimp in an Ecosphere—by keeping plants alive? Could you seal a man up in a sunlit bottle with enough living things so that their mutual exhalations would balance? It was a question worth doing something about.

Every school child knows animals consume the oxygen and food that plants generate, while plants consume the carbon dioxide and nutrients that animals generate. It's a lovely mirror, one side producing what the other needs, just as the shrimp and algae serve each other. Perhaps the right mix of plants and mammals in their symmetrical demands could support each other. Perhaps a human could find its proper doppelganger of organisms in a closed bottle.

The first person crazy enough to experimentally try this was a Russian researcher at the Moscow Institute for Biomedical Problems. In 1961, during the heady early years of space research, Evgenii Shepelev welded together a steel casket big enough to hold himself and eight gallons of green algae. Shepelev's careful calculations showed that eight gallons of chlorella algae under sodium lights should supply enough oxygen for one man, and one man should generate enough carbon dioxide for eight gallons of chlorella algae. The two sides of the equation should cancel each other out into unity. In theory it should work. On paper it balanced. On the blackboard it made perfect sense.

Inside the airtight iron capsule, it was a different story. You can't breathe theories. If the algae faltered, the brilliant Shepelev would follow; or, if he succumbed, the algae would do likewise. In the box the two species would become nearly symbiotic allies entirely dependent on each other, and no longer dependent upon the vast planetary web of support outside—the oceans, air, and creatures large and small. Man and algae sealed in the capsule divorced themselves from the wide net woven by the rest of life. They would be a separate, closed system. It was by an act of faith in his science that a trim Shepelev crawled into the chamber and sealed the door.

Algae and man lasted a whole day. For about 24 hours, man breathed into algae and algae breathed into man. Then the staleness of the air drove Shepelev out. The oxygen content initially produced by the algae plummeted rapidly by the close of the first day. In the final hour when Shepelev cracked open the sealed door to clamber out, his colleagues were bowled over by the revolting stench in his cabin. Carbon dioxide and oxygen had traded harmoniously, but other gases, such as methane, hydrogen sulfide, and ammonia, given off by algae and Shepelev himself, had gradually fouled the air. Like the mythological happy frog in slowly boiling water, Shepelev had not noticed the stink.

Shepelev's adventuresome work was taken up in seriousness by other Soviet researchers at a remote and secret lab in northern Siberia. Shepelev's own group was able to keep dogs and rats alive within the algae system for up to seven days. Unbeknownst to them, about the same time the United States Air Force School of Aviation Medicine linked a monkey to an algae-produced atmosphere for 50 hours. Later, by parking the tiny eight-gallon tub of chlorella in a larger sealed room, and tweaking the algae nutrients as well as the intensity of lights, Shepelev's lab found that a human could live in this airtight room for 30 days! At this extreme duration the researchers noticed that the respirations of man and algae were not exactly matched. To keep a balance of atmosphere, excess carbon dioxide needed to be removed by chemical filters. But the scientists were encouraged that stinky methane stabilized after 12 days.

By 1972, more than a decade later, the Soviet team, directed by Josepf Gitelson, constructed the third version of a small biologically based habitat that could support humans. The Russians called it Bios-3. It housed up to three men. The habitat was crowded inside. Four small airtight rooms enclosed tubs of hydroponically (soil-less) grown plants anchored under xenon lights. The men-in-a-box planted and harvested the kind of crops you might expect in Russia—potatoes, wheat, beets, carrots, kale, radishes, onions and dill. From the harvest they prepared about half of their own food, including bread from the grain. In this cramped, stuffy, sealed greenhouse, the men and plants lived on each other for as long as six months.

The box was not perfectly closed. While its atmosphere was sealed to air exchanges, the setup recycled only 95 percent of its water. The Soviet scientists stored half of their food (meat and proteins) beforehand. In addition, the Bios-3 system did not recycle human fecal wastes or kitchen scraps; the Bios-dwellers ejected these from the container, thereby ejecting some trace elements and carbon.

In order not to lose all carbon from the cycle, the inhabitants burned a portion of the inedible dead plant matter rendering it into carbon dioxide and ash. Over weeks the rooms accumulated trace gases generated by a number of sources: the plants, the materials of the room, and the men themselves. Some of these vapors were toxic, and methods to recycle them unknown then, so the men burned off the gases by simply "burning" the air inside with a catalytic furnace.

NASA, of course, was interested in feeding and housing humans in space. In 1977 they launched the still-going CELSS program (Controlled Ecological Life Support Systems). NASA took the reductionist approach: find the simplest units of life that can produce the required oxygen, protein, and vitamins for human consumption. It was in messing around with elemental systems that NASA's Joe Hanson stumbled on the interesting, but to NASA's eyes, not very useful shrimp/algae combo.

In 1986 NASA initiated the Breadboard Project. The program's agenda was to take what was known from tabletop experiments and implement them at a larger scale. Breadboard managers found an abandoned cylinder left over from the Mercury space shots. This giant tubular container had been built to serve as a pressure-testing chamber for the tiny astronaut capsule that would spearhead the Mercury rocket. NASA retrofitted the two-story cylinder with outside ductwork and plumbing, transforming the interior into a bottled home with racks of lights, plants, and circulating nutrients.

Just as the Soviet Bios-3 experiments did, Breadboard used higher plants to balance the atmosphere and provide food. But a human can only choke down so much algae each day. Even if algae was all one ate, chlorella only provides 10 percent of the daily nutrients a person needs. For this reason, NASA researchers drifted away from algae-based systems, and migrated toward plants that provided not only clean air but also food.

Ultra-intensive gardening seemed to be what everyone was coming up with. Gardening could produce really edible stuff, like wheat. Among the most workable setups were various hydroponic contraptions that delivered aqueous nutrients to plants as a mist, a foam, or a thin film dripping through plastic holding racks matted with lettuce or other greens. This highly engineered plumbing produced concentrated plant growth in cramped spaces. Frank Salisbury of Utah State University discovered ways to plant spring wheat at 100 times its normal density by precisely controlling the wheat's optimal environment of light, humidity, temperature, carbon dioxide, and nutrients. Extrapolating from his field results, Salisbury calculated the amount of calories one could extract from a square meter of ultradensely planted wheat sown, say, on an enclosed lunar base. He concluded that "a moon farm about the size of an American football field would support 100 inhabitants of Lunar City."

One hundred people living off a football field-size truck farm! The vision was Jeffersonian! One could envision a nearby planet colonized by a network of Superdome villages, each producing its own food, water, air, people, and culture.

But NASA's approach to inventing a living in a closed system struck many as being overly cautious, strangulatingly slow, and intolerably reductionistic. The operative word for NASA's Controlled Ecological Life Support Systems was "Controlled."

What was needed was a little "out-of-control."

THE APPROPRIATE out-of-controlness started on a ramshackle ranch near Santa Fe, New Mexico. During the commune heydays of the early 1970s, the ranch collected a typically renegade group of cultural misfits. Most communes then were freewheeling. This one, named Synergia Ranch, wasn't; it demanded discipline and hard work. Rather than lie back and whine while the apocalypse approached, the New Mexican commune worked on how it might build something to transcend the ills of society. They came up with several designs for giant arks of sanity. The more grandiose their mad ark visions got, the more interested in the whole idea they all became.

It was the commune's architect, Phil Hawes, who came up with the galvanizing idea. At a 1982 conference in France, Hawes presented a mock-up of a spherical, transparent spaceship. Inside the glass sphere were gardens, apartments, and a pool beneath a waterfall. "Why not look at life in space as *a life* instead of merely travel?" Hawes asked. "Why not build a spaceship like the one we've been traveling on?" That is, why not create a living satellite instead of hammering together a dead space station? Reproduce the holistic nature of Earth itself as a tiny transparent globe sailing through space. "We knew it would work," said John Allen, the ranch's charismatic leader, "because that's what the biosphere does every day. We just had to get the size right."

The Synergians stuck with the private vision of a living ark long after they left the ranch. In 1983, Ed Bass of Texas, one of the ranch's former members, used part of his extraordinary family oil fortune to finance a proof-of-concept prototype.

Unlike NASA, the Synergians wouldn't rely on technology as the solution. Their idea was to stuff as many *biological* systems—plants, animals, insects, fish, and microorganisms—as they possibly could into a sealed glass dome, and then rely on the emergent system's own self-stabilizing tendencies to self-organize a biospheric atmosphere. Life is in the business of making its environment agreeable for life. If you could get a bunch of life together and then *give it enough freedom to cultivate the conditions it needed to thrive*, it would go forever, and no one needed to understand how it worked.

Indeed, neither they nor biologists had any real idea of how one plant worked—what its exact needs and products were—and no idea at all of how a distributed miniecosystem sealed in a hut would work. Instead, they would rely on decentralized, uncontrolled life to sort itself out and come to some self-enhancing harmony.

No one had ever built any living thing that large. Even Gomez hadn't built his coral reef yet. The Synergians had only a vague notion of Clair Folsome's ecospheres and even vaguer knowledge of the Russian Bios-3 experiments.

The group, now calling itself Space Biosphere Ventures (SBV), and financed to the tune of tens of millions of dollars by Ed Bass, designed and built a tiny cottage-size test unit during the mid-1980s. The hut was crammed with a greenhouse-worth of plants, some fancy plumbing for recycling water, black boxes of sensitive environmental monitoring equipment, a tiny kitchenette and bathroom, and lots of glass.

In September 1988, for three days, John Allen sealed himself in for the unit's first trial run. Much like Evgenii Shepelev's bold step, this was an act of faith. The plants had been selected by rational guess, but there was nothing controlled about how well they would work as a system. Contrary to Gomez's hard-won knowledge about sequencing, the SBV folks just threw everything in together, at once. The sealed home depended on at least some of the individual plants being able to keep up with the lungs of one man.

The test results were very encouraging. Allen wrote in his journal for September 12: "It appears we are getting close to equilibrium, the plants, soil, water, sun, night and me." In the confined loop of a 100 percent recycled atmosphere, 47 trace gases, "all of which were probably anthropogenic in origin," fell to minute levels when the air of the hut was sent through the plant soil—an old technique modernized by SBV. Unlike Shepelev's case, when Allen stepped out, the air inside was fresh, ready for more human life. To someone outside, a whiff of the air inside was shockingly moist, thick, and "green."

The data from Allen's trial suggested a human could live in the hut for a while. Biologist Linda Leigh would later spend three weeks in the small glass shed. After her 21-day solo drive Leigh told me, "At first I was concerned whether I'd be able to stand breathing in there, but after two weeks I hardly noticed the moisture. In fact I felt invigorated, more relaxed, and healthier, probably because of the air-cleansing and oxygen-producing nature of close plants. The atmosphere even in that small space was stable. I felt that the test module could have gone on for the full two years and kept its atmosphere right."

During the three-week run, the sophisticated internal monitoring equipment indicated no buildup of gases either from building materials or biological sources. Although the atmosphere was stable overall, it was sensitive to perturbations which caused it to vacillate easily. While harvesting sweet potatoes out of their dirt beds in the hut, Leigh's digging disturbed $CO_2$-producing soil organisms. The rattled bugs temporarily altered the $CO_2$ concentration in the module's air. This was an illustration of the butterfly effect. In complex systems a small alteration in the initial conditions can amplify into wide-ranging effects throughout the rest of the system. The principle is usually illustrated by the fantasy of the flap of a butterfly's wings in Beijing triggering a hurricane in Florida. Here in SBV's sealed glass cottage the butterfly effect appeared in miniature: by wiggling her fingers Leigh upset the balance of the atmosphere.

John Allen and another Synergian, Mark Nelson, envisioned a near-future Mars station built as a mammoth closed-system bottle. Allen and Nelson gradually formulated a hybrid technology—called ecotechnics—based on a convergence of both machines and living organisms to support future human habitats.

They were dead serious about going to Mars and began working out the details. In order to journey to Mars or beyond, you needed a crew. How many people would you need? Military captains, expedition leaders, start-up managers, and crisis centers had long recognized that a team of eight was the ideal number for any complex hazardous project. More than eight people, and decisions got slow and squirrely; less than eight, accidents and ignorance became serious handicaps. Allen and Nelson settled on a crew of eight.

Next step: how big would you have to make a bottle-world to shelter, feed, water, and oxygenate eight people indefinitely?

Human requirements were well established. Each day a human adult needed about half a kilogram of food, a kilo of oxygen, 1.8 kilos of drinking water, FDA amounts of vitamins, and a couple of gallons of water for washing. Clair Folsome had extrapolated the results of his tiny ecospheres and calculated that you would need a sphere with a radius of 58 meters—half air and half microbial soup—to support the oxygen needs for one person indefinitely. Allen and Nelson then took the data from the Russian Bios-3 experiments and combined it with Folsome's, Salisbury's, and others' intensive farming harvest results. They estimated that right now—with the knowledge and technology of the 1980s—they could support eight adults on . . . three acres of land.

Three acres! The transparent container would have to be the size of the Astrodome. Such a span would demand at least a 50-foot ceiling. Clothed in glass, it would be quite a sight. And quite expensive.

But it would be magnificent! They would build it! And they did, with the further help of Ed Bass—to the tune of $100 million. Hard-hat construction of the 8-person ark began in 1988. The Synergians called the grand project Biosphere 2 (Bio2), a bonsai version of Biosphere 1, our Earth. It took three years to build.

Small compared to Earth, the completed self-contained terrarium was awesome at the human scale. Bio2 was a gigantic glass ark the size of an airport hangar. Think of an inverted ocean liner whose hull is transparent. The gigantic greenhouse was superairtight, sealed at the bottom, too, with a stainless steel tray 25 feet under the soil to prevent seepage of air from its basement. No gas, water, or matter could enter or leave the ark. It was a stadium-size Ecosphere—a big materially closed and energetically open system—but far more complex. Bio2 was the second only to Biosphere 1 (the Earth) as largest closed vivisystem.

The challenge of creating a living system of any size is daunting. Creating a living wonder at the scale of Bio2 could only be described as an experiment in sustained chaos. The challenge included: Select a couple of thousand parts out of several billion possibilities, and arrange them so that all the parts complemented and provided for each other, so that the whole mixture was self-sustaining over time, and that no single organism became dominant at the expense of others, so that the whole aggregate kept all the constituents in constant motion, without letting any ingredient become sequestered off to the side, while keeping the entire level of activity and atmospheric gases elevated at the point of perpetually almost-falling. Oh, and humans should be able to live, eat, and drink within and from it.

SBV decided to stake the survival of Bio2 on the design tenet that an extraordinarily diverse hodgepodge of living creatures would settle into a unified stability. If it proved nothing else, the experiment would at least shed some light on the almost universally held assumption in the last two decades: that diversity ensures stability. It would also test whether a certain level of complexity birthed self-sustainability.

As an architecture of maximum diversity, the final Bio2 floor plan had seven biomes (biogeographical habitats). Under the tallest part of the glass canopy, a rock-faced concrete mountain bulged. Planted with transplanted tropical trees and a misting system, the synthetic hill was transformed into a cloud forest—a high altitude rain forest. The cloud forest drained into an elevated hot grassland (the size of a big patio, but stocked with waist-high wild grasses). One edge of the rain forest stopped before a rocky cliff which fell to a saltwater lagoon, complete with coral, colorful fishes, and lobsters. The high savanna lowered into a lower, drier savanna, dark with thorny, tangled thickets. This biome is called thornscrub and is one of the most common of all habitats on Earth. In real life it is nearly impenetrable to humans (and thus ignored), but in Bio2, it served as a little hideaway for both wildlife and humans. The thicket leads into a compact marshy wetland, the fifth biome, which finally emptied into the lagoon. The low end of Bio2 was a desert, as big as a gymnasium. Since it was pretty humid inside, the desert was planted with fog desert plants from Baja California and South America. Off to one side was the seventh biome—an intensive agriculture and urban area where eight *Homo sapiens* grew all their own food. Like Noah's place, animals were aboard; some for meat, some for pets, and some on the loose: lizards, fish, birds roaming about the wild parts. There were honeybees, papaya trees, a beach, cable TV, a library, a gym, and a laundromat. Utopia!

The scale was stupendous. Once while I was visiting the construction site, an 18-wheeler semi-truck pulled up to the Bio2 office. The truck driver leaned out the window and asked where they wanted their ocean. He'd been hauling a full truckload of ocean salt and needed to unload it before dark. The office clerks pointed down to a very large hole in the center of the project. That's where Walter Adey from the Smithsonian Institution was building a one-million-gallon ocean, coral reef, and lagoon. There was enough elbow room in this gargantuan aquarium for all kinds of surprises to emerge.

Making an ocean is no cinch. Ask Gomez and the hobby saltwater aquarists. Adey had grown an artificial self-regenerating coral reef once before as a museum exhibit at the Smithsonian. But this one in Bio2 was huge; it had its own sandy beach. An expensive wave-making pump at one end would supply the turbulence coral love. The same machine created a half-meter tide on a lunar cycle.

The trucker unloaded the ocean: stacks of 50-pound bags of InstantOcean, the same stuff you buy at tropical aquarium stores. A starter solution harboring all the right microbeasties (sort of the yeast for the dough) was later hauled in on a different truck from the Pacific Ocean. Stir together well, and pour.

The ecologists building the wilderness areas of Bio2 were of the school that says: soil + bugs = ecology. To have the kind of tropical rain forest you want, you needed to have the right kind of jungle dirt. And to get that in Arizona you had to make it from scratch. Take a couple of bulldozer buckets of basalt, a few of sand, and a few of clay. Sprinkle in the right microorganisms. Mix in place. The underlying soils in each of the six wild biomes of Bio2 were manufactured in this painstaking way. "The thing we didn't realize at first," said Tony Burgess, "was that soils are alive. They breathe as fast as you do. You have to treat soil as a living organism. Ultimately it controls the biota."

Once you have soil, you can play Noah. Noah rounded up everything that moved for his ark, but that certainly wasn't going to work here. The designers of the Bio2 closed-system kept coming back to that most exasperating but thrilling question: what species should Bio2 include? No longer was it merely "Which organisms do we need to mirror the breath of eight humans?" The dilemma was "Which organisms do we need to mirror Gaia? Which combination of species would produce oxygen to breathe, plants to eat, plants to feed the animals to eat (if any), and species to support the food plants? How do we weave a self-supporting network out of random organisms? How do we launch a coevolutionary circuit?"

Take almost any creature as an example. Most fruit requires insects to pollinate it. So if you wanted blueberries in Bio2, you needed honeybees. But in order to have honeybees around when the blueberries are ready for pollination, you needed to provide the honeybees with flowers for the rest of the season. But in order to supply sufficient seasonal flowers to keep honeybees alive, there would be no room for other kinds of plants. So, perhaps another type of pollinating bee would work? You could use straw bees which can be supported with meager amounts of flowers, but they don't pollinate blueberry blossoms or several other fruits you wanted. How about moths? And so on down the catalog of living creatures. Termites are necessary to decompose old woody vegetation, but they were fond of eating the sealant around the windows. What's a benign termite substitute that would get along with the rest of the crowd?

"It's a sticky problem," said Peter Warshall, a consulting

ecologist for the project. "It's a pretty impossible job to pick 100 living things, even from the same place, and put them together to make a 'wilderness.' And here we're taking them from all over the world to mix together since we have so many biomes."

To cobble together a synthetic biome, the half-dozen Bio2 ecologists sat down at a table together and played this ultimate jigsaw puzzle. Each scientist had expertise in either mammals, insects, birds, or plants. But while they knew something about sedges and pond frogs, very little of their knowledge was systematically accessible. Warshall sighed, "It would have been nice if somewhere there was a database of all known species listing their food and energy requirements, their habitat, their waste products, their companion species, their breeding needs, etc., but there isn't anything remotely like that. We know very little about even common species. In fact, what this project shows is how little we know about *any* species."

The burning question for the summer the biomes were designed was "Well, how many moths does a bat really eat?" In the end, selecting the thousand or so higher species came down to informed guesses and biodiplomacy. Each ecologist wrote up a long list of possible candidates, including favorite species they thought would be the most versatile and flexible. Their heads were full of conflicting factors—pluses and minuses, likes to be near this guy but can't stand this one. The ecologists projected the competitiveness of rival organisms. They bickered for water or sunlight rights. It was as if they were ambassadors protecting the territory of their species from encroachments.

"I needed as much fruit as possible dropped from trees for my turtles to eat," said Bio2 desert ecologist Tony Burgess, "but the turtles would leave none for the fruit flies to breed on, which Warshall's hummingbirds needed to eat. Should we have more trees for leftover fruit, or use the space for bat habitat?"

So negotiations take place: If I can have this flower for the birds, you can keep the bats. Occasionally the polite diplomacy reverted to open subversion. The marsh-man wanted his pick of sawgrass, but Warshall didn't like his choice because he felt the species was too aggressive and would invade the dry land biome he was overseeing. In the end Warshall capitulated to the marsh-man's choice, but added, half in jest, "Oh, it doesn't make any difference because I'm just gonna plant taller elephant grass to shade out your stuff, anyway." The marsh-man retaliated by saying he was planting pine trees, taller than either. Warshall promised with a hearty laugh to plant a defense border of guava trees, which don't grow any taller, but grow much faster, staking out the niche early.

Everything was connected to everything. It made planning a nightmare. One approach the ecologists favored was building a redundancy of pathways into the food webs. With multiple foodchains in every web, if the sand flies died off, then something else became second choice food for the lizards. Rather than fight the dense tangle of interrelationships, they exploited them. The key was to find organisms with as many alternative roles as possible, so that if one didn't work out, it had another way or two to complete somebody's loop.

"Designing a biome was an opportunity to think like God," recalled Warshall. You, as a god, could create something by nothing. You could create something—some wonderful synthetic vibrant ecosystem—but you had no control over precisely what something emerged. All you could do was gather all the parts and let them *self-assemble into something that worked.* Walter Adey said, "Ecosystems in the wild are made up of patches. You inject as many species as you can into the system and let it decide what patch of species it wants to be in." Surrendering control became one of the "Principles of Synthetic Ecology." Adey continued, "We have to accept the fact that the amount of information contained in an ecosystem far exceeds the amount contained in our heads. We are going to fail if we only try things we can control and understand." The exact details of an emerging Bio2 ecology, he warned, were beyond predicting.

But details counted. Eight human lives rested on the details fusing into a whole. Tony Burgess, one of the Bio2 gods, ordered dune sand to be trucked in for the desert biome because construction sand, the only kind on hand at the Bio2 site, was too sharp for the land turtles; it cut their feet. "You've got to take care of your turtles, so they can take care of you," he said in a priestly way.

The number of free-roaming animals taking care of the system was pretty thin for the first two years in Bio2 because there wasn't enough wild food to support very many of them. Warshall almost didn't put any monkeylike galagos from Africa in because he wasn't sure the young acacia trees could produce enough gum to satisfy them. In the end he released four galagos and stored a couple hundred pounds of emergency monkeychow in the basement of the ark. Other wild animal occupants of Bio2 included leopard tortoises, blue-tongued skinks ("because they are generalists"—not picky what they eat), various lizards, small finches, and pygmy green hummingbirds, partially for pollination. "Most of the species will be pygmy," Warshall told a *Discover* reporter before closure, "because we really don't have that much space. In fact, ideally we'd have pygmy people, too."

The animals didn't go in two by two. "You want to have a higher ratio of females to males for reproduction insurance," Warshall told me. "Ideally we like to have at minimum five females per three males. I know director John Allen says that eight humans—four female, four male—is the minimum-size group needed for human colony start-up and reproduction, but from an ecologically correct rather than politically correct point of view, the Bio2 crew should be five females and three males."

For the first time biologists were being forced by the riddle of creating a biosphere to think like engineers: "Here is what we need, what materials will do that job?" At the same time, the engineers on the project were being forced to think like biologists: "That's not dirt, that's a living organism!"

A stubborn problem for the designers of Bio2 was making rain for the cloud forest. Rain is hard. The original plans optimistically called for cooling coils at the peak of the 85-foot glass roof over the jungle section. The coils would condense the jungle's moisture into gentle drops descending from the celestial heights—real artificial rain. Early tests proved the drops to be scarce, too large and destructive when they landed, and not at all the constant gentle mist the plants wanted. Second plan was for the rain to be pumped up into sprinklers bolted to the frame structure high overhead, but that proved to be a maintenance nightmare since over a two-year period the fine-holed mist heads were sure to need unclogging or replacements. The design they ended up with was "rain" squirted from misting nozzles fitted on the ends of pipes stationed here and there on the slopes.

One unexpected consequence of living in a small materially closed system is that rather than water becoming precious, it's in virtual abundance. In about one week 100 percent of the water is recycled, cleansed by microbiological activity in wetland treatment areas. When you use more water, it just goes around the loop a little faster.

Any field of life is a cloth woven with countless separate loops. The loops of life—the routes which materials, functions, and energy follow—double up, cross over and interweave as knots until it is impossible to tell one thread from another. Only the larger pattern knitted by the loops emerges. Each circle strengthens the others, until the whole is hard to unravel.

That is not to say there will be no extinctions in a tightly wrapped ecosystem. A certain extinction rate is essential for evolution. Walter Adey had about 1 percent attrition rate in his previous partially closed coral reef. He expected about a 30 to 40 percent drop-off in species within the whole of Bio2 by the end of its first two-year run. (The biologists from Yale University who are currently counting the species after reopening have not finished their studies of species attrition as of my writing.)

But Adey believes that he already has learned how to grow diversity: "What we are doing is cramming more species in than we expect to survive. So the numbers drop. Particularly the insects and lower organisms. Then, at the beginning of the next run we overstock it again, injecting slightly different species—our second guesses. What will probably happen is that there will still be a large loss again, maybe one quarter, but we reinject again next closure. Each time the numbers of species will stabilize at a higher level than the first. The more complex the system, the more species it can hold. We keep doing that, building up the diversity. If you loaded up Biosphere 2 with all the species it ends up with, it would collapse at the start." The huge glass bottle is a diversity pump that grows complexity.

The Bio2 ecologists were left with the large question of how best to jump start the initial variety, upon which further diverse growth would be leveraged. This was very much related to the practical problem of how to load all the animals onto the ark. How do you get 3,000 interdependent creatures into a cage, alive? Adey proposed moving an entire natural biome into Bio2's relatively miniature space by compressing it in the manner of a condensed book: selecting choice highlights here and there, and fusing these bits into a sampler.

He selected a fine 30-mile stretch of a Florida everglade mangrove swamp and had it surveyed into a grid. Every half mile or so along the salt gradient, a small cube (4-feet deep by 4-feet square) of mangrove roots was dug out. The block of leafy branches, roots, mud, and piggybacking barnacles was boxed and hauled ashore. The segments of the marsh, each one tuned to a slightly different salt content with slightly different microorganisms, were trucked to Arizona (after long negotiations with very confused agricultural custom agents who thought "mangroves" were "mangoes").

While the chunks of everglades were waiting to be placed in the Bio2 marsh, the Bio2 workers hooked the watertight boxes up into a network of pipes so that they became one distributed saltwater tide. Later the 30 or so cubes were reassembled into Bio2. Unboxed, the reconstituted marsh takes up only a micro 90-by-30 feet. But within this volleyball court-size everglade, each section harbors a gradually increasing salt-loving mixture of microorganisms. Thus, the flow of life from freshwater to brine is compressed into talking distance. The problem with the analog method is that scale is an important dimension of an ecosystem. As Warshall juggled the parts to manufacture a miniature savanna, he shook his head: "At best we are putting about one-tenth the variety of a system into Bio2. For the insect population it's more like one-hundredth. In a West African savanna there are 35 species of worms. At most we'll have three kinds. So the dilemma is: are we making a savanna or a lawn? It's surely better than a lawn . . . but how much better I don't know."

Constructing a wetlands or savanna by reassembling portions of a natural one is only one method of biome building—which the ecologists call the "analog" way. It seemed to work fine. But as Tony Burgess pointed out, "You can go two ways with this.

You can mimic an analog of a particular environment you find in nature, or you can invent a synthetic one based on many of them." Bio2 wound up being a synthetic ecosystem, with many analog parts, such as Adey's marshland.

"Bio2 is a synthetic ecosystem, but so is California by now," said Burgess. Warshall agrees: "What you see in California is a symbol of the future. A heavily synthetic ecology. It has hundreds of exotic species. A lot of Australia is going this way too. And the redwood/eucalyptus forest is also a new synthetic ecology." As are many other ecosystems in this world of jet travel, when species are jet-setted far from their home territories and introduced accidentally or deliberately in lands they would otherwise never reach. Warshall said, "Walter Adey first used the term 'synthetic ecology'. Then I realized that there were already huge amounts of synthetic ecology in Biosphere One. And that I wasn't inventing a synthetic ecology in Bio2, I was merely duplicating what already existed." Edward Mills of Cornell University has identified 136 species of fish from Europe, the Pacific and elsewhere now thriving in the Great Lakes. "Probably most of the biomass in the Great Lakes is exotic," Mills claims. "It's a very artificial system now."

We might as well develop a science of synthetic ecosystem creation since we've been creating them anyway in a haphazard fashion. Many archeo-ecologists believe that the entire spectrum of early humanoid activities—hunting, grazing, setting prairie fires, and selective herb gathering—forged an "artificial" ecology upon the wilderness, that is, an ecology greatly shaped by human arts. In fact, all that we think of as natural virgin wilderness is abundant with artificiality and the mark of human activity. "Many rain forests are actually pretty heavily managed by indigenous Indians," Burgess says. "But the first thing we do when we come in is wipe out the indigenous people, so the management expertise disappears. We assumed that this growth of old trees is pristine rain forest because the only way we know how to manage a forest is to clear the trees, and these weren't clear-cut." Burgess believes that the mark of human activity runs so deep that it cannot be undone easily. "Once you alter the ecosystem, and you get the right seeds dispersed in the ground and the essential climate window, then the transformation starts and it's irreversible. This does not require the presence of man to keep the synthetic ecosystem going. It runs undisturbed. All the people in California could die and its current synthetic flora and fauna will remain. It's a new meta-stable state that remains as long as the self-reinforcing conditions stay the same.

"California, Chile and Australia are converging very rapidly to become the same synthetic ecology," Burgess claims. "They were established by the same people, and shaped by the same goal: removal of the ancient herbivores to be replaced by the production of bovines: cow meat." As a synthetic ecology, Bio2 is a foreshadowing of ecologies to come. It is clear that we are not retreating from our influence on nature. Perhaps the bottle of Bio2 can teach us how to artificially evolve useful, less disruptive synthetic ecosystems.

As the ecologists began to assemble the first deliberately synthesized ecology they made an attempt to devise guidelines they felt would be important in creating any living closed biosystem. The makers of Bio2 called these the Principles of Biospherics. When creating a biosphere remember that:

- Microorganisms do most of the work.
- Soil is an organism. It is alive. It breathes.
- Make redundant food webs.
- Increase diversity gradually.
- If you can't provide a physical function, you need to simulate it.
- The atmosphere communicates the state of the whole system.
- Listen to the system; see where it wants to go.

Rain forests, tundras, and everglades are not themselves natural closed systems; they are open to each other. There is only one natural closed system we know of: the Earth as a whole, or Gaia. In the end our interest in fashioning new closed systems rests on concocting second examples of living closed systems so that we may generalize their behavior and understand the system of Earth, our home.

Closed systems are a particularly intense variety of coevolution. Pouring shrimp into a flask and pinching off the throat of the flask is like putting a chameleon in a mirrored bottle and pinching closed the entrance. The chameleon responds to the image it has generated, just as the shrimp responds to the atmosphere it has generated. The closed bottle—once the internal loops weave together and tighten—accelerates change and evolution within. This isolation, like the isolation in terrestrial evolution, breeds variety and marked differences.

But eventually all closed systems are opened or at least leak. We can be certain that whatever artificial closed systems we fabricate will sooner or later be opened. Bio2 will be closed and unsealed every year or so. And in the heavens, on the scale of galactic time, the closed systems of planets will be penetrated and shared in a type of cross-panspermia—a few exchanges of species here and there. The ecology of the cosmos is this type: a universe of isolated systems (planets), furiously inventing things in that mad way of a chameleon locked in a mirrored bottle. Every now and then marvels from one closed system will arrive with a shock into another.

On Gaia, the briefly closed miniature Gaias we construct are mostly instructional aides. They are models made to answer primarily one question: what influence do we, and can we, have over the unified system of life on Earth? Are there levels we can reach, or is Gaia entirely out of our control?

# 9

# Pop Goes the Biosphere

"I FEEL I AM FAR OUT IN SPACE," said Roy Walford, one of the people who lived inside Biosphere 2. Walford was speaking to a reporter via a video hookup during the first two-year closure of the ark, from September 26, 1991, to September 26, 1993. During that time eight humans, or biospherians as they are called, dramatically removed themselves from the direct touch of all other life on the planet, and from all the affirming flows of materials propelled by life, and lived instead in the tiny autonomous backwater swirl of life they had conjured up in a miniature surrogate Gaia. They could have been in space.

Walford was healthy but extremely skinny and underfed. For two years, all the biospherians were hungry. Their pocket-size farm had been plagued with insect infestations. Because they couldn't spray the beasties with poisons—since they would be drinking the evaporated runoff later in the week—they ate less. At one point the desperate biospherians crept down their rows of potato plants with portable hair dryers to drive the mites off the leaves, but without success. Altogether they lost five staple crops. One of the biospherians plummeted from 208 to 156 pounds. But he was prepared for this. He brought in clothes several sizes too small at the start.

Some scientists felt starting the Bio2 project with humans inside was not the most productive way. Peter Warshall, their consulting naturalist, said, "As a scientist, I would have preferred that we closed the whole thing up for one year with only the first two or three kingdoms in it: unicellular organisms and below. We could have seen how much the microbial cosmos controls the atmosphere. Then later we'd put everything in, close it up for the next year and compare the fluctuations." A few scientists felt the troublesome and difficult-to-support species of *Homo sapiens* shouldn't be in there at all, and that the humans became a mere entertainment factor. But many were sure the ecological study was pointless compared to the practical goal of developing technologies of human survival away from the Earth. To review the conflicting views of the scientific import and agenda of the project, an independent Scientific Advisory Committee was commissioned by Bio2's financier Ed Bass. They issued a report in July 1992 which acknowledged the dual nature of the experiment. It stated:

> The committee recognizes that there are at least two major areas of science to which Biosphere 2 can contribute significantly. One is the understanding of biogeochemical cycles of closed systems. From this perspective Biosphere 2 represents a much larger and more complex closed system than has ever been studied. For these studies the presence of human beings in the system is not essential except that they provide the capacity to make observations and measurements not initially regarded as important.
>
> The second is to gain the knowledge and experience to maintain humans within equilibrium in a closed ecological system. For these the presence of people is central to the experiment.

As an example of the latter case, within the first year people living inside the closed system yielded a completely unexpected medical result. Regular blood tests of the sequestered biospherians showed increased levels of pesticides and herbicides in their blood. Since every aspect of the environment within Bio2 was monitored constantly and precisely—it was the most monitored environment of all time—scientists knew that there were no pesticides or herbicides anywhere inside. One biospherian who had previously lived in third world countries had traces in her blood of a pesticide banned in the U.S. twenty years ago. What the medics guessed was that as the biospherians lost significant weight due to their restricted diet, they burnt up fat reserves stored in the past and flushed out toxins deposited in them decades ago. Until Bio2 was built, there was no scientific reason to precisely test people for internal toxins because there was no way to rigorously control what they ate, drank, breathed, or touched. Now there was. Just as Bio2 provided an experimental lab for meticulously tracking the flow of pollutants through an ecosystem, it also provided a lab for meticulously tracking the flow of pollutants through a human body.

Human bodies themselves are a vast complex system—despite our advanced medical knowledge, still unmapped—which can only be properly studied by isolating them from the greater complexity of life. Bio2 was an elegant way to do this. But the Science Advisory Committee missed another reason to have humans aboard, perhaps one of equal importance to getting ready for space. This justification was a matter of control and scaffolding. Humans were to serve as the "thumb on the way to thought," the chaperone present at the introduction, but not needed past that. People were not necessarv for a closed ecosystem to run once stable, but they might be helpful in stabilizing it.

For instance, there was the practical matter of time. No scientist could afford to run the emerging ecosystem for years and let it crash whenever it wanted, only to have to start over. As long as the humans inside measured and recorded what they did, they could steer the closed system away from the precipices of disaster and still be scientific about it. Within great latitudes, the artificial ecosystem of Bio2 ran its own course, but when it veered toward a runaway state, or stalled, the biospherians nudged it. They shared control with the emergent system itself. They were copilots.

One of the ways the biospherians shared control was by acting as "keystone predators"—biological checks of last resort. Populations of plants or animals that outran their niches were kept in reasonable range by human "arbitration." If the lavender shrub began to take over, the biospherians hacked it back. When the savanna grass shouldered out cactus, they weeded fiercely. In fact the biospherians spent several hours per day weeding in the wilderness areas (not counting the weeding they did on their crop plots). Adey said, "You can build synthetic ecosystems as small as you want. But the smaller you make it, the greater role human operators play because they must act out the larger forces of nature beyond the ecological community. The subsidy we get from nature is incredible."

Again and again, this was the message from the naturalists who assembled Bio2: *The subsidy we get from nature is incredible.* The ecological subsidy most missing from Bio2 was turbulence. Sudden, unseasonable rainfall. Wind. Lightning. A big tree falling over. Unexpected events. Just as in a miniature Ecosphere, nature both mild and wild demands variance. Turbulence is crucial to recycle nutrients. The explosive imbalance of fire feeds a prairie or starts a forest. Peter Warshall said, "Everything is controlled in Bio2, but nature needs wildness, a bit of chaos. Turbulence is an expensive resource to generate artificially. But turbulence is also a mode of communication, how different species and niches inform each other. Turbulence, such as wave action, is also needed to maximize the productivity of a niche. And we ain't got any turbulence here."

Humans in Bio2 were the gods of turbulence and the deputies of chaos. As pilots responsible for co-controlling the ark, they paradoxically were also *agents provocateurs* responsible for staging a certain amount of out-of-controllness.

Warshall was in charge of creating the minisavanna within Bio2 and its miniturbulence. Savannas, said Warshall, have evolved in conditions of periodic disturbance and require a natural kick every now and then. Any savanna's plants need a jolt by being burnt to the ground by fire or grazed by antelope. He said, "The savanna is so adapted to disturbances that it can not sustain itself without it," and then joked about putting a sign in the Bio2 savanna that says "Please Disturb."

Turbulence is an essential catalyst in ecology, but it was not cheap to replicate in a man-made environment like Bio2. The wave machine that sloshed the lagoon water was complicated, noisy, expensive, endlessly breaking down, and after all that, only made tiny highly regular waves—minimal turbulence. Huge fans in the basement of Bio2 pushed the air around for some semblance of wind, but it hardly moved pollen. Pollen-moving wind would have been prohibitively expensive to manufacture. And fire would have smothered the humans with captive smoke.

"If we were really doing this right, we would be piping in thunder for the frogs, who are stimulated to reproduce by rain splatters and thunder," said Warshall. "But we are not really modeling the Earth, we are modeling Noah. In reality the question we are asking is, How many links can we break and still have a species survive?"

"Well, we haven't had a crash yet!" Walter Adey chuckled. Both his analog coral reef in Bio2 and his analog swamp at the Smithsonian (which gets a thunderstorm when someone turns a gushing water hose onto it) thrived despite the sustained shock of being isolated and closed off from the big subsidies of nature. "They are hard to kill, given reasonable treatment. Or even occasional unreasonable treatment," Adey said. "One of my students forgot to remove a certain plug from the [Smithsonian] swamp one night, which flooded the main electrical panel with saltwater, which blew up the whole damn thing at 2 A.M. It wasn't until the next afternoon that we got its pumps running again, but it survived. We don't know how long we could have been down and still lived."

Life keeps rising. It rose again and again inside Bio2. The bottle was fecund, prolific. Of the many babies born in Bio2 during its first two years, the most visible was a galago born in the early months of closure. Two African pygmy goats birthed five kids, and an Ossabaw Island pig bore seven piglets. A checkered garter snake gave birth to three baby snakes in the ginger belt at the edge of the rain forest. And lizards hid lots of baby lizards under the rocks in the desert.

But all the bumblebees died. And so did the hummingbirds, all four of them. One species of coral in the lagoon (out of forty) went "extinct," but it was represented by only a single individual. All the cordon bleu finches died, still in their transition cages; maybe they were too cold during an unusually cloudy Arizona winter. Linda Leigh, who was Bio2's in-house biologist, wondered ruefully whether, if she had let them out earlier, they could have discovered a warm corner on their own. Humans make such remorseful gods. Furthermore, fate is always ironic. Three uninvited English sparrows who snuck into the structure before closure thrived merrily. Leigh complained that the sparrows were brash and noisy, even vulgar in their pushiness, while the finches were elegant, peaceful, and melodious singers.

Stewart Brand once needled Linda on the phone: "What's the matter with you guys that you don't want to go with success? Keep the sparrows and forget about the finches." Brand urged Darwinism: find what works, and let it reproduce; let the biosphere tell you where it wants to go. Leigh confessed, "I was horrified when Stewart first said that, but more and more I agree with him." The problem was not just sparrows. It was aggressive passion vines in the artificial savanna, and savanna grasses in the desert, ants everywhere, and other creatures not invited.

Urbanization is the advent of edge species. The hallmark of the modern world is its fragmentation, its division into patchworks. What wilderness is left is divided into islands and the species that thrive best thrive on the betweenness of patches. Bio2 is a compact package of edges. It has more ecological edges per square foot than anywhere else on Earth. But there is no heartland, no dark deepness, which is increasingly true of most of Europe, much of Asia, and eastern North America. Edge species are opportunists: crows, pigeons, rats, and the weeds found on the borders of urban areas all over the world.

Lynn Margulis, outspoken champion and coauthor of the Gaia Theory, told me her prediction of the Bio2 ecology before it closed. "It'll all go to Urban Weed," she said. Urban Weeds are those bully cosmopolitan varieties of both plants and animals that flourish in the edges of the patchwork habitats that people make. Bio2, after all, was a patchwork wilderness par excellence. According to Margulis's hypothesis, one expected to open the doors of Bio2 at the end and find it filled with dandelions, sparrows, cockroaches, and raccoons.

The human role was to prevent that from happening. Leigh said, "If we didn't tamper with it—that is, if no humans weeded the ones that were highly successful—I agree that Bio2 could go towards what Lynn Margulis is talking about: a world of Bermuda grass and mallard ducks. But since we are doing selective harvesting, I don't think that will happen, at least in the short run."

I harbor personal doubts about the ability of biospherians to steer the emergent ecology of 3,800 species. In the first two years, the fog desert became a fog thicket—it was wetter than expected, and grasses loved it. Weedy morning-glory vines overran the rain-forest canopy. The 3,800 species will sidestep, outmaneuver, burrow under, and otherwise wear down the "keystone predator" the biospherians hope to be, in order to go where they want to go. The cosmopolitan types are tenacious. They are in their element, and they want to stay.

Witness the curved-bill thrasher. One day an official from the U.S. Fish and Wildlife Department showed up outside a Bio2 window. The death of the finches had made the TV news and animal-rights activists had been calling his office. They wanted his service to check if the finches inside Bio2 had been collected from wild exotic places and brought in there to die. The biospherians showed the officer receipts and other paperwork that proved the late finches were mere captive-bred store pets, a status that was okay with the Wildlife Department. "By the way, what other birds do you have in there?" he asked them.

"Right now, only some English sparrows and a curved-bill thrasher."

"Do you have a permit for that curved-bill thrasher?"

"Uhhh, no."

"You know that under the Migratory Bird Treaty it's against federal law to contain a curved-bill thrasher. I'll have to give you a citation if you are holding him deliberately."

"Deliberately? You don't understand. He's a stowaway. We tried very hard to get him out of here. We tried trapping him every way we could think of. We didn't want him here before and we don't want him here now. He eats our bees, and butterflies, and as many insects as he can find, which isn't many by now."

The game warden and the biospherians were facing each other on either side of a thick airtight window. Although their noses were inches apart they talked on walkie-talkies. The surreal conversation continued. "Look," the biospherians said, "we couldn't get him out now even if we could catch him. We are *completely sealed* up in here for another year and a half."

"Oh. Umm. I see." The warden pauses. "Well, since you aren't keeping him intentionally, I'll issue you a permit for a curved-bill thrasher, and you can release him when you open up."

Anyone want to bet he won't ever leave?

Go with success. Unlike the fragile finches, both the hearty sparrows and the stubborn thrasher liked Bio2. The thrasher had his charms. His beautiful song wove through the wilderness in the morning and cheered the "key predators" during their sunrise routines.

The messy living thing knitting itself together inside Bio2 was pushing back. It was a coevolutionary world. The biospherians would have to coevolve along with it. Bio2 was specifically built to test how a *closed system* coevolves. In a coevolutionary world, the atmosphere and material environment in which beasties dwell become as adaptable and as lifelike as the beasties themselves. Bio2 was a test bench to find out how an environment governs the organisms immersed in it, and how the organisms in turn govern the environment. The atmosphere is the paramount environmental factor; it produces life, while life produces it. The transparent bottle of Bio2 turned out to be the ideal seat from which to observe an atmosphere in the act of conversing with life.

Inside this ultraairtight world—hundreds of times more airtight than any NASA space capsule—the atmosphere was full of surprises. It was unexpectedly clean for one thing. The trace gas buildup that was such a horrendous problem for earlier closed habitats and hi-tech closed systems such as NASA's space shuttle was eliminated by the collective respiration of a wilderness area. Scrubbed by some unknown balancing mechanism—most probably microbial—the air inside Bio2 was far cleaner than any space journey so far. Mark Nelson says, "Someone figured out it costs about $100 million a year to keep an astronaut in space, yet those guys are living in the worst environmental conditions you can imagine, worse than a ghetto." Mark told of an acquaintance who was honored to greet the returning space shuttle astronauts.

She was nervously waiting in front of cameras as they readied the door. They opened the hatch. She got a whiff. She puked. Mark says, "These guys really are heroes, because they are living a lousy life."

For two years in Bio2, carbon dioxide levels meandered up and down. At one point during a six-day sunless period, $CO_2$ reached a high of 3,800 parts per million (ppm). To give a sense of where that fits in, ambient carbon dioxide levels outside normally hover steadily at 350 ppm. The interior of a modern office building on a busy street may reach 2,000 ppm, and submarines let their $CO_2$ concentration rise to 8,000 ppm before they turn on $CO_2$ "scrubbers." Crew members of the NASA space shuttle work in a "normal" atmosphere of 5,000 ppm. Compare that to a very respectable 1,000 ppm average during a spring day in Bio2. The fluctuations, then, are well within the range of ordinary urban life and hardly noticeable to humans.

But the dance of atmospheric $CO_2$ does have consequences on plants and the ocean. During the tense days of higher $CO_2$, the biospherians worried that increased $CO_2$ in their air would dissolve in the mild ocean water, increasing the formation of carbonic acid ($CO_2$ + water), and lowering the water's pH, harming the newly transplanted corals. Discerning further biological effects of increased $CO_2$ is part of the Biosphere 2 mission.

People pay attention to the makeup of the Earth's atmosphere because it seems to be changing. We are sure it is changing, but beyond that we know almost nothing about its behavior. The only measurement of any historical accuracy we have relates to one component: carbon dioxide. The information on $CO_2$ concentration in the Earth's atmosphere shows an accelerating global rise over the past thirty years; that graph is due to a single, persistent scientist: Charles Keeling. In 1955, Keeling devised an instrument that could measure concentrations of carbon dioxide in all kinds of environments, from sooty city rooftops to pristine wilderness forests. Keeling obsessively measured $CO_2$ anywhere he thought the level might vary. He measured $CO_2$ at all times of the day and night. He initiated continuous measurements of $CO_2$ on a Hawaiian mountaintop and in the Antarctic. A colleague of Keeling told a reporter, "Keeling's outstanding characteristic is that he has an overwhelming desire to measure carbon dioxide. He wants to measure it in his belly. Measure it in all its manifestations, atmospheric and oceanic. And he's done this all his life." Keeling is still measuring carbon dioxide around the world.

Keeling discovered very early that $CO_2$ in the Earth's atmosphere cycles daily. $CO_2$ in the air increases measurably at night when plants shut down photosynthesis for the day, and then hits a low in the sunny afternoon as the plants go full steam turning $CO_2$ into vegetables. A few years later Keeling observed a second cycle: a hemispherical seasonal cycle of $CO_2$, low in summer and peaking in the winter for the same reason $CO_2$ peaks at night: no greens at work to eat it. But it is the third trend Keeling discovered that has focused attention on the dynamics of the atmosphere. Keeling noticed that the lowest level of $CO_2$, no matter where or when, would never sink beyond 315 ppm. This threshold was the ambient, global $CO_2$ level. And he noticed that every year it rose a little higher. By now, it's 350 ppm. Recently, other researchers have spotted in Keeling's meticulous recordings a fourth trend: the seasonal cycle is increasing in amplitude. It is as if the planet breathes yearly, summer (inhale) to winter (exhale), and its breath is getting deeper and deeper. Is Gaia hyperventilating or gasping?

Bio2 is a miniature Gaia. It is a small self-enclosed world with its own miniature atmosphere derived from living creatures. It is the first whole atmosphere/biosphere laboratory. And it has a chance to answer some of the tremendous questions science has about the workings of the Earth's atmosphere. Humans are inside the test tube to prevent the experiment from crashing, to divert the trials from overt crisis. The rest of us humans are outside, but inside the test tube of planet Earth. We are fiddling with Earth's atmosphere, yet haven't the slightest idea of how to control it, or where the dials are, or even if the system really is out of kilter and in crisis. The Bio2 experiment can offer clues to all those questions.

The atmosphere of Bio2 is so sensitive that the $CO_2$ needle rises when a cloud passes over. The shade momentarily slows green manufacturing, which momentarily lets the input flow of

$CO_2$ back up, which immediately registers as a blip at the $CO_2$ meter. On a partly cloudy day Bio2's $CO_2$ graph shows a string of little atmospheric hiccups.

Despite all the attention $CO_2$ levels have garnered in the past decade, and despite all the scrutiny agriculturists have given to the carbon cycle in plants, the fate of carbon in the Earth's atmosphere is a puzzle. It is generally agreed by climatologists that the curve of increasing $CO_2$ within modern times very roughly matches the rates of carbon-burning by industrial humans. That neat fit leaves out one astounding factor: when measured more precisely, only half of the carbon now burned on Earth remains in the atmosphere as increased $CO_2$ levels. The other half disappears!

Theories for the lost carbon abound. Three theories dominate: (1) it is being dissolved in the ocean, and then it precipitates to the sea bottom as carbon rain; or (2) it is being deposited in soils by microbes; or (3), most controversial, the lost carbon is fueling growth of the world's savanna grass, or being turned into tree wood, on an imperceptible but massive scale that we haven't yet been able to measure. $CO_2$ is the accepted limiting resource for the biosphere. At 350 ppm, the concentration of carbon dioxide is only a faint .03 percent—a mere trace gas. A field of corn in full sunshine will deplete the available trace $CO_2$ within a zone three feet above ground in under five minutes. Even small increases in $CO_2$ levels can boost biomass production significantly. Accordingly, says this hypothesis, wherever we aren't cutting down forests, trees are putting on extra weight due to the 15 percent of additional $CO_2$ "fertilizer" in the air, perhaps even at a rate greater than they are being destroyed elsewhere.

So far, the evidence is confusing. But in April of 1992, two studies published in *Science* claimed that the ocean and biosphere of Earth are indeed stockpiling carbon at the scale needed. One article showed that European forests have gained 25 percent or more treeflesh since 1971—despite the negative effects of acid rain and other pollutants. But hardly anyone has looked at the global carbon budget in detail. Our global ignorance of the global atmosphere makes the Biosphere experiment very promising. Here in the relatively controlled conditions of a sealed bottle, the links between an operating atmosphere and a living biosphere can be explored and mapped.

The amounts of carbon in the atmosphere, in the soil, in the plants, and in the ocean of Bio2 were carefully measured before closure. As the sun heated up photosynthesis, the carbon was moved from air to living things by measurable amounts. Each time any plant material was harvested, it was laboriously weighed and recorded by the biospherians. They could perturb the system slightly to see how it changed. For instance, when Linda Leigh "turned on the savanna" with artificial summer rains, the biospherians made simultaneous measurements of carbon levels in all domains of subsoil, topsoil, air, and water. They compiled a rich chart of where all carbon lies at the end of two years. By saving dried samples of leaf clippings, they also traced (somewhat) the route that carbon traveled within the surrogate world by following shifts in the ratio of naturally occurring carbon isotopes.

Carbon was only the first mystery. But the riddle deepened. Oxygen levels were lower inside Bio2 than outside. Oxygen dropped from 21 percent of the Bio2 atmosphere to 15 percent. A 6 percent drop in oxygen concentration was equivalent to Bio2 being transported to a site at a higher elevation, with a thinner atmosphere. The residents of Lhasa, Tibet, thrive at a similar, slightly reduced oxygen level. The biospherians experienced headaches, sleep loss, and fatigue. Though not catastrophic, the drop in oxygen levels was bewildering. In a sealed bottle, where does disappearing oxygen go?

Unlike the lost-carbon riddle, the mysterious oxygen vanishing act in Bio2 was completely unexpected. Speculation was that oxygen in Bio2 was tied up in the newly minted soil, maybe being captured into carbonates formed by microorganisms. Or, perhaps the fresh concrete absorbed it. In a quick survey of the scientific literature, biospheric researchers found little data concerning atmospheric oxygen levels in the Earth's atmosphere. The only known (but little-reported) fact is that oxygen in the atmosphere of the Earth is most likely also disappearing! Nobody knows why or even by how much. "I am surprised that the general public all over the world is not clamoring to know how fast we are using up the oxygen," said visionary physicist Freeman Dyson, one of the few scientists to even raise the problem.

And why stop there? Several experts watching the Bio2 experiment have suggested that tracing the comings and goings of atmospheric nitrogen should be next. Although nitrogen is the bulk component of the atmosphere, its role in the Great Cycle is known only broadly. Like carbon and oxygen, what is known has been extrapolated from reductionist experiments in the lab and computer modeling. Others have proposed that the biospherians map the element sodium or phosphorus next. Generating big questions about Gaia and the atmosphere may be Bio2's most important contribution to science.

When the $CO_2$ levels first began to rocket inside, the biospherians launched a countermove to limit the $CO_2$ rise. The chief tool to leverage the atmosphere was deployment of an "intentional season." Take a dry, dormant savanna, desert or thorn scrub and rouse it into spring with rising temperatures. Soon a thousand leaf buds swell. Then pour on the rain. Bam! In four days the plants explode into leaf and flower. The awakened biome sucks up $CO_2$. Once up, the biome can be kept awake past its normal retiring time by pruning old growth to stimulate new $CO_2$-consuming growth. As Leigh wrote in late fall of the first year, "With short days of winter approaching, we have to prepare for reduced light. Today we began to prune back the ginger belt on the north edge of the rain forest in order to stimulate rapid growth—a routine atmosphere management task."

The humans managed the atmosphere by turning the "$CO_2$ valve." Sometimes they reversed it. To flood the air with carbon dioxide, the biospherians hauled back the tons of dried grass clippings they had removed earlier. The clippings were piled on the soil as mulch and wetted. As bacteria decomposed it, they released $CO_2$ into the air.

Leigh called the biospherian interference in the atmosphere a "molecule economy." When they coordinated the atmosphere, they would "deposit the carbon into our account for safekeeping so that we can spend it next summer when we will need it for long days of plant growth." The underground areas where the plant clippings were dried served as a carbon bank. Carbon was lent as needed and primed with water. Water in Bio2 was diverted from one locality to another like so much federal spending meant to stimulate a regional economy. By channeling water onto the desert, $CO_2$ shrank; by channeling the water onto the dried mulch, $CO_2$ expanded. On Earth, our carbon bank is the black oil under Arabian sands, but all we do is spend it.

Bio2 compressed geological time into years. The biospherians twiddled with "geological" adjustments of carbon—storing and withdrawing carbon atoms in bulk—in the hope of roughly tuning the atmosphere. They tinkered with the ocean, lowering its temperature, adjusting the return of salty leachate, nudging its pH, and simultaneously guessing on a thousand other variables. "It's those few thousand other variables that make the Bio2 system challenging and controversial," said Leigh. "Most of us are taught not to mess with even two simultaneous variables." The biospherians hoped that if they were lucky, they could temper the initial wild oscillations of the atmosphere and ocean in the first years with a few well-chosen drastic actions. They would be the training wheels until the system could cycle through the year relying only on the natural action of sun, seasons, plants and animals to keep it in balance. At that point the system would "pop."

"Pop" is the term hobbyists in the saltwater aquarium trade use to describe what happens when a new fish tank suddenly balances after a long, meandering period of instability. Like Bio2, a saltwater fish tank is a delicate closed system that relies on an invisible world of microorganisms to process the waste of larger animals and plants. As Gomez, Folsome, and Pimm discovered in their microcosms, it can take 60 days for the microbes to settle into a stable community. In aquariums it takes several months for the various bacteria to develop a food web and to establish themselves in the gravel of the start-up tank. As more species of life are slowly added to the embryonic aquarium, the water becomes extremely sensitive to vicious cycles. If one ingredient drifts out of line (say, the amount of ammonia), it can kill off a few organisms, which decompose to release even more ammonia, killing more creatures, thus rapidly triggering the crash of the whole community. To ease the tank through this period of acute imbalance, the aquarist nudges the system gently with judicious changes of water, select chemical additives, filtration devices, and inoculations of bacteria from other successful aquariums. Then after about six weeks of microbial give-and-take—the nascent community teetering on the edge of chaos—suddenly, overnight, the system "pops" to zero ammonia. It's now ready for the long haul. Once the system has popped, it is more self-sustaining, self-stabilizing, not requiring the artificial crutches that set-up needed.

Wllat is interesting about a closed-system pop is that the conditions the day before the pop and the day after the pop hardly change. Beyond a little babysitting, there is often nothing one can do except wait. Wait for the thing to mature, to ripen, to grow, and develop. "Don't rush it," is the advice from saltwater hobbyists. "Don't hurry gestation as the system self-organizes. The most important thing you can give it is time."

Still green after two years, Bio2 is ripening. It suffers from wild, infantile oscillations that require "artificial" nurturing to soothe. It has not popped yet. It may be years (decades?) before it does, if it ever does, if it even can. That is the experiment.

We have not really looked yet, but we may find that all complex coevolutionary systems need to pop. Ecosystem restorationists such as Packard on the prairie and Wingate on Nonsuch Island seem to find that large systems can be assembled by ratchetting up complexity; once a system reaches a level of stability it tends not to easily fall back again, as if the system was "attracted" by the cohesion birthed by the new complexity. Human institutions, such as teams and companies, exhibit pop. Some little nudge—the additional right manager, a nifty new tool—can suddenly turn 35 competent hard-working people into a creative organism in the state of runaway success. Machines and machine systems, once we build them with sufficient complexity and flexibility, will also pop.

DIRECTLY BENEATH the wilderness of savanna and forest, the farm, and the modern apartments of the biospherians, lies the other face of Bio2: the mechanical "technosphere." The technosphere is the scaffolding put in place to help Bio2 pop. At several places in the wilderness, stairs wind down to a cavernous basement stuffed with basementish fixtures. Fifty miles of color-coded pipes as thick as an arm wind along the wall. There are huge ductworks right out of the movie *Brazil*; miles and miles of electrical wiring; workshops full of heavy-duty tools; hallways crowded with threshing and milling machines; shelves of spare parts; switchboxes, dials, vacuum blowers; over 200 motors, 100 pumps, and 60 fans. It could be the inside of a submarine or the backside of a skyscraper. The territory is industrial grunge.

The technosphere supports the biosphere. Huge blowers circulate the entire air of Bio2 several times in one day. Heavy pumps move the rainwater. The motors of the wave machine run day and night. Machines hum. This unabashedly manufactured world is not outside Bio2 but inside its tissue, like bone or cartilage, an integral part of the greater organism.

For example, Bio2's coral reef would not have worked without an eerie backroom in the basement where the algae scrubbers hide. The scrubbers were table-wide shallow plastic trays filled with a pool of algae. The whole room was flooded with the same type of halide sunlamps as illuminated artificial coral reefs in museums. The scrubbers were in fact the mechanical kidneys of the Bio2 coral reef. They performed the same function as a pool filter: to clear the water. The algae consumed waste products from the reef and under the intense artificial sunlight they proliferated in stringy green mats. The green strands soon clogged the scrubber; and just like a pool or aquarium filter, the scrubber needed to be scraped clean every ten days by some poor schmuck—another job for the eight humans. Cleaning the algae scrubbers (the harvest became compost) was the most despised assignment in Bio2.

The nerve center of the whole system was the computer room run by an artificial cortex of wires, chips, and sensors from around Bio2. Every valve, every pipe, and every motor of the infrastructure was simulated in a software network. Very little activity in the ark, either natural or man-made, happened without the distributed computer knowing about it. Bio2 responded as if it was one beast. About a hundred chemical compounds were continuously measured in the air, soil, and water throughout the whole structure. A potential profit-making technology that SBV imagined spinning off the project was sophisticated environmental-monitoring techniques.

Mark Nelson got it right when he said that Bio2 was the "marriage of ecology and technics." That's the beauty of Bio2—it's a fine example of ecotech, the symbiosis of nature and technology. We don't know enough yet about how to invent biomes without installing pumps. But by using the scaffolding of pumps now, we can try the system out and learn.

To a large degree it's a matter of learning a new form of control. Tony Burgess said, "NASA goes about by optimizing resource utilization. They take wheat and optimize the environment for the production of wheat. But the problem is when you put together a whole bunch of species you can't optimize each species separately, you have to optimize the whole thing. Doing this one at a time you become dependent on governance by engineering. SBV hopes that you can remove governance by engineering and switch it to governance by biology. Which ultimately should be cheaper. You may lose some optimization of production, but you gain independence from the technics."

Bio2 is a gigantic flask for ecological experiments that require more control over the environment than could (or should) be done in the wild. Individual lives can be studied in a laboratory. But ecological life and biospheric life require a more monumental room to view things in. For instance, in Bio2 a single species can be introduced or deleted with great confidence knowing that no other species have been altered—all in a space large enough for something "ecological" to happen. "Biosphere 2," said John Allen, "is a cyclotron for the life sciences."

Or maybe Bio2 is really a better Noah's ark. A futuristic zoo within one large cage where everything runs wild, including the observing *Homo sapiens*. The species are free to be themselves and to coevolve with others into anything they want.

At the same time, space cowboys see Bio2 as a pragmatic step on a spiritual journey off the planet into the galaxies. As space technology, Bio2 is the most thrilling news since the moon landings. NASA, after routinely poohpoohing the enterprise in its conceptual stages, refusing to help out at any time, has had to swallow their pride and acknowledge that, yes, there is something useful here. Out-of-control biology has a place.

All three spirits are really manifestations of the same metamorphosis best described by Dorion Sagan in his book *Biospheres*:

> The "man-made" ecosystems known as biospheres are ultimately "natural"—a planetary phenomenon that is part of the reproductive antics of life as a whole . . . We are at the first phase of a planetary metamorphosis, . . . [the] reappearance of individuality at a hitherto unsuspected scale: not of reproducing microorganisms, or plants or animals, but of the Earth as a living whole . . .
>
> Yes, human beings are involved in this reproduction, but are not insects involved in the reproduction of many flowers? That the living Earth now depends upon us and our engineering technology for its reproduction does not invalidate the proposition that biospheres, ostensibly built for human beings, represent the reproduction of the planetary biosystem . . .
>
> What is definitive success? Eight people living inside it for two years? How about ten years, or a century? In fact, biosphere reproduction, the building of dwellings that internally recycle all that is needed for human life, begins something whose end we cannot foresee.

When everything works, and free time loosens up daydreams, the biospherians can wonder, Where *does* all this lead? What's next? A Bio2 oasis at the South Pole? Or a bigger Bio2 with many more bugs and birds and berries? The most interesting question may be: how small can a Bio2 be? Those master miniaturists, the Japanese, are crazy over Biosphere 2. In one poll conducted in Japan, over 50 percent of the population recognized the project. To those used to claustrophobic living quarters and the isolation of island living, a mini-Bio2 seems positively charming. In fact, one government department in Japan has announced plans for a Biosphere J. The "J" stands not for Japan (they say), but for Junior, as in tinier. Official sketches show a small warren of rooms, lit by artificial lights and stuffed with compact biological systems.

The ecotechnicians who built Bio2 have figured out the basic techniques. They know how to seal the glass, schedule perpetual subsistence crops in a very small plot, recycle their wastes, balance their atmosphere, live without paper, and get along inside. That's a pretty good start for biospheres of any size. The future should birth Bio2s in all sizes and varieties, housing every combination of species. As Mark Nelson told me, "In the future there will be an enormous proliferation of niches for biospheres." Indeed, he sees varieties of biospheres of different sizes and composition, as if they were different species of biospheres, competing over territory, mingling to share genes, and hybridizing in the manner of biological organisms. Planets would be settled with them, and every city on Earth would have one, for experiments and education.

ONE EVENING in the spring of 1991, by some bureaucratic oversight, I found myself without an escort in the nearly completed Biosphere. The construction guys had gone home for the day, and the SBV staff were turning out lights up on the hill. I was alone in the first offspring of Gaia. It was eerily quiet. I felt I was standing in a cathedral. Loitering in the agricultural biome, I could barely hear the muffled thump of the distant wave machine in the ocean, as it exhaled a wave every twelve seconds. Near the machine—which sucks up ocean water and then releases it in a wave—it sounded, as Linda Leigh says, like the blow of a gray whale. Back in the garden where I stood, the distant deep guttural moan sounded like Tibetan monks chanting in the basement.

Outside, brown desert at dusk. Inside, a world thick with green life. Tall grass, seaweed adrift in tubs, ripe papaya, the splash of a fish jumping. I was breathing green, that heavy plant smell you get in jungles and swamps. The atmosphere moved slowly. Water cycled. The space-frame structure creaked as it cooled. The oasis was alive, yet everything was still. Quietly busy. No people. But something was happening together; I could sense the "co" in coevolutionary life.

The sun had nearly set. Its light was soft and warm on the white cathedral. I could live here a bit, I thought. There's a sense of place. A cave coziness. Yet open to the stars at night. A womb with a view. Mark Nelson said, "If we are really going to live in space like human beings, then we have to learn how to make biospheres." He said that the first thing macho, no-time-for-nonsense cosmonauts did after floating out of bed in the Soviet skylab was to tend their tiny pea seedling "experiments." Their kinship with peas became evident to them. We need other life.

On Mars, I would only want to live in an artificial biosphere. On Earth, living in an artificial biosphere is a noble experiment, suitable for pioneers. I could imagine it coming to feel like living inside a giant test tube after awhile. Great things will be learned inside Bio2 about our Earth, ourselves, and the uncountable other species we depend on. I have no doubt that someday what is learned here will land on Mars or the Moon. Already it has taught me, an outsider, that to *live as human beings means to live with other life*. The nauseating fear that machine technology will replace all living species has subsided in my mind. We'll keep other species, I believe, because as Bio2 helps prove, life is a technology. Life is the ultimate technology. Machine technology is a temporary surrogate for life technology. As we improve our machines they will become more organic, more biological, more like life, because life is the best technology for living. Someday the bulk of the technosphere in Bio2 will be replaced by engineered life and lifelike systems. Someday the difference between machines and biology will be hard to discern. Yet "pure" life will still have its place. What we know as life today will remain the ultimate technology because of its autonomy—it goes by itself, and more importantly, it learns by itself. Ultimate technologies, of any sort, inevitably win the allegiance of engineers, corporations, bankers, visionaries, and pioneers—all the agents who once were thought of as pure life's biggest threat.

The glass spaceship parked in the desert is called a biosphere because the logic of the Bios runs through it. The logic of Bios (bio-logic, biology) is uniting the organic and the mechanical. In the factories of bioengineering firms and in the chips of neural-net computers, the organic and the machine are merging. But nowhere is that marriage between the living and the manufactured so clear as in the pod of the Bio2. Where does the synthetic coral reef end and the chanting wave machine begin? Where does the waste-treatment marsh begin and the toilet plumbing end? Is it the fans or the soil bugs that control the atmosphere?

The bounty of a journey inside Bio2 is mostly questions. I sailed in it for only hours and got years of things to consider. That's enough. I turned the massive handle on the air lock doors in the quiet Biosphere 2 and debarked into a twilight desert. Two years in there would fill a lifetime.

# 10

# Industrial Ecology

Barcelona, Spain, is a city of die-hard optimists. Its citizens embrace not only trade and industry, art and opera, but also the Future, with a capital F. Twice, in 1888 and 1929, Barcelona hosted the Universal Exhibition, the then equivalent of a world's fair. Barcelona eagerly courted this future-friendly fiesta because, in one Spanish writer's opinion, the city ". . . really has no reason to be . . . so [it] is constantly re-inventing itself by creating great prospects." Barcelona's 1992 self-made great prospect was an Olympic vision, with a capital O. Young athletes, mass culture, new technology, big bucks—quite appealing prospects to this square town bustling with commonsense design and an earnest mercantile spirit.

Smack in the middle of this pragmatic place, the legendary Antonio Gaudi built several dozen of the strangest buildings on Earth. His structures are so futuristic and weird that Barcelonians and the world didn't know what to make of them until recently. His most famous creation is the unfinished cathedral known as the Sagrada Familia. Begun in 1884, the parts of the cathedral completed in Gaudi's time seethe with organic energy. The facade of stone drips, arcs, and blossoms as if it were vegetable. Four soaring steeples are honeycombed with cavities, revealing them to be the bony skeleton of support they are. One-third of the way up a second set of towers in the rear, massive thighbone braces lean up from the ground and steady the church. From a distance the braces look to be giant bleached drumsticks of a creature long dead.

All of Gaudi's work squirms with the flow of life. Ventilator chimneys sprouting on the roofs of his Barcelona apartments resemble a collection of mounted life forms from an alien planet. Window eaves and roof gutters curve in organic efficiency rather than follow a mechanical right angle. Gaudi captures that peculiar living response which cuts across a square campus lawn and traces a graceful curving shortcut. His buildings seem to be grown rather than constructed.

Imagine an entire city of Gaudi buildings, a human-made forest of planted homes and organic churches. Imagine if Gaudi did not have to stop with the static face of a stone veneer, but could endow his building with organic behavior over time. His building would thicken its hide on the side where the wind blows most or rearrange its interior as its inhabitants shifted their use of it. Imagine if Gaudi's city not only stood by organic design but adapted and flexed and evolved as living creatures do, forming an ecology of buildings. This is a future vision that not even optimistic Barcelona is ready for. But it is a future that is arriving now with the advent of adaptive technologies, distributed networks, and synthetic evolution.

You can browse through old *Popular Science* magazines from the early '60s and see that a living house has been in speculation for decades, not counting wonderful science-fiction stories even earlier. The animated Jetsons live in such a home, talking to it as if it were an animal or person. I think the metaphor is close but not quite correct. The adaptive house of the future will be more like an ecology of organisms than a single being, more like a jungle than a dog.

The ingredients for an ecological house are visible in an ordinary contemporary house. I can already program my home's thermostat to automatically run our furnace at different temperatures during weekdays and weekends. In essence, fire is networked to a clock. Our VCR knows how to tell time and talk to the TV. As computers continue to collapse into mere dots which find themselves wired into all appliances, it is reasonable to expect our washing machine, stereo, and smoke alarm to communicate in a householdwide network. Someday soon, when a visitor rings the doorbell, the doorbell will turn down the vacuum cleaner so that we can hear its chime. When the clothes are done in the washer, it will flash a message on the TV to let us know it's ready for the dryer. Even furniture will become part of the living forest. A microchip in a couch will sense the presence of a sitter and turn the heat up in the room.

The vehicle for this house-net, as it is presently envisioned by engineers in several research labs, is a universal outlet peppering the rooms in every home. You plug *everything* into it. Your telephone, computer, doorbell, furnace, and vacuum cleaner all insert into the same outlet to get both power and information. These smart outlets dispense 110-volt juice only to "qualified" appliances and only when they request it. When you plug a smart object into the house-net, its chip declares its identity ("I am a toaster"), status ("I am turned on"), and need ("Give me 10 watts of 110"). A child's fork or broken cord won't get power.

Outlets trade information all the time, powering-up things when needed. Most importantly, the networked outlets bundle many wires into one socket, so that intelligence, energy, information, and communication can be sucked from any point. You plug a doorbell button in a socket near the front door; you can then plug a doorbell chime into any socket in any room. Plug in a stereo in one room, and music is ready in all the other rooms as well. Likewise, the clock. Soon universal time signals will be transmitted through all power and telephone lines. Once something is plugged in anywhere, it will at least know the time and date and automatically recalibrate daylight savings when instructed by the master timekeeper in Greenwich, England, or the U.S. Naval Observatory. All information plugged into the household net will also be shared. The furnace's thermostat can feed a room's temperature to any appliance that would like to know, say, a fire alarm or a ceiling fan. Anything that can be measured—level of light, motion of inhabitants, noise level—can be broadcast into the home's network.

An intelligently wired house would be a lifesaver to the disabled and elderly. From a switch near the bed, they could control the lights, TV, and security gizmos in the rest of the house. An ecological building would also be moderately more energy efficient. Says Ian Allaby, a journalist reporting on the dawning smart-house trade, "You might not want to climb from bed to run the dishwasher at 2 A.M. to save 15 cents, but if you could pre-arrange the utility to switch the machine on for you, then great!" The prospect of decentralized efficiency is attractive to utility companies, since the profits in efficiency are greater than those in building a new power plant.

So far, nobody actually lives in a smart house. A grand partnership of electronic firms, building industry associations, and telephone companies banded together in 1984 under the umbrella of Smart House Partnership to develop protocols and hardware for an intelligent house. As of late 1992 the group had built about a dozen demo homes to distract reporters and garner investments. The partnership dropped their initial 1984 vision of a standard one-size-fits-all outlet as too radical on first pass. For interim technology, Smart House uses wiring that divides functions into three cables and three connections at the outlet box (AC power, DC power, and communications). This would allow "backward compatibility"—the opportunity to plug dumb ol' power tools and appliances into the house without having to scrap them for new smart objects. Competing agencies in the U.S., Japan, and Europe play with other ideas and other standards, including using a wireless infrared network to connect widgets. This would enable portable battery-powered devices, or nonelectric objects to be linked into the web. Doors could have small semi-intelligent chips that "plug in" via invisible signals in the air, to let the household ecology know that a room was closed or that a visitor was coming down the hall.

MY PREDICTION IN 1994: Smart offices will materialize before smart homes. Because of the intensive informational nature of business—its reliance on machines and its need to constantly adapt—wizardry that is merely marginal in a home can make an economic difference in an office. Time at home is often regarded as leisure, so saving a bit through the intelligence of a net isn't as valuable as accumulating small amounts of time on the job. Networked computers and phones are mandatory in offices now; networked lights and furniture will be next.

The research labs of Xerox in Palo Alto, California (PARC), invented, but unfortunately never exploited, the signature elements of the first friendly Macintosh computers. Not to be burned twice, PARC intends to fully exploit yet another radical (and potentially profitable) concept brewing in their labs now. Mark Weiser, young and cheerful, is director of a Xerox initiative to view the office as a superorganism—a networked being composed of many interlinked parts.

The glassy offices of PARC perch on a Bay Area hill overlooking Silicon Valley. When I visit Weiser he is wearing a loud yellow shirt flanked by red suspenders. He smiles constantly, as if inventing the future was a big joke and I'm in on it. I take the couch, an obligatory furnishing in hacker dens, even posh hacker dens like these at Xerox. Weiser is too animated to sit; he's waving his arms—a marker in one hand—in front of a huge white board that runs from the floor to the ceiling. This is complicated, his arms say, you are going to need to *see* it. The picture Weiser begins drawing on the white board looks like a diagram of a Roman army. Down at the bottom are one hundred small units. Above it are ten medium-size units. Perched at the top level is one large unit. The army that Weiser is drawing is a field of Room Organisms.

What I really want, Weiser is telling me, is a mob of tiny smart objects. One hundred small things throughout my office that have a uniform, dim awareness of each other, of themselves, and of me. My room becomes a supercolony of quasi-smart bits. What you want, he says, is every book on your shelf to have a chip embedded in it so that it keeps track of where it is in the room, when it was last open, and to what page. The chip might even have a dynamic copy of the book's index that will link itself to your computer database when you first bring the book into the room. The book now has a community presence. All information stored on a shelf as, say, books or videotapes is implanted with a cheap chip to communicate both where they are and what they are about.

In the ecological office stocked with swarmish things, the room will know where I am. If I'm not there, obviously it (they?) should turn the lights off. Weiser: "Instead of having a light switch in every room, everyone carries their own light switch with them. When they want the lights on, the smart switch in their pocket turns them on or dims them to a level you want, in the room you happen to be in. Rather than the room having a dimmer, you have a dimmer. Personal light control. Same with volume control. In an auditorium everyone has their personal volume control. The volume is often too loud or too low, so everyone sort of votes with their pocket devices. The sound settles at an average for those people."

In Weiser's vision of an intelligent office, ubiquitous smart things form a hierarchy. At bottom, an army of microorganisms act as a background sensory net for the room. They feed location and usage information directly to the upper levels. These front-line soldiers are cheap, disposable small fry attached to writing pads, booklets, and smart Post-it notes. You buy them by the dozen—like pads of paper or RAM chips. They work best massed into a mob.

Next, about ten mid-size (slightly bigger than a bread box) displays, such as furniture and appliances, interact more frequently and directly with the office holder. Linked into the superorganism of a smart room, my chair will recognize me when I sit in it, versus someone else. When I first plop down in the mornings, it will remember what I usually do in the A.M. It can then assist my routine, awakening appliances that need a warm-up, preparing the day's schedule.

Every room also has at least one electronic display that is a yard-wide or bigger—a window, a painting, or a computer/TV screen. In Weiser's world of environmental computing, the big display in every room is the smartest nonhuman in the room. You talk to it, point over it, write on it and it understands. The big screen does movies, text, super graphics, whatever. It almost goes without saying that it is interconnected with all the other objects in the room, knows exactly what they are up to, and can *represent them on its screen* with some faithfulness. So I can interact with a book two ways: by handling the actual object or by handling its image on the screen.

Every room becomes an environment of computation. The adaptive nature of computers recedes into the background until it is nearly invisible and ubiquitous. "The most profound technologies are those that disappear," says Weiser. "They weave themselves into the fabric of everyday life until they are indistinguishable from it." The technology of writing descended from elite status, steadily lowering itself out of our consciousness altogether until we now hardly notice words scribbled everywhere from logos stamped on fruit to movie subtitles. Motors began as huge noble beasts; they have since evaporated into micro-things fused (and forgotten) in most mechanical devices. George Gilder, writing in *Microcosm*, says, "The development of computers can be seen as the process of collapse. One component after another, once well above the surface of the microcosm, falls into the invisible sphere, and is never again seen clearly by the naked eye." The adaptive technologies that computers bring us started out as huge, conspicuous, and centralized. But as chips, motors, and sensors collapse into the invisible realms, their flexibility lingers as a distributed environment. The materials evaporate, leaving only their collective behavior. We interact with the collective behavior—the superorganism, the ecology—so that the room as a whole becomes an adaptive cocoon.

Gilder again: "The computer will ultimately collapse to a pinhead that can respond to the human voice. In this form, human intelligence can be transmitted to any tool or appliance, to any part of our environment. Thus the triumph of the computer does not dehumanize the world; it makes our environment more subject to human will." It is not machines we are creating but a mechanical environment permeated with our sense of learning. We are extending our life into our surroundings.

"You know the premise of virtual reality—to put you inside a computer world," says Mark Weiser. "Well, I want to do the opposite. I want to put the computer world around you on the outside. In the future, the smartness of computers will surround you." This is a nice switcheroo. Rather than have to don goggles and body suit to experience immersion in a computer-generated world, all you have to do to be completely surrounded by the magic of constant computation is to open a door.

Once you are in a net-ridden room, all smart rooms talk to each other. The big picture on the wall then is a portal into both my own room and into other folks' rooms. Say I hear about a book I should read. I do a data search for it in my building; my screen says a copy lives in Ralph's office, behind his desk on a shelf of company-bought books, and was used last week. There is also another copy in Alice's cubby, next to the computer manuals, that hasn't ever been read, even though it is her own personal purchase. I pick Alice and send her a loan plea on the net. She says okay. When I physically take the book from Alice's room, it reconfigures its display to match the rest of the books in my room as is my preference. (I like to have the pages I "dog-eared" displayed first.) The book's new location is recorded in its internal biography, and noted by everyone's databank. This book is unlikely to go the one-way journey of most borrowed books.

In the colony of a smart room, the telephone rings slightly louder if the stereo is on; the stereo lowers itself when you answer the phone. Your office voice-mail unit knows your car is not in the parking lot so it tells the caller you haven't arrived yet. When you pick up a book, it tells the lamp above your favorite reading chair to turn on. Your TV notifies you that the novel you've been reading is available this week as a movie. Everything is connected to everything. Clocks listen to the weather, refrigerators watch the time and order milk before the carton is empty, and books remember where they are.

Weiser writes that in Xerox's experimental office, "doors open only to the right badge wearer, rooms greet people by name, telephone calls can be automatically forwarded to wherever the recipient may be, receptionists actually know where people are, computer terminals retrieve the preferences of whoever is sitting at them, and appointment diaries write themselves." But what if I don't want everyone in my department to know what room I'm in? Workers participating in initial trials of ubiquitous computing at Xerox PARC often left their office in order to *get away* from the phone-blob. They felt imprisoned by always being findable.

Network culture cannot thrive without the technologies of privacy. Privacy in the form of personal encryption and unforgeable digital signatures are being rapidly developed (see following chapter). Privacy can also be secured in the anonymous nature of the mob.

WEISER'S BUILDINGS are a coevolutionary ecology of machines. Each device is an organism that reacts to stimulus and communicates with the others. Cooperation is rewarded. Alone, most of the electronic bits are wimpy and would die of nonuse. Together, they form a community that is attentive and robust. What each microbit lacks in depth is made up by the communal net which casts its collective influence wide over a building, outreaching even a human.

Not only would rooms and halls have embedded intelligence and ecological fluidity but entire streets, malls, and towns. Weiser uses the example of words. Writing, he says, is a technology that is ubiquitously embedded into our environment. Writing is everywhere, urban and suburban, passively waiting to be read. Now imagine, Weiser suggests, computation and connection embedded into the built environment to the same degree. Street signs would communicate to car navigation systems or a map in your hands (when street names change, all maps change too). Streetlights in a parking lot would flick on ahead of you in anticipation of your walk. Point to a billboard properly, and it would send you more information on its advertised product and let its sponsor know what part of the street most of the queries came from. The environment becomes animated, responsive, and adaptable. It responds not only to you but to all the other agents plugged in at the time.

One definition of a coevolutionary ecology is a collection of organisms that serve as their own environment. The flamboyant world of orchid flowers, ant colonies, and seaweed beds overflows with richness and mystery because the movie that each creature stars in features walk-ons and extras who are simultaneously acting as stars in their own movie filmed on the same lot. Every borrowed set is alive and liquid as the star is. Thus, the fate of a mayfly is primarily determined by the histrionics of neighboring frog, trout, alder, water spider, and the rest of stream life, each playing the environment for the other. Machines too will play on a coevolutionary stage.

The refrigerator you can purchase today is an arrogant snob. When you bring it home it assumes that it alone is the only appliance in the house. It has nothing to learn from other machines in the building, and nothing it will tell them. A wall clock will tell *you* the time of day but not its manufactured brethren. Each utensil haughtily serves only its buyer without regard to how much better it could serve in cooperation with the other items around it.

An ecology of machines, on the other hand, enhances the limited skills of dumb machines. The chips imbedded in book and chair have only the smartness of ants. They're no supercomputer; they could be manufactured now. But by the alien power of distributed being, sufficient numbers of antlike agents can be lifted into a type of colony intelligence by connecting them in bulk. More is different.

Collaborative efficiency, however, has a price. An ecological intelligence will penalize anyone new to the room, just as a tundra ecology will penalize anyone new to the arctic. Ecologies demand local knowledge. The only folks who know where the mushrooms bloom in the woods are native sons. To track wallabies through the Australian outback you want a local bush ranger as a guide.

Where there is an ecosystem, there are local experts. An outsider can muddle through an unfamiliar wilderness at some level, but to thrive or to survive a crisis, he'll require local expertise. Gardeners regularly surprise academic experts by growing things they aren't supposed to be able to grow because, as local experts, they tune into the neighborhood soil and climate.

The work of managing a natural environment is inescapably a work of local knowledge. A roomful of mechanical organisms improvising with each other demands a similar local knowledge. The one advantage snooty old Refrigerator had was that he ignored everyone equally, owner and visitor alike. In a room enlivened by a colony intelligence, visitors are at a disadvantage. Every room will be different; indeed, every telephone will be different. Because the new phones will merely be one node of a far larger organism—linking furnace, cars, TVs, computers, chairs, whole buildings—whose own behavior will hinge on the holistic sum of everything else going on in the room. The behavior of each will particularly depend on how its most frequent user employs it. To visitors, the indefinite beast of a room will seem to be out of control.

Adaptable technology means technology that will adapt *locally*. The logic of the network induces regionalism and localism. Or to put it another way, global behavior entails regional variety. We see evidence of the shift already. Try using someone else's "smart" phone. It is already either too smart, or not smart enough. Do you dial "9" to get out? Can you punch any button for a line? How do you (gulp!) transfer a call? Only the owner knows for sure. The local knowledge needed to fully operate a VCR is legendary. Just because you can preprogram yours to record *The Prisoner* reruns doesn't in any way mean you can handle your friend's.

Rooms and buildings will vary in their electronic ecology, as will appliances within a room, since they all will be aggregations of smaller distributed parts. No one will know the idiosyncrasies of my office's technology as well as I will. Nor will I be able to work another's technology as easily as my own. As computers become assistants, toasters become pets.

When the designers get it right, the coffee machine that an impatient visitor tries to use will default to "novice mode" when it senses desperate attempts to make it work. Mr. Coffee will cop to the situation by engaging only the five basic universal appliance functions that every school child will know.

But I find the emerging ecology in its earliest stages already daunting to strangers. Since computers are the locus where all these devices hail from and head toward, we can see in them now the alienness of unfamiliar complex machines. It doesn't matter how acquainted you are with a particular brand of computer. When you need to borrow someone else's, it feels like you're using their toothbrush. The instant you turn a friend's computer on, it's there: that strange arrangement of familiar parts (why do they do it like that?), the whole disorienting logic of a place you thought you knew. You kind of recognize it. There's an order here. Then, a moment of terror. You are . . . peering into someone else's mind!

The penetration goes both ways. So personal, so subtle, so minute is everyone's parochial intelligence of their own computer's ecology, that any disturbance is alarming. A pebble dislodged, a blade of grass bent, a file moved. "Someone has been in my compu-room! I know it!"

There will be nice-dog rooms and bad-dog rooms. Bad-dog rooms will bite intruders. Nice-dog rooms will herd visitors to someplace safe, away from places where real harm can be done. The nice-dog room may entertain guests. People will acquire reputations on how well-trained their computers are and how well-groomed their computational ecology is. And others will gain notoriety for how fiercely wild their machinery is. There are sure to be neglected areas in large corporations someday where no one wants to work or visit because the computational infrastructure has been neglected to the point that it is rude, erratic, swampy (although brilliant), and unforgiving, yet no one has time to tame or retrain it.

Of course there is a strong counterforce to keep the environment uniform. As Danny Hillis pointed out to me, "The reason we create artificial environments instead of accepting natural ones is that we like our environments to be constant and predictable. We used to have a computer editor that let everyone have a different interface. So we all did. Then we discovered it was a bad idea because we couldn't use each other's terminals. So we went back to the old way: a shared interface, a common culture. That's part of what brings us together as humans."

Machines will never go completely on their own way, but they will become more aware of other machines. To survive in the Darwinian marketplace, their designers must recognize that these machines inhabit an environment of other machines. They gather a history together, and in the manufactured ecology of the future, they will have to share what they know.

ON THE COUNTER of every American auto parts store sits a massive row of catalogs, a horizontal stack of pages as wide as a dump truck, spines down, page edges outward. Even from the other side of the Formica you can easily spot the dozen or so pages out of ten thousand that the mechanics use the most: their edges are smeared black by a mob of greasy fingers. The wear marks help the guys find things. Each soiled bald spot pinpoints a section they most often need to look up. Similar wear-indicators can be found in a cheap paperback. When you lay it down on your night table, its spine buckles open slightly at the page you were last reading. You can pick up your story the next evening at this spontaneous bookmark. Wear encodes useful information. When two trails diverge in a yellow wood, the one more worn tells you something.

Worn spots are emergent. They are sired by a mob of individual actions. Like most emergent phenomena, wear is liable to self-reinforce. A gouge in environment is likely to attract future gouges. Also, like most emergent properties, wear is communication. In real life "wear is tattooed directly on the object, appearing exactly where it can make an informative difference," says Will Hill, a researcher at Bellcore, the telephone companies' research consortium.

What Hill would like to do is transfer the environmental awareness communicated by physical wear into the ecology of objects in an office. As an example, Hill suggests that an electronic document can be enriched by a record of how others interact with it. "While using a spreadsheet to refine a budget, the count of edit changes per spreadsheet cell can be mapped onto a gray scale to give a visual impression of which budget numbers have been reworked the most and least." This gives an indication of where confusion, controversy, or errors lie. Another example: businesses with an efficiency bent can track what parts of documents acquire the most editorial changes as it bounces back and forth between various departments. Programmers call such hot spots of wheel-spinning change "churns." They find it useful to know where, in a million lines of group-written programming code, the areas of churn are. Software makers and appliance manufacturers would gladly pay for amalgamated information about which aspects of their products are used the most or least, since such explicit feedback can improve them.

Where Hill works, all the documents that pass through his lab keep track of how others (human or machine) interact with them. When you select a text file to read, a thin graph on your screen displays little tick marks indicating the cumulative time others have spent reading this part. You can see at a glance the few places other readers lingered over. Might be a key passage, or a promising passage that was a little unclear. Community usage can also be indicated by gradually increasing the type size. The effect is similar to an enlarged "pull quote" in a magazine article, except these highlighted "used" sections emerge out of an uncontrolled collective appreciation.

Wear is a wonderful metaphor for a commonwealth. A single wear mark is useless. But bunched and shared, they prove valuable to all. The more they are distributed, the more valuable. Humans crave privacy, but the fact is, we are more social than solitary. If machines knew as much about each other as we know about each other (even in our privacy), the ecology of machines would be indomitable.

IN MECHANICAL COMMUNITIES, or ecosystems, some machines are more likely to associate with certain other machines, just as red-winged blackbirds favor nesting in cattail swamps. Pumps go with pipes, furnaces go with air-conditioners, switches go with wires.

Machines form food webs. Viewed in the abstract, one machine "preys" upon another. One machine's input is another's output. A steel factory eats the effluent of an iron-mining machine. Its own extrusion of steel is in turn eaten by an automobile-making machine, and fashioned into a car. When the car dies it is consumed by a scrapyard crusher. The crusher's ejected iron cud is later swallowed by a recycling factory and excreted as, say, galvanized roofing.

If you were to follow an iron particle as it was dug out of the ground to be passed up the industrial food chain, it would trace a crisscrossing circuit for its path. The first time around the particle may appear in a Chevrolet; the second cycle around it may land in a Taiwanese ship hull; the third time around it shapes up as a railroad rail; and the fourth as a ship again. Every raw material meanders through such a network. Sugar, sulfuric acid, diamonds, and oil all follow different routes, but each navigates a web that touches various machines and may even cycle around again to its elemental form.

The tangled flow of manufactured materials from machine to machine can be seen as a networked community—an industrial ecology. Like all living systems, this interlocking human-made ecosystem tends to expand, to work around impediments, and to adapt to adversity. Seen in the right light, a robust industrial ecosystem is an extension of the natural ecosystem of the biosphere. As a splinter of wood fiber travels from tree to wood chip to newspaper and then from paper to compost to tree again, the fiber easily slips in and out of the natural and industrial spheres of a larger global megasystem. Stuff circles from the biosphere into the technosphere and back again in a grand bionic ecology of nature and artifact.

Yet, human-made industry is a weedy thing that threatens to overcome the natural sphere that ultimately supports it. The crabgrass character of industry sparks confrontations between advocates for nature and apologists of the artificial, both of whom believe only one side can prevail. However, in the last few years, a slightly romantic view that "the future of machines is biology" has penetrated science and flipped a bit of poetry into something useful. The new view claims: Both nature and industry can prevail. Employing the metaphor of organic machine systems, industrialists and (somewhat reluctantly) environmentalists can sketch out how manufacturing can repair its own messes, just as biological systems clean up after themselves. For instance, nature has no garbage problem because nothing becomes waste. An industry imitating this and other organic principles would be more compatible with the organic domain around it.

Until recently the mandate to "do as nature does" has been impossible to implement among isolated and rigid machines. But as we invest machines, factories, and materials with adaptive behavior, coevolutionary dynamics, and global connections, we can steer the manufactured environment into an industrial ecology. Doing so shifts the big picture from industry conquering nature to industry cooperating with nature.

Hardin Tibbs is a British industrial designer who picked up a sense of machines as whole systems while consulting on large engineering projects such as the NASA space station. To make a remote space station, or any other large system, utterly reliable requires steady attention to all the interacting, and at times conflicting, needs of each mechanical subsystem. Balancing several machines' opposing demands, while unifying common ones, instilled a holistic attitude in engineer Tibbs. As an avid environmentalist, Tibbs wondered why this holistic mechanical outlook—which stresses a systems approach to minimizing inefficiencies—could not be applied to industry in general as a way to solve the pollution it generated. The idea, said Tibbs, was to "take the pattern of the natural environment as a model for solving environmental problems." He and his fellow engineers were calling it "industrial ecology."

The term "industrial ecology" was a metaphor resurrected by Robert Frosch in a 1989 *Scientific American* article. Frosch, a scientist who runs GM's research laboratories and was once head of NASA, defined this fresh perspective: "In an industrial ecosystem . . . the consumption of energy and materials is optimized, waste generation is minimized, and the effluents of one process . . . serve as the raw material for another process. The industrial ecosystem would function as an analogue of biological ecosystems."

The term "industrial ecology" had been used since the 1970s as a way to think about workplace health and environmental issues, "stuff like whether you have mites living on dust particles in your factory or not," says Tibbs. Frosch and Tibbs expanded the concept of industrial ecology to include the environment formed by and among a web of machines. The goal according to Tibbs was "to model the systemic design of industry on the systemic design of the natural system" so that "we could not only improve the efficiency of industry but also find more acceptable ways of interfacing it with nature." In one daring step, engineers hijacked an age-old metaphor of machines as organisms and put the poetry to work.

One of the first ideas born out of the organic view of manufacturing was the notion of "design for disassembly." Ease of assembly has been the paramount factor in manufacturing for decades. The easier something was to assemble, the cheaper it could be made. Ease of repair and ease of disposal were almost wholly neglected. In the ecological vision, a product designed for disassembly would combine the tradeoffs of efficient disposal or repair as well as efficient assembly. The best-designed automobile, then, would not only be a joy to drive, and cheap to assemble, but would also easily break apart into common ingredients when dead. These technicians aim to invent devices that adhere better than glues or one-way fasteners, but are reversible, and materials as sound as Kevlar and molded polycarbonate, but easier to recycle.

The incentive for these inventions is imposed by requiring the manufacturer, rather than the consumer, to be responsible for disposal. It pushes the burden of waste "upstream" to the producer. Germany recently passed legislation that makes it mandatory for automobile manufacturers to design cars that dismantle easily into homogeneous parts. You can buy a new electric tea kettle featuring easy-to-dismember recyclable parts. Aluminum cans are already designed for recycling. What if everything else was? You couldn't make a radio, a running shoe, or a sofa without accounting for the destination of its dead body. You'd have to work with your ecological partners—those preying upon your machine's matter—to ensure someone took on your corpses. Every product would incorporate its engineered offal.

"I think that you can go a long way with the idea that any waste you can think of is a potential raw resource," Tibbs says. "And any material that might not have a use right now, we can eliminate upstream by design so that that material is not produced. We already know, in principle, how to make intrinsically zero-pollution processes. The only reason we aren't doing so is because we haven't decided to do it. It's a matter of volition rather than technology."

All evidence points to ecological technology being cost effective, if not shockingly profitable. Since 1975, the global conglomerate 3M has saved $500 million while reducing pollution 50 percent per unit of production. By reformulating products, modifying production processes (to use less solvents, say), or simply by recovering "pollutants," 3M has made money by applying technical innovations to its internal industrial ecology.

Tibbs told me of another example of an internal ecosystem that pays for itself: "In Massachusetts a metal refinishing plant had been discharging heavy metal solutions into the local waterways for years. And every year the environmental people were raising water-purity thresholds, until it got to the point where the plant would either have to stop what they were doing and farm out the plating to somewhere else, or install a very expensive state-of-the-art full-scale water treatment plant. Instead the refinishers did something radical—they invented a totally closed-loop system. Such a system did not exist in electroplating."

A closed-loop system constantly recycles the same materials over and over again, just as Bio2 does or a space capsule should. In practice small amounts leak in and out in industrial systems, but overall, the bulk of mass circles in a "closed loop." The Massachusetts plating company devised a way to take the tremendous amounts of water and toxic solvents demanded by the dirty process and recycle them entirely within the walls of the factory. Their innovative system, which reduced pollution output to zero, also paid for itself in two years. Tibbs says, "The water treatment plant would have cost them $½ million, whereas their novel closed-loop system cost only about $¼ million. They saved on water costs by no longer needing ½-million gallons per week. They reduced their chemical intake because they now reclaim the metals. At the same time they improved the quality of their plating product because their water filtration is so good that the reused water is cleaner than the local water they bought before."

Closed-loop manufacturing mirrors the natural closed-loop

production in living plant cells, which internally circulate the bulk of their materials during nongrowth periods. The same zero-pollution closed-loop principles in a plating factory can be designed into an industrial park or entire region. Add a global perspective and you up the scale to cover the entire planetary network of human activity. Nothing is thrown away in this grand loop because there is no "away." Eventually, all machines, factories, and human institutions will be members of the greater global bionic system that imitates biological manners.

Tibbs can already point to one ongoing prototype. Eighty miles west of Copenhagen, local Danish businesses have cultivated an embryonic industrial ecosystem. About a dozen industries cooperate in exploiting "wastes" from neighboring factories in an open-loop which is steadily "closing in" as they learn how to recycle each other's effluent. A coal-fired electric power plant supplies an oil refinery with waste heat from its steam turbines (previously released into a nearby fjord). The oil company removes polluting sulfur from gas released by the refining process which can then be burned by the power plant, saving 30,000 tons of coal per year. The removed sulfur is sold to a nearby sulfuric acid plant. The power plant also precipitates pollutants from its coal smoke in the form of calcium sulfate, which is consumed as a substitute for gypsum by a sheetrock company. Ash removed from the same smoke goes to a cement factory. Other surplus steam from the power plant warms a biotech pharmaceutical plant and 3,500 homes, as well as a seawater trout farm. Hi-nutrient sludge from both the fish farm and the pharmaceutical factory's fermentation vats are used to fertilize local farms, and perhaps someday soon, also horticulture greenhouses warmed by the power plant's waste heat.

Yet, to be realistic, no matter how cleverly manufacturing loops are closed, a tiny fraction of energy or unusable stuff will be wasted into the biosphere. The impact of this inevitable entropy *can* be absorbed by the organic sphere if the mechanical systems that generate it run at the pace and scope of natural systems. Living organisms such as water hyacinth can condense dilute impurities in water into a concentration with economic value. In '90s lingo, if industry interfaces well with nature, biological organisms can carry what minimal waste the industrial ecosystem generates.

The bugaboos in larger versions of this optimistic vision are highly variable flows of material, and decentralized, dilute concentrations of reclaimable stuff. Nature excels in dealing with variance and dilute being, while human artifacts do not. A multimillion-dollar paper recycling plant needs an unvarying stream of constant quality old paper to operate; it cannot afford to be down a day if volunteers tire of bundling their used newspapers. The usual solution, massive storage centers for recycled resources, burns up its slim profitability. Industrial ecology must grow into a networked just-in-time system that dynamically balances the flow of materials so that local overflows and shortages are shuttled around to minimize variable stocks. More net-driven "flex-factories" will be able to handle a more erratic quality of resources by running adaptable machinery or making fewer units of more different kinds of products.

Technologies of adaptation, such as distributed intelligence, flex-time accounting, niche economics, and supervised evolution, all stir up the organic in machines. Wired together into one megaloop, the world of the made slips steadily toward the world of the born.

As Tibbs studied what was needed to imitate "the world of the born" in manufacturing, he became convinced that industrial activities would become "sustainable," to use current jargon, as they become more organic. Imagine, Tibbs suggests, that we push grimy workaday industrial processes toward the character of biological processes. Instead of the high-pressure and high-temperature needs of most factories, lay out a factory operating within the everyday range of biological values. "Biological metabolism is primarily fueled by solar energy and operates at ambient temperatures and pressures," Tibbs writes in his landmark 1991 monograph *Industrial Ecology*. "If this were true of industrial metabolism, there could be significant gains in plant operating safety." Hot is fast, furious, and efficient. Cool is slow, safe, and flexible. Life is cool. Pharmaceutical companies are undergoing a revolution as bioengineered yeast cells replace toxic, solvent-intense chemicals to create medicinal drugs. While the pharmaceutical factory's hi-tech plumbing remains, genes spliced into a living yeasty soup take over as the engine. The use of bacteria to extract mineral ores from spent mine tailings—a job that in the industrial past required harsh and environmentally destructive methods—is another proven biological-scale process replacing a mechanical one.

Although life is built upon carbon, it is not powered by it. But carbon has fueled industrial development, as well as its accompanying atmospheric shock. $CO_2$ and other pollutants burn off into the air in direct proportion to the presence of complex hydrocarbons in fuel. The more carbon, the more mess. Yet the real energy gain in fuels does not come from burning the carbon component of hydrocarbons, but the hydrogen portion.

The best fuel of the ancients was wood. Expressed as the proportion of carbon to hydrogen, fuelwood is roughly 91 percent carbon. During the peak of the industrial revolution, the preferred fuel was coal, which is 50 percent carbon. Oil for the modern factory is 33 percent carbon, while natural gas, the upcoming favorite clean fuel, is 20 percent carbon. Tibbs notes that, "As the industrial system has evolved [fuels] have become increasingly hydrogen-rich. In theory at least, pure hydrogen would be the ideal 'clean fuel.' "

A future "hydrogen economy" would use sunlight to crack water into hydrogen and oxygen, and then pump the hydrogen around like natural gas, burning it for energy where needed. Such an environmentally benign carbonless energy system would ape the photon-based powerpacks in plant cells.

By pushing industrial processes toward the organic model, bionic engineers create a spectrum of ecosystem types. At one extreme are pure, natural ecosystems like an alpine meadow or a mangrove swamp. These systems can selfishly be thought to produce biomass, oxygen, foodstuffs, and thousands of fancy organic chemicals, a few of which we harvest. At the other extreme are pure, raw industrial systems, which synthesize compounds not found in nature, or not found in such large volumes. In between are a spectrum of hybrid ecosystems such as marshland sewage treatment plants (which use microbes to digest waste) or wineries (which use living yeast to make vintage brews), and soon, bioengineered processes that will use gene-spliced organisms to produce silk or vitamins or glues.

Both genetic engineering and industrial ecology promise the third category of bionic systems—part biology, part machine. We have only begun to imagine the varieties of ecotech systems that could create the things we desire.

Industry will inevitably adopt biological ways because:

- It takes less material to do the same job better. Cars, planes, houses, and, of course, computers, now consume less material than two decades ago, and give far better performance. Most of the processes that will generate our wealth in the future will shrink to biological scale and resolution, even when these processes make products as large as redwood trees. Manufacturers will perceive natural biological processes as competitive and inspirational, and this will drive manufactured processes toward biological-type solutions.
- The complexity of built things now reaches biological complexity. Nature, the master manager of complexity, offers priceless guidance in handling messy, counterintuitive webs. Artificial complex systems will be deliberately infused with organic principles simply to keep them going.
- Nature will not move, so it must be accommodated. Nature—which is larger than us and our contraptions—sets the underlying pace for industrial progress, so the artificial will have to conform to the natural in the long term.
- The natural world itself—genes and life forms—can be engineered (and patented) just like industrial systems. This trend narrows the gap between the two spheres of natural and artificial/industrial ecosystems, making it easier for industry to finance and appreciate the biological.

Anyone can see that our world is steadily paving itself over with human-made gadgets. Yet for every rapid step our society takes toward the manufactured, it is taking an equally quick step toward the biological. While electronic gizmos dazzle, they are here primarily to ferment the real revolution . . . in biology. The next century ushers in an era not of silicon—as everyone trumpets—but of biology: Mice. Viruses. Genes. Ecology. Evolution. Life.

Sort of. What the next century will really usher in is hyperbiology: Synthetic Mice. Computer Viruses. Engineered Genes. Industrial Ecology. Supervised Evolution. Artificial Life. (But they all are of one.) Silicon research is stampeding toward biology. Teams are in hot competition to design computers that not only assist the study of nature, but are natural themselves.

Note the woolly flavor of these recent technical conferences and workshops: Adaptive Computation (Santa Fe, April 1992), modeling organic flexibility into computer programs; Biocomputation (Monterey, June 1992), claiming that "natural evolution is a computational process of adaptation to an ever changing environment"; Parallel Problem Solving from Nature (Brussels, September 1992), treating nature as a supercomputer; The Fifth International Conference on Genetic Algorithms (San Diego, 1992), mimicking DNA's power of evolution; and uncountable conferences on neural networks, which focus on copying the distinctive structure of the brain's neurons as a model for learning.

Ten years from now the wowiest products in your living room, office, or garage will be based on ideas from these pioneering meetings.

Here in one paragraph is a pop-history of the world: The African savanna hatches human hunter-gatherers—raw biology; the hunter-gatherers hatch agriculture—domestication of the natural; the farmers hatch the industrial—domestication of the machine; the industrialists hatch the currently emerging postindustrial whatever. We are still figuring out what it is, but I'll call it the marriage of the born and the made.

To be precise, the flavor of the next epoch is *neo-biological* rather than bionic, because, although it may start symmetrically, biology always wins in any blending of organic and machine.

Biology always wins because the organic is not a sacred stance. It is not a holy status that living entities inherit by some mystical means. Biology is an inevitability—almost a mathematical certainty—that all complexity will drift towards. It is an omega point. In the slow mingling of the made and born, the organic is a dominant trait, while the mechanic is recessive. In the end, bio-logic always wins.

# 11

# Network Economics

John Perry Barlow's exact mission in life is hard to pin down. He owns a ranch in Pinedale, Wyoming. He once made a bid for a Republican seat in that state's Senate. He often introduces himself to boomer types as the B-string lyricist for that perennial underground cult band, the Grateful Dead. It's a role he relishes, particularly for the cognitive dissonance it serves up: A Republican Deadhead?

At any one moment Barlow may be working on getting a whaleboat launched in Sri Lanka (so environmentalists can monitor gray whale migrations), or delivering an address to an electrical engineers association on the future of privacy and freedom of speech. He is as likely to be sitting in a Japanese hot spring in Hokkaido with Japanese industrialists, brainstorming on ways to unify the Pacific Rim, as he would be soaking in a sweat lodge with the last of the space visionaries planning to settle Mars. I know Barlow from an experimental computer meeting place, the WELL, a place where no one has a body. There, he plays the role of "hippie mystic."

On the WELL, Barlow and I met and worked together years before we ever met in the flesh. This is the usual way of friendships in the information age. Barlow has about ten phone numbers, several different towns where he parks his cellular phone, and more than one electronic address. I never know where he is, but I can almost always reach him in a couple of minutes. The guy flies on planes with a laptop computer plugged into an in-flight phone. The numbers I hit to contact him might take me anywhere in the world.

I get discombobulated by this disembodiment. When I connect, I am confused if I can't picture at least what part of the globe I'm connected to. He might not mind being placeless, but I mind. When I dial what I think is him in New York City and I wind up with him over the Pacific, I feel flung.

"Barlow, where are you right now?" I demand impatiently during an intense phone call discussing some pretty hairy, nontrivial negotiations.

"Well, when you first called I was in a parking lot. Now I'm in a luggage store getting my luggage repaired."

"Gee," I said, "why don't you just get a receiver surgically wired into your brain? It'd be a lot more convenient. Free up your hands."

"That's the idea," he replies in total seriousness.

Barlow moved from the emptiness of Wyoming and is now homesteading in the vaster wilds of cyberspace, the frontier where our previous conversation technically took place. As originally envisioned by writer William Gibson, cyberspace encompasses the realm of large electronic networks which are invisibly spreading "underneath" the industrial world in a kind of virtual sprawl. In the near future, according to Gibson's science-fiction, cyberspace explorers would "jack in" to a borderless maze of electronic data banks and video-gamelike worlds. A cyberspace scout sits in a dark room and then plugs a modem directly into his brain. Thus jacked in, he cerebrally navigates the invisible world of abstracted information, as if he were racing through an infinite library. By all accounts, this version of cyberspace is already appearing in patches.

Cyberspace, as expanded by hippie mystic Barlow, is something yet broader. It includes not only the invisible matrix of databases and networks, and not only the three-dimensional games one can enter wearing computer-screen goggles, but also the entire realm of any disembodied presence and of all information in digital form. Cyberspace, says Barlow, is the place that you and a friend "are" when you are both talking on the phone.

"Nothing could be more disembodied than cyberspace. It's like having your everything amputated," Barlow once told a reporter. Cyberspace is the mall of network culture. It's that territory where the counterintuitive logic of distributed networks meets the odd behavior of human society. And it is expanding rapidly. Because of network economics, cyberspace is a resource that increases the more it is used. Barlow quips that it is "a peculiar kind of real estate which expands with development."

I BOUGHT MY FIRST computer to crunch a database of names for a mail order company I owned. But within several months of getting my first Apple II running, I hooked the machine up to a telephone and had a religious experience.

On the other side of the phone jack, an embryonic web stirred—the young Net. In that dawn I saw that the future of computers was not numbers but connections. Far more voltage crackled out of a million interconnected Apple IIs than within the most coddled million-dollar supercomputer standing alone. Roaming the Net I got a hit of network juice, and my head buzzed.

Computers, used as calculating machines, would, just as we all expected, whip up the next efficient edition of the world. But no one expected that once used as communication machines, networked computers would overturn the improved world onto an entirely different logic—the logic of the Net.

In the Me-Decades, the liberation of personal computers was just right. Personal computers were personal slaves. Loyal, bonded silicon brains, hired for cheap and at your command, even if you were only 13. It was plain as daylight that personal computers and their eventual high-powered offspring would reconfigure the world to our specifications: personal newspapers, video on demand, customized widgets. The focus was on you the individual. But in one of those quirks reality is famous for, the real power of the silicon chip lay not in its amazing ability to flip digits to think for us, but in its uncanny ability to use flipped switches to connect us. We shouldn't call them computers; we really should call them connectors.

By 1992 the fastest-growing segment of the computer industry was network technology. This reflects the light-speed rate at which every sector of business is electronically netting itself into a new shape. By 1993, both *Time* and *Newsweek* featured cover stories on the fast-approaching data superhighway that would connect television, telephones, and the Sixpack family. In a few years—no dream—you would pick up a gadget and get a "video dial-tone" which would enable you to send or receive a movie, a color photograph, an entire database, an album of music, some detailed blueprints, or a set of books—instantly—to or from anyone, anywhere, anytime.

Networking at that scale would truly revolutionize almost every business. It would alter:

- What we make
- How we make it
- How we decide what to make
- The nature of the economy we make it in.

There is hardly a single aspect of business not overhauled, either directly or indirectly, by the introduction of networking logic. Networks—not merely computers alone—enable companies to manufacture new kinds of innovative products, in faster and more flexible ways, in greater response to customers' needs, and all within a rapidly shifting environment where competitors can do the same. In response to these groundswell changes, laws and financing change, too, not to mention the incredible alterations in the economy due to global 24-hour networking of financial institutions. And not to mention the feverish cultural brew that will burst as "the Street" takes hold of this web and subverts it to its own uses.

Network logic has already shaped the products that are shaping business now. Instant cash, the product which is disgorged from ATM machines, could only be born in a network. Ditto for credit cards of any stripe. Fax machines, too. But also such things as the ubiquitous color printing in our lives. The high quality and low cost of modern four-color printing is made possible by a networked printing press which coordinates the hi-speed overlap of each color as it zips through the web of rollers. Biotech pharmaceuticals require networked intelligence to manage living soups as they flow by the barrelful from one vat to the next. Even processed snack foods are here to tempt us because the dispersed machines needed to cook them can be coordinated by a network.

Ordinary manufacturing becomes better when managed by netted intelligence. Networked equipment creates not only purer steel and glass, but its adaptive nature allows more varieties to be made with the same equipment. Small differences in composition can be maintained during manufacturing, in effect creating new kinds of precise materials where once there was only one fuzzy, imprecise material.

Networking will also inform the maintenance of products. Already, in 1993, some business equipment (Pitney Bowes's fax machines, Hewlett-Packard's minicomputers, General Electric's body scanners) can be diagnosed and repaired from a distance. By plugging a phone line into a machine, operators at the factory can peek inside its guts to see if it is working properly and often fix it if not. The technique of remote diagnostics was developed by satellite makers who had no choice but to do repairs at a distance. Now the methods arc being used to fix a fax machine, to dissect a hard disk, or to speed repair of an X-ray machine thousands of miles away. Sometimes new software can be uploaded into the machine to create a fix; at the very least, the repairman can learn beforehand what parts and tools he'll need if he visits and thus speed up the on-site repair. In essence, these networked devices become nodes of a larger distributed machine. In time all machines will be wired into a net so that they warn repairmen when they are flaking out, and so that they can receive updated intelligence and thus improve while on the job.

The Japanese perfected the technique of combining well-educated human beings and networked computer intelligence into one seamless companywide network to ensure uncompromised quality. Intense coordination of critical information in

Japanese manufacturing corporations gave the world palm-size camcorders and durable cars. While the rest of the industrialized sector frantically installs network-driven manufacturing machinery, the Japanese have moved on to the next frontier in network logic: flexible manufacturing and mass customization. For instance the National Bicycle Industrial Company in Kokubu, Japan, builds custom bicycles on an assembly line. You can order any one of 11 million variations of its models to suit your taste, at prices only 10 percent higher than mass-produced noncustomized models.

The challenge is simply stated: Extend the company's internal network outward to include all those with whom the company interacts in the marketplace. Spin a grand web to include employees, suppliers, regulators, and customers; they all become part of your company's collective being. *They* are the company.

Cases in both Japan and America where corporations have started building an extended distributed company demonstrate the immense power it releases. For example, Levi Strauss, makers of jeans for the whole world, has networked a large portion of its being. Continuous data flows from its headquarters, its 39 production plants, and its thousands of retailers into an economic superorganism. As stone-washed jeans are bought at the mall in, say, Buffalo, a message announcing those sales flies that night from the mall's cash register into Levi's net. The net consolidates the transaction with transactions from 3,500 other retail stores and within hours triggers the order for more stone-washed jeans from a factory in Belgium, or more dye from Germany, or more denim cloth from the cotton mills in North Carolina.

The same signal spurs the networked factory into action. Here bundles of cloth arrive from the mills decked in bar codes. As the stacks of cloth become pants, their bar-coded identity will be followed with hand-held laser readers, from fabric to trucker to store shelf. A reply is sent back to the mall store saying the restocking pants are on their way. And they will be, in a matter of days.

So tight is this loop of customer purchase/order materials/make, that other highly networked clothiers such as Benetton boast that they don't dye their sweaters until they are on their way out the door. When customers at the local chains start ringing up turquoise jumpers, in a few days Benetton's network will begin dyeing more jumpsuits in that color. Thus, the cash registers, not fashion mavens, choose the hues of the season. In this way, hip Benetton stays abreast of the unpredictable storms of fashion.

If you link computer-assisted design tools, and computer-assisted manufacturing, then not only can colors be nimbly manipulated but entire designs as well. A new outfit is quickly drawn up, made in low volume, distributed to stores, and then rapidly modified or multiplied if successful. The whole cycle is measured in days. Up until recently, the cycle of a far more limited choice was measured in seasons and years. Kao, a detergent and toiletry manufacturer in Japan, has developed a distribution system so tightly networked that it delivers even the smallest order within 24 hours.

Why not make cars or plastics this way? In fact, you can. A truly adaptable factory must be modular. Its tools and workflow can be quickly modified and reassembled to manufacture a different version of car or a different formula plastic. One day the assembly line is grinding out station wagons or Styrofoam, the next day jeeps or Plexiglas. Technicians call it flexible manufacturing. The assembly line adapts to fit the products needed. It's a hot field of research with immense potential. If you can alter the manufacturing process on the fly without stopping the flow, you then have the means to make stuff in batches of one.

But this flexibility demands tiptoe agility from multi-ton machines that are presently bolted to the floor. To get them to dance requires substituting a lot of mass with a lot of networked intelligence. Flexibility has to sink deep into the system to make flexible manufacturing work. The machine tools must themselves be adjustable, the schedules of material delivery must turn on a dime, the labor force must coordinate as a unit, the suppliers of packaging must be fluid, the trucking lines must be adaptable, the marketing must be in sync. That's all done with networks.

Today my factory needs 21 flatbed trucks, 73 tons of acetate resin, 2,000 kilowatts, and 576 man hours. The next day I may not need any of those. So if you are the acetate or electric company, you'll need to be as nimble as I am if we are to work together. We'll coordinate as a network, sharing information and control, decentralizing functions between us. It will be hard at times to tell who is working for whom.

Federal Express used to deliver key parts for IBM computers. Now they warehouse them too. By means of networks, Federal Express locates the just-finished part recently arrived in a FedEx warehouse from some remote overseas IBM supplier. When you order an item from an IBM catalog, FedEx brings it to you via their worldwide delivery service. An IBM employee may never touch the piece. So when the Federal Express man delivers the part to your door, who sent it, IBM or Federal Express? Schneider National, the first national trucking company to have all its trucks fully networked in real time by satellite, has some major customers who deposit their orders directly into Schneider's dispatching computers and who are billed by the same method. Who is in charge? Where does the company end and the supplier start?

Customers are being roped into the distributed company just as fast. Ubiquitous 800-numbers just about ring on the factory floor, as the feedback of users shape how and what the assembly line makes.

One can imagine the future shape of companies by stretching them until they are pure network. A company that was pure network would have the following traits: *distributed, decentralized, collaborative*, and *adaptive*.

***Distributed***—There is no single location for the business. It dwells among many places concurrently. The company might not even be headquartered in one place. Apple Computer, Inc., has numerous buildings spread thickly over two towns. Each one is a "headquarter" for a different function of the company. Even small businesses may be distributed within the same locality. Once networked, it hardly matters whether you are on the floor below, or across town.

Open Vision, based in Pleasanton, California, is an example of a rather ordinary, small software company, molded in the new pattern. "We are operating as a true distributed company," said CEO Michael Fields. Open Vision has clients and *employees* in most U.S. cities, all served on computer networks, but "most of them don't even know where Pleasanton is," Fields told the *San Francisco Chronicle*.

Yet in this stretch toward ultimate networks, companies will not break down into a network of individuals working alone. The data collected so far, as well as my own experience, says that the natural resolution of a purely distributed company coalesces into teams of 8 to 12 people working in a space together. A very large global company in the pure network form could be viewed as a system of cells of a dozen people each, including minifactories manned by a dozen people, a "headquarters" staffed with a dozen, profit centers managed by eight and suppliers run by ten people.

***Decentralized***—How can any large-scale project ever get anything done with only ten people? For most of the industrial revolution, serious wealth was made by bringing processes under central control. Bigger was more efficient. The "robber barons" of yesteryear figured out that by controlling every vital and auxiliary aspect of their industry, they could make millions. Steel companies proceeded to control the ore deposits, mine their own coal, set up their own railways, make their own equipment, house their own workers, and strive for self-containment within the borders of a gigantic company. That worked magnificently when things moved slowly.

Now, when the economy shifts daily, owning the whole chain of production is a liability. It is efficient only while the last hours of its relevancy lasts. Once that moment of power recedes, control has to be traded in for speed and nimbleness. Peripheral functions, like supplying your own energy, are quickly passed on to another company.

Even supposedly essential functions are subcontracted out. For instance, Gallo Winery no longer grows the specialized grapes required for its wines; it farms that chore out to others and focuses on brewing and marketing. A car rental company subcontracts out the repair and maintenance of its fleet, and focuses on renting. One passenger airline subcontracted its cargo space on transcontinental flights (a vitally important profit center) to an independent freight company, figuring they would manage it better and earn the airline more than it could itself.

Detroit automobile manufacturers were once famous for doing everything themselves. Now they subcontract out about half of their functions, including the rather important job of building engines. General Motors even hired PPG Industries to handle the painting of auto bodies—a critical job in terms of sales—within GM's factories. In the business magazines this pervasive decentralization by means of subcontracting is called "outsourcing."

The coordination costs for large-scale outsourcing have been reduced to bearable amounts by electronic trading of massive amounts of technical and accounting information. In short, networks make outsourcing feasible, profitable, and competitive. The jobs one company passes off to another can be passed back several times until they rest upon the shoulders of a small, tightly knit group, who will complete the job with care and efficiency. That group will most likely be a separate company, or they may be an autonomous subsidiary.

Research shows that the transactional costs needed to maintain the quality of a task as it stretches across several companies *are* higher than if the job stayed within one company. However: (1) those costs are being lowered every day with network technology such as electronic data transfers (EDI) and videoconferencing, and (2) those costs are *already* lower in terms of the immense gains in adaptability—not having to manage jobs you no longer need, and being able to start jobs you will need—that centralized companies lack.

Extending outsourcing to its logical conclusion, a 100 percent networked company would consist solely of one office of professionals linked by network technology to other independent groups. Many invisible million-dollar businesses are being run from one office with two assistants. And some don't have an office at all. The large advertising firm of Chiat/Day is working on dismantling its physical headquarters. Project team members will rent hotel conference rooms for the duration of the project, working on portable computers and call-forwarding. They'll disband and regroup when the project is done. Some of those groups might be "owned" by the office; others would be separately controlled and financed.

Let's imagine the office of the future in a hypothetical Silicon Valley *automobile* manufacturer that I'll call Upstart Car, Inc. Upstart Car intends to compete with the big three Japanese automobile giants.

Here's Upstart's blueprint: A dozen people share a room in a sleek office building in Palo Alto, California. Some finance people, four engineers, a CEO, a coordinator, a lawyer, and a marketing guy. Across town in a former warehouse, crews assemble 120-mpg, nonpolluting cars made from polychain composite materials, ceramic engines, and electronic everything else. The hi-tech plastics come from a young company with whom Upstart has formed a joint venture. The engines are purchased in Singapore; other automobile parts arrive each day in bar-coded profusion from Mexico, Utah, and Detroit. The shipping companies deal with temporary storage of parts; only what is needed that day appears at the plant. Cars, each one customer-tailored, are ordered by a network of customers and shipped the minute they are done. Molds for the car's body are rapidly shaped by computer-guided lasers, and fed designs generated by customer response and targeted marketing. A flexible line of robots assemble the cars.

Robot repair and improvement is outsourced to a robot company. Acme Plant Maintenance Service keeps the factory sheds going. Phone reception is hired out to a small outfit physically located in San Mateo. The clerical work is handled by a national agency who services all the other groups in the company. Same with computer hardware. The marketing and legal guys each oversee (of course) the marketing and legal services which Upstart also hires out. Bookkeeping is pretty much entirely computerized, but an outside accounting firm, operating from remote terminals, tends to any accounting requests. In total about 100 workers are paid directly by Upstart, and they are organized into small groups with varying benefit plans and pay schedules. As Upstart's cars soar in popularity, it grows by helping its suppliers grow, negotiating alliances, and sometimes investing in their growth.

Pretty far out, huh? It's not so farfetched. Here's how a real pioneering Silicon Valley company was launched a decade ago. James Brian Quinn writes in the March-April 1990 *Harvard Business Review*:

> Apple bought microprocessors from Synertek, other chips from Hitachi, Texas Instruments, and Motorola, video monitors from Hitachi, power supplies from Astec, and printers from Tokyo E:lectric and Qume. Similarly, Apple kept its internal service activities and investments to a minimum by outsourcing application software development to Microsoft, promotion to Regis McKenna, product styling to Frogdesign, and distribution to ITT and ComputerLand.

Businesses aren't the only ones to tap the networked benefits of outsourcing. Municipalities and government agencies are rapidly following suit. As one example out of many, the city of Chicago hired EDS, the computer outsourcing company Ross Perot founded, to handle its public parking enforcement. EDS devised a system based on hand-held computers that print out tickets and link into a database of Chicago's 25,000 parking meters to increase fine collection. After EDS outsourced this service for the city, parking tickets that were paid off jumped from 10 percent to 47 percent, raising $60 million in badly needed income.

***Collaborative***—Networking internal jobs can make so much economic sense that sometimes vital functions are outsourced to competitors, to mutual benefit. Enterprises may be collaborators on one undertaking and competitors on another, at the same time.

Many major domestic airlines in the U.S. outsource their complex reservation and ticketing procedures to their competitor American Airlines. Both MasterCard and Visa credit card companies sometimes delegate their vital work of processing customer charges and transactions to arch-competitor American Express. "Strategic Alliances" is the buzz word for corporations in the 1990s. Everyone is looking for symbiotic partners, or even symbiotic competitors.

The borders between industries, between transportation, wholesaling, retailing, communications, marketing, public relations, manufacturing, warehousing all disappear into an indefinite web. Airlines run tours, sell junk by direct mail, arrange hotel reservations, while computer companies hardly even handle computer hardware.

It may get to the point that wholly autonomous companies become rare. The metaphor for corporations is shifting from the tightly coupled, tightly bounded organism to the loosely coupled, loosely bounded ecosystem. The metaphor of IBM as an organism needs overhauling. IBM is an ecosystem.

***Adaptive***—The shift from products to service is inevitable because automation keeps lowering the price of physical reproduction. The cost of copying a disk of software or a tape of music is a fraction of the cost of the product. And as things continue to get smaller, their cost of reproduction continues to shrink because less material is involved. The cost of manufacturing a capsule of drug is a fraction of the cost it sells for.

But in pharmaceutical, computer, and gradually all hi-tech industries, the cost of research, development, stylizing, licenses, patents, copyrights, marketing and customer support—the service component—are increasingly substantial. All are information and knowledge intensive.

Even a superior product is not enough to carry a company very long these days. Things churn so fast that innovative substitutions (wires built on light instead of electrons), reverse engineering, clones, third party add-ons that make a weak product boom, and quickly shifting standards (Sony lost badly on Beta VCRs but may yet prevail with 8-mm tapes) all conspire to bypass the usual routes to dominance. To make money in the new era, follow the flow of information.

A network is a factory for information. As the value of a product is increased by the amount of knowledge invested in it, the networks that engender the knowledge increase in value. A factory-made widget once followed a linear path from design to manufacturing and delivery. Now the biography of a flexibly processed widget becomes a net, distributed over many departments in many places simultaneously, and spilling out beyond the factory, so that it is difficult to say what happens first or where it happens.

The whole net happens at once. Marketing, design, manufacturing, suppliers, buyers are all involved in the creation of the successful product. Designing a product concurrently entails having marketing, legal, and engineering teams all design the product at once, instead of sequentially as in the past.

Retail products (cans of soda, socks) have communicated their movement at the cash register to the back office since the 1970s when the UPC bar code became popular in stores. However in a full-bore network economy, the idea is to have these items communicate to the front office and customer as well by adding weak communication abilities. Manufacturing small items with active microchips instead of passive bar codes embedded into them means you now have hundreds of items with snail-minds sitting on a shelf in a discount store by the thousands. Why not turn them on? They are now smart packages. They can display their own prices, thank you, easily adjusting to sales. They can recalculate their prices if the store owner wants to sell them at a premium or if you the shopper are carrying a coupon or discount card of some sort. And a product would remember if you passed it over even after seeing the sale price, much to the interest of the store owner and manufacturer. At least you looked, boasts the product's ad agency. When shelf items acquire awareness of each other and themselves and interact with their consumers, they rapidly erupt into a different economy.

DESPITE MY SUNNY FORECAST for the network economy, there is much about it that is worrisome. These are the same concerns that accompany other large, decentralized, self-making systems:

- You can't understand them.
- You have less control.
- They don't optimize well.

As companies become disembodied into some Barlowian cyberspace, they take on the character of software. Clean, massless, quick, useful, mobile, and interesting. But also complicated and probably full of bugs no one can find.

If the companies and products of the future become more like software of today, what does that promise? Televisions that crash? Cars that freeze up suddenly? Toasters that bomb?

Large software programs are about the most complex things humans can make right now. Microsoft's new operating system had 4 million lines of code. Naturally Bill Gates claims there will be no bugs in it after the 70,000 beta-test sites are done checking it.

Is it possible for us to manufacture extremely complex things without defects (or even with merely a few defects)? Will network economics help us to create complexity without any bugs, or just complexity with bugs?

Whether or not companies become more like software themselves, it is certain that more and more of what they make depends on more complex software, so the problems of creating complexity without defects becomes essential.

And in an age of simulations, the problem of verifying the truthfulness of a simulation is the same type of problem as testing massive complex software to determine whether it is or is not flawless.

David Parnas, a Canadian computer scientist, developed a set of eight criticisms of President Reagan's "Star Wars" (SDI) initiative. He based his criteria on the inherent instabilities of extremely complex software, which is what SDI essentially was. The most interesting of Parnas's points was that there are two kinds of complex systems: continuous, and discontinuous.

When GM tests a new car on its track field, it puts the car through its paces at different speeds. It will test how it handles a sharp curve going at 50, 60, 70 mph. To no one's surprise, the car's performance varies continuously with the speed. If the car passed the curve test at 50, 60, and 70 mph, then the GM engineers know—without explicit testing—that it will also pass at all the intermediate speeds of 55 and 67 mph.

They don't have to worry that at 55 mph the car will sprout

wings or go into reverse. How it behaves at 55 will be some interpolated function of what it does at 50 and 60 mph. A car is a continuous system.

Computer software, distributed networks, and most vivisystems are discontinuous systems. In complex adaptive systems you simply can't rely on interpolated functions. You can have software that has been running reliably for years, then suddenly, at some particular set of values (63.25 mph), *kaboom!*, the system bombs, or something novel emerges.

The discontinuity was always there. All the neighboring values had been tested but this particular exact set of circumstances had not. In retrospect it is obvious why the bug caused the system to crash and perhaps even why one should have looked for it. But in systems with astronomical numbers of possibilities, it is impossible to test every case. Worse, you can't rely on sampling because the system is discontinuous.

A tester can have no confidence that unexamined values in extremely complex systems will perform continuously with examined values. Despite that hurdle there is a movement toward "zero-defect" software design. Naturally it's happening in Japan.

For small programs, zero is 0.000. For extremely large programs, zero is 0.001, and falling. That's the number of defects per thousand lines of code (KLOC), and it is just one crude measure of quality. The methods for attaining zero-defect software borrow heavily from the Japanese engineer Shigeo Shingo's pioneering work on zero-defect manufacturing. Of course, computer scientists claim, "software is different." It duplicates perfectly in production, so the only problem is making the first copy.

In network economics the major expense of new product development stems from *designing the manufacturing process* and not designing the product. The Japanese have excelled at designing and improving processes; Americans have excelled at designing and improving products. The Japanese view software as a process rather than product. And in the dawning network culture, more and more of what we make—certainly more and more of our wealth—is tangled up in symbol processing which resembles code more than corn.

Software reliability guru C. K. Cho admonished the industrialist not to think of software as a product but as a portable factory. You are selling—or giving—a factory (the program code) to others who will use it to manufacture an answer when they need one. Your problem is to make a factory that will generate zero-defect answers. The methods of making a factory that produces perfectly reliable widgets can be easily applied to creating a factory that makes perfectly reliable answers.

Ordinarily, software is constructed according to three centralized milestones. It is first designed as one big picture, then coded in detail, then finally, near the end of the project, it is tested as an interacting whole. Zero-defect quality design proceeds by thousands of distributed "inchstones," instead of a few milestones. Software is designed, coded, and tested daily, in a hundred cubicles, as each person works on it.

The zero-defect evangelists have a slogan that summarizes network economics: "Every person *within* a company has a customer." Usually that customer is the coworker you hand your work off to. And you don't hand off your work until you've done the milestone cycle in miniature—specifying, coding, and testing what you made as if you were shipping it.

When you ship your work to your customer/coworker, she immediately checks it and lets you know how you did, reporting errors back to you, which you correct. In essence, software is grown from the bottom up in a manner not unlike Rodney Brooks's subsumption architecture. Each inchstone is a small module of code that works for sure, and from which more complex layers are added and tested.

Inchstones alone won't get you zero-defect software. Underlying the zero goal is a key distinction. A defect is an error shipped. An error corrected before shipping in not a defect. Shingo says, "What we absolutely cannot prevent are errors, but we can keep those errors from generating defects." Therefore, the task of zero-defect design is to detect errors early and rectify them early.

But that much is obvious. The real progress comes from identifying the cause of the error early and then eliminating the cause early. If a worker is inserting the wrong bolt, institute a system that prevents the incorrect bolt from being inserted. To err is human; to manage error is system.

The classic Japanese invention for error prevention is a "poka-yoke" system—making things foolproof. On assembly lines, cleverly simple devices prevent mistakes. A holding tray may have a specific hole for every bolt so that if there are any bolts left the operator knows he missed one. An example of poka-yoke for software production is a spell-checker that doesn't allow the programmer to type a misspelled command or even to enter an illegal (illogical) command. Software developers have an ever widening choice of amazingly sophisticated "automatic program correction" packages that check ongoing programming to prevent typical errors.

State-of-the-art developer tools perform meta-evaluations on a program's logic—"Hey, that step doesn't make sense!" it says—eliminating logical errors at the first chance. A software industry trade magazine recently listed almost a hundred error test and removal tools for sale. The most elegant of these programs offer the creator a legitimate alternative, just as a good spell-checker does, to correct the error.

Another poka-yoke of great importance is the modularization of complex software. A 1982 study published in the *IEEE Transactions on Software Engineering* revealed how the same number of lines of code broken up into smaller subprograms would decrease the number of faults, all other things being equal. A 10,000-line program in one hunk would exhibit 317 faults; when broken into three subprograms, 10,000 lines would total a lesser 265 faults. The decrease per subdivision follows a slight linear function, so fragmenting is not a cure-all, but it is a very reliable trick.

Furthermore, below a certain threshold, a subprogram can be small enough to be absolutely free of defects. IBM's code for their IMS series was written in modules of which three-quarters were entirely defect free. That is, 300 out of 420 modules had zero defects. Over half of the faults were found in only 31 of the modules. The move toward modular software, then, is a move in the direction of reliability.

The hottest frontier right now in software design is the move to object-oriented software. Object-oriented programming (OOP) is relatively decentralized and modular software. The pieces of an OOP retain an integrity as a standalone unit; they can be combined with other OOP pieces into a decomposable hierarchy of instructions. An "object" limits the damage a bug can make. Rather than blowing up the whole program, OOP effectively isolates the function into a manageable unit so that a broken object won't disrupt the whole program; it can be swapped for a new one just like an old brake pad on a car can be swapped for a better one. Vendors can buy and sell libraries of prefabricated "objects" which other software developers can buy and reassemble into large, powerful programs very quickly, instead of writing huge new programs line by line. When it comes time to update the massive OOP, all you have to do is add upgraded or new objects.

Objects in OOP are like Lego blocks, but they also carry a wee bit of intelligence with them. An object can be similar to a file folder icon on a Macintosh screen, but one that knows it's a folder and would respond to a program's query for all file folders to list their contents. An OOP object could also be a tax form, or an employee in a firm's database, or an e-mail message. Objects know what tasks they can and can't do, and they communicate laterally with each other.

Object-oriented programs create a mild distributed intelligence in software. Like other distributed beings, it is resilient to errors, it heals faster (remove the object), and it grows by incrementally assembling working subunits.

The 31 error-filled modules mentioned earlier that were found in IBM's code beautifully illustrate one characteristic of software that can be used to achieve sigma-precision quality. Errors tend to cluster. *Zero Defect Software*, the bible of the movement says, "The next error you find is far more likely to be found in the module where eleven other errors have already been found, rather than in the module where no errors have been found." Error clustering is so prevalent in software that it is known as the cockroach rule of thumb: where there is one error seen, another twenty-three lurk unnoticed.

Here's the remedy, according to the *Zero* bible: "Do not spend money on defect-prone code, get rid of it. Coding cost is nearly irrelevant compared to the cost of repairing error-prone modules.

If a software unit exceeds an error threshold, throw it out, and have a different developer do the recoding. Discard work in progress that shows a tendency toward errors because early errors predict late errors."

As software programs mushroom in complexity, it becomes impossible to exhaustively test them at the end. Because they are discontinuous systems, they will always harbor odd corners or a fatal response triggered by a one-in-a-million combination of input that eluded detection of both systematic and sample-based testing. And while statistical sampling can tell if there are likely to be faults left, it can't locate them.

The neo-biological approach is to assemble software from working parts, while continuously testing and correcting the software as it grows. One still has the problems of unexpected "emergent behaviors" (bugs) arising from the aggregation of bugless parts. But there is hope that as long as you only need to test at the new emergent level (since the lower orders are already proven) you have a chance—and you are far ahead of where you'd be if you had to test for emergent bugs along with deeper sub-bugs.

Ted Kaehler invents new kinds of software languages for his living. He was an early pioneer of object-oriented languages, a codeveloper of SmallTalk and HyperCard. He's now working on a "direct manipulation" language for Apple Computers. When I asked him about zero-defect software at Apple he waved it off. "I think it is possible to make zero defects in production software, say if you are writing yet another database program. Anywhere you really understand what you are doing, you can do it without defects."

Ted would never get along in a Japanese software mill. He says, "A good programmer can take anything known, any regularity, and cleverly reduce it in size. In creative programming then, anything completely understood disappears. So you are left writing down what you don't know . . . So, yeah, you can make zero-defect software, but by writing a program that may be thousands of lines longer than it needs to be."

This is what nature does: it sacrifices elegance for reliability. The neural pathways in nature continue to stun scientists with how non-optimized they are. Researchers investigating the neurons in a crayfish's tail reported astonishment at how clunky and inelegant the circuit was. With a little work they could come up with a more parsimonious design. But the crayfish tail circuit, more redundant than it perhaps needed to be, was error free.

The price of zero-defect software is that it's over-engineered, overbuilt, a bit bloated, and never on the edge of the unknown where Ted and friends hang out. It trades efficiency of execution for efficiencies of production.

I asked Nobel Laureate Herbert Simon how zero-defect philosophy squared with his concept of "satisficing"—don't aim for optimization, aim for good enough. He laughed and said, "Oh, you *can* make zero-defect products. The question is, can you do it profitably? If you are interested in profits, then you need to satisfice your zero defects." There's that complexity tradeoff again.

The future of network economics is in devising reliable processes, rather than reliable products. At the same time the nature of this economy means that processes become impossible to optimize. In a distributed, semiliving world, goals can only be satisficed, and then for only a moment. A day later the landscape has changed, and another upstart is shaping the playing field.

## Characteristics of the Emerging Network Economy: Executive Summary

As I see it, a few general systemic patterns will prevail in the economy of the near future. And what economic plan would be without its executive summary? Certainly not this one. Cataloged below are some traits I believe a networked-based economy would exhibit:

• *Distributed Cores*—The boundaries of a company blur to obscurity. Tasks, even seemingly core tasks like accounting or manufacturing, are jobbed out via networks to contractors, who subcontract the tasks further. Companies, from one-person to Fortune 500, become societies of work centers distributed in ownership and geography.

• *Adaptive Technologies*—If you are not in real time, you are dead. Bar codes, laser scanners, cellular phones, 700-numbers, and satellite uplinks which are directly connected to cash registers, polling devices, and delivery trucks steer the production of goods. Heads of lettuce, as well as airline tickets, have shifting prices displayed on an LED on the grocery shelf.

• *Flex Manufacturing*—Smaller numbers of items can be produced in smaller time periods with smaller equipment. Film processing used to happen in a couple of national centers and take weeks. It's now done in a little machine on every street corner in an hour. Modular equipment, no standing inventory, and computer-aided design shrink product development cycles from years to weeks.

• *Mass Customization*—Individually customized products produced on a mass scale. Cars with weather equipment for your local neighborhood; VCRs preprogrammed to your habits. All products are manufactured to personal specifications, but at mass production prices.

• *Industrial Ecology*—Closed-loop, no-waste, zero-pollution manufacturing; products designed for disassembly; and a gradual shift to biologically compatible techniques. Increasing intolerance for transgressions against the rule of biology

• *Global Accounting*—Even small businesses become global in perspective. Unexploited, undeveloped economic "frontiers" disappear geographically. The game shifts from zero-sum, where every win means someone else's loss, to positive-sum, where the economic rewards go to those able to play the system as a unified whole. Alliances, partnerships, collaboration, even if temporary or paradoxical, become essential and the norm.

• *Coevolved Customers*—Customers are trained and educated by the company, and then the company is trained and educated by the customer. Products in a network culture become updatable franchises that coevolve in continuous improvement with customer use. Think software updates and subscriptions. Companies become clubs or user groups of coevolving customers. A company cannot be a learning company without also being a teaching company.

- *Knowledge Based*—Networked data makes any job faster, better, easier. But data is cheap, and in the large volumes on networks, a nuisance. The advantage no longer lies in "how you do a job" but in "which job do you do?" Data can't tell you that; knowledge does. Coordination of data into knowledge becomes priceless.

- *Free Bandwidth*—Connecting is free; switching is expensive. You can send anyone anything anytime; but choosing who, what, and when to send, or what and when to get is the trick. Selecting what *not* to connect to is key.

- *Increasing Returns*—Them that has, gets. Them that gives away and shares, gets. Being early counts. A network's value grows faster than the number of members added to it. A 10 percent increase in customers for a company in a nonnetworked economy may increase its revenue 10 percent. But adding 10 percent more customers to a networked company, such as a telephone company, could increase revenues by 20 percent because of the exponentially greater numbers of conversations between each member, both new and old.

- *Digital Money*—Everyday digital cash replaces batch-mode paper money. All accounts become real-time.

- *Underwire Economies*—The dark side: the informal economy booms. Creative edges and fringe areas expand, but now they are invisibly connected on encrypted networks. Distributed cores and electronic money drive economic activity underwire.

In network economics the customer can expect increased speed and choice, and *more responsibility as a customer*. The provider can expect increased decentralization of all functions and increased symbiotic relationships with customers. Finding the right customer in the chaotic web of infinite communications will be a new game.

The central act of the coming era is to connect everything to everything. All matter, big and small, will be linked into vast webs of networks at many levels. Without grand meshes there is no life, intelligence, and evolution; with networks there are all of these and more.

My friend Barlow—at least Barlow's disembodied voice—has already connected his everything to his everything. He lives and works in a true network economy. He gives away information—for free—and he is given money. The more he gives away, the more money he gets. He had something to say about the emerging network in an e-mail message to me:

> Computers—the gizmos themselves—have far less to do with techie enthusiasm than some half-understood resonance to The Great Work: hardwiring collective consciousness, creating the Planetary Mind. Teilhard de Chardin wrote about this enterprise many years ago and would be appalled by the prosaic nature of the tools we will use to bring it about. But I think there is something sweetly ironic that the ladder to his Omega Point might be built by engineers and not mystics.

The boldest scientists, technologists, economists, and philosophers of this day have taken the first steps to interconnect all things and all events into a vast complex web. As very large webs penetrate the made world, we see the first glimpses of what emerges from that net—machines that become alive, smart, and evolve—a neo-biological civilization.

There is a sense in which a global mind also emerges in a network culture. The global mind is the union of computer and nature—of telephones and human brains and more. It is a very large complexity of indeterminate shape governed by an invisible hand of its own. We humans will be unconscious of what the global mind ponders. This is not because we are not smart enough, but because the design of a mind does not allow the parts to understand the whole. The particular thoughts of the global mind—and its subsequent actions—will be out of our control and beyond our understanding. Thus network economics will breed a new spiritualism.

Our primary difficulty in comprehending the global mind of a network culture will be that it does not have a central "I" to appeal to. No headquarters, no head. That will be most exasperating and discouraging. In the past, adventurous men have sought the holy grail, or the source of the Nile, or Prester John, or the secrets of the pyramids. In the future the quest will be to find the "I am" of the global mind, the source of its coherence. Many souls will lose all they have searching for it—and many will be the theories of where the global mind's "I am" hides. But it will be a never-ending quest like the others before it.

12

# E-Money

In Tim May's eyes a digital tape is a weapon as potent and subversive as a shoulder-mounted Stinger missile. May (fortyish, trim beard, ex-physicist) holds up a $9.95 digital audio tape, or DAT. The cassette—just slightly fatter than an ordinary cassette—contains a copy of Mozart equivalent in fidelity to a conventional digital compact disc. DAT can hold text as easily as music. If the data is smartly compressed, one DAT purchased at K-Mart can hold about 10,000 books in digital form.

One DAT can also completely cloak a smaller library of information interleaved within the music. Not only can the data be securely encrypted within a digital tape, but the library's existence on the tape would be invisible even to powerful computers. In the scheme May promotes, a computer hard disk's-worth of coded information could be made to disappear inside an ordinary digital tape of Michael Jackson's *Thriller*.

The vanishing act works as follows. DAT records music in 16 binary digits, but that precision is beyond perception. The difference contained in the 16th bit of the signal is too small to be detected by the human ear. An engineer can substitute a long message—a book of diagrams, a pile of data spreadsheets (in encrypted form)—into the positions of the 16th bits of music. Anyone playing the tape would hear Michael Jackson crooning in the exact digital quality they would hear on a purchased

*Thriller* tape. Anyone examining the tape with a computer would see only digital music. Only by matching an untampered-with tape with the encrypted one bit by bit on a computer could someone detect the difference. Even then, the random-looking differences would appear to be noise acquired while duping a digital tape through an analog CD player (as is normally done). Finally, this "noise" would have to be decrypted (not likely) to prove that it was something other than noise.

"What this means," says May, "is that already it is totally hopeless to stop the flow of bits across borders. Because anyone carrying a single music cassette bought in a store could carry the entire computerized files of the stealth bomber, and it would be completely and totally imperceptible." One tape contains disco music. The other tape contains disco and the essential blueprints of a key technology.

Music isn't the only way to hide things, either. "I've done this with photos," says May. "I take a digitized photo posted on the Net, download it into Adobe Photoshop, and then strip an encrypted message into the least significant bit in each pixel. When I repost the image, it is essentially indistinguishable from the original."

The other thing May is into is wholly anonymous transactions. If one takes the encryption methods developed by military agencies and transplants them into the vast terrain of electronic networks, very powerful—and very unbreakable—technologies of anonymous dealing become possible. Two complete strangers could solicit or supply information to each other, and consummate the exchange with money, without the least chance of being traced. That's something that cannot be securely done with phones and the post office now.

It's not just spies and organized crime who are paying attention. Efficient means of authentication and verification, such as smart cards, tamper-proof networks, and micro-size encryption chips, are driving the cost of ciphers down to the consumer level. Encryption is now affordable for the everyman.

The upshot of all this, Tim believes, is the end of corporations in their current form and the beginning of more sophisticated, untaxed black markets. Tim calls this movement Crypto

Anarchy. "I have to tell you I think there is a coming war between two forces," Tim May confides to me. "One force wants full disclosure, an end to secret dealings. That's the government going after pot smokers and controversial bulletin boards. The other force wants privacy and civil liberties. In this war, encryption wins. Unless the government is successful in banning encryption, which it won't be, encryption always wins."

A couple of years ago May wrote a manifesto to alert the world to the advent of widespread encryption. In this electronic broadside published on the Net, he warned of the coming "specter of crypto anarchy":

> . . . The State will of course try to slow or halt the spread of this technology, citing national security concerns, use of the technology by drug dealers and tax evaders, and fears of societal disintegration. Many of these concerns will be valid; crypto anarchy will allow national secrets to be traded freely and will allow illicit and stolen materials to be traded. An anonymous computerized market will even make possible abhorrent markets for assassinations and extortion. Various criminal and foreign elements will be active users of CryptoNet. But this will not halt the spread of crypto anarchy.
>
> Just as the technology of printing altered and reduced the power of medieval guilds and the social power structure, so too will cryptologic methods fundamentally alter the nature of corporations and of government interference in economic transactions. Combined with emerging information markets, crypto anarchy will create a liquid market for any and all material which can be put into words and pictures. And just as a seemingly minor invention like barbed wire made possible the fencing-off of vast ranches and farms, thus altering forever the concepts of land and property rights in the frontier West, so too will the seemingly minor discovery out of an arcane branch of mathematics come to be the wire clippers which dismantle the barbed wire around intellectual property.

The manifesto was signed:

> Timothy C. May, Crypto Anarchy: encryption, digital money, anonymous networks, digital pseudonyms, zero knowledge, reputations, information markets, black markets, collapse of government.

I asked Tim May, a retired Intel physicist, to explain the connection between encryption and the collapse of society as we know it. May explained, "Medieval guilds would monopolize information. When someone tried to make leather or silver outside the guilds, the King's men came in and pounded on them because the guild paid a levy to the King. What broke the medieval guilds was printing; someone could publish a treatise on how to tan leather. In the age of printing, corporations arose to monopolize certain expertise like gunsmithing, or making steel. Now encryption will cause the erosion of the current corporate monopoly on expertise and proprietary knowledge. Corporations won't be able to keep secrets because of how easy it will be to sell information on the nets."

The reason crypto anarchy hasn't broken out yet, according to May, is that the military has a monopoly on the key knowledge of encryption—just as the Church once tried to control printing. With few exceptions, encryption technology has been invented by and for the world's military organizations. To say that the military is secretive about this technology would be an understatement. Very little developed by the U.S. National Security Agency (NSA)—whose mandate it is to develop crypto systems—has ever trickled down for civilian use, unlike technologies spun off from the rest of the military/industrial alliance.

But who needs encryption, anyway? Only people with something to hide, perhaps. Spies, criminals, and malcontents. People whose appetite for encryption may be thwarted righteously, effectively, and harshly.

The ground shifted two decades ago when the information age arrived, and intelligence became the chief asset of corporations. Intelligence was no longer the monopoly of the Central Intelligence Agency, but the subject of seminars for CEOs. Spying meant corporate spying. Illicit transfer of corporate know-how, rather than military plans, became the treasonous information the state had to worry about.

In addition, within the last decade, computers became fast and cheap; enciphering no longer demanded supercomputers and the superbudgets needed to run them. A generic brand PC picked up at a garage sale could handle the massive computations that decent encryption schemes consumed. For small companies running their entire business on PCs, encryption was a tool they wanted on their hard disks.

And now, within the last few years, a thousand electronic networks have blossomed into one highly decentralized network of networks. A network is a distributed thing without a center of control, and with few clear boundaries. How do you secure something without boundaries? Certain types of encryption, it turns out, are an ideal way to bring security to a decentralized system while keeping the system flexible. Rather than trying to seal out trouble with a rigid wall of security, networks can tolerate all kinds of crap if a large portion of its members use peer-to-peer encryption.

Suddenly, encryption has become incredibly useful to ordinary people who have "nothing to hide" but their privacy. Peer-to-peer encryption, sown into the Net, linked with electronic payments, tied into everyday business deals, becomes just another business tool like fax machines or credit cards.

Just as suddenly, tax-paying citizens—whose dollars funded the military ownership of this technology—want the technology back.

But the government (at least the U.S. government) may not give encryption back to the people for a number of antiquated reasons. So, in the summer of 1992, a loose federation of creative math hackers, civil libertarians, free-market advocates, genius programmers, renegade cryptologists, and sundry other frontier folk, began creating, assembling, or appropriating encryption technology to plug into the Net. They called themselves "cypherpunks."

On a couple of Saturdays in the fall of 1992, I joined Tim May and about 15 other crypto-rebels for their monthly cypherpunk meeting held near Palo Alto, California. The group meets in a typically nondescript office complex full of small hi-tech start-up companies. It could be anywhere in Silicon Valley. The room has corporate gray carpeting and a conference table. The moderator for this meeting, Eric Hughes, tries to quiet the cacophony of loud, opinionated voices. Hughes, with sandy hair halfway down his back, grabs a marker and scribbles the agenda on a whiteboard. The items he writes down echo Tim May's digital card: reputations, PGP encryption, anonymous re-mailer update, and the Diffie-Hellmann key exchange paper.

After a bit of gossip the group gets down to business. It's class time. One member, Dean Tribble, stands up front to report on his research on digital reputations. If you are trying to do business with someone you know only as a name introducing some e-mail, how can you be sure they are legit? Tribble suggests that you can buy a reputation from a "trust escrow"—a company similar to a title or bond company that would guarantee someone for a fee. He explains the lesson from game theory concerning iterated negotiation games, like the Prisoner's Dilemma; how payoffs shift when playing the game over and over instead of just once, and how important reputations become in iterated relationships. The potential problems of buying and selling reputations online are chewed on, and suggestions of new directions for research are made, before Tribble sits down and another member stands to give a brief talk. Round the table it goes.

Arthur Abraham, dressed in heavy studded black leather, reviews a recent technical paper on encryption. Abraham flicks on an overhead projector, whips out some transparencies painted with equations, and walks the group through the mathematical proof. It is clear that the math is not easy for most. Sitting around the table are programmers (many self-taught), engineers, consultants—all very smart—but only a single member is equipped with a background in mathematics. "What do you mean by that?" questions one quiet fellow as Abraham talks. "Oh, I see, you forgot the modulus," chimes in another guy. "Is that '*a* to the *x*' or '*a* to the *y*'?" The amateur crypto-hackers challenge each statement, asking for clarification, mulling it over until each understands. The hacker mind, the programmer's drive to whittle things down to an elegant minimum, to seek short cuts, confronts the academic stance of the paper. Pointing to a large hunk of one equation, Dean asks, "Why not just scrap all this?" A voice from back: "That's a great question, and I think I know why not." So the voice explains. Dean nods. Arthur looks around to be sure everyone got it. Then he goes on to the next line in the paper; those who understand help out those who don't. Soon the room is full of people saying, "Oh, that means you can serve this up on a network configuration! Hey, cool!" And another tool for distributed computing is born; another component is transferred from the shroud of military secrecy to the open web of the Net; and another brick is set into the foundation of network culture.

The main thrust of the group's efforts takes place in the virtual online space of the cypherpunk electronic mailing list. A growing crowd of cryptohip folks from around the world interact daily via an Internet "mailing list." Here they pass around code-in-progress as they attempt to implement ideas on the cheap (such as digital signatures), or discuss the ethical and political implications of what they are doing. Some anonymous subset of them has launched the Information Liberation Front. The ILF locates scholarly papers on cryptology appearing in very expensive (and very hard-to-find) journals, scans them in by computer, and "liberates" them from their copyright restrictions by posting the articles anonymously to the Net.

Posting anything anonymously to the Net is quite hard: the nature of the Net is to track everything infallibly, and to duplicate items promiscuously. It is theoretically trivial to monitor transmission nodes in order to backtrack a message to its source. In such a climate of potential omniscience, the crypto-rebels yearn for true anonymity.

I confess my misgivings about the potential market for anonymity to Tim: "Seems like the perfect thing for ransom notes, extortion threats, bribes, blackmail, insider trading, and terrorism." "Well," Tim answers, "what about selling information that isn't viewed as legal, say about pot growing, do-it-yourself abortion, cryonics, or even peddling alternative medical information without a license? What about the anonymity wanted for whistleblowers, confessionals, and dating personals?"

Digital anonymity is needed, the crypto-rebels feel, because anonymity is as important a civil tool as authentic identification is. Pretty good anonymity is offered by the post office; you don't need to give a return address and the post office doesn't verify it if you do. Telephones (without caller ID) and telegrams are likewise anonymous to a rough degree. And everyone has a right (upheld by the Supreme Court) to distribute anonymous handbills and pamphlets. Anonymity stirs the most fervor among those who spend hours each day in networked communications. Ted Kaehler, a programmer at Apple Computer, believes that "our society is in the midst of a privacy crisis." He sees encryption as an extension of such all-American institutions as the Post Office: "We have always valued the privacy of the mails. Now for the first time, we don't have to trust in it; we can enforce it." John Gilmore, a crypto-freak who sits on the board of the Electronic Frontier Foundation, says, "We clearly have a societal need for anonymity in our basic communications media."

A pretty good society needs more than just anonymity. An online civilization requires online anonymity, online identification, online authentication, online reputations, online trust holders, online signatures, online privacy, and online access. All are essential ingredients of any open society. The cypherpunk's agenda is to build the tools that provide digital equivalents to the interpersonal conventions we have in face-to-face society, and hand them out for free. By the time they are done, the cypherpunks hope to have given away free digital signatures, as well as the opportunity for online anonymity.

To create digital anonymity, the cypherpunks have developed about 15 prototype versions of an anonymous re-mailer that would, when fully implemented, make it impossible to determine the source of an e-mail message, even under intensive monitoring of communication lines. One stage of the re-mailer works today. When you use it to mail to Alice, she gets a message from you that says it is from "nobody." Unraveling where it came from is trivial for any computer capable of monitoring the entire network—a feat few can afford. But to be mathematically untraceable, the re-mailers have to work in a relay of at least two (more is better)—one re-mailer handing off a message to the next re-mailer, diluting information about its source to nothing as it is passed along.

Eric Hughes sees a role for digital pseudonymity—your identity is known by some but not by others. When cloaked pseudonymously "you could join a collective to purchase some information and decrease your actual cost by orders of magnitude—that is, until it is almost free." A digital co-op could form a private online library and collectively purchase digital movies, albums, software, and expensive newsletters, which they would "lend" to each other over the net. The vendor selling the information would have absolutely no way of determining whether he was selling to one person or 500. Hughes sees these kinds of arrangements peppering an information-rich society as "increasing the margins where the poor can survive."

"One thing for sure," Tim says, "long-term, this stuff nukes tax collection." I venture the rather lame observation that this may be one reason the government isn't handing the technology back. I also offer the speculation that an escalating arms race with a digital IRS might evolve. For every new avenue the digital underground invents to disguise transactions, the digital IRS will counter with a surveillance method. Tim pooh-poohs the notion. "Without a doubt, this stuff is unbreakable. Encryption *always* wins."

And this is scary because pervasive encryption removes economic activity—one driving force of our society—from any hope of central control. Encryption breeds out-of-controllness.

ENCRYPTION ALWAYS WINS because it follows the logic of the Net. A given public-key encryption key can eventually be cracked by a supercomputer working on the problem long enough. Those who have codes they don't want cracked try to stay ahead of the supercomputers by increasing the length of their keys (the longer a key, the harder it is to crack)—but at the cost of making the safeguard more unwieldy and slow to use. However, *any* code can be deciphered given enough time or money. As Eric Hughes often reminds fellow cypherpunks, "Encryption is economics. Encryption is always possible, just expensive." It took Adi Shamir a year to break a 120-digit key using a network of distributed Sun workstations working part-time. A person could use a key so long that no supercomputer could crack it for the foreseeable future, but it would be awkward to use in daily life.

A building-full of NSA's specially hot-rodded supercomputers might take a day to crack a 140-digit code today. But that is a full day of big iron to open just one lousy key!

Cypherpunks intend to level the playing field against centralized computer resources with the Fax Effect. If you have the only fax machine in the world it is worth nothing. But for every other fax installed in the world, your fax machine increases in value. In fact, the more faxes in the world, the more valuable everybody's fax becomes. This is the logic of the Net, also known as the law of increasing returns. It goes contrary to classical economic theories of wealth based on equilibratory tradeoff. These state that you can't get something from nothing. The truth is, you can. (Only now are a few radical economics professors formalizing this notion.) Hackers, cypherpunks, and many hi-tech entrepreneurs already know that. In network economics, more brings more. This is why giving things away so often works, and why the cypherpunks want to pass out their tools gratis. It has less to do with charity than with the clear intuition that network economics reward the more and not the less—and you can seed the "more" at the start by giving the tools away. (The cypherpunks also talk about using the economics of the Net for the reverse side of encryption: to crack codes. They could assemble a people's supercomputer by networking together a million Macintoshes, each one computing a coordinated little part of a huge, distributed decryption program. In theory, such a decentralized parallel computer would in sum be the most powerful computer we can now imagine—far greater than the centralized NSA's.)

The idea of choking Big Brother with a deluge of petty, heavily encrypted messages so tickles the imagination of crypto-rebels that one of them came up with a freeware version of a highly regarded public-key encryption scheme. The software is called PGP, for Pretty Good Privacy. The code has been passed out on the nets for free and made available on disks. In certain parts of the Net it is quite common to see messages encrypted with PGP, with a note that the sender's public-key is "available upon request."

PGP is not the only encryption freeware. On the Net,

cypherpunks can grab RIPEM, an application for privacy-enhanced mail. Both PGP and RIPEM are based on RSA, a patented implementation of encryption algorithms. But while RIPEM is distributed as public domain software by the RSA company itself, Pretty Good Privacy software is home-brew code concocted by a crypto-rebel named Philip Zimmermann. Because Pretty Good Privacy uses RSA's patented math, it's outlaw-ware.

RSA was developed at MIT—partly with federal funds—but was later licensed to the academic researchers who invented it. The researchers published their crypto-methods before they filed for patents out of fear that the NSA would hold up the patents or even prevent the civilian use of their system. In the U.S., inventors have a year after publication to file patents. But the rest of the world requires patents before publication, so RSA could secure only U.S. patents on its system. PGP's use of RSA's patented mathematics is legitimate overseas. But PGP is commonly exchanged in the no-place of the Net (what country's jurisdiction prevails in cyberspace?) where the law on intellectual property is still a bit murky and close to the beginnings of crypto anarchy. Pretty Good Privacy deals with this legal tar baby by notifying its American users that it is their responsibility to secure from RSA a license for use of PGP's underlying algorithm. (Sure. Right.)

Zimmermann claims he released the quasi-legal PGP into the world because he was concerned that the government would reclaim all public-key encryption technology, including RSA's. RSA can't stop distribution of existing versions of PGP because once something goes onto the Net, it never comes back. But it's hard for RSA to argue damages. Both the outlawed PGP and the officially sanctioned RIPEM infect the Net to produce the Fax Effect. PGP encourages consumer use of encryption—the more use, the better for everyone in the business. Pretty Good Privacy is freeware; like most freeware, its users will sooner or later graduate to commercially supported stuff. Only RSA offers the license for that at the moment. Economically, what could be better for a patent holder than to have a million people use the buddy system to teach themselves about the intricacies and virtues of your product (as pirated and distributed by others), and then wait in line to buy your stuff when they want the best?

The Fax Effect, the rule of freeware upgrade, and the power of distributed intelligence are all part of an emerging network economics. Politics in a network economy will also definitely require the kind of tools the cypherpunks are playing with. Glenn Tenney, chairman of the annual Hackers' Conference, ran for public office in California last year using the computer networks for campaigning, and came away with a realistic grasp of how they will shape politics. He notes that digital techniques for establishing trust are needed for electronic democracy. He writes online, "Imagine if a Senator responds to some e-mail, but someone alters the response and then sends it on to the *NY Times*? Authentication, digital signatures, etc., are essential for protection of all sides." Encryption and digital signatures are techniques to expand the dynamics of trust into a new territory. Encryption cultivates a "web of trust," says Phil Zimmermann, the very web that is the heart of any society or human network. The short form of the cypherpunk's obsession with encryption can be summarized as: Pretty good privacy means pretty good society.

One of the consequences of network economics, as facilitated by ciphers and digital technology, is the transformation of what we mean by pretty good privacy. Networks shift privacy from the realm of morals to the marketplace; privacy becomes a commodity.

A telephone directory has value because of the energy it saves a caller in finding a particular phone number. When telephones were new, having an individual number to list in a directory was valuable to the lister and to all other telephone users. But today, in a world full of easily obtained telephone numbers, an unlisted phone number is more valuable to the unlisted (who pay more) and to the phone company (who charge more). Privacy is a commodity to be priced and sold.

Most privacy transactions will soon take place in the marketplace rather than in government offices because a centralized government is handicapped in a distributed, open-weave network, and can no longer guarantee how things are connected or

. connected. Hundreds of privacy vendors will sell bits of privacy at market rates. You hire Little Brother, Inc., to demand maximum payment from junk mail and direct marketers when you sell your name, and to monitor uses of that information as it tends to escape into the Net. On your behalf, Little Brother, Inc., negotiates with other privacy vendors for hired services such as personal encrypters, absolutely unlisted numbers, bozo filters (to hide the messages from known "bozos"), stranger ID screeners (such as caller ID on phones that only accept calls from certain numbers), and hired mechanical agents (called network "know-bots" to trace addresses, and counter-knowbots that unravel traces of your own activities.

Privacy is a type of information that has its polarity reversed; I imagine it as anti-information. The removal of a bit of information from a system can be seen as the reproduction of a corresponding bit of anti-information. In a world flooded with information ceaselessly replicating itself to the edges of the Net, the absence or vaporization of a bit of information becomes very valuable, especially if that absence can be maintained. In a world where everything is connected to everything—where connection and information and knowledge are dirt cheap—then disconnection and anti-information and no-knowledge become expensive. When bandwidth becomes free and entire gigabytes of information are swapped around the clock, what you don't want to communicate becomes the most difficult chore. Encryption systems and their ilk are technologies of disconnection. They somewhat tame the network's innate tendency to connect and inform without discrimination.

WE MANAGE THE DISCONNECTION of domestic utilities, such as water or electricity, through metering. But metering is neither obvious nor easy. Thomas Edison's dazzling electrical gizmos were of little use to anyone until people had easy access to electricity in their factories and homes. So at the peak of his career Edison diverted his attention away from designing electrical devices to focus on the electrical delivery network itself. At first, very little was settled about how electricity should be created (DC or AC?), carried, or billed. For billing, Edison favored the approach that most information providers today favor: charge a flat fee. Readers pay the same for a newspaper no matter how much of it they read. Ditto for cable TV, books and computer software. All are priced flat for all you can use.

Edison pushed a flat fee for electricity—a fixed amount if you are connected, nothing if you aren't—because he felt that the costs of accounting for differential usage would exceed the cost of variances in electricity usage. But mostly Edison was stymied about how to meter electricity. For the first six months of his General Electric Lighting Company in New York City, customers paid a flat fee. To Edison's chagrin, that didn't work out economically. Edison was forced to come up with a stop-gap solution. His remedy, an electrolytic meter, was erratic and impractical. It froze in winter, it sometimes ran backwards, and customers couldn't read it (nor did they trust the company's meter readers). It wasn't until a decade after municipal electrical networks were up and running that another inventor came up with a reliable watt-hour meter. Now we can hardly imagine buying electricity any other way.

A hundred years later the information industry still lacks an information meter. George Gilder, hi-tech gadfly, puts the problem this way: "Rather than having to pay for the whole reservoir every time you are thirsty, what you want is to only pay for a glass of water."

Indeed, why buy an ocean of information when all you want is a drink? No reason at all, if you have an information meter. Entrepreneur Peter Sprague believes he has just invented one. "We use encryption to force the metering of information," says Sprague. His spigot is a microchip that doles out small bits of information from a huge pile of encrypted data. Instead of selling a CD-ROM crammed with a hundred thousand pages of legal documents for $2,000, Sprague invented a ciphering device that would dispense the documents off the CD-ROM at $1 per page. A user only pays for what she uses and can use only what she pays for.

Sprague's way of selling information per page is to make each page unreadable until decrypted. Working from a catalog of contents, a user selects a range of information to browse. She reads the abstracts or summaries and is charged a minuscule amount. Then she selects a full text, which is decrypted by her dispenser. Each act of decryption rings up a small charge (maybe 50 cents). The charge is tallied by a metering chip in her dispenser that deducts the amount from a prepaid account (also stored on the metering chip), much as a postage meter deducts credit while dispensing postage tapes. When the CD-ROM credit runs out, she calls a central office, which replenishes her account via an encrypted message sent on a modem line running into her computer's metering chip. Her dispenser now has $300 credit to spend on information by the page, by the paragraph, or by the stock price, depending on how fine the vendor is cutting it.

What Sprague's encryption metering device does is *decouple* information's fabulous ease in being copied from its owner's need to have it selectively disconnected. It lets information flow freely and ubiquitously—like water through a town's plumbing—by metering it out in usable chunks. Metering converts information into a utility.

The cypherpunks note, quite correctly, that this will not stop hackers from siphoning off free information. The Videocipher encryption system, used to meter satellite-delivered TV programs such as HBO and Showtime, was compromised within weeks of its introduction. Despite claims by the meter's manufacturer that the encrypto-metering chip was unhackable, big moneymaking scams capitalized on hacks around the codes. (The scams were set up on Indian reservations—but that's a whole 'nother story.) Pirates would find a descrambler box with a valid subscription—in a hotel room, for instance—and then clone the identity into other chips. A consumer would send their box to the reservation for "repairs" and it would come back with a new chip cloned with the identity of the hotel box. The broadcasting system couldn't perceive clones in the audience. In short, the system was hacked not by cracking the code but by subverting places where the code tied into the other parts of the system.

No system is hack-proof. But disruptions of an encrypted

system require deliberate creative energy. Information meters can't stop thievery or hacking, but meters can counteract the effects of lazy mooching and the natural human desire to share. The Videocipher satellite TV system eliminates user piracy *on a mass scale*—the type of piracy that plagued the satellite TV outback before scrambling and that still plagues the lands of software and photocopying. Encryption makes pirating a chore and not something that any slouch with a blank disk can do. Satellite encryption works overall because encryption always wins.

Peter Sprague's crypto-meter permits Alice to make as many copies of the encrypted CD-ROMs as she likes, since she pays for only what she uses. Crypto-metering, in essence, disengages the process of payment from the process of duplication.

Using encryption to force the metering of information works because it does not constrain information's desire to reproduce. All things being equal, a bit of information will replicate through an available network until it fills that network. With an animate drive, every fact naturally proliferates as many times as possible. The more fit—the more interesting or useful—a fact is, the wider it spreads. A pretty metaphor compares the spread of genes through a population with the similar spread of ideas, or memes, in a population. Both genes and memes depend on a network of replicating machines—cells or brains or computer terminals. A network in this general sense is a swarm of flexibly interconnected nodes each of which can copy (either exactly or with variation) a message taken from another node. A population of butterflies and a flurry of e-mail messages have the same mandate: replicate or die. Information wants to be copied.

Our digital society has built a supernetwork of copiers out of hundreds of millions of personal faxes, library photocopiers, and desktop hard disks. It is as if our information society is one huge aggregate copying machine. But we won't let this supermachine copy. Much to everyone's surprise, information created in one corner finds its way into all the other corners rather quickly. Because our previous economy was built upon scarcity of goods, we have so far fought the natural fecundity of information by trying to control every act of replication as it occurs. We take a massively parallel copy machine and try to stifle most acts of reproduction. As in other puritanical regimes, this doesn't work. Information wants to be copied.

"Free the bits!" shouts Tim May. This sense of the word "free" shifts Stewart Brand's oft-quoted maxim, "Information wants to be free"—as in "without cost"—to the more subtle "without chains or imprisonment." Information wants to be free to wander and reproduce. Success, in a networked world of decentralized nodes, belongs to those plans that do not resist either the replication or roaming urges of information.

Sprague's encrypted meter capitalizes on the distinction between pay and copy. "It is easy to make software count how many times it has been invoked, but hard to make it count how many times it has been copied," says software architect Brad Cox. In a message broadcast on the Internet, Cox writes:

> Software objects differ from tangible objects in being fundamentally unable to monitor their copying but trivially able to monitor their use . . . So why not build an information age market economy around this difference between manufacturing-age and information -age goods? If revenue collection were based on monitoring the use of software inside a computer, vendors could dispense with copy protection altogether.

Cox is a software developer specializing in object-oriented programming. In addition to the previously mentioned virtue of reduced bugs which OOP delivers, it offers two other magnificent improvements over conventional software. First, OOP provides the user with applications that are more fluid, more interoperable with various tasks—sort of like a house with movable "object" furniture instead of a house saddled with built-in furniture. Second, OOP provides software developers the ability to "reuse" modules of software, whether they wrote the modules themselves or purchased them from someone else. To build a database, an OOP designer like Cox takes a sort routine, a field manager, a form generator, an icon handler, etc., and assembles the program instead of rewriting a working whole from scratch. Cox developed a set of cool OOP objects that he sold to Steve Jobs to use in his Next machine, but selling small bits of modular code as a regular business has been slow. It is similar to trying to peddle limericks one by one. To recoup the great cost of writing an individual object by selling it outright would garner too few sales, but selling it by copy is too hard to monitor or control. But if objects could generate revenue each time a user activated one, then an author could make a living creating them.

While contemplating the possible market for OOP objects that were sold on a "per use" plan, Cox uncovered the natural grain in networked intelligence: Let the copies flow, and pay per use. He says, "The premise is that copy protection is exactly the wrong idea for intangible, easily copied goods such as software. You want information-age goods to be freely distributed and freely acquired via whatever distribution means you want. You are positively encouraged to download software from networks, give copies to your friends, or send it as junk mail to people you've never met. Broadcast my software from satellites. Please!"

Cox adds (in echo of Peter Sprague, although surprisingly the two are unfamiliar with each other's work), "This generosity is possible because the software is actually 'meterware.' It has strings attached that make revenue collection independent of how the software was distributed."

"The approach is called superdistribution," Cox says, using a term given by Japanese researchers to a similar method they devised to track the flow of software through a network. Cox: "Like superconductivity, it lets information flow freely, without resistance from copy protection or piracy."

The model is the successful balance of copyright and use rights worked out by the music and radio industries. Musicians earn money not only by selling customers a copy of their work but by selling broadcast stations a "use" of their music. The copies are supplied free, sent to radio stations in a great unmonitored flood by the musicians' agents. The stations sort through this tide of free music, paying royalties only for the music they *broadcast*, as metered (statistically) by two agencies representing musicians, ASCAP and BMI.

JEIDA, a Japanese consortium of computer manufacturers, developed a chip and a protocol that allows each Macintosh on a network to freely replicate software while metering use rights. According to Ryoichi Mori, the head of JEIDA, "Each computer is thought of as a station that broadcasts, not the software itself, but the use of the software, to an audience of a single 'listener.'" Each time your Mac "plays" a piece of software or a software component from among thousands freely available, it triggers a royalty. Commercial radio and TV provide an "existence proof" of a working superdistribution system in which the copies are disseminated free and the stations only pay for what they use. Musicians would be quite happy if one radio station made copies of their tapes and distributed them to other stations ("Free the bits!") because it increases the likelihood of some station using their music.

JEIDA envisions software percolating through large computer networks unencumbered by restrictions on copying or mobility. Like Cox, Sprague, and the cypherpunks, JEIDA counts on public-key encryption to keep these counts private and untampered as they are transmitted to the credit center. Peter Sprague says plainly, "Encrypted metering is an ASCAP for intellectual property."

Cox's electronically disseminated pamphlet on superdistribution sums up the virtues very nicely:

> Whereas software's ease of replication is a liability today, superdistribution makes it an asset. Whereas software vendors must spend heavily to overcome software's invisibility, superdistribution thrusts software out into the world to serve as its own advertisement.

A hoary ogre known as the Pay-Per-View Problem haunts the information economy. In the past this monster ate billions of dollars in failed corporate attempts to sell movies, databases, or music recordings on a per view or per use basis. The ogre still lives. The problem is, people are reluctant to pay in advance for information they haven't seen because of their hunch that they might not find it useful. They are equally unwilling to pay after they have seen it because their hunch usually proves correct: they could have lived without it. Can you imagine being asked to pay after you've seen a movie? Medical knowledge is the only type of information that can be easily sold sight unseen because the buyers believe they can't live without it.

The ogre is usually slain with sampling. Moviegoers persuaded to pay beforehand by lapel-grabbing trailers. Softwa is loaned among friends for trial; books and magazines are browsed in the bookstore.

The other way to slay the problem is by lowering the price of admission. Newspapers are cheap; we pay before looking. The ingenious thing about information metering is that it delivers two solutions: it provides a spigot to record how much data is used, and it provides a spigot that can be turned down to a cheap trickle. Encryption-metering chops big expensive data hunks into small inexpensive doses of data. People will readily pay for bits of cheap information before viewing, particularly if the payment invisibly deducts itself from an account.

The fine granularity of information-metering gets Peter Sprague excited. When asked for an example of how fine it could get, he volunteers one so fast it's obvious that he has been giving it some thought: "Say you want to write obscene limericks from your house in Telluride, Colorado. If you could write one obscene limerick a day, we can probably find 10,000 people in the world who want to pay 10 cents a day to get it. We'll collect $365,000 per year and pay you $120,000, and then you can ski for the rest of your life." In no other kind of marketplace would one measly limerick, no matter how bawdy and clever, be worth selling on its own. Maybe a book of them—an ocean of limericks—but not one. Yet in an electronic marketplace, a single limerick—the information equivalent of a stick of gum—is worth producing and offering for sale.

Sprague ticks off a list of other fine-grained items that might be traded in such a marketplace. He catalogs what he'd pay for right now: "I want the weather in Prague for 25 cents per month, I want my stocks updated for 50 cents a stock, I want the *Dines Letter* for $12 a week, I want the congestion report from O'Hare Airport updated continuously because I'm always getting stuck in Chicago, so I'll pay a buck per month for that, and I want 'Hagar the Horrible' cartoon for a nickel a day." Each of these products is currently either given away scattershot or peddled in the aggregate very expensively. Sprague's electronically mediated marketplace would "unbundle" the data and deliver a narrowly selected piece of information to your desktop or mobile palmtop for a reasonable price. Encryption would meter it out, preventing you from filching other tiny bits of data that would hardly be worth protecting (or selling) in other ways. In essence, the ocean of information flows through you, but you only pay for what you drink.

At the moment, this particular technology of disconnection exists as a $95 circuit board that can slide into a personal computer and plug into a phone line. To encourage established computer manufacturers such as Hewlett-Packard to hardwire a similar board into units coming off their assembly line, Sprague's company, Waves, Inc., offers manufacturers a percentage of the revenue the encryption system generates. Their first market is lawyers, "because," he says, "lawyers spend $400 a month on information searches." Sprague's next step is to compress the encrypto-metering circuits and the modem down into a single $20 microchip that can be tucked into beepers, video recorders, phones, radios, and anything else that dispenses information. Ordinarily, this vision might be dismissed as the pipe dream of a starry-eyed junior inventor, but Peter Sprague is chairman and founder of National Semiconductor, one of the major semiconductor manufacturers in the world. He is sort of a Henry Ford of silicon chips. A cypherpunk, not. If anyone knows how to squeeze a revolutionary economy onto the head of a pin, it might be him.

THIS ANTICIPATED INFORMATION ECONOMY and network culture still lacks one vital component—an ingredient that, once again, is enabled by encryption, and a key element that, once again, only long-haired crypto-rebels are experimenting with: electronic cash.

We already have electronic money. It flows daily in great invisible rivers from bank vault to bank vault, from broker to broker, from country to country, from your employer to your bank account. One institution alone, the Clearing House Interbank Payment System, currently moves an average of a trillion dollars (a million millions) each day via wire and satellite.

But that river of numbers is *institutional* electronic money, as remote from electronic cash as mainframes are from PCs. When pocket cash goes digital—demassified into data in the same transformation that institutional money underwent—we'll experience the deepest consequences of an information economy. Just as computing machines did not reorganize society until individuals plugged into them outside of institutions, the full effects of an electronic economy will have to wait until everyday petty cash (and check) transactions of individuals go digital.

We have a hint of digital cash in credit cards and ATMs. Like most of my generation, I get the little cash I use at an ATM, not having been inside a bank in years. On average, I use less cash every month. High-octane executives fly around the country purchasing everything on the go—meals, rooms, cabs, supplies, presents—carrying no more than $50 in their wallets. Already, the cashless society is real for some.

Today in the U.S., credit card purchases are used for one-tenth of all consumer payments. Credit card companies salivate while envisioning a near future where people routinely use their cards for "virtually every kind of transaction." Visa U.S.A. is experimenting with card-based electronic money terminals (no slip to sign) at fast-food shops and grocery stores. Since 1975, Visa has issued over 20 million debit cards that deduct money from one's bank account. In essence, Visa moved ATMs off of bank walls and onto the front counters of stores.

The conventional view of cashless money thus touted by banks and most futurists is not much more than a pervasive extension of the generic credit card system now operating. Alice has an account at National Trust Me Bank. The bank issues her one of their handy-dandy smart cards. She goes to an ATM and loads the wallet-size debit card with $300 cash deducted from her checking account. She can spend her $300 from the card at any store, gas station, ticket counter, or phone booth that has a Trust Me smart-card slot.

What's wrong with this picture? Most folks would prefer this system over passing around portraits of dead presidents. Or over indebtedness to Visa or MasterCard. But this version of the cashless concept slights both user and merchant; therefore it has slept on the drawing boards for years, and will probably die there.

Foremost among the debit (or credit) card's weaknesses is its nasty habit of leaving every merchant Alice buys from—newsstand to nursery—with a personalized history of her purchases. The record of a single store is not worrisome. But each store's file of Alice's spending is indexed with her bank account number or Social Security number. That makes it all too easy, and inevitable, for her spending histories to be combined, store to store, into an exact, extremely desirable marketing profile of her. Such a monetary dossier holds valuable information (not to mention private data) about her. She has no control over this information and derives no compensation for it.

Second, the bank is obliged to hand out whiz-bang smart cards. Banks being the legendary cheapskates they are, you know who is going to pay for them, at bank rates. Alice will also have to pay the bank for the transaction costs of using the money card.

Third, merchants pay the system a small percentage whenever a debit card is used. This eats into their already small profits and discourages vendors from soliciting the card's use for small purchases.

Fourth, Alice can only use her money at establishments equipped with slots that accept Trust Me's proprietary technology. This hardware quarantine has been a prime factor in the nonhappening of this future. It also eliminates person-to-person payments (unless you want to carry a slot around for others to poke into). Furthermore, Alice can only refill her card (essentially purchase money) at an official Trust Me ATM branch. This obstacle could be surmounted by a cooperative network of banks using a universal slot linked into an internet of all banks; a hint of such a network already exists.

The alternative to debit card cash is true digital cash. Digital cash has none of the debit or credit card's drawbacks. True digital cash is real money with the nimbleness of electricity and the privacy of cash. Payments are accountable but unlinkable. The cash does not demand proprietary hardware or software.

Therefore, money can be received or transferred from and to anywhere, including to and from other individuals. You don't need to be a store or institution to get paid in nonpaper money. Anyone connected can collect. And any company with the right reputation can "sell" electronic money refills, so the costs are at market rates. Banks are only peripherally involved. You use digital cash to order a pizza, pay for a bridge toll, or reimburse a friend, as well as to pay the mortgage, if you want. It is different from plain old electronic money in that it can be anonymous and untraceable except by the payer. It is fueled by encryption.

The method, technically known as blinded digital signatures, is based on a variant of a proven technology called public-key encryption. Here's how it works at the consumer level. You use a digicash card to pay Joe's Meat Market for a prime roast. The merchant can verify (by examining the digital signature of the bank issuing the money) that he was paid with money that had not been "spent" before. Yet, he'll have no record of *who* paid him. After the transaction, the bank has a verifiable account that you spent $7, and spent it only once, and that Joe's Meat Market did indeed receive $7. But those two sides of the transaction are not linked and cannot be reconstructed unless you the payer enable them to be. It seems illogical at first that such blind but verifiable transactions can occur, but the integrity of their "disconnection" is pretty watertight.

Digital cash can replace every use of pocket cash except flipping a coin. You have a complete record of all your payments and to whom they were made. "They" have a record of being paid but not by whom they were made. The reliability of both impeccably accurate accounting and 100 percent anonymity is ranked mathematically "unconditional"—without exceptions.

The privacy and agility of digital cash stems from a simple and clever technology. When I ask a digicash card entrepreneur if I could see one of his smart cards, he says that he is sorry. He thought he had put one in his wallet but can't find it. It looks like a regular credit card, he says, showing me his very small collection of them. It looks like . . . why, here it is! He slips out a blank, very thin, flexible card. The plastic rectangle holds math money. In one corner is a small gold square the size of a thumbnail. This is a computer. The CPU, no larger than a soggy cornflake, contains a limited amount of cash, say, $500 or 100 transactions, whichever comes first. This one, made by Cylink, contains a coprocessor specifically designed to handle public-key encryption mathematics. On the tiny computer's gold square are six very minute surface contacts which connect to an online computer when the card is inserted into a slot.

Less smart cards (they don't do encryption) are big in Europe and Japan, where 61 million of them are already in use. Japan is afloat in a primitive type of electronic currency—prepaid magnetic phone cards. The Japanese national phone company, NTT, has so far sold 330 million (some 10 million per month) of them. Forty percent of the French carry smart cards in their wallets today to make phone calls. New York City recently introduced a cashless phone card for a few of its 58,000 public phone booths. New York is motivated not by futurism but by thieves. According to *The New York Times*, "Every three minutes, a thief, a vandal, or some other telephone thug breaks into a coin box or yanks a handset from a socket. That's more than 175,000 times a year," and costs the city $10 million annually for repairs. The disposable phone card New York uses is not very smart, but it's adequate. It employs an infrared optical memory, common in European phone cards, which is hard to counterfeit in small quantities but cheap to manufacture in large numbers.

In Denmark, smart cards substitute for the credit cards the Danes never got. So everyone who would tote a credit card in America, packs a smart debit card in Denmark. Danish law demanded two significant restrictions: (1) that there be no minimum purchase amount; (2) that there be no surcharge for the card's use. The immediate effect was that the cards began to replace cash in everyday use even more than checks and credit cards have replaced cash in the States. The popularity of these cards is their undoing because unlike cheap, decentralized phone cards, these cards rely on realtime interactions with banks. They are overloading the Danish banking system, hogging phone lines as the sale of each piece of candy is transmitted to the central bank, flooding the system with transactions that cost more than they are worth.

David Chaum, a Berkeley cryptographer now living in Holland, has a solution. Chaum, head of the cryptography group at the center for Mathematics and Computer Science in Amsterdam, has proposed a mathematical code for a distributed, true digital cash system. In his solution, everyone carries around a refillable smart card that packs anonymous cash. This digicash seamlessly intermingles with electronic cash from home, company, or government. And it works offline, freeing the phone system.

Chaum looks like a Berkeley stereotype: gray beard, full mane of hair tied back in a professional ponytail, tweed jacket, sandals. As a grad student, Chaum got interested in the prospects and problems of electronic voting. For his thesis he worked on the idea of a digital signature that could not be faked, an essential tool for fraud-proof electronic elections. From there his interest drifted to the similar problem in computer network communications: how can you be sure a document is really from whom it claims to be from? At the same time he wondered: how can you keep certain information private and untraceable? Both directions—security and privacy—led to cryptography and a Ph.D. in that subject.

Sometime in 1978, Chaum says, "I had this flash of inspiration that it was possible to make a database of people so that someone could *not* link them all together, yet you could prove everything about them was correct. At the time, I was trying to convince myself that it was not possible, but I saw a loophole, how you might do it and I thought, gee . . . But it wasn't until 1984 or '85 that I figured out how to actually do that."

"Unconditional untraceability" is what Chaum calls his innovation. When this code is integrated with the "practically unbreakable security" of a standard public-key encryption code, the combined encryption scheme can provide anonymous electronic money, among other things. Chaum's encrypted cash (to date none of the other systems anywhere are encrypted) offers several important practical improvements in a card-based electronic currency.

First, it offers the bonafide privacy of material cash. In the past, if you bought a subversive pamphlet from a merchant for a dollar, he had a dollar that was definitely a dollar and could be paid to anyone else; but he had no record of who gave him that dollar or any way to provably reconstruct who gave it to him. In Chaum's digital cash, the merchant likewise gets a digital dollar transferred from your card (or from an online account), and the bank can prove that indeed he definitely has one dollar there and no more and no less, but no one (except you if you want) can prove where that dollar came from.

One minor caveat: the smart-card versions of cash implemented so far are, alas, as vulnerable and valuable as cash if lost or stolen. However, encrypting them with a PIN password would make them substantially more secure, though also slightly more hassle to use. Chaum predicts that users of digicash will use short (4-digit) PINs (or none at all) for minor transactions and longer passwords for major ones. Speculating a bit, Chaum says, "To protect herself from a robber who might force her to give up her passwords at gunpoint, Alice could use a 'duress code' that would cause the card to appear to operate normally, while hiding its more valuable assets."

Second, Chaum's card-based system works offline. It does not require instant verification via phone lines as credit cards do, so the costs are minimal and perfect for the numerous small-time cash transactions people want them for—parking meters, restaurant meals, bus rides, phone calls, groceries. Transaction records are ganged together and zapped once a day, say, to the central accountant computer.

During this day's delay, it would theoretically be possible to cheat. Electronic money systems dealing in larger amounts, running online in almost real time, have a smaller window for cheating—the instant between sending and receiving—but the minute opportunity is still there. While it is not theoretically possible to break the privacy aspect of digital cash (who paid whom) if you were desperate enough for small cash, you could break the security aspect—has this money been spent?—with supercomputers. By breaking the RSA public-key code, you could use the compromised key to spend money more than once. That is, until the data was submitted to the bank and they caught you. For in a delicious quirk, Chaum's digital cash is untraceable except if you try to cheat by spending money more than once. When that happens, the extra bit of information the twice-spent money now carries is enough to trace the payer. So electronic money is as anonymous as cash, except for cheaters!

Because of its cheaper costs, the Danish government is making plans to switch from the Dencard to the Dencoin, an offline system suited to small change. The computational overhead needed to run a system like this is nano-small. Each encrypted transaction on a smart card consumes only 64 bytes. (The previous sentence contains 67 bytes.) A household's yearly financial record of all income and all expenditure would easily fit on one hi-density floppy disk. Chaum calculates that the existing mainframe computers in banks would have more-than-adequate computational horsepower to handle digital cash. The encryption safeguards of an offline system would reduce much of the transactional computation that occurs online over phone lines (for ATMs and credit card checks), enabling the same banking computers to cover the increase in electronic cash. Even if we assume that Chaum guessed wrong about the computational demands of a scaled-up system, and he is off by a factor of ten, computer speed is accelerating so fast that this defers the feasibility of using existing bank power by only a few years.

In variations on Chaum's basic design, people may also have computer appliances at home, loaded with digital cash software, which allow them to pay other individuals, and get paid, over phone lines. This would be e-money on the networks. Attached to your e-mail message to your daughter is an electronic $100 bill. She may use that cash to purchase via e-mail an airplane ticket home. The airline sends the cash to one of their vendors, the flight's meal caterer. In Chaum's system nobody has any trace of the money's path. E-mail and digital cash are a match made in heaven. Digital cash could fail in real life, but it is almost certain to flourish in the nascent network culture.

I asked Chaum what banks think of digital cash. His company has visited or been visited by most of the big players. Do they say, gee, this threatens our business? Or do they say, hmm, this strengthens us, makes us more efficient? Chaum: "Well, it ranges. I find the corporate planners in $1,000 suits and private dining halls are more interested in it than the lower-level systems guys because the planners' job is to look to the future. Banks don't go about building stuff themselves. They have their systems guys buy stuff from vendors. My company is the first vendor of electronic money. I have a very extensive portfolio of patents on electronic money, in the U.S., Europe, and elsewhere." Some of Chaum's crypto-anarcho friends still give him a hard time about taking out patents on this work. Chaum tells me in defense, "It turns out that I was in the field very early so I wiped out all the basic problems. So most of the new work now [in encrypted electronic money] are extensions and applications of the basic work I did. The thing is, banks don't want to invest into something that is unprotected. Patents are very helpful in making electronic money happen."

Chaum is an idealist. He sees security and privacy as a tradeoff. His larger agenda is providing tools for privacy in a networked world so that privacy can be balanced with security. In the economics of networks, costs are disproportionately dependent on the number of other users. To get the Fax Effect going, you need a critical mass of early adopters. Once beyond the threshold, the event is unstoppable because it is self-reinforcing. Electronic cash shows all the signs of having a lower critical mass threshold than other implementations of data privacy. Chaum is betting that an electronic cash system inside an e-mail network, or a card-based electronic cash for a local public transportation network, has the lowest critical mass of all.

The most eager current customers for digital cash are European city officials. They see card-based digital cash as the next step beyond magnetic fastpasses now issued regularly by most cities' bus and subway departments. One card is filled with as much bus money as you want. But there are added advantages: the same card could fit into parking meters when you did drive or be used on trains for longer-distance travel.

Urban planners love the idea of automatic tolls charging vehicles for downtown entry or crossing a bridge without having the car stop or slow down. Bar-code lasers can identify moving cars on the road, and drivers will accept purchasing vouchers.

What's holding up a finer-grain toll system is the Orwellian fear that "*they* will have a record of my car's travels." Despite that fear, automatic tolls that record car identities are already operating in Oklahoma, Louisiana, and Texas. Three states in the busy Northeast have agreed to install one compatible system starting with experimental setups on two Manhattan/New Jersey bridges. In this system, a tiny card-size radio taped to the car windshield transmits signals to the toll gate which deducts the toll from your account at the gate (not from the card). Similar equipment running on the Texas turnpike system is 99.99 percent reliable. These proven toll mechanisms could easily be modified to Chaum's untraceable encrypted payments, and true electronic cash, if people wanted.

In this way the same cash card that pays for public transportation can also be used to cover fees for private transportation. Chaum relates that in his experience with European cities, the Fax Effect—the more people online, the more incentive to join—takes hold, quickly drawing other users. Officials from the phone company get wind of what's up and make it known that they would like to use the card to rid themselves of a nasty plague called "coins" that bog public phones down. Newspaper vendors call to inquire if they can use the card . . . Soon the economics of networks begin to take over.

Ubiquitous digital cash dovetails well with massive electronic networks. It's a pretty sound bet that the Internet will be the first place that e-money will infiltrate deeply. Money is another type of information, a compact type of control. As the Net expands, money expands. Wherever information goes, money is sure to follow. By its decentralized, distributed nature, encrypted e-money has the same potential for transforming economic structure as personal computers did for overhauling management and communication structure. Most importantly, the privacy/security innovations needed for e-money are instrumental in developing the next level of adaptive complexity in an information-based society. I'd go so far as to say that truly digital money—or, more accurately, the economic mechanics needed for truly digital cash—will rewire the nature of our economy, communications, and knowledge.

The consequential effects of digital money upon the hive mind of our network economy are already underway. Five we can expect are:

• *Increased velocity.* When money is disembodied—removed from any material basis at all—it speeds up. It travels farther, faster. Circulating money faster has an effect similar to circulating more money. When satellites went up, enabling near-the-speed-of-light, round-the-clock world stock trade, they expanded the *amount* of global money by 5 percent. Digital cash used on a large scale will further accelerate money's velocity.

• *Continuity.* Money that is composed of gold, precious materials, or paper comes in fixed units that are paid at fixed times. The ATM spits out $20 bills; that's it. You pay the phone company once a month even though you use the phone everyday. This is batch-mode money. Electronic money is continuous-flow. It allows recurring expenses to be paid, in Alvin Toffler's phrase, by "bleeding electronically from one's bank account in tiny droplets, on a minute-by-minute basis." Your e-money account pays for each phone call as soon as you hang up, or—how about this?—*as you are talking.* Payment coincides with use. Together with its higher velocity, continuous electronic money can approach near instantaneity. This puts a crimp on banks which derive a lot of their current profit on the "float"—which instantaneity erases.

• *Unlimited fungibility.* Finally, *really* plastic money. Once completely disembodied, digitized money escapes from a single transmission form and merrily migrates to whatever medium is handiest. Separate billing fades away. Accounts can be interleaved with the object or service itself. The bill for a video comes incorporated into the video. Invoices reside alongside of bar codes and can be paid with the zap of a laser. Anything that can hold an electronic charge can hold a fiscal charge. Foreign currencies become a matter of changing a symbol. Money is as malleable as digitized information. This makes it all the easier to monetize exchanges and interactions that were never part of an economy before. It opens the floodgates of commerce onto the Net.

• *Accessibility.* Until now, sophisticated manipulations of money have been the private domain of professional financial institutions—a financial priesthood. But just as a million Macs broke the monopoly of the high priests guarding access to mainframe computers, so e-money will break the monopoly of financial Brahmins. Imagine if you could charge (and get) interest on any money due you by dragging an icon over that electronic invoice. Imagine if you could factor in the "interest due" icon and give it variable interest, ballooning as it aged. Or maybe you would charge interest by the minute if you sent a payment in early. Or program your personal computer to differentially pay bills depending on the prime rate—programmed bill-trading for amateurs. Or perhaps you would engineer your computer to play with exchange rates, paying bills in whatever currency is least valuable at the time. All manner of clever financial instruments will surface once the masses can drink from the same river of electronic money as the pros. To the list of things to hack, we may now add finance. We are headed toward programmed capitalism.

• *Privatization.* The ease with which e-money is caught, flung, and shaped makes it ideal for private currencies. The 214 billion yen tied up by Japan's NTT's phone cards is one limited type of private currency. The law of the Net is: he who owns a computer not only owns a printing press, but also a mint, when that computer is linked to e-money. Para-currencies can pop up anywhere there is trust (and fail there, too).

Historically, most modern barter networks rapidly slide into exchanges of real currency; one could expect the same in electronic barter clubs, but the blinding efficiency of an e-money system may not tend that way. The $350 billion tax question is whether para-currency networks would ever rise above unofficial status.

The minting and issuing of currency has been one of the few remaining functions of government that the private sector has not encroached upon. E-money will lower this formidable barrier. By doing so it will provide a powerful tool to private governance systems, such as might be established by renegade ethnic groups, or the "edge cities" proliferating near the world's megacities. The use of institutional electronic money transfers to launder money on a global scale is already out of anyone's control.

THE NATURE OF E-MONEY —invisible, lightning quick, cheap, globally penetrating—is likely to produce indelible underground economies, a worry way beyond mere laundering of drug money. In the net-world, where a global economy is rooted in distributed knowledge and decentralized control, e-money is not an option but a necessity. Para-currencies will flourish as the network culture flourishes. An electronic matrix is destined to be an outback of hardy *underwire* economies. The Net is so amicable to electronic cash that once established interstitially in the Net's links, e-money is probably ineradicable.

In fact, the legality of anonymous digital cash is in limbo from the start. There are now strict limits to the size of transactions U.S. citizens can make with physical cash; try depositing $10,000 in greenbacks in a bank. At what amount will the government limit anonymous digital cash? The drift of all governments is to demand fuller and fuller disclosures of financial transactions (to make sure they get their cut of tax) and to halt unlawful transactions (as in the War on Drugs). The prospect of allowing untraceable commerce to bloom on a federally subsidized network would probably have the U.S. government seriously worried if they were thinking about it. But they aren't. A cashless society smells like stale science-fiction, and the notion reminds every bureaucrat drowning in paper of the unfulfilled predictions of a paperless society. Eric Hughes, maintainer of the cypherpunks' mailing list, says, "The Really Big Question is, how large can the flow of money on the nets get before the government requires reporting of every small transaction? Because if the flows can get large enough, past some threshold, then there might be enough aggregate money to provide an economic incentive for a transnational service to issue money, and it wouldn't matter what one government does."

Hughes envisions multiple outlets for electronic money springing up all over the global net. The vendors would act like traveler's check companies. They would issue e-money for, say, a 1 percent surcharge. You could then spend Internet Express Checks wherever anyone accepts them. But somewhere on the global Net, underwire economies would dawn, perhaps sponsored by the governments of struggling developing countries. Like the Swiss banks of old, these digital banks would offer unreported transactions. Paying in online Nigerian nairas from a house in Connecticut would be no more difficult than using U.S. dollars. "The interesting market experiment," Hughes says, "is to see what the difference in the charge for anonymous money is, once the market equalizes. I bet it'll be on the order of 1–3 percent higher, with an upper limit of about 10 percent. That amount will be the first real measure of what financial privacy is worth. It might also be the case that anonymous money will be the *only* kind of money."

Usable electronic money may be the most important outcome of a sudden grassroots takeover of the formerly esoteric and forbidden field of codes and ciphers. Everyday e-money is one novel use for encryption that never would have occurred to the military. There are certainly many potential uses of encryption that the cypherpunks' own ideological leanings blind them to, and that will have to wait until encryption technology enters the mainstream—as it certainly will.

To date encryption has birthed the following: digital signatures, blind credentials (you have a diploma that says, yes, you have a Ph.D., yet no one can link that diploma with the other diploma in your name from traffic school), anonymous e-mail, and electronic money. These species of disconnection thrive as networks thrive.

Encryption wins because it is the necessary counterforce to the Net's runaway tendency to link. Left to itself, the Net will connect everyone to everyone, everything to everything. The Net says, "Just connect." The cipher, in contrast, says, "Disconnect." Without some force of disconnection, the world would freeze up in an overloaded tangle of unprivate connections and unfiltered information.

I'm listening to the cypherpunks not because I think that anarchy is a solution to anything but because it seems to me that encryption technology civilizes the grid-locking avalanche of knowledge and data that networked systems generate. Without this taming spirit, the Net becomes a web that snares its own life. It strangles itself by its own prolific connections. A cipher is the yin for the network's yang, a tiny hidden force that is able to tame the explosive interconnections born of decentralized, distributed systems.

Encryption permits the requisite out-of-controllness that a hive culture demands in order to keep nimble and quick as it evolves into a deepening tangle.

# 13

# God Games

Populous II is a state-of-the-art computer god game. You play god. A son of Zeus to be exact. Through the portal of the computer screen you spy down upon a patch of Earth where the tiny figures of men scurry about farming, building, and wandering around. With a shimmering blue hand (the hand of god) you can reach down and touch the land, transforming it. You can either gradually level mountains or gradually build up valleys. In both cases, you try to create flat farmland for people. Except for the power to deliver a spectrum of disasters such as earthquakes, tidal waves, and tornadoes, your direct influence over the people of your world is limited to this geological hand.

Good farmland makes happy people. You can see them prosper and bustle about. They build farmhouses first; then as their numbers increase, they build red-tile roofed town houses, and if things continue to bode well, eventually they construct complex walled cities, whitewashed and gleaming in the Mediterranean sun. The more the little beings prosper, the more they worship you, and the more manna (power) you, the god, accumulate.

Here's your problem, though. Elsewhere in the greater landscape other sons of Zeus are contesting for immortality. These gods can be played by other humans, or by the game's own AI agent. The other gods will rain the seven plagues on your populace, wiping out your base of support and worship. They can send a crashing blue tidal wave which not only drowns your citizenry but submerges their farmland, endangering your own divine existence. No people, no worship, no god.

Of course, you can do the same—if you have enough manna in store. Using your destructive powers consumes manna by the barrelful. Besides, there are other ways to defeat your enemies and gain manna without sending a zigzagging crack through an area, a crack which swallows groaning people as they fall in. You can devise Pan figures that roam the countryside luring newcomers to your religion with magic flutes. Or you can erect a "Papal Magnet," a granite ankh monument which acts as a shrine, attracting worshipers and pilgrims.

Meanwhile your own citizens are dodging fire storms from your scheming half-brothers. And after those minor-league gods are through trashing one of your countries, you've got to decide whether to rebuild it or go after their populations with your arsenal. You could use a tornado which sucks up houses and people alike and visibly tosses them across the land. Or a biblical column of fire which scorches the earth into barrenness (until a god restores it by sowing healing wildflowers). Or, you can send burning flows of lava from a well-placed volcano.

I got an expert tour of this world from a metagod's point of view on a visit to the office of Electronic Arts, the game's publisher, where I was taken through the paces of god powers. Jeff Haas is one of the developers of the game. You could call Haas a supergod who created the other gods. He pointed to a gathering dark mass of clouds over one village that suddenly erupted into a shower of lightning. The bolts shimmied down to Earth. When a white bolt struck a person, the figure fried to a blackened crisp. Haas chortled in delight at the exquisitely rendered graphic but caught my raised eyebrow. "Yes," he admitted sheepishly, "the point of the game is destruction—total slash and burn."

"There *are* a few positive things you can do as a god," Haas volunteered, "but not many. Making trees is one of them. Trees always make people happy. And you can bless the land with wildflowers. But mostly it's destroy or be destroyed." Aristotle might have understood. In his day, gods were entities to be feared. God as a buddy, or even an ally, is hopelessly modern. You kept out of the gods' way, appeased them when needed, and prayed that your god would vanquish the other gods. The world was dangerous and capricious.

"Let me put it this way," Haas says, "you definitely do not want to be one of the people in this world." You bet. It's godhood for me.

To win Populous, you've got to think like a god. You cannot live many small individual lives and succeed. Nor can you manipulate every individual simultaneously and hope to remain sane. Control must be surrendered to a populous mob. Individuals of Populous land, who are no more than a few bits of code, have a certain amount of autonomy and anonymity. Their pandemonium must be harnessed collectively in an intelligent way. That's your job.

As god, you have only indirect control. You can offer incentives, play with global events, make calculated tradeoffs, and hope that you get the mix right so that your underlings follow you. Cause and effect in this game is coevolutionarily fuzzy; changing one thing always changes many things, often in the direction you wanted least. All management is done laterally.

Software stores sell other god games: Railroad Tycoon, A-Train, Utopia, Moonbase. They all enable you, the neo-god, to entice citizens to create a self-sustaining empire. In the game Power Monger you are one of four godlike kings hoping to rule supreme over a large region of a planet. The population below, which numbers in the hundreds, is not faceless. Each citizen has a name, an occupation, and a biography. As deity, your job is to urge the citizenry to explore the land, mine ore, make plows, or hammer them into swords. All you can do is adjust the society's parameters and then set the beings loose. It's hard for a god to guess what will emerge. If your folks manage to rule over the most land, you win.

In the brief annals of classic god games, the game of Civilization ranks pretty high. Here the goal is to steer your bottom-up population through the evolution of culture. You can't tell them how to build a car, but you can set them up so that they can make the "discoveries" needed to build one. If they invent a wheel, then they can make chariots. If they acquire masonry skills, then they can make arithmetic. Electricity needs metallurgy and magnetism; corporations first require banking skills.

This is a new way of steering. Pushing too hard can backfire. The denizens in Civilization might revolt at any time, and occasionally they do. All the while you are racing against other cultures being tweaked by your opponent. Lopsided contests are quite common. I once heard an avid Civilization player boast that he overran the other society with stealth bombers while they were still working on chariots.

It's only just a game, but Populous embodies the subtle shift in our interactions with all computers and machines. Artifacts no longer have to be inert homogeneous lumps. They can be liquid, adaptable, slippery webs. These collectivist machines run on myriad tiny agents interacting in ways we can't fathom, generating results we can only indirectly control. Getting a favorable end result is a challenge in coordination. It feels like herding sheep, managing an orchard, or raising kids.

In the development of computers, games come first, work later. Kids who become comfortable relating to machines as if they behave organically, later expect the same from machines at work when they are older. MIT psychologist Sherry Turkle describes the readiness of children to perceive complicated devices as organic as an affinity for a "second self"—a projection of themselves onto their machines. Toy worlds certainly encourage that personification.

SimEarth, yet another god game, bills itself, somewhat tongue in cheek, as "the ultimate experience in planet management." An acquaintance of mine told a story of making a long car trip with three 10–12-year-old boys in the back seat, the trio equipped with a laptop computer running SimEarth. He drove while eavesdropping on the boys' conversation. He gathered that the boys had decided their goal was to evolve intelligent snakes. The kids:

"Do you think we can start the reptiles now?"

"Oh shoot. The *mammals* are taking over."

"We better add more sunlight."

"How can we make the snakes smarter?"

SimEarth has no narrative or fixed goals—a nonstarter for many adults. Kids, on the other hand, fall into the game without hesitation or instruction. "We are as gods, and might as well get good at it," declared Stewart Brand in 1968, who had personal computers (a term he later coined) and other vivisystems in mind when he said it.

Stripped of all secondary motives, all addictions are one: to make a world of our own. I can't imagine anything more addictive than being a god. A hundred years from now nothing will keep us away from artificial cosmos cartridges we can purchase and pop into a world machine to watch creatures come alive and interact on their own accord. Godhood is irresistible. The hemorrhaging expense of yet another hero will not keep us away. World-makers could charge us anything they want for a daily fix of a few hours immersed in the interactive saga of our characters' lives, and to keep our world going we will pay it. Organized crime will make billions of dollars peddling crude artificial calamities—first class hurricanes or high priced tornadoes—to addicts compelled to buy. Over time, god-customers will evolve fairly sturdy and endearing populations, which they will be eager to test with yet another fully rendered natural disaster. For the poor there will surely be underground exchanges of generic mutant beings and pilfered scenarios. The headlong high of substituting for Jehovah, and the genuine, overwhelming, sheer *love* for one's private world, will suck in any and all who near it.

Because simulated worlds behave—in a tiny but measurable way—similarly to worlds of living organisms, the ones that survive will grow in complexity and value. The organic ambiance of distributed, parallel world-games is not mere anthropomorphism, despite the second self projected upon them.

SimEarth was meant to model Lovelock's and Margulis's Gaia hypothesis, which it succeeded in doing to a remarkable degree. Fairly serious changes in the simulated Earth's atmosphere and geology are compensated by convoluted feedback loops in the system itself. For instance, overheating the planet increases biomass production, which reduces $CO_2$ levels, which cools the planet.

Scientists debate whether the evidence of self-correcting cohesion seen in the Earth's global geochemistry qualify Earth as a large organism (Gaia), or merely a large vivisystem. Applying the same test to SimEarth we get a more certain answer: SimEarth, the game, is not an organism. But it is a step in the direction of the organic. By playing SimEarth and other god games we can get a feel of what it will be like to parry with autonomous vivisystems.

In SimEarth, a mind-boggling web of factors impinge on each other, making it impossible to sort out what does what. Players sometimes complain that SimEarth appears to run without regard to human control. It's as if the game has its own agenda and you are just watching.

Johnny Wilson, a gaming expert and author of a SimEarth handbook, says that the only way to derail Gaia (SimEarth) is to launch a cataclysmic alteration such as tilting the axis of the Earth to horizontal. He says there is an "envelope" of limits within which the SimEarth system will always bounce back; one must bump the system beyond that envelope to crash it. As long as SimEarth runs inside the envelope, it follows its own beat; outside of it, it follows no beat. As a comparison, Wilson points out that SimCity, SimEarth's older sister, "is much more satisfying as a game, because you get more instant and clear feedback on changes, and because you feel like you are more in control."

Unlike SimEarth, SimCity is the paramount example of an underling-driven god game. This award-winning simulation of a city is so convincing that professional urban planners use it to demonstrate the dynamics of real cities, which are also driven by underlings. SimCity succeeds, I believe, because it is based on the swarm, the same foundation that all vivisystems are based on: a collective of richly linked, autonomous, localized agents working in parallel. In SimCity a working city bubbles up from a swarm of hundreds of ignorant Sims (or Simpletons) doing their simple-minded tasks.

SimCity obeys the usual tail-swallowing logic of god games. Sims won't take up residence in your city unless there are factories, but factories generate pollution which drives away residents. Roads help commuters but also raise taxes, which drive down your ratings as a mayor, which you need to survive politically. The maze of interrelated factors required to construct a sustainable SimCity can unfold along the lines of the following fairly typical account from a heavy SimCity-using friend of mine: "In one city which I built up over many Sim-years I had a 93 percent approval in the public opinion polls. Things were going great! I had a nice balance of tax-producing commerce and citizen-retaining beauty. To lessen pollution in my great metropolis I ordered a nuclear power plant built. Unfortunately I inadvertently placed it in my airport's flight path. One day a plane crashed into the generators, causing a meltdown. This set fire to the town. But since I hadn't built enough fire stations in the vicinity (way too costly), the fires spread and eventually burnt down the whole city. I'm rebuilding now, differently."

Will Wright, the author of SimCity and coauthor of SimEarth, is thirtyish, bookish, and certainly one of the most innovative programmers working today. Because Sim games are so hard to control, he likes to call them Software Toys. You diddle with them, explore, try out fantasies, and learn. You don't win, any more than you might win at gardening. Wright sees his robust simulation toys as the initial baby steps toward a full march of "adaptive technologies." These technologies are not designed, improved upon, or adjusted by the creator; rather, they—on their own accord—adapt, learn, and evolve. It shifts a bit of power from the user to the used.

The origins of SimCity trace Will's own path to this vision. In 1985 Will wrote what he calls "a really, and I mean *really*, stupid video game" entitled Raid on Bungling Bay. It was a typical shoot-'em-up starring a helicopter that bombed everything in sight.

"To create this game I had to draw all these islands that the helicopter would go bomb," recalls Will. Normally the artist/author modeled the complete fantasy in minute pixelated detail, but Will got bored. "Instead," Will says, "I wrote a separate program, a little utility, that would let me go around and build these islands real quick. I also wrote some code that could automatically put roads on the islands."

By engaging his land-making or road-making module the program would—on its own!—fill in land or roads in the simulated world. Will remembers, "Eventually I finished the shoot-'em-up game part, but for some reason I kept going back to the darn thing and making the building utilities more and more fancy. I wanted to automate the road function. I made it so that when you added each connecting piece of island, the road parts on them would connect up automatically to form a continuous road. Then I wanted to put down buildings automatically, so I built a little menu choice for buildings.

"I started asking myself, why am I doing this since the game is finished? The answer was that I found that I had a lot more fun building the islands than I had destroying them. Pretty soon I realized that I was fascinated by bringing a city to life. At first I just wanted to do a traffic simulation. But then I realized that traffic didn't make a lot of sense unless you had places where the people drove to . . . and that led layer upon layer to a whole city; SimCity."

A player building a SimCity recapitulates Will Wright's sequence in inventing it. First, he makes the lower geographical foundation of land and water which support the road traffic and telephone infrastructure which support residential homes which support the Sims which support the mayor.

To get a feel for the dynamics of a city, Wright studied a simulation of an average city done in the 1960s at MIT by Jay Forrester. Forrester summarized city life into quantitative relations rendered as mathematical equations. They were almost rules of thumb: it takes so many residents to support one firefighter; or, you need so many parking spaces for each car. Forrester published his findings as *Urban Dynamics*, a book which influenced many aspiring computer modelers. Forrester's own computer simulation was entirely numerical with no visual interface. He ran the simulation and got a stack of printouts on lined paper.

Will Wright put flesh onto Jay Forrester's equations, and gave them a decentralized, bottom-up existence. Cities assembled themselves (according to the laws and theories of the god Will Wright) on the computer screen. In essence, SimCity is an urban theory provided with a user interface. In the same sense, a dollhouse is a theory of the household. A novel is theory told as story. A flight simulator is an interactive theory of aviation. Simulated life is a theory of biology left to fend for itself.

A theory abstracts the complicated pattern of real things into the facsimile pattern—a model, or a simulation. If done well, the miniature captures some integrity of the larger whole. Einstein, working at the peak of human talent, reduced the complexity of the cosmos to five symbols. His theory, or simulation, works. If done well, an abstraction becomes a creation.

There are many reasons to create. But what we create is always a world. I believe we may be unable to create anything less. We can create hurriedly, in fragments, in thumbnail sketches, and streams of consciousness, but always we are filling in an unfinished world of our own. Of course we sometimes doodle, literally and metaphorically. But we immediately see this for what it is: theory-free gibberish, and model-less nonsense. In essence, every creative act is no more or less than the reenactment of the Creation.

A FEW YEARS AGO, right before my eyes, a man with matted hair created an artificial world, a simulation of swaying fernlike arches rising off of an arabesque floor of maroon tiles and a tall red chimney going nowhere in particular. This world had no material form. It was a nether world that only two hours earlier had been a daydream in the man's imagination. Now it was a daydream circulating on a pair of Silicon Graphics computers.

The man donned magical goggles and climbed into his simulacra. I climbed in after him.

As far as I know, this descent into a man's daydream in the summer of 1989 was the first time a human created an instant fantasy and let others crawl in to share it.

The man was Jaron Lanier, a round guy with a mop of rastafarian dreadlocks and a funny giggle, who always reminds me of Big Bird. He was nonchalant about entering and exiting a dreamland and talked about the travel like someone who had been exploring "the other side" for years. The walls of Jaron's company's office displayed fossils of past experimental magic goggles and gloves. The usual computer hardware and software paraphernalia littered the rest of the lab: soldering irons, floppy disks, soda cans, and in this case, ripped body suits woven with wires and bejeweled with connector plugs.

Jaron's hi-tech method of generating visitable worlds had been pioneered years earlier by institutional researchers including NASA. Scores of people had already entered into disembodied imaginary worlds. Research worlds. But Jaron devised a low-rent system that worked even better than the university setups, and he built wildly unscientific "crazy worlds" on the fly. And Jaron coined a catchy name for the result: "virtual reality."

To participate in a virtual reality, a visitor suits up into a uniform that is wired to monitor major body movements. The costume includes a face mask that can signal the movement of the head. Inside the mask are two tiny color video monitors which deliver the participant a vision of stereoscopic realism. From behind the mask it appears to the visitor that he inhabits a 3-D virtual reality.

The general concept of a computer-generated reality is probably familiar to most readers because in the years following Jaron's demonstrations, the prospect of everyday virtual reality (VR) became a regular staple of magazine and TV news features. The surreal aspect was always emphasized. Eventually the *Wall Street Journal* headlined virtual reality as "An Electronic LSD."

I must confess that "drugs" were exactly my first thought watching Jaron disappear into his world. Here's a 29-year-old company founder wearing an electrified scuba mask. While I and other friends watch soberly, Jaron rolls slowly on the floor, mouth agape. He writhes into a new position, one arm pushing against the air, grasping nothing. Like a man possessed in slo-mo, he bends from one contortion to another as he explores hidden aspects of his newly minted universe. He carefully crawls across the carpet, stopping every so often to inspect some unseen wonder in the air before him. Watching him is eerie. His maneuvers follow a distant, internal logic, a separate reality. Occasionally, Jaron disturbs the quiet with a yelp of delight.

"Hey, the chalk pedestals are hollow! You can go up inside them and see the bottom of the rubies!" he squeals. Jaron himself had created the pedestals topped with red gems, but when he imagined them he hadn't bothered to consider their bottoms. A whole world is too complex to hold in one's head. But a simulation can play out those complexities. Again and again, Jaron reported back details in the world that he, the god, had not foreseen. Jaron's virtual world was like other simulations; the only way to predict what would happen was to run it.

Simulations are not new. Nor is visiting them. Toy worlds are a very early human invention, perhaps even a sign of humanity's emergence, since toys and games in a burial site are recognized by archaeologists as evidence of human culture. Certainly the urge to create toys arises very early in individual development. Children immerse themselves in their own artificial worlds of miniatures. Dolls and choo-choo trains properly belong to the microcosms of simulation. So does much of the great art in our culture: Persian miniatures, painterly landscape realism, Japanese tea gardens, and perhaps all novels and theater. Tiny worlds.

But now in the computer age—the age of simulations—we are making tiny worlds in larger bandwidths, with more interaction, and with deeper embodiment. We've come from inert figurines to SimCity. Some simulations, like Disneyland, are no longer so tiny.

Anything at all, in fact, is a candidate for a simulation when it is given energy, possible behaviors, and room to grow. We live in a culture that is rapidly animating a million objects into simulations by electrifying them with smartness. A telephone switchboard becomes a simulated operator voice, a car becomes a tiger in a commercial, fake trees and robotic alligators become a simulated jungle in an amusement park. We don't even blink anymore.

In the early 1970s the Italian novelist Umberto Eco drove around America visiting as many low-brow roadside attractions as he could get to. Eco was a semiotician—a decipherer of unnoticed signs. He found America trafficking in subtle messages about simulations and degrees of reality. The national icon, Coca-Cola, as an example, advertised itself as "the real thing." Wax museums were Eco's favorite text. The more kitsch-laden they were with altarlike velvet drapes and soft narrations, the better. Eco found wax museums to be populated with exquisite copies of real people (Brigitte Bardot in a bikini) and exquisite fakes of fictional characters (Ben Hur in a chariot race). Both history and fantasy were sculptured in equally realistic and neurotic detail so that there was no boundary between the real and faked. Tableau artists spared no effort in rendering an unreal character in supreme realism. Mirrors reflected one period room's figures into another time period to further blur the distinction of real and not. Between San Francisco and Los Angeles, Eco was able to visit seven wax versions of Leonardo's *Last Supper*. Each you'll-never-be-the-same-afterwards waxwork tried to outdo the other in degree of faithful realism to a fictionalized painting.

Eco wrote that he was on a "journey into hyperreality, in search of instances where the American imagination demands the real thing and, to attain it, must fabricate the absolute fake." The reality of the *absolute fake* Eco called hyperreality. In hyperreality, as Eco puts it, "absolute unreality is offered as real presence."

A perfect simulation and a computer toy world are works of hyperreality. They fake so wholly that as a whole they have a reality.

French pop-philosopher Jean Baudrillard opens his small book, *Simulations* (1983), with these two tightly wound paragraphs:

> If we were able to take as the finest allegory of simulation the Borges tale where the cartographers of the Empire draw up a map so detailed that it ends up exactly covering the territory (but where the decline of the Empire sees this map become frayed and finally ruined, a few threads still discernible in the deserts . . .) then this fable has come full circle for us . . .
>
> Abstraction today is no longer that of the map, the double, the

> mirror or the concept. Simulation is no longer that of a territory, a referential being, or a substance. It is the generation of models of a real without origin or reality: a hyperreal. The territory no longer precedes the map, nor survives it. Henceforth, it is the map that precedes the territory—PRECESSION OF SIMULACRA—it is the map that engenders the territory and if we were to revive the fable today, it would be the territory whose shreds are slowly rotting across the map. It is the real, and not the map, whose vestiges subsist here and there, in the deserts which are no longer of the Empire, but our own. *The desert of the real itself.*

In the desert of the real, we are busy building paradises of the hyperreal. It is the model (the map) that we prefer. Steven Levy, author of *Artificial Life* (1991), a book that celebrates the advent of simulations so rich that we can only declare them alive, rephrases Baudrillard's point this way: "The map is not the territory, but a map *is* a territory."

However, the territory of the simulacra is blank. The absolute fake is so obvious that it is still invisible to us. We have no taxonomy yet to differentiate subtle types of simulations. Take simulacra's long list of indistinct synonyms: fake, phony, counterfeit, replication, artificial, second grade, phantom, image, reproduction, deception, camouflage, pretense, imitation, false appearance, pretended, effigy, an enactment, shadow, shade, insincerity, a mask, disguise, substitute, surrogate, feign, parody, a copy, something bluffed, a sham, a lie. The word "simulacra" is a word loaded with heavy karma.

The Greek Epicureans, a school of radical philosophers who figured out there must be atoms, had an unusual theory of vision. They believed every object gave off an "idol" (*eidola*). The same concept came to be called *simulacra* in Latin. Lucretius, a Roman Epicurean, says you can think of simulacra as "images of things, a sort of outer skin perpetually peeled off the surfaces of objects and flying about this way and that through the air."

These simulacra were physical, but ethereal, things. Invisible simulacra emanated from an object and impinged upon the eye causing vision. A thing's reflection assembled in a mirror demonstrated the existence of simulacra; how else could there be two of them, and one so diaphanous? Simulacra, the Epicureans believed, could enter into people's senses through their pores while they slept, thus conveying the idols (images) carried in dreams. Art and paintings captured the idols radiated by the original subject, just as flypaper might catch bugs.

A simulacra then was a derived entity, second to the original, a parallel image—or to use modern words, a virtual reality.

In the Roman vernacular simulacrum came to mean a statue or image that was animated by a ghost or spirit. Thus its Greek predecessor, the term "idol", crept into the English language in 1382, when the first English Bible needed a word to describe the hyperreality of animated, and sometimes talking, statues that were presented as gods.

Some of these ancient temple automatons were quite elaborate. They had moving heads and limbs, and tubes to channel voices from behind them. Ancient people were far more sophisticated than we often give them credit for. No one mistook the idols for the real god they represented. But no one ignored the idol's presence, either. The idol really moved and said things; it had its own behavior. The idols were neither real nor faked—they were real idols. In Eco's terms, they were hyperreal, just as Murphy Brown, a virtual character on TV, is treated as kind of real.

We post-modern urbanites spend a huge portion of our day immersed in hyperrealities: phone conversations, TV viewing, computer screens, radio worlds. We value them highly. Try to have a dinner conversation without referencing something you saw or heard via the media! Simulacra have become the terrain we live in. In most ways we care to measure, the hyperreal is real for us. We enter and leave hyperreality with ease.

Take, for instance, a hyperreality that Jaron Lanier built months after his first instant world. Not long after he was done, I immersed myself in his world of idols and simulacra. This artificial reality included a circle of railway track about a block in diameter and a locomotive about chest high. The ground was pink, the train light gray. Other blocky figures lay about like so many dropped toys. The shape of the choo-choo train and toys were aggregations of polygons—no graceful curves. Colors were uniform and bright. When I turned my head, the scene shifted in a stuttered way. Shadows were stark. The sky was an empty dark blue with no hint of distance or space. I had the impression of being a toon in Toontown.

A gloved hand—roughly rendered in tiny polygonal blocks—floated in front of me. It was my hand. I flexed the disembodied thing. When I mentally willed the hand into a point, I began to fly in the direction of my finger. I flew over to the small train engine and sat on it or above it, I couldn't tell. I reached out my floating hand and yanked a lever on the train. The train began to circle and I could watch the pink landscape go by. At some point I hopped off the train near an inverted top hat. I stood and watched the train chug around the loop of track without me. I bent to grab the top hat and the instant I touched it, it turned into a white rabbit.

I heard someone outside the world laugh, a heavenly chuckle. That was the god's little joke.

The disappearance of the top hat was real, in a hyperreality way. The trainy thing really started and eventually really stopped. It was really going around in circles. When I flew I really transposed a distance *of some sort.* To anyone watching me on the outside, I was a guy stiffly gyrating in a carpeted office in the same odd way that Jaron did. But inside, hyperreal events really happened. Anyone else visiting could corroborate; there was consensual evidence. In the parallel world of the simulacra, they were real.

HAND-WRINGING about the reality of simulations would be an appropriate academic exercise for French and Italian philosophers, if simulacra didn't turn out to be so useful.

In the Entertainment & Information Systems Group at the MIT Media Lab, Andy Lippman is developing an approach to television transmission that "lets the audience drive." A major objective of the Media Lab's research is to allow the consumer to personalize the presentation of information. Lippman invented a scheme to deliver video in an ultracompact form which can then be unpacked in a thousand different ways. He does this by transmitting not a staid image but a simulacra.

In the demo that he shows, Lippman's group took an early episode of "I Love Lucy" and extracted a visual model of Lucy's living room from the footage. Lucy's living room becomes a virtual living room on a hard disk. Any part or view of it can be displayed on cue. Lippman then used a computer to remove Lucy's moving image from the background scenes. When he wants to transmit the entire episode, he sends two kinds of data: the background as a virtual model and the film of Lucy moving. The viewer's computer reassembles Lucy's character moving against a background produced by the model. Thus Lippman can broadcast the living room set data only once in a single burst—not continuously as is normal—updating only when the scene or light shifts. Says Lippman, "Conceivably, we might choose to store all of the background sets from a TV serial at the front of a single optical disk, while the action and camera motion instructions needed to reconstitute 25 episodes could fit on the remaining tracks."

Nicholas Negroponte, director of the Media Lab, speaks of this method as "transmission of models rather than content, so content is something the receiver derives from the model." He extrapolates from the simple "I Love Lucy" experiment to a future when entire scenes, figures and all, are modeled into simulacra to be transmitted. Rather than broadcast a two-dimensional picture of a ball, send a simulacra of the ball. The broadcasting machine says "Here is a simulacra of a ball: shiny blue, with a dimension of 50 centimeters, moving at this velocity and direction." The receiving machines says, "Umm yes, a simulacra of a bouncing ball. Oh, I *see* it," and displays the hopping blue ball as a moving hologram. Now the home viewer can visually examine the ball from any perspective he wants.

As a commercial example, Negroponte suggests broadcasting a holographic image of a football game into living rooms. Rather than merely sending the data for the game's two-dimensional image, the sports station transmits a simulacra of the game; the stadium, players, and plays are abstracted into a model which can be compressed for transmission. The receiving machine in the home unpacks the model into visual form. The couch potato with a six-pack sees a dynamic mirage of the players as they rush, pass, and punt in 3-D. He chooses the angle he wants to watch it from. His kids can horse around by watching the game from the ball's point of view.

Besides being able to "break the tyranny of video as prepackaged frames," the purpose of transmitting simulacra is primarily data compression. Real-time holography requires astronomical amounts of bits. Using all the smart processing tricks in the foreseeable future, a state-of-the-art supercomputer would spend hours computing a few seconds of a real-time holograph the size of a TV console. The ball game would be over before you saw the last of the amazing (and terrifying in three dimensions) opening flying logos.

What better way to compress a complication than to model it, mail it, and let the recipient supply the intelligent details? Transmitting a simulacra is not a step down from transmitting reality. It is a step up from transmitting data.

The military is keen on simulacra as well.

IN AN UNNAMED STRETCH of desert, in the spring of 1991, Captain H. R. McMaster of the U.S. 2d Armored Cavalry Regiment paced over the quiet battlefield. Hardly a month had passed since he had last been there. The rocky sand was quiet and still now. Iraqi tanks lay in twisted wrecks just as he had left them a few weeks ago, although now they no longer burned like an inferno. Thank God he and his troops had all survived; the Iraqis had not done as well. A month ago neither side knew they were engaged in the pivotal battle of the Desert Storm war. Things moved fast; thirty days after their fateful skirmish, historians already had a name for it: The Battle of 73 Easting.

Now Captain McMaster was at this desolate site again. He had reconvened at the behest of some crazy analysts back in the States. The Pentagon wanted all troop officers gathered at the battlefield while the U.S. still controlled the territory, and while their memories were fresh. The Army was going to recreate the entire 73 Easting battle as a fully three-dimensional simulated reality which any future cadet could enter and relive. "A living history book," they called it. A simulacra of war.

On the plains of Iraq, the real soldiers sketched out the month-old battle. They walked off the action as best their feverish memories of the day could remind them. A few soldiers supplied diaries to reconstruct their actions. A couple even consulted personal tape recordings taken during the chaos. Tracks in the sand gave the simulators precise traces of movement. A black box in each tank, programmed to track three satellites, confirmed the exact position on the ground to eight digits. Every missile shot left a thin wire trail which lay undisturbed in the sand. Headquarters had a tape recording of radio voice communications from the field. Sequenced overhead photos from satellite cameras gave the big view. Soldiers paced the sun-baked ground in hot arguments sorting out who shot whom. A digital map of the terrain was captured by lasers and radar. When the Pentagon left, they had all the information they needed to recreate history's most documented battle.

Back at the Simulation Center, a department at the Institute for Defense Analysis in Alexandria, Virginia, technicians spent nine months digesting this overdose of information and compiling a synthetic reality from a thousand fragments. A few months into the project, they had the actual desert troops, then stationed in Germany, review a preliminary version of the recreation. The simulacra were sufficiently fleshed out that the soldiers could sit in tank simulators and enter the virtual battle. They reported corrections of the simulated event to the techies, who modified the model. Just about one year after the confrontation, following the final review by Captain McMaster, the recreated Battle of 73 Easting premiered for the military brass. McMaster laconically understates that the simulacra give "a very realistic sensation of being in a vehicle in that battle." Every vehicle and soldier's movements, gun fire, and fall were captured in facsimile. A four-star general, who was far from the battlefield but close to the human consequences of war, entered the virtual battle and came out with the hair on his arm on end. What did he see?

A panoramic view on three 50-inch TV screens at the resolution of a very good video game. The sky is jet black with oil-fire smoke. A floor of ashen gray desert, wet from rain earlier, recedes to the black horizon. Steel blue hulks of demolished tanks spew tongues of yellow-orange fire which lean and drift in the steady wind. Over 300 vehicles—tanks, jeeps, fuelers, water trucks, even two Iraqi Chevy pickups—roam the landscape. Late in the day a wicked forty-knot Shamal sandstorm kicks up, cutting visibility to a yellow haze of 1,000 meters. Individual infantry soldiers march on the screen. Likewise hundreds of Iraqi soldiers who scramble from their muddy spider holes to hop into their tanks when they realize the shelling is not a precision air attack. Helicopters show up for about six minutes, but the blowing sand shoos them away. Fixed-wing aircraft are deep into another battle behind Iraqi lines.

To enter the battle, the general can pick any vehicle and see what that driver would see. As in the real battle, a low hill might hide a tank. Views are blocked, important things hidden, nothing is clear, everything is happening at once. But in the virtual world you can mount every soldier's dream of a flying carpet and zoom around high above the action. Go up far enough and you get a maplike God's-eye point of view. The truly demented can enter the simulation sitting astride a missile madly arching toward its target.

It's just a three-dimensional movie right now. But here's the next step: allow future cadets to take on the Republican Guard by unleashing what-ifs into the simulation. What if the Iraqis had infrared night vision? What if their missiles had twice the range? What if they weren't out of their tanks at first? Would you still win?

Without the ability to what-if, the Battle of 73 Easting simulation is a very expensive and fanatical documentary. But animated with the tiniest liberty to run in unplanned directions, the simulation takes on a soul and becomes a powerful teacher. It becomes something real in itself. It is no longer just the Battle of 73 Easting. Tuned to different values, equipped with different powers, the model war begins in the same place with the same formation, but quickly runs into its own future. The cadets immersed in the simulation are fighting a hyperreal war, a war only they know about and which only they can fight. The alternative battles they wage are as real as the simulated 73 Easting battle is real, or perhaps even realer, because these battles have unknown endings, much as real life does.

On an everyday basis, the U.S. military thrusts troops into the realm of the hyperreal. At a dozen U.S. Army bases around the world, top-gun tank and aircraft pilots compete in simulated AirLand battles, woven together by a military system called SIMNET, the same window through which the four-star general entered the recreated 73 Easting battle. In the words of *National Defense* columnist Douglas Nelms, SIMNET "transports crews of land and aerial vehicles from planet Earth to a surrogate world where they can do battle without the constraints of safety, cost, environmental impact or geographical boundaries." The first place the SIMNET warriors explore is their backyard. At Fort Knox, Tennessee, 80 crews of M1 tank simulators drive through an amazing virtual reconstruction of Fort Knox's outdoor wargaming arena. Every tree, every building, every creek, every telephone pole, every dip in the land for hundreds of square miles is digitized and represented inside the three-dimensional land of the SIMNET model. The virtual space is huge enough to easily get lost in. One day the troops may ride their greasy real tanks over the real course, and the next day they may traverse the same terrain in facsimile. Only the simulation doesn't smell like burning diesel. When the troops master Fort Knox they can beam themselves to another location by choosing from the computer's menu. Up comes one of two dozen other immaculately rendered places: Fort Irwin's famous National Training Grounds, parts of rural Germany, hundreds of thousands of empty square miles of the oil-rich Gulf States, and (why not?) downtown Moscow.

Standard M1 tanks are the most common entity in the virtual land of SIMNET. Seen from the outside, an M1 simulator never moves: it's a big fiberglass box about the shape of an oversize dumpster that is bolted to the floor. A crew of four men squat, sit, and recline at their cramped stations. The inside is molded in plastic to resemble the gadget-filled interior of the

M1. The men twirl hundreds of facsimile dials and switches and peer into monitors. When the pilot puts a tank simulator into gear, it rumbles, groans, and shakes much like the ride in a real tank.

Eight or more of these fiberglass boxes are electronically linked in the drab Fort Knox warehouse. One M1 can play against the other M1s in SIMNET-land. Long-haul telephone lines link the other 300 existing simulator boxes worldwide into one network, so that 300 vehicles can be hurling through the same virtual battle, even though some of the crew may be at Fort Irwin, California, and others in Graffenvere, Germany.

To boost the realism of SIMNET, military hackers devised vehicles steered by artificial intelligence which are loosely herded by one computer operator. Launching these "semi-automated forces" onto the virtual battlefield, the army can get a bigger, more realistic engagement of forces beyond the 300 simulator boxes built. Says Neale Cosby, who runs the Simulation Center, "We once had a thousand entities on SIMNET at the same time. One guy at a console can throw out 17 semi-automated vehicles, or a company of tanks." Cosby explains the practical virtues of semi-automated forces: "Let's say you are the captain of a national guard unit. You're in charge of an armory of 100 guys coming in on Saturday morning. You want to run your company in a defensive posture, and you want to be attacked by a battalion of 500 people. Well, where are you going to get 500 people Saturday morning in downtown San Diego? So the idea is you can call up SIMNET and have three other guys, each operating a couple of consoles, run those forces against you. You send a message: tonight at 2100 meet us on the Panama database and be ready to go. You could be talking to guys in Germany, Panama, Kansas, and California, and we'd all meet on the same piece of virtual map-sheet. The thing about semi-automated vehicles is that you wouldn't know if they were real or Memorex."

He obviously meant you wouldn't know if they were real simulations or fake simulations (the hyperreal), a modern distinction the military is only now coming to appreciate. The slippery fuzz between the real, the faked, and the hyperreally faked can be used to some advantage in war. U.S. Forces in the Gulf War overturned popular opinion of the relative expertise of both sides. Conventional wisdom said Iraq's forces were older, experienced, and battle hardened; the U.S.'s were young, inexperienced, and couch potatoes with joy sticks. Conventional wisdom was right; only about 1 out of 15 U.S. pilots had previous combat experience; most were fresh out of flight school. Yet the lopsided victory of the U.S. could not be accounted for merely by the absence of gumption from Iraq. Military insiders point to simulation training. A retired colonel asked one commander of the Battle of 73 Easting, "How do you account for your dramatic success, when not a single officer or man in your entire outfit ever had combat experience, and yet you beat Republican Guards who were operating on their own combat training maneuver grounds?" The troop leader answered, "But we *were* experienced. We had fought such engagements six times before in complete battle simulations at the National Training Center and in Germany. It was no different than practice."

Participants of the Battle of 73 Easting were not unique. Ninety percent of the U.S. Air Force units in Desert Storm, and 80 percent of the leaders of the ground forces had intensive training in battle simulations beforehand. The National Training Center (NTC) polished a soldier's SIMNET experience with another level of simulation. NTC, a Rhode Island-size blank spot on the map in the western deserts of California, uses a $100 million hi-tech laser and radio network to simulate battle with real tanks in a real desert. Cocky U.S. veterans dress in Russian uniforms, fight to Russian rules, and occasionally communicate in Russian as they play the home team opposing force (Opfor). They have a reputation of being unbeatable. But not only did U.S. trainees play against mock Iraqi forces drilled in Soviet tactics, but in some cases they simulated specific battle tactics until "they were second nature." For instance, the attack program for the awesome air blitz against Baghdad's targets had been rehearsed in simulated detail for months by U.S. pilots. As a result, only one out of 600 allied aircraft failed to return that first night. Colonel Paul Kern, the commander of a Gulf infantry brigade, told the electrical engineer's journal *IEEE Spectrum*, "Almost every commander I talked to said the combat situations they found in Iraq were not as hard as what they'd encounter at NTC."

What the military is groping towards is "embedded training"—training simulation so real it is indistinguishable from actual combat. It is no leap of faith for the gunner of a modern tank, or a modern jetfighter, to imagine gaining more combat experience in SIMNET simulators than in an Iraqi war. A real tank gunner in a real tank reclines in a tiny windowless burrow tucked into the bowels of a multimillion-dollar steel capsule. He is surrounded by electronics and dials and LED readouts. His only portal to the outside battlefield is on the tiny TV monitor in front of his face which he can swivel like a periscope with his hands. His only link to the rest of his crew is through a headset. For all practical purposes a real gunner in a real tank operates a simulation. For all he knows, the numbers on his dials and the picture on his screen, even the image of the explosion his missiles generate, could be fantasized by a computer. What difference does it make for his job whether the one-inch-tall tanks on his monitor are "real" or not?

For a combatant of the Battle of 73 Easting, simulations came as a trinity. The soldier fought the battle first as a simulation, secondly for real via the simulation of monitors and sensors, and thirdly in the recreated simulation for history. Perhaps someday he wouldn't really be able to tell the difference between them.

That worrisome notion came up once at a NATO-sponsored conference on "Embedded Training," convened to examine this problem. As Michael Moshell, of the Institute for Simulation and Training, recalls, someone read the punch line of a memorable 1985 science fiction novel called *Ender's Game*, written by Orson Scott Card. Card originally wrote *Ender's Game* inside the virtual space of the GEnie teleconferencing system, for an audience who appreciated the hyperreal aspects of online life. In this tale, young boys are trained from childhood to be generals. They play nonstop tactical and strategic games in a zero-gravity space station. Their military training culminates as serious computer war games. Eventually, the most brilliant player and born leader, Ender, supervises a group of teammates in a massive and complex video war game against his adult mentor.

Unbeknownst to them the mentor switches the inputs so that the Nintendo kids in reality are commanding galactic star ships (full of real people) fending off real hostile aliens invading the solar system. The kids win by blowing up the aliens' planet. Later they are told the truth: That wasn't just practice.

A reality switch could be made at other points, too. If there is little difference between simulated tank practice and real war, why not use simulated practice to fight a real war? If you can drive a tank through simulated Iraq from a plastic box connected in Kansas, why not drive a tank through real Iraq from the same safe place? That dream, which meshes so nicely with the Pentagon number-one mandate to lessen U.S. casualties, flitters all across the military these days. Prototype passengerless roving jeeps driven by "telepresent" operators back at the base already zip down real roads. These robo-soldiers keep "humans in the loop" but out of harm's way as the Army prefers. Unmanned but human-piloted aircraft played an immense part in the recent Gulf War. Imagine a very big model airplane loaded with video cameras and computers. These remotely guided planes, steered from bases in Saudi Arabia, served as spy platforms or command relays hovering directly over hostile territory. At the back end, a human leaned into a simulation.

The military's forward vision is big but slow. The power of cheap smart chips is ballooning faster than the Pentagon can think ahead. As far as I can discern, as of 1992, military simulations and war games are only marginally advanced over commercial versions for the public.

JORDAN WEISMAN and buddy Ross Babcock were naval cadets at the Merchant Marine Academy, and deep into dungeons-and-dragons fantasy games. Once on a naval tour they got a peek at a supertanker bridge simulator, a wall of monitors that could fake the color details of a passage through 50 different harbors around the world. They were dying to play. Sorry, this is not a toy, the brass told them. Yes it is, they knew. So they decided to build their own. A simulated world that would let others into their secret of fantasy worlds. They'd use plywood, Radio Shack electronic parts, some homegrown software. And, they would charge admission.

Weisman and Babcock launched BattleTech in 1990. Funded by their lucrative success in the role-playing game business, and based on one of their game's premises, the $2.5 million center runs seven days a week in a mall on the North Pier in downtown Chicago. (With new investment from Tim Disney, Walt's grandson, other centers are opening up around the country.) "Just follow the noise," the attendant on the phone says when I asked for directions. Rowdy teenagers linger at the Star Trek-styled storefront where T-shirts stamped "No Guts, No Galaxy" hang for sale.

BattleTech bears an uncanny resemblance to SIMNET: a set of twelve cramped boxes bolted to a concrete floor linked in an electronic network. Each box is detailed with futuristic nonsense of the outside ("Beware of Blast") and inside stuffed with glorious "switchology"—knobs, meters, flashing lights—a sliding seat, two computer screens, a microphone by which to communicate with teammates, and a few working controls. You steer with foot pedals (as on a tank), you accelerate with a throttle, and you fire with a joystick. At the whistle, the game flickers to life. You are immersed in a redsand desert world chasing other legged tanks (à la *Return of the Jedi*) and being chased in return. The rules are war simple: it's kill or be killed. Driving through the red desert world is cool. The other "mechs," as they are called, dashing about madly in this simulated world are steered by 11 other customers crouched in adjacent boxes. Half are supposed to be on your side, but in the booming mayhem its hard to tell who's who. I see on my readout that my teammates (whom I've not really met) are Doughboy, Ratman, and Genghis. Apparently I'm just "Kevin" on their monitors since I neglected to supply a "handle" before setting off. We are all novices dying early. I am a journalist doing research. Who are they?

Predominantly unmarried males in their twenties, according to a Michigan State University study on fanatical users of the game. The report surveys veterans who have played at least 200 games (at $6 a pop!). Some masters live and work at BattleTech Center calling it "home." I talked to several who've played over a thousand games. Masters of BattleTech claim that it took them about 5 games merely to get used to driving the mech and firing basic weapons, and about 50 games to master cooperating with others. Team-playing is the whole point. Masters see BattleTech primarily as a social contract. To a man (and every master but one is male) they believe that wherever new networked virtual worlds would emerge, special communities of people would come to live in them. When asked what compels them to return to the BattleTech simulated world, the masters mention "the other people," "being able to find competent foes," "fame and glory," "compatible teammates."

The survey queried 47 maniacal players and asked them what BattleTech should change; only two replied that the management should work on "improving reality." Rather the majority wanted lower costs, less crashable software, more of the same ("more mechs, more terrain, more missiles"). Most of all, they wanted more players inside the simulation.

This is the call of the Net. Keep adding players. The more they are connected, the more valuable my connection becomes. It is revealing that these obsessive game players realize they get more "reality" by increasing the fullness of the network than they get by increasing the visual resolution of the environment. Reality is first coevolutionary dynamics, only secondly is it six million pixels.

More is different. Keep adding grains of sand to the first grain and you'll get a dune, which is altogether different than a single grain. Keep adding players to the Net and you get . . . what? . . . something very different . . . a distributed being, a virtual world, a hive mind, a networked community.

While the behemoth size of the military quells innovation, its gigantic scale allows the military to attempt the grand—which nimble commercial entrepreneurs cannot. DARPA, the highly regarded creative research and development branch of the defense department, has drawn up an ambitious next step beyond SIMNET. DARPA would like a 21st century style of simulation. When Col. Jack Thorpe from DARPA gives military briefings promoting this new kind of simulation, he throws up a couple of slides on the overhead projector. One says, Simulation: a Strategic U.S. Technology. Another proclaims,

Simulate Before You Build!
Simulate Before You Buy!
Simulate Before You Fight!

Thorpe is trying to sell the top brass and the military industrialists the key idea that they can get better weapons per buck applying simulation at every point in the process. By designing technology via simulations, testing them via simulated action before committing money for them, and then training users and officers via simulations before actually unwrapping the hardware, they gain a strategic advantage.

"Simulate Before You Build" is already happening to a degree. Northrop built the B-2 stealth bomber without paper. It was simulated in a computer instead. Some industrial experts call the B-2 "the most complex system ever to be simulated." The entire project was designed as a computer simulacra so intricate and precise that Northrop didn't bother fabricating a mechanical mock-up before actually building the billion-dollar plane. Normally a system consisting of 30,000 parts entails redesigning 50 percent of the parts during the course of actual construction. Northrop's "simulate-first" approach reduced that number of refitted parts to 3 percent.

Boeing explored the idea of a hypothetical tilting-rotor aircraft, called the VS-X, by constructing it in virtual reality first. Once built as a simulacra, Boeing sent more than 100 of its engineers and staff inside the simulated aircraft to evaluate it. As one small example of the advantage of simulated building, Boeing's engineers discovered that a critical pressure gauge in the maintenance hatch was obscured from view no matter how hard the crew tried to look at it. So the hatch was redesigned before building, saving millions.

The elaborate platform for this pervasive simulation is codenamed ADST, an awkward acronym that stands for Advanced Distributed Simulation Technology. The key word is "Distributed." Col. Thorpe's distributed simulation technology is nothing less than visionary: a seamless distributed military/industrial complex. A seamless distributed army. A seamless distributed war hyperreality. Imagine a thin film of optical fibers spanning the globe opening a portal to real-time, broadband, multiuser, 3-D simulation. Any soldier who wants to plug into a hyperreal battle, or any defense manufacturer who wants to test a possible product in a virtual reality, need only jack into the great international superhighway-in-the-sky known as Internet. Ten thousand decentralized simulators linked into a single virtual world. Thousands of different kinds of simulators—virtual jeeps, simulated ships, Marines with head-mounts, and shadow forces generated by artificial intelligences—are all summed together into one seamless consensual simulacra.

ARMIES WIN and mobs lose. And the lone Rambo always dies. The most important thing the military knows more about than anyone else is in how to make teams work. Teams are what transform mobs into armies and Rambos into soldiers. Col. Thorpe rightly proclaims that distributed intelligence—not firepower—wins wars. Other visionaries say the same about the future of corporations. "The next breakthrough won't be in the individual interface but in the team interface," says John Seely Brown, the research director of Xerox's PARC.

If Col. Thorpe has his way, the four divisions of the U.S. military and hundreds of industrial contractors become a single interconnected superorganism. The immediate step to this world of distributed intelligence and distributed presence is an engineering protocol developed by a consortium of defense simulation centers in Orlando, Florida. Known as the DSI (Distributed Simulation Internet) protocol, this standard permits independent bits of simulation (a tank here, a building there) to be interleaved into a unified simulation when sent over the existing Internet. In effect, a scene emerges in this virtual space as sufficient parts of it are supplied from afar and assembled in the marvelous decentralized way of swarms. The entire hyperreality of a 10,000-piece battle scene is distributed across many computers through the optic fibers of Internet. The outfit supplying detailed virtual mountains may not supply surging rivers or creeks and may not know whether creeks are flowing down its mountains at all.

Distributed intelligence is the way to go. Students on the Internet (which was developed by DARPA but now is global and demilitarized) can't wait. They see the promise of distributed simulations and have begun making their own versions in quiet corners of the Net.

DAVID SPENDS TWELVE HOURS a day as a swashbuckling explorer in a subterranean world of dungeons and elves. He plays a character called Lotsu. He should be in class getting A grades. Instead he has succumbed to the latest fad sweeping college campuses: total immersion into multiuser fantasy games.

Multiuser fantasy games are electronic adventures run on a large network fed by university and personal computers. Players commonly spend four or five hours a day logged into fantasy worlds based on Star Trek, the Hobbit, or Anne McCaffrey's popular novels about dragon-riders and wizards.

Students like David use school computers, or their own personal machine, to log onto the Internet. This mega-network, now collectively funded by governments, universities, and private corporations around the world, subsidizes all ordinary passengers traveling across it. Colleges freely issue Internet accounts to any student wanting to do "research." By logging on from a dorm in Boston, a student can "drive" to any participating computer in the world, link up for free and stay connected for as long as he or she wishes.

What can one do with such virtual travel, besides downloading papers on genetic algorithms? If 100 other students were to suddenly show up in the same virtual place, it might be pretty cool. You could: throw a party, devise pranks, role-play, scheme, and plot to build a better world. All at the same time. The only thing you'd need is a multiuser place to meet. A place to swarm online.

In 1978, Roy Trubshaw wrote an electronic role-playing game similar to Dungeons and Dragons while he was in his final undergraduate year at Essex College in England. The following year, his classmate Richard Bartle took over the game, expanding the number of potential players and their options for action. Trubshaw and Bartle called the game MUD, for Multi-User Dungeons, and put it onto the Internet.

MUD is very much like the classic game ZORK, or any of the hundreds of text-based adventure video games that have flourished on personal computers since day one. The computer screen says: "You are in a cold, damp dungeon lit by a flickering torch. There is a skull on the stone floor. One hallway leads to the north, the other south. There is a grate on the grimy floor."

Your job is to explore the room and its objects and eventually discover treasures hidden in the labyrinth of other rooms connected to it. You'll probably need to find a small collection of treasures and clues along the way in order to win the motherlode booty, which is usually to break a spell, or become a wizard, or kill the dragon, or escape the dungeon.

You explore by typing something like: "Look skull." The computer replies: "The skull says, 'Beware of the rat.'" You type: "Look grate" and the computer replies: "This way lies Death." You type: "Go north," and you exit through the tunnel on your way into the unknown in the next room.

MUD and its many improved offspring (known generically as MUDs, MUSEs, TinyMUDs, etc.) are very similar to classic 1970s-style adventure games but with two powerful improvements. First, MUDs can handle up to 100 other human players immersed in the dungeon along with you. This is the distributed, parallel characteristic of MUDs. The others can be playing alongside you as jolly partners, or against you as wicked adversaries, or above you as capricious gods creating miracles and spells.

Secondly, and most significantly, the other players (and yourself) can be at work adding rooms, modifying passages, or inventing new and magical objects. You say to yourself, "What this place needs is a tower where a bearded elf can enslave the unwary." So you make one. In short, the players invent the world as they live in it. The game is to create a cooler world than you had yesterday.

MUDs then become a parallel, distributed platform for a consensual superorganism to emerge. Someone tinkers up a virtual holodeck for the heck of it. Later, someone else adds a captain's bridge and maybe an engine room. Next thing you know you have built the *Starship Enterprise* in text. Over the course of months, several hundred other players (who should be doing calculus homework) jack in and build a fleet of rooms and devices until you wind up with fully staffed Klingon battleships, Vulcan planets, and the interconnected galaxies of a *Star Trek* MUD. (Such a place exists on the Internet.) You can log on at any time, 24 hours a day, greet fellow members of the crew—all in role-playing characters—to collectively obey orders broadcast by the captain, and battle enemy ships built and managed by a different set of players.

The more hours one spends exploring and hacking the MUD-world, the more status one earns from the rulers overseeing that world. A player who assists newcomers, or who takes on janitorial chores in keeping the database going, can earn increasing rank and power, such as being able to teleport for free or being exempt from certain everyday laws. Ultimately every MUDer dreams of achieving local god or wizard status. Some become better gods than others. Ideally, gods promote fair play, keep the system going, and help those "below." But stories of abusive and deranged gods are legendary on the Internet.

Real-life events are recapitulated within MUDs and Tiny-MUDs. Players will hold funerals and wakes for characters who die. There have been TinyWeddings for virtual and real people. The slipperiness between real life and virtual life is one of MUD's chief attractions, particularly for teenage kids who are wrestling with their identity.

On a MUD, you define who you are. As you enter a room, others read your description: "Judi enters. She is a tall, dark-haired Vulcan woman, with small pointed ears, and a lovely reddish tinge to her skin. She walks with a gymnast's bounce.

Her green eyes seem to flirt." The author may be petite female with a bad case of acne, or she may be a bearded male masquerading as a women. So many female-presenting characters are actually males pretending at this point that most savvy MUDers now assume all players to be male unless proven otherwise. This has led to a weird prejudice against true female players who are subject to the harassment of "proving" their gender.

Most players live out virtual life with more than one character, as if they are trying out various facets of their persona. "MUDs are a workshop for the concept of identity," says Amy Bruckman, a MIT researcher who studies the sociological aspects of MUDs and TinyMUDs. "Many players notice that they are somehow different on the net than off, and this leads them to reflect on who they are in real life." Flirting, infatuation, romance, and even TinySex are as ubiquitous in MUD worlds as on real campuses. Only the characters vary.

Sherry Turkle, who calls the computer an occasion for a "second self", goes further. She says, "On a MUD, the self is multiplied and decentralized." It is no coincidence that a multiple, decentralized structure is the emerging model for understanding real-life, healthy human selves.

Pranks are also rampant. One demented player devised an invisible "spud" that, when accidentally picked up by another player we'll call Visitor, would remove Visitor's limbs. Others in the room would read: "Visitor rolls about on the floor, twitching excitedly." The gods were summoned to fix player Visitor. But as soon as they "looked" at him, they too got spudded, so that everyone would read, "Wizard rolls about on the floor, twitching excitedly." Ordinary objects can be booby-trapped to do almost anything. A favorite pastime is to manufacture a neat object and get others to copy it without knowing its true powers. For example, when you innocently inspect a "Home Sweet Home" cross-stitch hanging on someone's wall, it might instantly and forcibly teleport you home (while it flashes "There is no place like home").

Since most MUDers are 20-year-old males, violence often permeates these worlds. Elaborate slash-'n'-hack universes repel all but the most thick skinned. But one experimental world running at MIT outlaws all killing and has gathered a huge following of elementary and high school kids. The world, Cyberion City, is modeled on a cylindrical space station. On any one day about 500 kids beam up into Cyberion City to roam or build without ceasing. So far the kids have built 50,000 objects, characters, and rooms. There's a mall with multiplex cinema (and text movies written by kids), a city hall, science museum, a Wizard of Oz theme park, a CB radio network, acres of housing suburbs, and a tour bus. A robot real estate agent roams around making deals with anyone who wants to buy a house.

There is deliberately no map of Cyberion City. To explore is the thrill. Not to be told how things work is the teacher. You are expected to do what the kids do: ask another kid. As Barry Kort, the real-life administrator of the project, says, "One of the charms of entering an unfamiliar environment and culture such as Cyberion City is that it tends to put adults and children back on an equal footing. Some adults would say it reverses the balance of power." The main architects of Cyberion City are 15 years old, or younger. The sheer bustle and intricacy of the land they have built is intimidating to the lone, over-educated immigrant trying to get somewhere, or build *anything*. As *San Francisco Chronicle* columnist Jon Carroll exclaimed on his first visit, "The *psychological* size of the place, all those rooms, and the 'puppets' flitting about, makes it seem like being dropped into downtown Tokyo with a Tootsie Roll and a screwdriver." To survive is the only task.

Kids get lost, then find their way, then they get lost in another sense and never leave. The continuous telecommunication traffic due to nonstop MUDing can cripple a computer center. The college of Amherst outlawed all MUDing from its campus. Australia, linked to the rest of the world by a limited number of precious satellite datalines, banned all international MUDs from the continent. Student-constructed virtual worlds were crowding out bank note updates and calls from Aunt Sheila. Other institutions are sure to follow the ban on unlimited virtual worlds.

Until now, every MUD going (and there are about 200 of them) has been written by fanatical students in their spare time with no one's approval. A couple of pseudo-MUDs have a large following on commercial online services. These almost-MUDs, such as Federation 2, Gemstone, and ImagiNation's Yserbius permit multiusers but give them only limited power to alter their worlds. Xerox PARC is nurturing an experimental MUD running on its company computer. This trial, code-named the Jupiter Project, explores MUDs as a possible environment in which to run a business. An experimental Scandinavian system and a start-up called the Multiplayer Network (running a game called Kingdom of Drakkar) both boast a prototype visual MUD. The dawn of commercial profit-making MUDs is not far away.

Children of the 22nd century will marvel at Nintendo games of the 1990s and wonder why anyone bothered to play a simulation where only one person could enter. It's sort of like having one telephone in the world and no one to talk to.

The future of MUDs, then, converges upon the future of SIMNET, the future of SimCity, and the future of virtual reality. Somewhere in that mix is the ultimate god game. I imagine it as a vast world set into motion with a few well-chosen rules. It is populated by myriad autonomous critters and other creatures who are mere simulacra of distant human players. Characters unfold over time. Tangles grow.

Eventually the simulated world quickens with palpable energy as the interrelations deepen and the entities alter and shape their world. The participants—real, fake, and hyperreal—coevolve the system into a game different than it began. Then, the god himself dons a pair of magic goggles, suits up, and descends into his creation.

The god who lowered himself into his own creation is an old theme. Stanislaw Lem once wrote a great science-fiction classic about a tyrant who kept his world in a box. But another version predates it by millennia.

AS MOSES TELLS THE STORY, on the sixth day of creation, that is at the eleventh hour of a particularly frantic creative bout, the god kneaded some clayey earth and in an almost playful gesture, crafted a tiny model to dwell in his new world. This god, Yahweh, was an unspeakably mighty inventor who built his universe merely by thinking aloud. He had been able to do the rest of his creation in his head, but this part required some fiddling. The final hand-tuned model—a blinking, dazed thing, a "man" as Yahweh called him—was to be a bit more than the other creatures the almighty made that week.

This one was to be a model in imitation of the great Yahweh himself. In some cybernetic way the man was to be a simulacra of Yahweh.

As Yahweh was a creator, this model would also create in simulation of Yahweh's creativity. As Yahweh had free will and loved, this model was to have free will and love in reflection of Yahweh. So Yahweh endowed the model the same type of true creativity he himself possessed.

Free will and creativity meant an open-ended world with no limits. Anything could be imagined, anything could be done. This meant that the man-thing could be creatively hateful as well as creatively loving (although Yahweh attempted to encode heuristics in the model to help it decide).

Now Yahweh himself was outside of time, beyond space and form, and unlimited in scope—ultimate software. So making a model of himself that could operate in bounded material, limited in scale, and constrained by time was not a cinch. By definition, the model wasn't perfect.

To continue where Moses left off, Yahweh's man-thing has been around in creation for millennia, long enough to pick up the patterns of birth, being, and becoming. A few bold man-things have had a recurring dream: to do as Yahweh did and make a model of themselves—a simulacra that will spring from their own hands and in its turn create novelty freely as Yahweh and man-things can.

So by now some of Yahweh's creatures have begun to gather minerals from the earth to build their own model creatures. Like Yahweh, they have given their created model a name. But in the cursed babel of man-things, it has many designations: automata, robot, golem, droid, homunculus, simulacra.

The simulacra they have built so far vary. Some species, such as computer viruses, are more spirit than flesh. Others species of simulacra exist on another plane of being—virtual space. And some simulacra, like the kind marching forward in SIMNET, are terrifying hybrids between the real and the hyperreal.

The rest of the man-things are perplexed by the dream of the model builders. Some of the curious bystanders cheer: how wonderful to reenact Yahweh's incomparable creation! Others are worried; there goes our humanity. It's a good question. Will creating our own simulacra *complete* Yahweh's genesis in an act of true flattery? Or does it commence mankind's demise in the most foolish audacity?

Is the work of the model-making-its-own-model a sacrament or a blasphemy?

One thing the man-creatures know for sure: making models of themselves is no cinch.

The other thing the man-things should know is that their models won't be perfect, either. Nor will these imperfect creations be under godly control. To succeed at all in creating a creative creature, the creators have to turn over control to the created, just as Yahweh relinquished control to them.

To be a god, at least to be a creative one, one must relinquish control and embrace uncertainty. Absolute control is absolutely boring. To birth the new, the unexpected, the truly novel—that is, to be genuinely surprised—one must surrender the seat of power to the mob below.

The great irony of god games is that letting go is the only way to win.

# 14

# In the Library of Form

My path to the fiction section on the third floor of the university library meandered through hundreds of thousands of books sleeping on shelves. Have these books ever been read? Way in the back of the library, where the dark fluorescent lights must be turned on by the browser, I searched the international literature section for the work of the Argentinean author Jorge Luis Borges.

I found three shelves packed with books Borges wrote or that were written about him. Borges's stories are famously surreal. They are so absolutely fake that they appear real; they are literate hyperreality. Some of the books were in Spanish, some were biographies, some were full of poems, some were anthologies of his minor essays, some were duplicate copies of other books on the shelf, some were commentaries upon the commentaries on his essays.

I ran my hand over the volumes, thick, thin, slim, oversize, old, and newly bound. On a whim I slid out a worn chestnut-covered book. I opened it. It was an anthology of interviews Borges did in his eighties. The interviews were conducted in English, which Borges wielded more gracefully than most native speakers. I was stunned to find that the last 24 pages contained an interview with Borges, based on his writings in *Labyrinths*, which properly could only exist in my book, this book, *Out of Control*.

The interview began with my question: "I read in one of your essays about a labyrinthine maze of books. This library contained all possible books. It was clear that this library was born as a literary metaphor, but such a library now appears in scientific thought. Can you describe the origin of this hall of books to me?"

BORGES: The universe (which others call the Library) is composed of an indefinite and perhaps infinite number of hexagonal galleries, with vast air shafts between, surrounded by very low railings. There are five shelves for each of the hexagon's walls; each shelf contains thirty-five books of uniform format; each book is of four hundred and ten pages; each page, of forty lines, each line, of some eighty letters which are black in color.

ME: What do the books say?

BORGES: For every sensible line of straightforward statement in the books there are leagues of senseless cacophonies, verbal jumbles and incoherence. Nonsense is normal in the Library. The reasonable (and even humble and pure coherence) is an almost miraculous exception.

ME: You mean all the books are full of random letters?

BORGES: Nearly. One book which my father saw in a hexagon on circuit 1594 was made up of the letters *MCV*, perversely repeated from the first line to the last. Another (very much consulted, by the way) is a mere labyrinth of letters, but the next-to-the-last page says *Oh time thy pyramids*.

ME: But there must be some books in the Library which make sense!

BORGES: A few. Five hundred years ago, the chief of an upper hexagon came upon a book as confusing as the others, but which had nearly two pages of homogeneous lines. The content was deciphered: some notions of combinative analysis, illustrated with examples of variation with unlimited repetition.

ME: That's it? Two pages of rational sense discovered in five hundred years of searching? What did the two pages say?

BORGES: The text of the two pages made it possible for a librarian to discover the fundamental law of the Library. This thinker observed that all the books, no matter how diverse they might be, are made up of the same elements: the space, the period, the comma, the twenty-two letters of the alphabet. He also alleged a fact which travelers have confirmed: In the vast Library there are no two identical books. From these two incontrovertible premises he deduced that the Library is total and that its shelves register all the possible combinations of the twenty-odd orthographical symbols (a number which, though extremely vast, is not infinite).

ME: So, in other words, any book you could possibly write, in any language, could be found (theoretically) in the Library. It contains all past and future books!

BORGES: Everything: the minutely detailed history of the future, the archangels' autobiographies, the faithful catalog of the Library, thousands and thousands of false catalogs, the demonstration of the fallacy of the true catalogue, the Gnostic gospel of the Basilides, the commentary on that gospel, the commentary on the commentary on that gospel, the true story of your death, the translation of every book in all languages, the interpolations of every book in all books.

ME: One would have to guess, then, that the Library holds immaculate books—books of the most unimaginably beautiful writing and penetrating insight—books better than the best literature that anyone has written so far.

BORGES: It suffices that a book be possible for it to exist in the Library. On some shelf in some hexagon there must exist a book which is the formula and perfect compendium *of all the rest.* I pray to the unknown gods that a man—just one, even though it were thousands of years ago!—may have examined and read it.

Borges then went on at great length about a blasphemous sect of librarians who believed it was crucial to eliminate useless books: "They invaded the hexagons, showed credentials which were not always false, leafed through a volume with displeasure and condemned whole shelves."

He caught the curiosity in my eyes and said, "Those who deplored the 'treasures' destroyed by this frenzy neglect two notable facts. One: the Library is so enormous that any reduction of human origin is infinitesimal. The other: every copy is unique, irreplaceable, but (since the Library is total) there are always several hundred thousand imperfect facsimiles: works which differ only in a letter or comma."

ME: But how would one discern the difference between the real and the almost? Such proximity means that this book I hold in my hands not only exists in the Library, but so does a similar one, differing only by an alternative word in a previous sentence. Perhaps the related book reads: "every copy is *not* unique, irreplaceable." How would you know if you ever found the book you were looking for?

There was no reply. When I looked up I noticed I was surrounded by dusty shelves in an eerily lit hexagonal room. By some fantastical logic, I was standing in Borges's Library. Here were the twenty shelves, and the receding layers upon layers of upper and lower floors visible between the low railing, and the labyrinth of corridors lined with books.

Borges's Library was as marvelous as it was a temptation. For two years I had been working on the book you now hold. At that time I was one year past my deadline. I couldn't afford to finish it, and I couldn't afford to not finish it. A grand resolution to my dilemma lay somewhere in this Library of all possible books. I would search Borges's Library until I found on some shelf the best of all possible books I could write, one entitled *Out of Control.* This would be a book already written, edited, and proofed. It would spare me another year of tortuous work, work I was not sure I was even up to. It certainly seemed worth a try looking for it.

So I set off down the endless corridors of book-filled hexagons.

After passing through the fifth hexagon, I paused and on a whim I reached out and dislodged a stiff green book from a cramped upper shelf. Inside it was utter chaos.

So was the one next to it, and the next after that. I fled this hexagon and walked quickly through identical corridors of hexagons for about a half mile, until I stopped again and plucked a book from a nearby shelf without deliberation. The book was rotten with the same gibberish. I checked the entire row and found the same rot. I inspected several other spots in the hexagon and could not distinguish any improvement among them. For several more hours I wandered changing directions, checking hundreds of books, some on lower shelves near my feet and some perched almost at the ceiling, but all contained the same undistinguished garbage. There appeared to be billions of books of nonsense. A book entirely full of the letters *MCV*, as Borges's father found, would have been quite exhilarating.

Yet the temptation lingered. I figured I could spend days, or even weeks, searching for the completed *Out of Control* book by Kevin Kelly, at a profitable gamble. I might even find a better *Out of Control* book by Kevin Kelly than I could write myself, for which I would be thankful to spend a year hunting.

I stopped to rest upon the small landing on one of the spiral staircases that wound between floors. I reflected on the design of the Library. From where I sat I could see nine stores up the air shaft and nine below, and about a mile in the six directions of the honeycombed floors. If this Library contained all possible books, my reasoning went, then any volumes that fit the rules of grammar (let alone were interesting) would be so tiny a fraction of the total books, that my coming upon one by random search would be miraculous. Five hundred years sounded about right as the time needed to find two sensible pages—any two sensible pages. To find a readable book would take several millennia, with luck.

I decided to take a different tack.

There were a constant number of books per shelf. There were a constant number of shelves per hexagon. All the hexagons were uniform, lit by a grapefruit-size bulb of light, interspersed by hallways with two closet doors and a mirror in each. The Library was ordered.

If the Library was ordered that meant (most likely) the books it contained were also ordered. If the volumes were arranged so that books that differed only slightly were placed near each other, and books that differed greatly were separated widely, then this organization would yield a way for me to fairly quickly find a readable book somewhere in this Library of all possible books. If this vastness of the Library was so ordered, there was even a chance I could put my hands on a completed *Out of Control*, a book embossed with my name on the title page, but which I did not have to write.

I commenced my shortcut to achievement by selecting a book from the nearest stack. I spent ten minutes studying its nonsense.

I strode a hundred yards away to the seventh nearest hexagon and picked another book. I did the same in turn for each of the six radiating directions. I scanned the six new texts and then I selected the one that held the most "sense" compared to the first. In one I found a sensible three word sequence: "or bog and." Then I repeated the search routine using this "bog" volume as the base, comparing texts in the six directions around it. After several iterations I uncovered a book whose noisy pages contained *two* phraselike sequences. I was getting warmer. After many iterations of this ritual I found a book with four English phrases hidden among the detritus of garbled letters.

I quickly learned to search very wide—about 200 hexagons in each direction—spreading out from the last "best" book in order to explore the Library faster. I kept progressing in this fashion until I found books with many English phrases, although the clauses were scattered among the pages.

My hours turned to days. The topological pattern of "good" books formed an image in my mind. Every complete grammatical book in the Library sat in a disguised epicenter. At the center was the book; immediately surrounding it were shelves of close facsimiles of the book; each facsimile contained a mere alteration in punctuation—an inserted comma, a deleted period. Ringing these books were shelves of lesser counterfeits that altered a word or two. Surrounding this second ring was a further broad ring of books that differed by whole sentences, most of them degraded illogical statements.

I imagined the rings of grammar as a map of contour lines circling round a mountain. The map represented a geography of coherence. A single celestial, readable book resided on a summit's peak; below it lay ever greater masses of baser books. The lower the books, the more base they were, and the greater was the circumference of their bulk. The entire mountain of "almost" books stood in an enormous plain of undifferentiated nonsense.

To find a book then was a matter of scaling the summit of order. As long as I made sure that I was always climbing uphill—always marching toward books that contained more sense—I would inevitably arrive at the apex of a readable book. As long as I moved through the Library across the contour of increasingly better grammar, then I would inevitably arrive at the hexagon harboring a wholly grammatical book—the peak.

After several days of using what I began to call the Method, I found a book. Such a book could not have been found by aimless rambling of the kind that produced the two pages Borges's father found. Only the Method could have guided me to this center of coherence. I justified my investment of time by reminding myself that I found more with the Method than generations of librarians had uncovered by their unorganized rambles.

As forecasted by the Method, the book I found (entitled *Hadal*) was surrounded by broad concentric rings of similar pseudo-books. But the text itself, although grammatically correct, was disappointingly bland, flat, characterless. The most interesting parts read like very bad poetry. There was one line alone that shone with remarkable intelligence and has stuck with me: "The present is hidden from us."

However, I never did find a copy of *Out of Control*. Nor did I find a book that could steal an evening from me. I see now that would have taken years, even with the Method. Instead, I exited from Borges's Library into the university library and then returned home to conclude *Out of Control* by writing it myself.

The Method tickled my curiosity and distracted me from my writing. Was it widely known among travelers and librarians? I was prepared for the probability that others must have uncovered it in the past. Returning to the university library (finite and cataloged), I searched for a book with an answer. I bounced from index to footnote, from footnote to book, landing far from where I began. What I found amazed me. The truth seemed farfetched: Scientists believe the Method has saturated our world since time immemorial. It was not invented by man; by God perhaps. The Method is a variety of what we now call evolution.

If we can accept this analysis, then the Method is how we have all been found.

More amazing yet: I had taken Borges's Library to be the private dream (a virtual reality) of an imaginative author, yet I read with growing fascination that his Library was real. I believe the sly Borges had known this all along; he had cast his account as fiction, for who would have believed him? (Others say his fiction was a way to jealously guard his access to this most awesome space.)

Two decades ago nonlibrarians discovered Borges's Library in silicon circuits of human manufacture. The poetic can imagine the countless rows of hexagons and hallways stacked up in the Library corresponding to the incomprehensible microlabyrinth of crystalline wires and gates stamped into a silicon computer chip. A computer chip, blessed by the proper incantation of software, creates Borges's Library on command. The initiated chip employs its companion screen to display the text of any book in Borges's Library; first a text from block 1594, the next from the little visited section 2CY. Pages from the books appear on the screen one after another without delay. To search Borges's Library of all possible books, past, present, and future, one needs only to sit down (the modern solution) and click the mouse.

Neither the model, the speed, the soundness of design, or the geographical residence of the computer makes any difference while generating a portal to Borges's Library. This Borges himself did not know, although he would have appreciated it: that whatever artificial means are used to get there, all travelers arrive at exactly the same Library. (Which is to say all libraries of possible books are identical; there are no counterfeit Libraries of Borges; *all copies of the Library are original.*) The consequence of this universality is that any computer can create a Borgian Library of all possible books.

THE MOST POWERFUL COMPUTER made in 1993, the Connection Machine 5 (CM5), can effortlessly generate Borges's Library of books. But the CM5 can also generate equally vast and mysterious Borgian Libraries of complex things other than books.

Karl Sims, who works for Thinking Machines, the maker of the CM5, has made a Borgian Library of art and pictures. Sims first wrote special software for the Connection Machine and then constructed a universe (which others call a Library) of all possible pictures. The same machinery that can generate a possible book can generate a possible picture. In the former case the output are letters printed in linear sequence; in the latter, a rectangle of pixels displayed on a screen. Sims hunts for patterns of pixels instead of patterns of letters.

I visit Sims in his dark office cubicle at Thinking Machines's Cambridge, Massachusetts, offices. Two extra-large, bright monitors sit on Sims's desk. His largest monitor is divided into a matrix of 20 small projected rectangles, 4 down and 5 across. Each rectangle is a window that at the moment shows a realistically marbled doughnut. Each of the 20 pictures is slightly varied in pattern.

Sims uses his mouse to click on the lower right corner rectangle. In a blink all 20 rectangles are refreshed with newly marbled doughnuts, each new image a slight variation of the formerly selected corner pattern. By clicking on a sequence of images, Sims can walk through a Borgian Library of visual patterns using the Method. Instead of bodily running ahead seven yards (in many directions) to reach a stored pattern, Sims's software calculates what the pattern would logically be seven yards away (since it turns out the Borgian Library is extremely ordered). He then paints the newfound pattern on the screen. The Connection Machine does this in milliseconds, simultaneously figuring the new patterns in 20 different directions away from the last selection.

There is no limit to what picture could possibly appear from the Library. In true Borgian fashion, this total universe contains all shades of rose, all stripes; it contains the *Mona Lisa*, and all *Mona Lisa* parodies; every swirl, the blueprints of the Pentagon, all of Van Gogh's sketches, every frame from *Gone With the Wind*, all speckled scallop shells. These are desires, though; on whimsical rambles through this Library, Sims harvests chiefly windows filled with amorphous blotches, streaks, and psychedelic swirls of color.

The Method—as evolution—can be conceived of not as traveling but as breeding. Sims describes the twenty new images as twenty children of an original parent. The twenty pictures vary just as offspring do. Then he selects the "best" offspring, which in turn immediately sires twenty new variations. He'll pick the best of that lot, and that best will sire twenty more variations. He can begin with a simple sphere and by cumulative selection end with a cathedral.

Watching the forms appear, multiply in variation, get selected, ramify in form, winnow again, and begin to drift over generations to ever more complicated shapes, neither mind nor gut can escape the impression that Sims is really breeding images. Richer, wilder, more esthetically fit images unfold over generations. Sims and fellow computationalists call it artificial evolution.

The mathematical logic of breeding pictures is indistinguishable from the mathematical logic of breeding pigeons. Conceptually the two processes are equivalent. Although we may call it artificial evolution, there is nothing about it that is more or less artificial than breeding dachshunds. Both methods are equally artificial (of the art) and natural (true to nature).

In Sims's universe evolution has been yanked from the living world and left naked in mathematics. Stripped of its cloak of tissue and hair, stolen from its womb of moist wet flesh, and then spirited into circuits, the vital essence of evolution has moved from the world of the born to the world of the made, from its former sole domain of carbon ring to the manufactured silicon world of algorithmic chips.

The shock is not that evolution has been transported from carbon to silicon; silicon and carbon are actually very similar elements. The shock of artificial evolution is that it is fundamentally natural to computers.

Within ten cycles, Sims's artificial breeding will produce something that is "interesting." Often as few as five hops will land Sims someplace that is greater than mere chaotic splatters. While he clicks from picture to picture, Sims talks, as Borges did, of "traveling through the Library," or "exploring the space." The pictures exist "out there" even though they are not rendered into visual form until found or selected.

The electronic version of Borges's Library of books can be considered in the same way. The book texts exist abstractly, independent of form. Each sleeps in its assigned spot on a virtual shelf in the virtual Library. When selected, the cabalistic silicon chip breathes form into a book's virtual self to awaken the text onto the screen. A conjurer travels to a place in the space (which is ordered) and there awakens the particular book that must rest there. Every coordinate has a book; every book a coordinate. Just as for the traveler, one vista opens up many new possible locations for yet more vistas; in the Library one coordinate begets many subsequent related coordinates. An initiated librarian travels through the space in sequential hops; the path is a chain of selections.

Thus the six texts derived from the original text are six relatives; they share a familial form and informational seed. In the scale of the Library their variation is on the order of siblings. Since they are relatives derived in a following generation, they can thus be called offspring. The single chosen "best" offspring text becomes the parent in the next round; one of its six grand-offspring variations will become the parent in that generation.

While I was within Borges's Library, I saw myself hunting for a readable book over a trail that began at gibberish. But another looking in would see me breeding a nonsense book into a viable book, just as one might domesticate a disorganized wildflower into the elegant cup of a rose through many generations of selection.

Karl Sims breeds gray noise into jubilant images of plant life on the CM5. "There is no limitation to what evolution can come up with. It can surpass the design capabilities of humans," he claims. He devised a way to rope off the immense Library so that his wanderings would stay within the range of all possible plant forms. As he evolved his way through this space, he copied "seeds" of those forms he found most intriguing. Later Sims reconstituted his harvest and rendered them into fantastical three-dimensional plant shapes that he could animate. His domesticated forest included a giant unrolling fern frond, spindly pine things with a Christmas ball on top, grass with crab-claw blades, and twisty oak trees. Eventually these bizarre, evolved plants populated a video of his creations called *Panspermia*. In this animation, alien trees and strange giant grasses sprouted from seeds, eventually carpeting a barren planet with an unearthly jungle of rooted things. The evolved (now animated) plants produced their own seeds which were blasted from a bulbous cannon of a plant into space and onto the next barren world (the process of Panspermia).

Karl Sims is not the only explorer of the architecture of the Borgian universe (which some call the Library), nor was he the first. As far as I can tell, the first librarian of a synthetic Borgian world was the British zoologist Richard Dawkins. In 1985, Dawkins invented a universe he called "Biomorph Land." Biomorph Land is the space of possible biological shapes constructed with short straight lines and branches. It was the first computer-generated library of possible forms that could be searched by breeding.

Dawkins wrote Biomorph Land as an educational program to illustrate how designed things could be created without a designer. He wanted to demonstrate visually that while random selection and aimless wandering would never produce a coherent design, cumulative selection (the Method) could.

Despite a prestigious reputation in biology, Dawkins was experienced in programming mainframe computers. Biomorph is a fairly sophisticated computer program. It draws a stick of a certain length, and in a growthlike pattern, adds branches to it, and branches to the branches. How the branches fork, how many are added, and at what length they are added are all values that can vary independently by small amounts from form to form. In Dawkins's program these values also "mutate" at random. Every form it draws differs by one mutation of nine possible variables.

Dawkins hoped to traverse a library of tree shapes by artificial selection and breeding. A form was born in Biomorph Land as a line so short it was a dot. Dawkins's program generated eight offspring of the dot, much as Sims's later program would do. The dot's children varied in length depending on what value the random mutation assigned. The computer projected each offspring, plus the parent, in a nine-square display. In the now familiar style of selective breeding Dawkins selected the most pleasing form (his choice) and evolved a succession of ever more complex variant forms. By the seventh generation, offspring were accelerating in filigreed detail.

That was Dawkins's hope as he began writing the code in

BASIC. If he was lucky in his programming he'd get a universe of wonderfully diverse branching trees.

The first day he got the program running, Dawkins spent an exhilarating hour rummaging through the nearest shelves of his Borgian Library. Progressing a mutation at a time, he came upon unexpected arrangements of stem, stick, and trunk. Here were odd trees nature had never claimed. And line drawings of bushes, grass, and flowers that never were. Echoing the dual metaphor of evolution and libraries, Dawkins wrote in *The Blind Watchmaker*, "When you first evolve a new creature by artificial selection in the computer model, it feels like a creative process. So it is, indeed. But what you are really doing is *finding* the creature, for it is, in a mathematical sense, already sitting in its own place in the genetic space of Biomorph Land."

As the hours passed, he noticed he was entering a space in the Library where the branching structures of his trees began to cross back upon themselves, filling in areas with crisscrossing lines until they congealed into a solid mass. The recursive branches closed upon themselves forming little bodies rather than trunks. Auxiliary branches still sprouting from these bodies looked surprisingly like legs and wings. He had entered the part of the Library where insects dwelled (despite the fact that he as God had not intended there be such a country!). He discovered all sorts of weird bugs and butterflies.

Dawkins was astonished: "When I wrote the program I never imagined it would evolve anything but treelike shapes. I had hoped for weeping willows, poplars, and cedars of Lebanon."

Now there were insects everywhere. Dawkins was too excited to eat that evening. He spent more hours discovering amazingly complex creatures looking like scorpions and water spiders and even frogs. He said later, "I was almost feverish with excitement. I cannot convey the exaltation I felt of exploring a land which I had supposedly made. Nothing in my biologist's background, nothing in my 20 years of programming computers, and nothing in my wildest dreams, prepared me for what actually emerged on the screen."

That night he couldn't sleep. He kept pressing on, dying to survey the extent of his universe. What other surprises did this supposedly simple world contain? When he finally fell asleep in the early morning, images of "his" insects swarmed in his dreams.

Over the following months, Dawkins tramped the backwaters of Biomorph Land hunting for nonplant and abstract shapes. The short list of forms he encountered included: "fairy shrimps, Aztec temples, Gothic church windows, and aboriginal drawings of kangaroos." Making the best use of an idle minute here and there, Dawkins eventually used the evolutionary method to locate many letters of the alphabet. (These letters were bred into visibility, not drawn.) His goal was to capture the letters in his name, but he never could find a passable *D* or a decent *K*. (On the wall of my office I have a wonderful poster of the 26 letters and 10 numerals found shimmering on living butterfly wings—including a marvelous *D* and *K*. But although these letters evolved, they were not found by the Method. The photographer, Kjell Sandved, told me he inspected more than a million wings to gather all 36 symbols.)

Dawkins was on a quest. He later wrote, "There are computer games on the market in which the player has the illusion that he is wandering about in an underground labyrinth, which has a definite if complex geography and in which he encounters dragons, minotaurs or other mythic adversaries. In these games the monsters are rather few in number. They are all designed by a human programmer, and so is the geography of the labyrinth. In the evolution game, whether the computer version or the real thing, the player (or observer) obtains the same feeling of wandering metaphorically through a labyrinth of branching passages, but the number of possible pathways is all but infinite, and the monsters that one encounters are undesigned and unpredictable."

Most magically the monsters in this space were seen once and then were lost. The earliest versions of Biomorph Land did not have a function for saving the coordinates of every biomorph. The shapes appeared on the screen, roused from their shelf in the Library, and when the computer was turned off, they returned to their mathematical place. The probability of encountering them again was infinitesimal.

When Dawkins first arrived in the district of insects he

desperately wanted to keep one so he could find it again. He printed out a picture of it, and a picture of all the 28 ancestral forms he evolved along the way to get to it, but at that time his prototype program would not let him save the underlying numbers enabling him to reconstruct the form. He knew that once he flicked his computer off that night, the insect biomorphs would be gone except for the wisp of their souls held by their portraits. Could he ever reevolve identical forms? He killed the power. He had proof, at least, that they existed somewhere in his Library. Knowing they were there haunted him.

Despite the fact that Dawkins had both the starting point and the sequence of 28 "fossils" leading up to the specific insect he was trying to recapture, the biomorphs remained elusive. Karl Sims, too, once bred a dazzling, luminescent image of colorful loopy strings on his CM5—very reminiscent of a painting by Jackson Pollock—before he wrote a coordinate-saving feature; he too was never able to rediscover the image, although he owns a slide of it to serve as a trophy.

Borgian space is vast. Deliberately relocating a point in this space is as difficult as replaying an identical game of chess. A tiny, almost undetectable error of choice at any turn can carry one to a destination miles from one's aim. In Biomorph space the complexity of the forms, the complexity of choices at each juncture, and the subtlety of their differences, guarantees that every evolved form is probably the first and last visit.

Perhaps in the Library of Borges there is a book called *Labyrinths* that holds the following miraculous story (not contained in the book *Labyrinths* found on the shelf in the university library). In this book Jorge Luis Borges tells how his father, who was a traveler in the universe of all possible books, once came upon a sensible book in this confusing vastness. All four hundred and ten pages of the tome, including the table of contents, were filled with two sentence palindromes. The first 33 palindromes were both riddles and profound. That's all his father had time to read before an unusual fire in the basement forced the evacuation of the librarians working in this section. In the semi-orderly panic of exit, his father forgot the location of this volume. Out of shame the existence of the Book of Palindromes has never been mentioned outside the Library. For eight generations, a somewhat secretive association of exlibrarians has been meeting regularly to methodically retrace the old traveler's steps so that they might rediscover this book in the Library's enormity. There is little hope they will ever find their holy grail.

To demonstrate how vast such Borgian spaces are, Dawkins offered a prize to anyone who could rebreed (or find by hit or miss!) an image of a chalice that Dawkins had come upon by chance on one of his rambles in Biomorph Land. He called it the Holy Grail. So sure was Dawkins of its deep concealment that he offered $1,000 to the first person presenting him with the genes to the Holy Grail. "Offering my own money," said Dawkins, "was my way of saying nobody was going to find it." Much to his astonishment, within one year of his challenge, Thomas Reed, a software engineer in California, reencountered the cup. This appears akin to retracing the elder Borges's steps to locate the lost palindrome book, or the feat of finding *Out of Control* in the Library of Borges.

But Biomorph Land supplies assistance. Because its genesis reflects Dawkins's professional interests as a biologist, it was built on organic principles in addition to evolution. The secondary biological nature of biomorphs permitted Reed to find the chalice.

Dawkins saw that in order to make a practical biological universe, he would have to restrict the possibilities of forms to those that held some biological sense. Otherwise, the sheer vastness of all shapes would overwhelm any ordinary chance of finding enough biological morphs to play with—even using the cumulative selection method. After all, he reasoned, the embryonic development of living creatures limits the possibilities of what they can mutate into. For instance, most biological creatures display left-right symmetry; by instituting left-right symmetry as a fundamental element of every biomorph, Dawkins could reduce the overall size of the Library, thus making it easier to find a biomorph. He called this reduction a "constrained embryology." The task he set for himself was to design an embryology that was restricted, but in "biologically interesting directions."

"Very early I had a strong intuitive conviction that the embryology I wanted should be recursive. My intuition was based partly upon the fact that embryology in real life can be thought of as recursive," Dawkins told me. By recursive embryology, Dawkins meant that simple rules iterated over and over again (including rules that play upon their own results) would furnish much of the complexity of the final form. For instance, as the recursive rule "grow one unit then fork into two" is applied over successive generations to a starting stick, it will produce a bushy many-forked thing after about five iterations.

Secondly, Dawkins introduced the idea of gene and body into the Library. He saw that a string of letters (as in a book) is directly analogous to biological genes. (A gene is even represented as a string of letters in the formal notation of biochemistry.) The genes produce the tissues of the body. "But," says Dawkins, "biological genes don't control small fragments of the body, which would be the equivalent of controlling pixels on a monitor. Instead, genes control growing rules—embryological developmental processes—or in Biomorph Land, drawing algorithms." Thus, a string of numbers or text acts as a string of genes (a chromosome), which represents a formula, which then draws the image (body) in pixels.

The consequences of this indirect way of generating forms was that almost any random place in the Library—that is, almost any genes—produced a coherent biological shape. By having genes control algorithms rather than pixels, Dawkins built an inherent grammar into his universe which prevented any old nonsense from appearing. Even a wild mutation would not arrive at a flat gray blob. The same transformation could be done to the Library of Borges. Rather than each shelf place in the Library representing a possible arrangement of letters, each place could represent a possible arrangement of *words*, or even of possible *sentences*. Then, any book you picked out would at least be close to readable. This enhanced space of word strings is much smaller than the space of letter strings, but also, as Dawkins suggested, restricted in a more interesting direction: you are more likely to come across something comprehensible.

Dawkins's introduction of genes that behaved in a biological manner—each mutation affecting many pixels in a structured way—not only shrunk the biomorph library's size, distilling it to functional forms, but also provided an alternative way for human breeders to find a form. Any subtle shift made in the biomorph gene space would amplify into a noticeable ***and dependable*** shift in graphic image.

This gave Thomas Reed, freelance knight of the Holy Grail, a second way of breeding. Reed repeatedly altered genes of a parent form while observing the visual changes in forms the genes produced in order to learn how to steer a shape by altering individual genes. In this way he could steer to various biomorph forms by twiddling the gene dial. In an obvious analogy, Dawkins called this mode in his program "genetic engineering." As in the real world, it holds uncanny power.

In effect, Dawkins lost his $1,000 to the first genetic engineer of artificial life. Thomas Reed spent his lunch hours at work hunting for the chalice in Dawkins's program. Six months after Dawkins announced his contest, Reed converged upon the lost treasure by a combination of breeding images and genetically engineering their genes. Breeding is a way to brainstorm fast and loose; engineering is a way to fine-tune and control. Of the forty hours Reed estimated he spent hunting for the cup, he spent 38 of them engineering. "There is no way I could have found it by breeding," he said. As he closed in on the cup, Reed couldn't get the last pixel to budge without getting everything else to move. He spent many hours trying to control that single pixel in the penultimate form.

In a coincidence that completely astonished Dawkins, two other finders independently submitted correct gene solutions to the Holy Grail within weeks after Reed. They too were able to pinpoint his chalice in an astronomically large space of possibilities, not by breeding alone, but primarily by genetic engineering and, in one case, by reverse engineering.

Perhaps because of the visual nature of Biomorph Land, the first people to incorporate Dawkins's idea of computational breeding were artists. The first was a fellow Brit, William Latham; later Karl Sims in Boston would take artificial evolution further.

The exhibited work of William Latham in the early 1980s resembled a parts catalog from some unfathomable alien contraption. On a wall of paper, Latham drew a simple form, such as a cone, at the top center, and then filled the rest of the space with gradually complexifying cone shapes. Each new shape was generated by rules that Latham had devised. Thin lines connected one shape to its modified descendant shapes. Often, multiple variations would split off one form. By the bottom of these giant pages, the cone forms had metamorphosed into ornate pyramids and art-deco mounds. The logical structure of the drawing was a family tree, but with many common cross-marriages. The entire field was packed; it looked more like a network or circuit.

Latham called this "obsessive, rule-based process" of generating varieties of forms and selecting certain offspring to develop further, "FormSynth." Originally he used FormSynth as a tool to brainstorm ideas for possible sculptures. He would select a particularly pleasing form lifted from the map of his sketches and then sculpt the intricate shape in wood or plastic. One of Latham's gallery catalogs shows a modest black statue with a resemblance to an African mask that Latham created (or found) using FormSynth. But sculpting was so time-consuming, and in a way superfluous, that he ceased doing it. What most interested him was that vast uncharted Library of possible forms. Latham: "My focus shifted from producing a single sculpture to producing millions of sculptures, each spawning a further million sculptures. My work of art was now the whole *evolutionary tree of sculptures*."

Inspired by an avalanche of dazzling 3-D computer graphics in the U.S. in the late 1980s, Latham took up computing as a way to automate his form generation. He collaborated with programmers at an IBM research station in Hampshire, England. Together they modified a 3-D modeling program to produce mutant forms. For about a year artist Latham manually typed in or edited gene values in his shape-generating program to produce wonderfully complete trees of possible forms. By modifying a form's code by hand, Latham could search the space at random. With understatement Latham recalls this manual search as being "laborious."

In 1986, after encountering the newly published Biomorph program, Latham merged the heart of Dawkins's evolutionary engine with the sophisticated skin of his three-dimensional forms. This union birthed the idea of an evolutionary art program. Latham dubbed his method "the Mutator." The Mutator functioned almost identically to Dawkins's mutating engine. The program generated offspring of a current form, each with slight differences. However, instead of stick figures, Latham's forms were fleshy and sensual. They popped into one's consciousness in three dimensions, with shadows. Whole eye-riveting beasts were drummed up by the hi-octane IBM graphics computer. The artist then selected the best of the 3-D progeny. That best form became the next parent, begetting other mutations. Over many generations, the artist would evolve a completely new three-dimensional body in a true Borgian Library. Biomorph Land—huge as it was—was only a subset of Latham's space.

Echoing Dawkins, Latham states, "I had not anticipated the variety of sculpture types which my software could create. There appears no limit to the wealth of different forms that can be created using this method." The forms Latham retrieved, rendered in mind-boggling detail, include elaborately woven baskets, marbled giant eggs, double mushroom-things, twisty antlers from another planet, gourds, fantastical microbial beasts, starfish gone punk, and a swirling multi-arm Shiva god from outer space that Latham calls "Mutation Y1."

"A garden of unearthly delights," Latham calls his collection of forms. Rather than try to imitate the motif of earthly life, Latham is after alternative organic forms, "something more savage" than life on Earth. He remembers visiting a county fair and stopping by an artificial insemination tent and seeing photographs of gigantic mutant superbulls and other kinds of "useless" freaks. He finds these bizarre forms inspiring.

The printouts are surrealistically clear, as if photographed in the vacuum of the moon. Every form possesses a startling organic feel to it. These things are not copies of nature but natural shapes that do not exist on Earth. Latham: "The machine gave me freedom to explore forms which previously had not been accessible to me, as they had been beyond my imagination."

Deep in the recesses of the Borgian Library, racks of graceful antlers, shelves of left-handed snails, rows of dwarf flowering trees, and trays of lady bugs await their first visitor, whether that be nature or artist. As yet, neither nature nor artist has reached them. They remain unthought of, unseen, unmaterialized, mere possible forms. As far as we know, evolution is the only way to reach them.

The Library contains all the forms of life past and life future and even, perhaps, the shape of life present on other planets. We are blocked by our own natural prejudices from contemplating these alternative life forms in any detail. Our minds quickly drift back to what we know as natural. We can give it a momentary thought, but we balk at filling in much detail on so whimsical a fantasy. But evolution can be harnessed to serve as a wild bronco to carry us where we can't go by ourselves. On this untamed transport we arrive at a place stuffed with odd bodies, fully imagined (not by us) down to the last hair.

Karl Sims, CM5's artist, told me, "I use evolution for two reasons. One, to breed things I would have never thought of, nor would have found any other way. And, two, to create things in great detail that I might have thought of, but would never have time to draw."

Both Sims and Latham stumbled upon discontinuities in the Library. "You develop a feel for what kinds of things can happen in an evolutionary space," Sims claims. He reported that he often would be evolving away, making satisfactory progress—sort of whistling happily while things noticeably improved—when suddenly he'd hit a wall and the improvements would plateau. Even drastic choices would not "move" the sluggish form away from the rut it seemed stuck in. Generation after generation of progeny seemed to get no better. It was as if he were trapped on a large local desert basin where one step was identical to the next and the interesting peaks were far away.

As Thomas Reed stalked the lost chalice in Biomorph Land, he often needed to back up. He would be near the cup but getting nowhere. He often saved intermediate forms on his long chase. Once he needed to retreat hundreds of steps back to the sixth archived form in order to get out of a dead end.

Latham reported similar experiences while exploring his space. He often ran into what he called a territory of instabilities. In some regions of possible forms, significant changes in genes would effect only insignificant shifts in forms—Sims's basin of stagnation. He'd have to really push the genes miles around to move an inch in form. Yet, in other regions, minute changes in genes would produce huge alterations in form. In the former, Latham's progress through the space was glacial; in the latter, his tiniest move would send him rapidly careening through the Library at a zoom.

To avoid overshooting a destination of possible form, and to accelerate its discovery, Latham would purposefully twirl a mutation knob as he explored. At first he'd set the mutation rate high, to skip through the space. As the shapes became more interesting, he'd turn the mutation rate down so that each generation sliced thinner, and he'd slowly creep up to a concealed shape. Sims wired his system to perform a similar trick automatically. As the image he was evolving became more complex, his software would crank down the mutation rate for a soft landing on the final form. "Otherwise," Sims says, "things can get crazy as you are trying to fine-tune an image."

These frontiersmen developed a couple of other tricks for traveling through the Library. The most important trick was sex. Dawkins's Biomorph Land was a fertile, but puritanical, place that hadn't a hint of sex. All variation in Biomorph Land occurred by asexual mutations from a single parent. Sims's and Latham's worlds, in contrast, were driven by sex. A major lesson the frontiersmen realized was that you could do sex in an evolutionary system in any number of ways!

There was of course the orthodox missionary position: two parents, with genes from each. But even that plain vanilla mating can be accomplished in several ways. In the Library, breeding is analogous to taking two books and merging their text to form a child-book. You can beget two kinds of progeny: in-betweener books or outsider books.

In-betweener offspring inherit a position in between Mommy and Daddy. Imagine a beeline in the Library bridging Book A and Book B. Any child (Book C) would be found somewhere in the Library on that imaginary line. In-between offspring can be exactly halfway in between as they would be if they inherited exactly half of their genes from Pop and half from Mom. Or, they can be in-between at some other proportion, say 10 percent Mom and 90 percent Dad. In-betweeners can also inherit alternating chapters from Book A and Book B, or alternating clumps of genes from Mommy and Daddy. This method retains genes that may be linked to each other by a proximal function, making it more likely to accumulate "good stuff."

Another way to think of in-betweeners is to imagine creature A morphing—in the Hollywood term—into creature B. All the creatures it morphs through on its way from A to B are the pair's possible in-betweener offspring.

Outsider offspring inherit a position outside of the morph-line between Mommy and Daddy. Rather than some random halfway stage between a lion and snake, they are a chimera boasting a lion's head with a snake's tail and forked tongue. There are several different ways to generate chimera, including the pretty basic one of fishing in a potluck stew of random traits possessed either by Mom or Dad. Outsider offspring are wilder, less expected, more out of control.

But that's not the end of the weirdness feasible in evolutionary systems. Mating can also be perverse. William Latham is currently playing around with polygamy in his system. Why limit mating to two parents? Latham coded his system to allow him to choose up to five parents and assign each parent varying weights of inheritability. So he says to his brood of children forms: next time give me something very much like this one, that one, and that one, and somewhat like this one, and a little bit like that one. Then he marries them together and they co-procreate the next brood. Latham can also assign negative values: as in, *not* like this one. In effect he has made an antiparent. When an antiparent mates in multiple marriages it sires (or not-sires) children as *unlike* it as possible.

Moving further still from natural biology (at least as far as we know it) Latham hacked a program for Mutator which follows the breeder's progress through the Library. Genes that persist over a particular breeding course, the Mutator assumes the breeder likes. It makes those genes dominant. Genes that keep changing, the Mutator reads as "experimental" and unsatisfying to the breeder, so it reduces their impact by declaring them recessive in any mating.

The idea of tracking evolution in order to anticipate its future course is bewitching. Both Sims and Latham dream about an artificial intelligence module that could analyze a breeder's progress through form space. The AI program would deduce the common element shared by the selections and then reach far ahead into the Library to retrieve a form that encapsulates that trait.

At the Pompidou Center in Paris, and at the Ars Electronica Festival in Linz, Austria, Karl Sims installed a public version of his artificial evolution universe. In the middle of a long gallery space, a Connection Machine hummed on a platform. The jet-black cube was vested in flickering red lights, which syncopated as the machine thought. A heavy cable connected the supercomputer to an arc of 20 large monitors. A footpad on the floor sat in front of each color screen in the crescent. By stepping on a footpad (which covers a switch) a museum-goer chose a particular image out of the row.

I had a chance to breed CM2 images in Linz. To start, I selected what looked like an impression of poppies in a garden. Instantly, Sims's program bred 20 new offspring of the flowers. Two screens filled with gray rubbish, the other 18 displayed new "flowers," some fragmented, some in new colors. At each turn I tried to see how flowery I could push the image. I quickly worked up a sweat running from pad to pad in the computer-heated room. The physical work felt like gardening—nurturing shapes into existence. I kept evolving more elaborate floral patterns, until another visitor shifted the direction toward wild fluorescent plaids. I was dumbfounded by the range of beautiful images that the system uncovered: geometric still lifes, hallucinogenic landscapes, alien textures, eerie logos. One after another elaborate, brilliantly colored composition would appear on the monitors and then, unchosen, retreat forever.

Sims's installation breeds all day, every day, bending its evolution to the fancy of the passing mob of international museum visitors. The Connection Machine records every choice, and every choice leading up to the choice. Sims now has a database of what humans (at least art museum humans) find beautiful or interesting. He believes that these inarticulate qualities can be abstracted from such a rich trove of data and then used as selection criteria for future breeding in other regions of the Library.

Or, we may be very surprised to find that nothing unifies the selection criteria. It may be that *any highly evolved form is beautiful.* We find beauty in all biological creatures, although individual people have individual favorites. Overall, a monarch butterfly is no more or less striking than its host, the milkweed pod. If inspected without prejudice, parasitic beasts are beautiful. My suspicion is that the beauty of nature resides in the process of getting there by evolution and in the important fact that the form must work biologically as a whole.

Still, something distinguishes the selected forms, no matter what they are, from the speckled gray noise that surrounds them. Comparing the chosen to the random may tell us much about beauty and even help us figure out what we mean by "complexity."

THE RUSSIAN PROGRAMMER Vladimir Pokhilko reminds me that evolving for beauty alone may be a sufficient goal. Pokhilko and partner Alexey Pajitnov (who wrote the famously addictive computer game Tetris) designed a very powerful selection program that breeds virtual aquarium fish. Pokhilko told me during early work on the game, "When we started we didn't want to use the computer to make something very practical, but to make something very beautiful." Pokhilko and Pajitnov did not set out to make an evolutionary world. "Our starting point was ikebana, the Japanese art of arranging flowers. We wanted to make some kind of computer ikebana. But we wanted something alive, moving. And which never repeats itself." Since the computer screen "looks like an aquarium, we decided to make a customizable aquarium."

Users become artists by filling the aquarium with the right combinations of colored fishes and quantities of swaying seaweeds. Users would need a large variety of organisms. Why not let the aquarist breed their own? Thus "El-Fish" was hatched, and the Russians found themselves in the evolution game.

El-Fish became a monster of a program. It was mostly written in Moscow during a time when smart U.S. entrepreneurs could hire an entire unemployed Russian university math department for the salary of one U.S. hacker. Up to 50 Russian programmers, ignorant of Dawkins, Latham, and Sims, wrote code for El-Fish, rediscovering the power and method of computational evolution.

The commercial version of El-Fish, released by the U.S. software publisher Maxis in 1993, compresses the kind of flamboyant visual breeding done by Latham on large IBMs and Sims on a Connection Machine into a small desktop home computer.

Each El-Fish has 56 genes which define 800 parameters (a huge Library). The colorful fish swim in a virtual underwater world realistically, turning with the flick of a fin as fish do. They weave between strands of kelp (also bred by the program). They pace back and forth endlessly. They school around food when you "feed" them. They never die. When I first saw an El-Fish aquarium from ten paces away, I took it to be a video of a real aquarium.

The really fun part is breeding fish. I got started by dipping a net somewhat randomly into a map of the hypothetical El-Fish ocean, fishing for a couple of exotic parent fish. Different areas shelter different fish. The ocean is, in fact, the Library. I hauled up two fish which I kept: a plump yellow fish spotted in green with a thin dorsal fin and an overbite (the Mama) and a puny blue torpedo-shaped guy with a Chinese junk sail of a top fin (the Pa). I could evolve from either, that is, I could asexually mutate new fish from either the fat yellow one or the tiny blue one, or I could mate the pair and select from their joint offspring. I chose sex.

As in the other artificial evolution programs, about a dozen mutated offspring appeared on the screen. I could slide a knob to adjust the mutation rate. I was into fins. I chose a large-finned one, pushing its shape each generation toward increasingly ornate, heavy-duty fins. I got one fish that seemed to be all fins, top, bottom, side. I moved it from the incubator and animated it before plunking it into the aquarium (the animation procedure can take minutes or hours depending on the computer). After many generations of increasingly weird finny fish, I evolved a fish so freakish that it wouldn't breed anymore. This is the El-Fish program's way of keeping the fish, fish. I had entered the outer boundaries in the Library beyond which the forms are less than fishlike. El-Fish won't render nonfish creatures, and it won't animate unorthodox fish because it's too hard to make a monster move. (The code relies on standard fish proportions to keep a creature's movements convincing.) Part of the game is users trying to figure out where those fishy limits lie and whether there are any loopholes.

Storing full fish consumes far too much disk memory, so only the bare genes of the fish are filed. These tiny seeds of genes are called "roe." Roe are 250 times more compact than the fish they grow into. El-Fish aficionados swap the roe of selected creations over modem lines or stock them in digital public libraries.

One of the programmers at Maxis in charge of testing El-Fish discovered an interesting way to explore the outer limits of the fish Library. Instead of breeding or fishing the pool for sample stock, he inserted the text of his name (Roger) into a roe. Out came a short black tadpole. Pretty soon everyone in the office had a tadpole in their El-Fish tank. Roger wondered what else he could transform into fish roe. He took the digital text of the Gettysburg Address and grew the digits into a ghostly creature—a pale face trailing a deformed batwing. The wags dubbed it a "Gettyfish." Hacking around they discovered that a sequence of about 2,000 digits of any sort can be shanghaied into serving as roe for a possible fish. Getting into the swing of things, the project manager for El-Fish loaded the spreadsheet of his budget into El-Fish and birthed the bad omen of a fish skull, fangy mouth, and dragon body.

Breeding was once a craft belonging solely to the gardener. It is now available to the painter, the musician, the inventor. William Latham predicts evolutionism as the next stage in modern art. In evolutionism, the borrowed concepts of mutation and sexual reproduction spawn the art. Instead of painting or creating textures for computer graphic models, artist Sims evolves them. He drifts into a region of woodlike patterns and then evolves his way to the exact grainy, knot-ridden piney look which he can use to color a wall in a video he is making.

You can now do this on a Macintosh with a commercial template for Adobe Photoshop software. Written by Kai Krause, the Texture Mutator lets ordinary computer owners breed textures from a choice of eight offspring every generation.

Evolutionism reverses the modern trend in the design of artist's tools that bends toward greater analytical control. The ends of evolution are more subjective ("survival of the most aesthetic"), less controlled, more related to art generated in a dream or trance; more found.

The evolutionary artist creates twice. First, the artist acts as god by concocting a world, or a system for generating beauty. Second, he is the gardener and curator of this made world, interpreting and presenting the chosen works he nurtures. He fathers rather than molds a creation into existence.

At the moment the tools of exploratory evolution restrict an artist to begin with a random or primitive start. The next advance in evolutionism is to be able to begin with a human-designed pattern and then arbitrarily breed from there. Ideally, you would like to be able to pick up, say, a colorful logo or label that needed work (or mind-altering modification) and progressively evolve from that.

The outlines of such a commercial software are pretty clear. Will Wright, SimCity author and founder of Maxis, the innovative software publisher behind El-Fish, even came up with the perfect jazzy title: DarwinDraw. In DarwinDraw you sketch a new corporate logo. Every line, curve, dot, or paint stroke of the image you create is rendered into mathematical functions. When you are done, you have a logo on a screen and a mutable set of functions as genes in the computer. Then you breed the logo. You let it evolve outlandish designs you could never have thought of, in detail you don't have time to do. You jump around randomly at first, just to brainstorm. Then you home in on an unusual and striking arrangement. You turn the mutation rate down, use multiple marriages and antiparenting to fine-tune it to its final version. You now have an obsessively detailed evolved artwork with cross-hatching and filigrees you wouldn't believe. Because the image is based on algorithms, it has infinite resolution; you can blow it up as large as you like with unexpected detail to spare. Print it!

As a demo of this power of evolutionism, Sims scanned the logo of CM5 into his program and used it as a starting image to breed an "improved" CM5 logo. Rather than the sterile modern look, it had frilly organic lines around the edges of the letters. Folks in the office liked the evolved artwork so much that they decided to make a T-shirt out of it. "I'd really love to evolve neckties," Sims says. His other suggestions: "How about evolving textile patterns, wallpaper designs, or type fonts?"

IBM has been supporting artist William Latham's evolution experiments because the global corporation realizes there is commercial potential here. While Sims's evolution machine is, according to Latham, "a grammar that is more ragged, more uncontrolled," Latham's is more controlled and useful to engineers. IBM is turning the evolutionary tools Latham developed over to automobile designers and having them mutate car body shapes. One of the questions they are trying to answer is whether evolutionary design techniques are more useful in the beginning of rough ideas or later in fine tuning, or both. IBM intends to make a profitable project out of it. And not only for cars. They imagine evolutionary "steering" tools useful for all kinds of design problems entailing large numbers of parameters which require a user to "back up" to a stored previous solution. Latham pictures evolution taking root in packaging design, where the outer parameters are firmly fixed (size and shape of the container), but where what happens within that space is wide open. Here evolution can bring in multiple levels of detail that a human artist would never have the time, energy, or money to do. The other advantage of evolutionary industrial design, Latham has slowly come to realize, is that it is perfectly suited to design by committee. The more people that play, the better.

The copyright status of an artificially evolved creation is in legal limbo. Who gets the protection, the artist who bred or the artist who created the program? In the future, lawyers may demand a record of the evolutionary path an artist followed to arrive at an evolved creation as evidence that such work belongs to him and was not copied, or due to the creator of the Library. As Dawkins showed, in a truly large Library it's improbable to find a pattern more than once. Owning an evolutionary pathway to a particular point demonstrates irrefutable proof that the artist found that destination originally, since evolution doesn't strike twice.

In the end, breeding a useful thing becomes almost as miraculous as creating one. Richard Dawkins echoes this when he asserts that "effective searching procedures become, when the search-space is sufficiently large, indistinguishable from true creativity." In the Library of all possible books, finding a particular book is equivalent to writing it.

This sentiment was recognized centuries ago, long before the advent of computers. As Denis Diderot wrote in 1755:

> The number of books will grow continually, and one can predict that a time will come when it will be almost as difficult to learn anything from books as from the direct study of the whole universe. It will be almost as convenient to search for some bit of truth concealed in nature as it will be to find it hidden away in an immense multitude of bound volumes.

William Poundstone, author of *The Recursive Universe*, contrived an analogy to illustrate why searching huge Borgian libraries of knowledge is as difficult as searching the huge Borgian library of nature itself. Imagine, Poundstone said, that there is a library with all possible videos. Like all Borgian spaces, most of the items in this library are full of noise and random grayness. A typical tape would be two hours of snow. The main problem with searching for a viewable video is that no title, call name, or symbol of any sort could represent a random tape in any less space or time than the tape itself. Most of the items in a Borgian library are incompressible into anything shorter than the work itself. (This irreducibility is the current definition of randomness.) To search the tapes, they must be watched, and therefore the information, time, and energy needed to sort through all the tapes would exceed the information, time, and energy needed to create the tape you wanted, *no matter what the tape was*.

Evolution is a slow-witted way to outsmart this conundrum, but what we call intelligence is nothing more (and nothing less) than a tunnel through it. If I had been especially astute in my search in the Library for my book *Out of Control*, after several hours I might have discerned a cardinal direction to my wanderings through the library stacks. I might have noticed that in general, "sense" lay to the left of the last book I held. I could have anticipated many generations of slow evolution by running ahead miles to the left. I might have learned the architecture of the library and predicted where sense would hide, outrunning both random guessing and creeping evolution. I could have found *Out of Control* by a combination of evolution and by learning the inherent order of the Library.

Some students of the human mind make a strong argument that thinking is a type of evolution of ideas within the brain. According to this argument, all created things are evolved. As I write these words, I have to agree. I began this book not with a sentence formed in my mind but with an arbitrarily chosen phrase, "I am." Then in unconsciously rapid succession I evaluated a headful of possible next words. I picked one that seemed esthetically fit, "sealed." After "I am sealed," I went on to the next word, choosing from among 100,000s of possible ones. Each selected word bred the choices for the next until I had evolved almost a sentence of words. Toward the end of the sentence my choices were constrained somewhat by the words I had already chosen at the beginning, so learning helped the breeding go more quickly.

But the first word of the next sentence could have been any word. The end of my book, 150,000 choices away, looked as distant and improbable as the end of the galaxy. A book *is* improbable. Out of all the books written or to be written in the world, only this book, for instance, would have found the preceding two sentences in a row.

Now that I'm in the middle of the book, I'm still evolving the text. What will the next words be that I write in this chapter? In a real sense I don't know. There are probably billions of possibilities of what they might be, even taking into account the restriction that they must logically follow from the last sentence. Did you guess this sentence as the next one? I didn't either. But that's the sentence I found at the end of the sentence.

I wrote this book by finding it. I found it in the Library of Borges by evolving it at my desk. Word by word, I traveled through the Library of Jorge Luis Borges. By some kind of weird combination of learning and evolution that our heads do, I found my book. It was on the middle shelf, almost at eye level, in the seventh hexagon of region 52427. Who knows if it is my book or merely one that is almost my book (differing by a paragraph here or there, or maybe even by the omission of a few critical facts)?

The great satisfaction of the long search for me—no matter how the book fares—was that only I could find it.

15

# Artificial Evolution

The first time Tom Ray released his tiny handmade creature into his computer, it reproduced rapidly until hundreds of copies occupied the available memory space. Ray's creature was an experimental computer virus of sorts; it wasn't dangerous because the bugs couldn't replicate outside his computer. The idea was to see what would happen if they had to compete against each other in a confined world.

Ray cleverly devised his universe so that out of the thousands of clones from the first ancestral virus, about ten percent replicated with small variations. The initial creature was an "80"—so named because it had 80 bytes of code. A number of 80s "flipped a bit" at random and became creatures 79 or 81 bytes long. Some of these new mutant viruses soon took over Ray's virtual world. In turn, they mutated into further varieties. Creature 80 was nearly overwhelmed to the point of extinction by the mushrooming ranks of new "organisms." But the 80s never completely died, and long after the new arrivals 79, 51, and 45 emerged and peaked in population, the 80s rebounded.

After a few hours of operation, Tom Ray's electric-powered evolution machine had evolved a soup of nearly a hundred types of computer viruses, all battling it out for survival in his isolated world. On his very first try, after months of writing code, Ray had brewed artificial evolution

When he was a shy, soft-spoken Harvard undergraduate, Ray had collected ant colonies in Costa Rica for the legendary ant-man, E. O. Wilson. Wilson needed live leafcutting ant colonies for his Cambridge labs. Ray hired on in the lush tropics of Central America to locate and capture healthy colonies in the field, and then ship them to Harvard. He found that he was particularly good at the task. The trick was to dig into the jungle soil with the deftness of a surgeon in order to remove the guts of a colony. What was needed was the intact inner chamber of the queen's nest, along with the queen herself, her nurse ants, and a mini-ant-garden stocked with enough food to support the chamber for shipping. A young newborn colony was perfect. The heart of such a colony might fit into a tea cup. That was the other essential trick: to locate a really small nest hidden under the natural camouflaged debris of the forest floor. From a minuscule core that could be warmed in one's hands, the colony could grow in a few years to fill a large room.

While collecting ants in the rain forest, Ray discovered an obscure species of butterfly that would tag along the advancing lines of army ants. The army ants' ruthless eating habits—devouring any animal life in their path—would flush a cloud of flying insects eager to get out of the way. A kind of bird evolved to follow the pillaging army, happily picking off the agitated fleeing insects in the air. The butterfly, in turn, followed the birds who followed the army ants. The butterflies tagged along to feast on the droppings of the ant-birds—a much needed source of nitrogen for egg laying. The whole motley crew of ants, ant-birds and ant-bird-butterflies, and who knew what else, would roam across the jungle like a band of gypsies in cahoots.

Ray was overwhelmed by such wondrous complexity. Here was an entirely nomadic community! Most attempts to understand ecological relations seemed laughable in light of these weird creations. How in the universe did these three groups of species (one ant, three butterflies, and about a dozen birds) ever wind up in this peculiar codependency? And why?

By the time he had finished his Ph.D., Ray felt that the science of ecology was moribund because it could not offer a satisfying answer to such big questions. Ecology lacked good theories to generalize the wealth of observations piling up from every patch of wilderness. It was stymied by extensive local knowledge: without an overarching theory, ecology was merely a library of fascinating just-so stories. The life cycles of barnacle communities, or the seasonal pattern of buttercup fields, or behavior of bobcat clans were all known, but what principles, if any, guided all three? Ecology needed a science of complexity that addressed the riddles of form, history, development—all the really interesting questions—yet was supported by field data.

Along with many other biologists, Ray felt that the best hope for ecology was to shift its focus from ecological time (the thousand-year lifetime of a forest) to evolutionary time (the million-year lifetime of a tree species). Evolution at least had a theory. Yet, the study of evolution too was caught up with the same fixation on specifics. "I was frustrated," Ray told me, "because I didn't want to study the products of evolution—vines and ants and butterflies. I wanted to study evolution itself."

Tom Ray dreamed of making an electric-powered evolution machine. With a black box that contained evolution he could demonstrate the historical principles of ecology, how a rain forest descends from earlier woods, and how in fact ecologies emerge from the same primordial forces that spawn species. If he could develop an evolution engine, he'd have a test-bed with which to do real ecological experiments. He could take a community and run it over and over again in different combinations, making ponds without algae, woods without termites, grasslands without gophers, or just to cover the bases, jungles with gophers and grasslands with algae. He could start with viruses and see where it all would lead him.

Ray was a bird watcher, insect collector, plantsman—the farthest thing from a computer nerd—yet he was sure such a machine could be built. He remembered a moment ten years earlier when he was learning the Japanese game of Go from an MIT hacker who used biological metaphors to explain the rules. As Ray tells it, "He said to me, 'Do you know that it is possible to write a computer program that can self-replicate?' And right at that moment I imagined all the things I'm doing now. I asked him how to do it, and he said, 'Oh, it's trivial,' but I didn't remember what he said, or whether in fact he actually knew. When I remembered that conversation I stopped reading novels and started reading computer manuals."

Ray's solution to the problem of making an electronic evolution machine was to start with simple replicators and give them a cozy habitat and plenty of energy and places to fill. The closest real things to these creatures were bits of self-replicating RNA. But the challenge seemed doable. He would cook up a soup of computer viruses.

About this time in 1989, the news magazines were chock-full of cover stories pronouncing computer viruses worse than the plague and as evil as technology could get. Yet Ray saw in the simple codes of computer viruses the beginnings of a new science: experimental evolution and ecology.

To protect the outside world (and to keep his own computer from crashing), Ray devised a virtual computer to contain his experiments. A virtual computer is a bit of clever software that emulates a pretend computer deep within the operating subconscious of the real computer. By containing his tiny bits of replicating code inside this shadow computer, Ray sealed them from the outside world and gave himself room to mess with vital functions, such as computer memory, without jeopardizing the integrity of his host computer. "After a year of reading computer manuals, I sat down and wrote code. In two months the thing was running. And in the first two minutes of running without a crash, I had evolving creatures."

Ray seeded his world (which he called "Tierra") with a single creature he programmed by hand—the 80-byte creature—inserted into a block of RAM in his virtual computer. The 80 creature reproduced by finding an empty RAM block 80 bytes big and then filling it with a copy of itself. Within minutes the RAM was saturated with copies of 80.

But Ray had added two key features that modified this otherwise Xerox-like copying machine into an evolution machine: his program occasionally scrambled the digital bits during copying, and he assigned his creatures a priority tag for an executioner. In short he introduced variation and death.

Computer scientists had told him that if he randomly varied

bits of a computer code (which is all his creatures really are), the resulting programs would break and then crash the computer. They felt that the probability of getting a working program by randomly introducing bugs into code was so low as to make his scheme a waste of time. This sentiment seemed in line with what Ray knew about the fragile perfection needed to keep computers going; bugs killed progress. But because his creature programs would run in his shadow computer, whenever a mutation would birth a creature that was seriously broken, his executioner program—he named it "the Reaper"— would kill it while the rest of his Tierra world kept running. In essence, Tierra spotted the buggy programs that couldn't reproduce and yanked them out of the virtual computer.

Yet, the Reaper would pass over the very rare mutants that worked, that is, those that happened to form a bona fide alternative program. These legitimate variations could multiply and breed other variants. If you ran Tierra for a billion computer cycles or so, as Ray did, a startling number of randomly generated creatures formed during those billion chances. And just to keep the pot boiling, Ray also assigned creatures an age stamp so that older creatures would die. "The Reaper kills either the oldest creature or the most screwed-up creature," Ray says with a smile.

On Ray's first run of Tierra, random variation, death, and natural selection worked. Within minutes Ray witnessed an ecology of newly created creatures emerge to compete for computer cycles. The competition rewarded creatures of smaller size since they needed less cycles, and in Darwinian ruthlessness, terminated the greedy consumers, the infirm, and the old. Creature 79 (one byte smaller than 80) was lucky. It worked productively and soon outpaced the 80s.

Ray also found something very strange: a viable creature with only 45 very efficient bytes which overran all other creatures. "I was amazed how fast this system would optimize," Ray recalls. "I could graph its pace as the system would generate organisms surviving on shorter and shorter genomes."

On close examination of 45's code, Ray was amazed to discover that it was a parasite. It contained only a part of the code it needed to survive. In order to reproduce, it "borrowed" the reproductive section from the code of an 80 and copied itself. As long as there were enough 80 hosts around, the 45s thrived. But if there were too many 45s in the limited world, there wouldn't be enough 80s to supply copy resources. As the 80s waned, so did the 45s. The pair danced the classic coevolutionary tango, back and forth endlessly, just like populations of foxes and rabbits in the north woods.

"It seems to be a universal property of life that all successful systems attract parasites," Ray reminds me. In nature parasites are so common that hosts soon coevolve immunity to them. Then eventually the parasites coevolve strategies to circumvent that immunity. And eventually the hosts coevolve defenses to repel them again. In reality, these actions are not alternating steps but two constant forces pressing against one another.

Ray learned to run ecological experiments in Tierra using parasites. He loaded his "soup" with 79s which he suspected were immune to the 45 parasite. They were. But as the 79s prospered, a second parasite evolved that could prey on them. This one was 51 bytes long. When Ray sequenced its genes he found that a single genetic event had transformed a 45 into a 51. "Seven instructions of unknown origin," Ray says, "had replaced one instruction somewhere near the middle of the 45," transforming a disabled parasite into a newly potent one. And so it went. A new creature evolved that was immune to 51s, and so on.

Poking around in the soups of long runs, Ray discovered parasites that preyed on other parasites—hyperparasites: "Hyperparasites are like neighbors who steal power from your lines to the power plant. You sit in the dark while they use your power *and* you pay the bill." In Tierra, organisms such as the 45s discovered that they didn't need to carry a lot of code around to replicate themselves because their environment was full of code—of other organisms. Quips Ray, "It's just like us using other animals' amino acids [when we eat them]." On further inspection Ray found hyper-hyperparasites thriving, parasites raised to the third. He found "social cheaters"—creatures that exploit the code of two cooperating hyperparasites (the "cooperating" hyperparasites were stealing from each other!). Social cheaters require a fairly well developed ecology. They can't be seen yet, but there are probably hyper-hyper-hyperparasites and no end to elaborate freeloading games possible in his world.

And Ray found creatures that surpassed the programming skills of human software engineers.

"I started with a creature 80 bytes large," Ray remembers, "because that's the best I could come up with. I figured that maybe evolution could get it down to 75 bytes or so. I let the program run overnight and the next morning there was a creature—not a parasite, but a fully self-replicating creature—that was only 22 bytes! I was completely baffled how a creature could manage to self-replicate in only 22 instructions without stealing instructions from others, as parasites do. To share this novelty, I distributed its basic algorithm onto the Net. A computer science student at MIT saw my explanation, but somehow didn't get the code of the 22 creature. He tried to recreate it by hand, but the best he could do was get it to 31 instructions. He was quite distressed when he found out I came up with 22 instructions in my sleep!"

What humans can't engineer, evolution can. Ray puts it nicely as he shows off a monitor with traces of the 22s propagating in his soup: "It seems utterly preposterous to think that you could randomly alter a computer program and get something better than what you carefully crafted by hand, but here's living proof." It suddenly dawns on the observer that there is no end to the creativity that these mindless hackers can come up with.

Because creatures consume computer cycles, there is an advantage to smaller (shorter sets of instructions) creatures. Ray reprogrammed Tierra's code so the system assigned computer resources to creatures in proportion to their size; large ones getting more cycles. In this mode, Ray's creatures inhabited a size-neutral world, which seemed more suited for long runs since it wasn't biased to either the small or large. Once Ray ran a size-neutral world for 15 billion cycles of his computer. Somewhere around 11 billion cycles, a diabolically clever 36 creature evolved. It calculated its true size, then behind its back so to speak, shifted all the bits in the measurement to the left one bit, which in binary code is equal to doubling the number. So by lying about its size, creature 36 sneakily garnered the resources of a 72 creature, which meant that it got twice the usual CPU time. Naturally this mutation swept through the system.

Perhaps the most astounding thing about Tom Ray's electrically powered evolution machine is that it created sex. Nobody told it about sex, but it found it nonetheless. In an experiment to see what would happen if he turned the mutation function off, Ray let the soup run without deliberate error. He was flabbergasted to discover that even without programmed mutation, evolution pushed forward.

In real natural life, sex is a much more important source of variation than mutations. Sex, at the conceptual level, is genetic recombination—a few genes from Dad and a few genes from Mom combined into a new genome for Junior. Sometimes in Tierra a parasite would be in the middle of asexual reproduction, "borrowing" the copy function of some other creature's code, when the Reaper would happen to kill the host midway in the process. When this happens the parasite uses some copy code of the new creature born in the old creature's space, and part of the "dead" creature's interrupted reproduction function. The resultant junior was a wild, new recombination created without deliberate mutation. (Ray also says this weird reproduction "amounts to sex with the dead!") Interrupted sex had happened all the time in his soup, but only when Ray turned off his "flip-a-bit" mutator did he notice its results. It turned out that inadvertent recombination alone was enough to fuel evolution. There was sufficient irregularity in the moment of death, and where creatures lived in RAM, that this complexity furnished the variety that evolution required. In one sense, the system evolved variation.

To scientists, the most exhilarating news to come out of Ray's artificial evolution machine is that his small worlds display what seems to be punctuated equilibrium. For relatively long periods of time, the ratio of populations remains in a steady tango of give and take with only the occasional extinction or birth of a new species. Then, in a relative blink, this equilibrium is punctuated by a rapid burst of roiling change with many newcomers and eclipsing of the old. For a short period change is rampant. Then things sort out and stasis and equilibrium reign again. The current interpretation of fossil evidence on Earth is that this pattern predominates in nature. Stasis is the norm; change occurs in bouts. The same punctuated equilibrium pattern has been seen in other evolutionary computer models as well, such as Kristian Lindgren's coevolutionary Prisoner's Dilemma world. If artificial evolution mirrors organic evolution, one has to wonder what would happen if Ray let his world run forever? Would his viral creatures invent multicellularity?

Unfortunately, Ray has never turned his world on marathon mode just to see what would happen over months or years. He's still fiddling with the program, gearing it up to collect the immense store of data (50 megabytes per day) such a marathon run would generate. He admits that "sometimes we're like a bunch of boys with a car. We've always got the hood up and pieces of the engine out on the garage floor, but we hardly ever drive the car because we're too interested in souping it up."

In fact, Ray has his sights fixed on a new piece of hardware, a technology that ought to be. Ray figures that he could take his virtual computer and the fundamental language he wrote for it and "burn" it into a computer chip—a slice of silicon that did evolution. This off-the-shelf Darwin Chip would then be a module you could plug into any computer, and it would breed stuff for you, fast. You could evolve lines of computer code, or subroutines, or maybe even entire software programs. "I find it rather peculiar," Ray confides, "that as a tropical plant ecologist I'm now designing computers."

The prospects that a Darwin Chip might serve up are delicious. Imagine you have one in your PC where you use Microsoft Word as a word processor. With resident Darwinism loaded into your operating system, Word would evolve as you worked. It would use your computer's idle CPU cycles to improve, and learn, in a slow evolutionary way, to fit itself to your working habits. Only those alterations that improved the speed or the accuracy would survive. However Ray feels strongly that messy evolution should happen away from the job. "You want to divorce evolution from the end user," he says. He imagines "digital husbandry" happening offline in back rooms, so to speak, so that the common failures necessary for evolution are never seen by its customer. Before an evolving application is turned over to an end user, it is "neutered" so that it can't evolve while in use.

Retail evolution is not so farfetched. Today you can buy a spreadsheet module that does something similar in software. It's called, naturally enough, "Evolver." Evolver is a template for spreadsheets on the Macintosh—very complicated spreadsheets spilling over with hundreds of variables and "what-if" functions. Engineers and database specialists use it.

Let's say you have the medical records of thirty thousand patients. You'd probably like to know what a typical patient looks like. The larger the database, the harder it is to see what you have in there. Most software can do averaging, but that does not extract a "typical" patient. What you would like to know is what set of measurements—out of the thousands of categories collected by the records—have similar values for the maximum number of people? It's a problem of optimizing huge numbers of interacting variables. The task is familiar to any living species: how does it maximize the results of thousands of variables? Raccoons have to ensure their own survival, but there are a thousand variables (foot size, night vision, heart rate, skin color, etc.) that can be changed over time, and altering one parameter will alter another. The only way to tread through this vast space of possible answers, and retain some hope of reaching a peak, is by evolution.

The Evolver software optimizes the broadest possible profile for the largest number of patients by trying a description of a typical patient, then testing how many fit that description, then tweaking the profile in a multitude of directions to see if more patients fit it, and then varying, selecting, and varying again, until a maximum number of patients fit the profile. It's a job particularly suited for evolution.

"Hill climbing," computer scientists call the process. Evolutionary programs attempt to scale the peak in the libraries of form where the optimal solution resides. By relentlessly pushing the program toward better solutions, the programs climb up until they can't climb any higher. At that point, they are on a peak—a maximum—of some sort. The question always is: is their summit the tallest peak around, or is the program stuck on a local peak adjacent to a much taller peak across the valley, with no way to retreat?

Finding a solution—a peak—is not difficult. What evolution in nature and evolutionary programs in computers excel at is hill climbing to global summits—the highest peaks around—when the terrain is rugged with many false summits.

JOHN HOLLAND is a gnomic figure of indeterminate age who once worked on the world's earliest computers, and who now teaches at the University of Michigan. He was the first to invent a mathematical method of describing evolution's optimizing ability in a form that could be easily programmed on a computer. Because of the way his math mimicked the effects of genetic information, Holland called them genetic algorithms, or GAs for short.

Holland, unlike Tom Ray, started with sex. Holland's genetic algorithms took two strings of DNA-like computer code that did a job fairly well and recombined the two at random in a sexual swap to see if the new offspring code might do a little better. In designing his system, Holland had to overcome the same looming obstacle that Ray faced: any random generation of a computer program would most likely produce not a program that was either slightly better or slightly worse, but one that was not sensible at all. Statistically, successive random mutations to a working code were bound to produce successive crashes.

Mating rather than mutating was discovered by theoretical biologists in the early 1960s to make a more robust computer evolution—one that birthed a higher ratio of sensible entities. But sexual mating alone was too restrictive in what it could come up with. In the mid-1960s Holland devised his GAs; these relied chiefly on mating and secondarily on mutation as a background instigator. With sex and mutation combined, the system was both flexible and wide.

Like many other systems thinkers, Holland sees the tasks of nature and the job of computers as similar. "Living organisms are consummate problem solvers," Holland wrote in a summary of his work. "They exhibit a versatility that puts the best computer programs to shame. This observation is especially galling for computer scientists, who may spend months or years of intellectual effort on an algorithm, whereas organisms come by their abilities through the apparently undirected mechanism of evolution and natural selection."

The evolutionary approach, Holland wrote, "eliminates one of the greatest hurdles in software design: specifying in advance all the features of a problem." Anywhere you have many conflicting, interlinked variables and a broadly defined goal where the solutions may be myriad, evolution is the answer.

Just as evolution deals in *populations* of individuals, genetic algorithms mimic nature by evolving huge churning populations of code, all processing and mutating at once. GAs are swarms of slightly different strategies trying to simultaneously hill-climb over a rugged landscape. Because a multitude of code strings "climb" in parallel, the population visits many regions of the landscape concurrently. This ensures it won't miss the Big Peak.

Implicit parallelism is the magic by which evolutionary processes guarantee you climb not just any peak but the tallest peak. How do you locate the global optima? By testing bits of the entire landscape at once. How do you optimally balance a thousand counteracting variables in a complex problem? By sampling a thousand combinations at once. How do you develop an organism that can survive harsh conditions? By running a thousand slightly varied individuals at once.

In Holland's scheme, the highest performing bits of code anywhere on the landscape mate with each other. Since high performance increases the assigned rate of mating in that area, this focuses the attention of the genetic algorithm system on the most promising areas in the overall landscape. It also diverts computational cycles away from unpromising areas. Thus parallelism sweeps a large net over the problem landscape while reducing the number of code strings that need manipulating to locate the peaks.

Parallelism is one of the ways around the inherent stupidity and blindness of random mutations. It is the great irony of life that a mindless act repeated in sequence can only lead to greater depths of absurdity, while a mindless act performed in parallel by a swarm of individuals can, under the proper conditions, lead to all that we find interesting.

John Holland invented genetic algorithms while studying the mechanics of adaptation in the 1960s. His work was ignored until the late 1980s by all but a dozen wild-eyed computer grad students. A couple of other researchers, such as the engineers Lawrence Fogel and Hans Bremermann, independently played around with mechanical evolution of populations in the 1960s; they enjoyed equal indifference from the science community. Michael Conrad, a computer scientist now at Wayne State University, Michigan, also drifted from the study of adaptation to modeling evolving populations in computers in the 1970s, and met the same silence that Holland did a decade earlier. The totality of this work was obscure to computer science and completely unknown in biology.

No more than a couple of students wrote theses on GA until Holland's book *Adaptation in Natural and Artificial Systems* about GAs and evolution appeared in 1975. The book sold only 2,500 copies until it was reissued in 1992. Between 1972 and 1982, no more than two dozen articles on GAs were published in all of science. You could not even say computational evolution had a cult following.

The lack of interest from biology was understandable (but not commendable); biologists reasoned that nature was far too complex to be meaningfully represented by computers *of that time*. The lack of interest from computer science is more baffling. I was often perplexed in my research for this book why such a fundamental process as computational evolution could be so wholly ignored? I now believe the disregard stems from the messy parallelism inherent in evolution and the fundamental conflict it presented to the reigning dogma of computers: the von Neumann serial program.

The first functioning electronic computer was the ENIAC, which was booted up in 1945 to solve ballistic calculations for the U.S. Army. The ENIAC was an immense jumble of 18,000 hot vacuum tubes, 70,000 resistors, and 10,000 capacitors. The instructions for the machine were communicated to it by setting 6,000 switches by hand and then turning the program on. In essence the machine calculated all its values simultaneously in a parallel fashion. It was a bear to program.

The genius von Neumann radically altered this awkward programming system for the EDVAC, the ENIAC's successor and the first general-purpose computer with a stored program. Von Neumann had been thinking about systemic logic since the age of 24 when he published his first papers (in 1927) on mathematical logic systems and game theory. Working with the EDVAC computer group, he invented a way to control the slippery calculations needed to program a machine that could solve more than one problem. Von Neumann proposed that a problem be broken into discrete logical steps, much like the steps in a long division problem, and that intermediate values in the task be stored temporarily in the computer in such a way that those values could be considered input for the next portion of the problem. By feeding back the calculation through a coevolutionary loop (or what is now called a subroutine), and storing the logic of the program in the machine so that it could interact with the answer, von Neumann was able to take any problem and turn it into a series of steps that could be comprehended by a human mind. He also invented a notation for describing this step-wise circuit: the now familiar flow chart. Von Neumann's serial architecture for computation—where one instruction at a time was executed—was amazingly versatile and extremely suited to human programming. He published the general outlines for the architecture in 1946, and it immediately became the standard for every commercial computer thereafter, without exception.

In 1949, John Holland worked on Project Whirlwind, a follow-up to the EDVAC. In 1950 he joined the logical design team on what was then called IBM's Defense Calculator, later to become the IBM 701, the world's first commercial computer. Computers at that point were room-size calculators consuming a lot of electricity. But in the mid-fifties Holland participated in the legendary circle of thinkers who began to map out the possibility of artificial intelligence.

While luminaries such as Herbert Simon and Alan Newall thought of learning as a noble, high-order achievement, Holland thought of it as a polished type of lowly adaptation. If we could understand adaptation, especially evolutionary adaptation, Holland believed, we might be able to understand and maybe imitate conscious learning. But although the others could appreciate the parallels between evolution and learning, evolution was the low road in a fast-moving field.

Browsing for nothing in particular in the University of Michigan math library in 1953, Holland had an epiphany. He stumbled upon a volume, *The Genetical Theory of Natural Selection*, written by R. A. Fisher in 1929. It was Darwin who led the consequential shift from thinking about creatures as individuals to thinking about populations of individuals, but it was Fisher who transformed this population-thinking into a quantitative science. Fisher took what appeared to be a community of flittering butterflies evolving over time and saw them as a whole system transmitting differentiated information in parallel through a population. And he worked out the equations that governed that diffusion of information. Fisher single-handedly opened a new world of human knowledge by subjugating nature's most potent force—evolution—with humankind's most potent tool—mathematics. "That was the first time I realized that you could do significant mathematics on evolution," Holland recalled of the encounter. "The idea appealed to me tremendously." Holland was so enamored of treating evolution as a type of math that in a desperate attempt to get a copy of the out-of-print text (in the days before copiers) he begged the library (unsuccessfully) to sell it to him. Holland absorbed Fisher's vision and then leaped to a vision of his own: butterflies as coprocessors in a field of computer RAM.

Holland felt artificial learning at its core was a special case of adaptation. He was pretty sure he could implement adaptation on computers. Taking the insights of Fisher—that evolution was a class of probability—Holland began the job of trying to code evolution into a machine.

Very early in his efforts, he confronted the dilemma that evolution is a parallel processor while all available electronic computers were von Neumann serial processors.

In his eagerness to wire up a computer as a platform for evolution, Holland did the only reasonable thing: he designed a massively parallel computer to run his experiments. During parallel computing, many instructions are executed concurrently, rather than one at a time. In 1959 he presented a paper which, as its title says, describes "A Universal Computer Capable of Executing an Arbitrary Number of Sub-programs Simultaneously," a contraption that became known as a "Holland Machine." It was almost thirty years before one was built.

In the interim, Holland and the other computational evolutionists had to rely on serial computers to grow evolution. By various tricks they programmed their fast serial CPUs to simulate a slow parallelism. The simulations worked well enough to hint at the power of true parallelism.

It wasn't until the mid-1980s that Danny Hillis began building the first massively parallel computer. Just a few years earlier Hillis had been a wunderkind computer science student. His pranks and hacks at MIT were legendary, even on the campus that invented hacking. With his usual clarity, Hillis summed up for writer Steven Levy the obstacle the von Neumann bottleneck had become in computers: "The more knowledge you gave them, the slower computers got. Yet with a person, the more knowledge you give him, the faster he gets. So we were in this paradox that if you tried to make computers smart, they got stupider."

Hillis really wanted to be a biologist, but his knack for understanding complex programs drew him to the artificial intelligence labs of MIT, where he wound up trying to build a thinking computer "that would be proud of me." He attributes to John Holland the seminal design notions for a swarmy, thousand-headed computing beast. Eventually Hillis led a group that invented the first parallel processing computer, the Connection Machine. In 1988 it sold for a cool $1 million apiece, fully loaded. Now that the machines are here, Hillis has taken up computational biology in earnest.

"There are only two ways we know of to make extremely

complicated things," says Hillis. "One is by engineering, and the other is evolution. And of the two, evolution will make the more complex." If we can't engineer a computer that will be proud of us, we may have to evolve it.

Hillis's first massively parallel Connection Machine had 64,000 processors working in unison. He couldn't wait to get evolution going. He inoculated his computer with a population of 64,000 very simple software programs. As in Holland's GA or in Ray's Tierra, each individual was a string of symbols that could be altered by mutation. But in Hillis's Connection Machine, each program had an entire computer processor dedicated to running it. The populalion, therefore, would react extremely quickly and in numbers that were simply not possible for serial computers to handle.

Each bug in his soup was initially a random sequence of instructions, but over tens of thousands of generations they became a program that sorted a long string of numbers into numerical order. Such a sort routine is an integral part of most larger computer programs; over the years many hundreds of man hours have been spent in computer science departments engineering the most efficient sort algorithms. Hillis let thousands of his sorters proliferate in his computer, mutate at random, and occasionally sexually swap genes. Then in the usual evolutionary maneuver, his system tested them and terminated the less fit so that only the shortest (the best) sorting programs would be given a chance to reproduce. Over ten thousand generations of this cycle, his system bred a software program that was nearly as short as the best sorting programs written by human programmers.

Hillis then reran the experiment but with this important difference: He allowed the *sorting test* itself to mutate while the evolving sorter tried to solve it. The string of symbols in the test varied to become more complicated in order to resist easy sorting. Sorters had to unscramble a moving target, while tests had to resist a moving arrow. In effect Hillis transformed the test list of numbers from a harsh passive environment into an active organism. Like foxes and hares or monarchs and milkweed, sorters and tests got swept up by a textbook case of coevolution.

A biologist at heart, Hillis viewed the mutating sorting test as a parasitic organism trying to disrupt the sorter. He saw his world as an arms race—parasite attack, host defense, parasite counterattack, host counter-defense, and so on. Conventional wisdom claimed such locked arms races are a silly waste of time or an unfortunate blind trap to get stuck in. But Hillis discovered that rather than retard the advance of the sorting organisms, the introduction of a parasite *sped up* the rate of evolution. Parasitic arms races may be ugly, but they turbocharged evolution.

Just as Tom Ray would discover, Danny Hillis also found that evolution can surpass ordinary human skills. Parasites thriving in the Connection Machine prodded sorters to devise a solution more efficient than the ones they found without parasites. After 10,000 cycles of coevolution, Hillis's creatures evolved a sorting program previously unknown to computer scientists. Most humbling, it was only a step short of the all-time shortest algorithm engineered by humans. Blind dumb evolution had designed an ingenious, and quite useful, software program.

A single processor in the Connection Machine is very stupid. It might be as smart as an ant. On its own, a single processor could not come up with an original solution to anything, no matter how many years it spent. Nor would it come up with much if 64,000 processors were strung in a row.

But 64,000 dumb, mindless, ant-brains wired up into a vast interconnected network become a field of evolving populations and, at the same time, look like a mass of neurons in a brain. Out of this network of dumbness emerge brilliant solutions to problems that tax humans. This "order-emerging-out-of-massive-connections" approach to artificial intelligence became known as "connectionism."

Connectionism rekindled earlier intuitions that evolution and learning were deeply related. The connectionists who were reaching for artificial learning latched onto the model of vast webs interconnecting dumb neurons, and then took off with it. They developed a brand of connected concurrent processing—running in either virtual or hardwired parallel computers—that performed simultaneous calculations en masse, similar to genetic algorithms but with more sophisticated (smarter) accounting systems. These smartened up networks were called neural networks. So far neural nets have achieved only limited success in generating partial "intelligence," although their pattern-recognition abilities are useful.

But that *anything at all* emerges from a field of lowly connections is startling. What kind of magic happens inside a web to give it an almost divine power to birth organization from dumb nodes interconnected, or breed software from mindless processors wired to each other? What alchemic transformation occurs when you connect everything to everything? One minute you have a mob of simple individuals, the next, after connection, you have useful, emergent order.

There was a fleeting moment when the connectionists imagined that perhaps all you needed to produce reason and consciousness was a sufficiently large field of interlinked neurons out of which rational intelligence would assemble itself. That dream vanished as soon as they tried it.

But in an odd way, the artificial evolutionists still pursue the dream of connectionism. Only they, in sync with the slow pace of evolution, would be more patient. But it is the slow, very slow, pace of evolution that bothers me. I put my concern to Tom Ray this way: "What worries me about off-the-shelf evolution chips and parallel evolutionary processing machines is that evolution takes an incredible amount of time. Where is this time going to come from? Look at the speed at which nature is working. Consider all the little molecules that have just been snapped together as we talk here. Nature is *incredibly* speedy and vast and humongously parallel, and here we are going to try to beat it. It seems to me there's simply not enough time to do it."

Ray replied: "Well, I worry about that too. On the other hand, I'm amazed at how fast evolution has occurred in my system with only one virtual processor churning it. Besides, time is relative. In evolution, a generation sets the time scale. For *us* a generation is thirty years, but for my creatures it is a fraction of a second. And, when I play god I can crank up the global mutation rate. I'm not sure, but I may be able to get more evolution on a computer."

There are other reasons for doing evolution in a computer. For instance, Ray can record the sequence of every creature's genome and keep a complete demographic and genealogic record of every creature's birth and death. It produces an avalanche of data that is impossible to compile in the real world. And though the complexity and cost of extracting the information will surge as the complexity of the artificial worlds surge, it will probably remain easier to do than in the unwired organic world. As Ray told me, "Even if my world gets as complex as the real world, I'm god. I'm omniscient. I can get information on whatever attracts my attention without disturbing it, without walking around crushing plants. That's a crucial difference."

BACK IN THE 18TH CENTURY, Benjamin Franklin had a hard time convincing his friends that the mild electrical currents produced in his lab were identical in their essence to the thundering lightning that struck in the wild. The difference in scale between his artificially produced microsparks and the sky-splitting, tree-shattering, monstrous bolts generated in the heavens was only part of the problem. Primarily, observers found it unnatural that Franklin could recreate nature, as he claimed.

Today, Tom Ray has trouble convincing his colleagues that the evolution he has synthesized in his lab is identical in essence to the evolution shaping the animals and plants in nature. The difference in time scale between the few hours his world has evolved and the billions of years wild nature has evolved is only part of the problem. Primarily, skeptics find it unnatural that Ray can recreate such an intangible and natural process as he claims.

Two hundred years after Franklin, artificially generated lightning—tamed, measured, and piped through wires into buildings and tools—is the primary organizing force in our society, particularly our digital society. Two hundred years from now, artificial adaptation—tamed, measured and piped into every type of mechanical apparatus we have—will become the central organizing force in our society.

No computer scientist has yet synthesized an artificial intelligence—as desirable and immensely powerful and life-changing as that would be. Nor has any biochemist created an artificial life. But evolution captured, as Ray and others have done, and recreated on demand, is now seen by many technicians as the subtle spark that can create both our dreams of artificial life and artificial intelligence, unleashing their awesome potential. We can grow rather than make them.

We have built machines as complicated as is possible with unassisted engineering. The kind of projects we now have on the drawing boards—software programs reckoned in tens of millions of lines of code, communication systems spanning the planet, factories that must adapt to rapidly shifting global buying habits and retool in days, cheap Robbie the Robots—all demand a degree of complexity that only evolution can coordinate.

Because it is slow, invisible, and diffuse, evolution has the air of a hardly believable ghost in this fast-paced, in-your-face world of humanmade machines. But I prefer to think of evolution as a natural technology that is easily moved into computer code. It is this supercompatibility between evolution and computers that will propel artificial evolution into our digital lives.

ARTIFICIAL EVOLUTION is not merely confined to silicon, however. Evolution will be imported wherever engineering balks. Synthetic evolution technology is already employed in the frontier formerly called bioengineering.

Here's a real-world problem. You need a drug to combat a disease whose mechanism has just been isolated. Think of the mechanism as a lock. All you need is the right key molecule—a drug—that triggers the active binding sites of the lock.

Organic molecules are immensely complex. They consist of thousands of atoms that can be arranged in billions of ways. Simply knowing the chemical ingredients of a protein does not tell us much about its structure. Extremely long chains of amino acids are folded up into a compact bundle so that the hot spots—

the active sites of the protein—are held on the outside at just the right position. Folding a protein is similar to the task of pushing a mile-long stretch of string marked in blue at six points, and trying to fold the string up into a bundle so that the six points of blue all land on different outside faces of the bundle. There are uncountable ways you could proceed, of which only a very few would work. And usually you wouldn't know which sequence was even close until you had completed most of it. There is not enough time in the universe to try all of the variations.

Drug makers have had two traditional manners for dealing with this complexity. In the past, pharmacists relied on hit or miss. They tried all existing chemicals found in nature to see if any might work on a given lock. Often, one or two natural compounds activated a couple of sites—a sort of partial key. But now in the era of engineering, biochemists try to decipher the pathways between gene code and protein folding to see if they can engineer the sequence of steps needed to create a molecular shape. Although there has been some limited success, protein folding and genetic pathways are still far too complex to control. Thus this logical approach, called "rational drug design," has bumped the ceiling of how much complexity we can engineer.

Beginning in the late 1980s, though, bioengineering labs around the world began perfecting a new procedure that employs the only other tool we have for creating complex entities: evolution.

In brief, the evolutionary system generates billions of random molecules which are tested against the lock. Out of the billion humdrum candidates, one molecule contains a single site that matches one of, say, six sites on the lock. That partial "warm" key sticks to the lock and is retained. The rest are washed down the drain. Then, a billion new variations of that surviving warm key are made (retaining the trait that works) and tested against the lock. Perhaps another warm key is found that now has two sites correct. That key is kept as a survivor while the rest die. A billion variations are made of it, and the most fit of that generation will survive to the next. In less than ten generations of repeating the wash/mutate/bind sequence, this molecular breeding program will find a drug—perhaps a lifesaving drug—that keys all the sites of the lock.

Almost any kind of molecule might be evolved. An evolutionary biotechnician could evolve an improved version of insulin, say, by injecting insulin into a rabbit and harvesting the antibodies that the rabbit's immune system produced in reaction to this "toxin." (Antibodies are the complementary shape to a toxin.) The biotechnician then puts the extracted insulin antibodies into an evolutionary system where the antibodies serve as a lock against which new keys are tested. After several generations of evolution, he would have a complementary shape to the antibody, or in effect, an alternative working shape to the insulin shape. In short, he'd have another version of insulin. Such an alternative insulin would be extremely valuable. Alternative versions of natural drugs can offer many advantages: they might be smaller; more easily delivered in the body; produce fewer side effects; be easier to manufacture; or be more specific in their targets.

Of course, the bioevolutionists could also harvest an antibody against, say, a hepatitis virus and then evolve an imitation hepatitis virus to match the antibody. Instead of a perfect match, the biochemist would select for a surrogate molecule that lacked certain activation sites that cause the disease's fatal symptoms. We call this imperfect, impotent surrogate a vaccine. So vaccines could also be evolved rather than engineered.

All the usual reasons for creating drugs lend themselves to the evolutionary method. The resulting molecule is indistinguishable from rationally designed drugs. The only difference is that while an evolved drug works, we have no idea of how or why it does so. All we know is that we gave it a thorough test and it passed. Cloaked from our understanding, these invented drugs are "irrationally designed."

Evolving drugs allows a researcher to be stupid, while evolution slowly accumulates the smartness. Andrew Ellington, an evolutionary biochemist at Indiana University, told *Science* that in evolving systems "you let the molecule tell you about itself, because it knows more about itself than you do."

Breeding drugs would be a medical boon. But if we can breed software and then later turn the system upon itself so that software breeds itself, leading to who knows what, can we set molecules too upon the path of open-ended evolution?

Yes, but it's a difficult job. Tom Ray's electric-powered evolution machine is heavy on the heritable information but light on bodies. Molecular evolution programs are heavy on bodies but skimpy on heritable information. Naked information is hard to kill, and without death there is no evolution. Flesh and blood greatly assist the cause of evolution because a body provides a handy way for information to die. Any system that can incorporate the two threads of heritable information and mortal bodies has the ingredients for an evolutionary system.

Gerald Joyce, a biochemist at San Diego whose background is the chemistry of very early life, devised a simple way to incorporate the dual nature of information and bodies into one robust artificial evolutionary system. He accomplished this by recreating a probable earlier stage of life on Earth—"RNA world"—in a test tube.

RNA is a very sophisticated molecular system. It was not the very first living system, but life on Earth at some stage almost certainly became RNA life. Says Joyce, "Everything in biology points to the fact that 3.9 billion years ago, RNA was running the show."

RNA has a unique advantage that no other system we know about can boast. It acts at once as both body and info, phenotype and genotype, messenger and message. An RNA molecule is at once the flesh that must interact in the world and the information that must inherit the world, or at least be transmitted to the next generation. Though limited by this uniqueness, RNA is a wonderfully compact system in which to begin open-ended artificial evolution.

Gerald Joyce runs a modest group of graduates and postdocs at Scripps Institute, a sleek modern lab along the California coast near San Diego. His experimental RNA worlds are tiny drops that pool in the bottom of plastic micro-test tubes hardly the volume of thimbles. At any one time dozens of these pastel-colored tubes, packed in ice in styrofoam buckets, await being warmed up to body temperature to start evolving. Once warmed, RNA will produce a billion copies in one hour.

"What we have here," Joyce says pointing to one of the tiny tubes, "is a huge parallel processor. One of the reasons I went into biology instead of doing computer simulations of evolution is that no computer on the face of the Earth, at least for the near future, can give me $10^{15}$ microprocessors in parallel." The drops in the bottom of the tubes are about the size of the smart part of computer chips. Joyce polishes the image: "Actually, our artificial system is even better than playing with *natural* evolution because there aren't too many natural systems that come close to letting us turn over $10^{15}$ individuals in an hour, either."

In addition to the intellectual revolution a self-sustaining life system would launch, Joyce sees evolution as a commercially profitable way to create useful chemicals and drugs. He imagines molecular evolution systems that run 24 hours, 365 days a year: "You give it a task, and say don't come out of your closet until you've figured out how to convert molecule A to molecule B."

Joyce rattles off a list of biotech companies that are today dedicated solely to research in directed molecular evolution (Gilead, Ixsys, Nexagen, Osiris, Selectide, and Darwin Molecule). His list does not include established biotech companies, such as Genentech, which are doing advanced research into directed evolutionary techniques, but which also practice rational drug design. Darwin Molecule, whose principal patent holder is complexity researcher Stuart Kauffman, raised several million dollars to exploit evolution's power to design drugs. Manfred Eigen, Nobel Prize-winning biochemist, calls directed evolution "the future of biotechnology."

But is this really evolution? Is this the same vital spirit that brought us insulin, eyelashes, and raccoons in the first place? It is. "We approach evolution with a capital D for Darwin," Joyce told me. "But since the selection pressure is determined by us, rather than nature, we call this directed evolution."

Directed evolution is another name for supervised learning, another name for the Method of traversing the Library, another name for breeding. Instead of letting the selection emerge, the breeder directs the choice of varieties of dogs, pigeons, pharmaceuticals, or graphic images.

DAVID ACKLEY is a researcher of neural nets and genetic algorithms at Bellcore, the R&D labs for the Baby Bells. Ackley has some of the most original ways of looking at evolutionary systems that I've come across.

Ackley is a bear of a guy with a side-of-the-mouth wisecracking delivery. He broke up 250 serious scientists at the 1990 Second Artificial Life Conference with a wickedly funny video of a rather important artificial life world he and colleague Michael Littman had made. His "creatures" were actually bits of code not too different from a classical GA, but he dressed them up with moronic smiley faces as they went about chomping each other or bumping into walls in his graphical world. The smart survived, the dumb died. As others had, Ackley found that his world was able to evolve amazingly fit organisms. Successful individuals would live Methuselahian lifetimes—25,000 day-steps in his world. These guys had the system all figured out. They knew how to get what they needed with minimum effort. And how to stay out of trouble. Not only would individuals live long, but the populations that shared their genes would survive eons as well.

Noodling around with the genes of these streetwise creatures, Ackley uncovered a couple of resources they hadn't taken up. He saw that he could improve their chromosomes in a godlike way to exploit these resources, making them even better adapted to the environment he had set up for them. So in an early act of virtual genetic engineering, he modified their evolved code and set them back again into his world. As individuals, they were superbly fitted and flourished easily, scoring higher on the fitness scale than any creatures before them.

But Ackley noticed that their population numbers were always lower than the naturally evolved guys. As a group they were anemic. Although they never died out, they were always endangered. Ackley felt their low numbers wouldn't permit the species to last more than 300 generations. So while handcrafted genes suited individuals to the max, they lacked the robustness of organically grown genes, which suited the species to the max. Here, in the home-brewed world of a midnight hacker, was the first bit of testable proof for hoary ecological wisdom: that what is best for an individual ain't necessarily best for the species.

"It's tough accepting that we can't figure out what's best in the long run," Ackley told the Artificial Life conference to great applause, "but, hey, I guess that's life!"

Bellcore allowed Ackley to pursue his microgod world because they recognized that evolution is a type of computation. Bellcore was, and still is, interested in better computational methods, particularly those based on distributed models, because ultimately a telephone network is a distributed computer. If evolution is a useful type of distributed computation, what might some other methods be? And what improvements or variations, if any, can we make to evolutionary techniques? Taking up the usual library/space metaphor, Ackley gushes, "The space of *computational* machinery is unbelievably vast and we have only explored very tiny corners of it. What I'm doing, and what I want to do more of, is to expand the space of what people recognize as computation."

Of all the possible types of computation, Ackley is primarily interested in those procedures that underpin learning. Strong learning methods require smart teachers; that's one type of learning. A smart teacher tells a learner what it should know, and the learner analyzes the information and stores it in its memory. A less smart teacher can also teach by using a different method. It doesn't know the material itself, but it can tell when the learner guesses the right answer—as a substitute teacher might grade tests. If the learner guesses a partial answer the weak teacher can give a hint of "getting warm," or "getting cold" to help the learner along. In this way, a weaker teacher can potentially generate information that it itself doesn't own. Ackley has been pushing the edge of weak learning as a way of maximizing computation: leveraging the smallest amount of information in, to get the maximum information out. "I'm trying to come up with the dumbest, least informative teacher as possible," Ackley told me. "And I think I found it. My answer is: death."

Death is the *only* teacher in evolution. Ackley's mission was to find out: what can you learn using only death as a teacher? We don't know for sure, but some candidates are: soaring eagles, or pigeon navigation systems, or termite skyscrapers. It takes a while, but evolution *is* clever. Yet it is obviously blind and dumb. "I can't imagine any dumber type of learning than natural selection," says Ackley.

In the space of all possible computation and learning, then, natural selection holds a special position. It occupies the extreme point where information transfer is minimized. It forms the lowest baseline of learning and smartness, below which learning doesn't happen and above which smarter, more complicated learning takes place. Even though we still do not fully understand the nature of natural selection in coevolutionary worlds, natural selection remains the elemental melting point of learning. If we could measure degrees of evolution (we can't yet) we would have a starting benchmark against which to rate other types of learning.

Natural selection plays itself out in many guises. Ackley was right; computer scientists now realize that many modes of computation exist—many of them evolutionary. For all anyone knows, there may be hundreds of styles of evolution and learning. All such strategies, however, perform a search routine through a library or space. "Discovering the notion of the 'search' was the one and only brilliant idea that traditional AI research ever had," claims Ackley. A search can be accomplished in many ways. Natural selection—as it is run in organic life—is but one flavor.

Biological life is wedded to a particular hardware: carbon-based DNA molecules. This hardware limits the versions of search-by-natural-selection that can successfully operate upon it. With the new hardware of computers, particularly parallel computers, a host of other adaptive systems can be conjured up, and entirely different search strategies set out to shape them. For instance, a chromosome of biological DNA cannot broadcast its code to DNA molecules in other organisms in order for them to receive the message and alter their code. But in a computer environment you can do that.

David Ackley and Michael Littman, both of Bellcore's Cognitive Science Research Group, set out to fabricate a non-Darwinian evolutionary system in a computer. They chose a most logical alternative: Lamarckian evolution—the inheritance of acquired traits. Lamarckism is very appealing. Intuitively such a system would seem deeply advantageous over the Darwinian version, because presumably useful mutations would be adopted into the gene line more quickly. But a look at its severe computational requirements quickly convinces the hopeful engineer how unlikely such a system would be in real life.

If a blacksmith acquires bulging biceps, how does his body reverse-engineer the exact changes in his genes needed to produce this improvement? The drawback for a Lamarckian system is its need to trace a particular advantageous change in the body back through embryonic development into the genetic blueprints. Since any change in an organism's form may be caused by more than one gene, or by many instructions interacting during the body's convoluted development, unraveling the tangled web of causes of any outward form requires a tracking system almost as complex as the body itself. Biological Lamarckian evolution is hampered by a strict mathematical law: that it is supremely easy to multiply prime factors together, but supremely hard to derive the prime factors out of the result. The best encryption schemes work on this same asymmetrical difficulty. Biological Lamarckism probably hasn't happened because it requires an improbable biological decryption scheme.

But computational entities don't require bodies. In computer evolution (as in Tom Ray's electric-powered evolution machine) the computer code doubles as both gene and body. Thus, the dilemma of deriving a genotype from the phenotype is moot. (The restriction of monolithic representation is not all that artificial. Life on Earth must have passed through this stage, and perhaps any spontaneously organizing vivisystem must begin with a genotype that is restricted to its phenotype, as simple self-replicating molecules would be.)

In artificial computer worlds, Lamarckian evolution works. Ackley and Littman implemented a Lamarckian system on a parallel computer with 16,000 processors. Each processor held a subpopulation of 64 individuals, for a grand total of approximately one million individuals. To simulate the dual information lines of body and gene, the system made a copy of the gene for each individual and called the copy the "body." Each body was a slightly different bit of code trying to solve the same problem as its million siblings.

The Bellcore scientists set up two runs. In the Darwinian run, the body code would mutate over time. By chance a lucky guy might become code that provides a better solution, so the system chooses it to mate and replicate. But in Darwinism when it mates, it must use its original "gene" copy of the code—the code it inherited, not the improved body code it acquired during its lifetime. This is the biological way; when the blacksmith mates, he uses the code for the body he inherited, not the body he acquired.

In the Lamarckian run, by contrast, when the lucky guy with the improved body code is chosen to mate, it can use the improved code acquired during its lifetime as the basis for its mating. It is as if a blacksmith could pass on his massive arms to his offspring.

Comparing the two systems, Ackley and Littman found that, at least for the complicated problems they looked at, the Lamarckian system discovered solutions almost twice as good as the Darwinian method. The smartest Lamarckian individual was far smarter than the smartest Darwinian one. The thing about Lamarckian evolution, says Ackley, is that it "very quickly squeezes out the idiots" in a population. Ackley once bellowed to a roomful of scientists, "Lamarck just blows the doors off of Darwin!"

In a mathematical sense, Lamarckian evolution injects a bit of learning into the soup. Learning is defined as adaptation within an individual's lifetime. In classical Darwinian evolution, individual learning doesn't count for much. But Lamarckian evolution permits information acquired during a lifetime (including how to build muscles or solve equations) to be incorporated into the long-term, dumb learning that takes place over evolution. Lamarckian evolution produces smarter answers because it is a smarter type of search.

The superiority of Lamarckism surprised Ackley because he felt that nature did things so well: "From a computer science viewpoint it seems really stupid that nature is Darwinian and not Lamarckian. But nature is stuck on chemicals. We're not." It got

him thinking about other types of evolution and
that might be more useful if you weren't restrictec
on molecules.

A GROUP OF RESEARCHERS in Milan, Italy, have come up with a few new varieties of evolution and learning. Their methods fill a few holes in Ackley's proposed "space of all possible types of computation." Because they were inspired by the collective behavior of ant colonies, the Milan group call their searches "Ant Algorithms."

Ants have distributed parallel systems all figured out. Ants are the history of social organization and the future of computers. A colony may contain a million workers and hundreds of queens, and the entire mass of them can build a city while only dimly aware of one another. Ants can swarm over a field and find the choicest food in it as if the swarm were a large compound eye. They weave vegetation together in coordinated parallel rows, and collectively keep their nest at a steady temperature, although not a single ant has ever lived who knows how to regulate temperature.

An army of ants too dumb to measure and too blind to see far can rapidly find the shortest route across a very rugged landscape. This calculation perfectly mirrors the evolutionary search: dumb, blind, simultaneous agents trying to optimize a path on a computationally rugged landscape. Ants *are* a parallel processing machine.

Real ants communicate with each other by a chemical system called pheromones. Ants apply pheromones on each other and on their environment. These aromatic smells dissipate over time. The odors can also be relayed by a chain of ants picking up a scent and remanufacturing it to pass on to others. Pheromones can be thought of as information broadcasted or communicated within the ant system.

The Milan group (Alberto Colorni, Marco Dorigo, and Vittorio Maniezzo) constructed formulas modeled on ant logic. Their virtual ants ("vants") were dumb processors in a giant community operating in parallel. Each vant had a meager memory, and could communicate locally. Yet the rewards of doing well were shared by others in a kind of distributed computation.

The Italians tested their ant machine on a standard benchmark, the traveling salesman problem. The riddle was: what is the shortest route between a large number of cities, if you can only visit each city once? Each virtual ant in the colony would set out rambling from city to city leaving a trail of pheromones. The shorter the path between cities, the less the pheromone evaporated. The stronger the pheromone signal, the more other ants followed that route. Shorter paths were thus self-reinforcing. Run for 5,000 rounds or so, the ant group-mind would evolve a fairly optimal global route.

The Milan group played with variations. Did it make any difference if the vants all started at one city or were uniformly distributed? (Distributed was better.) Did it make any difference how many vants one ran concurrently? (More was better until you hit the ratio of one ant for every city, when the advantage peaked.) By varying parameters, the group came up with a number of computational ant searches.

Ant algorithms are a type of Lamarckian search. When one ant stumbles upon a short route, that information is indirectly broadcast to the other vants by the trail's pheromone strength. In this way learning in one ant's lifetime is indirectly incorporated into the whole colony's inheritance of information. Individual ants effectively broadcast what they have learned into their hive. Broadcasting, like cultural teaching, is a part of Lamarckian search. Ackley: "There are ways to exchange information other than sex. Like the evening news."

The cleverness of the ants, both real and virtual, is that the amount of information invested into "broadcasting" is very small, done very locally, and is very weak. The notion of introducing weak broadcasting into evolution is quite appealing. If there is any Lamarckism in earthly biology it is buried deep. But there remains a universe full of strange types of potential computation that might employ various modes of Lamarckian broadcasting. I know of programmers fooling around with algorithms to mimic "memetic" evolution—the flow of ideas (memes) from one mind to another, trying to capture the essence and power of cultural evolution. Out of all the possible ways to connect the nodes in distributed computers, only a very few, such as the ant algorithms, have even been examined.

As late as 1990, parallel computers were derided by experts as controversial, specialized, and belonging to the lunatic fringe. They were untidy and hard to program. The lunatic fringe disagreed. In 1989, Danny Hillis boldly made a widely publicized bet with a leading computer expert that as early as 1995, more bits per month would be processed by parallel machines than by serial machines. He is looking right. As serial computers audibly groaned under the burden of pushing complex jobs through the tiny funnel of von Neumann's serial processor, a change in expert opinion suddenly swept through the computer industry. Peter Denning signaled the new perspective when he wrote in a paper published by *Science* ("Highly Parallel Computation," November 30, 1990), "Highly parallel computing architectures are the only means to achieve the computational rates demanded by advanced scientific problems." John Koza of Stanford's Computer Science Department says flatly, "Parallel computers are the future of computing. Period."

But parallel computers remain hard to manage. Parallel software is a tangled web of horizontal, simultaneous causes. You can't check such nonlinearity for flaws since it's all hidden corners. There is no narrative to step through. The code has the integrity of a water balloon, yielding in one spot as another bulges. Parallel computers can easily be built but can't be easily programmed.

Parallel computers embody the challenge of all distributed swarm systems, including phone networks, military systems, the planetary 24-hour financial web, and large computer networks. Their complexity is taxing our ability to steer them. "The complexity of programming a massively parallel machine is probably beyond us," Tom Ray told me. "I don't think we'll ever be able to write software that fully uses the capacity of parallelism."

Little dumb creatures in parallel that can "write" better

software than humans can suggests to Ray a solution for our desire for parallel software. "Look," he says, "ecological interactions are just parallel optimization techniques. A multicellular organism essentially runs massively parallel code of an astronomical scale. Evolution can 'think' of parallel programming ways that would take us forever to think of. If we can evolve software, we'll be way ahead." When it comes to distributed network kinds of things, Ray says, "Evolution is the *natural* way to program."

The natural way to program! That's an ego-deflating lesson. Humans should stick to what they do best: small, elegant, minimal systems that are fast and deep. Let natural evolution (artificially injected) do the messy big work.

Danny Hillis has come to the same conclusion. He is serious when he says he wants his Connection Machine to evolve commercial software. "We want these systems to solve a problem we don't know how to solve, but merely know how to state." One such problem is creating multimillion-line programs to fly airplanes. Hillis proposes setting up a swarm system which would try to evolve better software to steer a plane, while tiny parasitic programs would try to crash it. As his experiments have shown, parasites encourage a faster convergence to an error-free, robust software navigation program. Hillis: "Rather than spending uncountable hours designing code, doing error-checking, and so on, we'd like to spend more time making better parasites!"

Even when technicians do succeed in engineering an immense program such as navigation software, testing it thoroughly is becoming impossible. But things grown, not made, are different. "This kind of software would be built in an environment full of *thousands* of full-time adversaries who specialize in finding out what's wrong with it," Hillis says, thinking of his parasites. "Whatever survives them has been tested ruthlessly." In addition to its ability to create things that we can't make, evolution adds this: it can also make them more flawless than we can. "I would rather fly on a plane running software evolved by a program like this, than fly on a plane running software I wrote myself," says Hillis, programmer extraordinaire.

The call-routing program of long-distance phone companies

tallies up to about 2 million lines of code. Three faulty lines in those 2 million caused the rash of national telephone system outages in the summer of 1990. And 2 million lines is no longer large. The combat computers aboard the Navy's Seawolf submarine contain 3.6 million lines of code. "NT," the new workstation computer operating system released by Microsoft in 1993, required 4 million lines of code. One-hundred-million-line programs are not far away.

When computer programs swell to billions of lines of code, just keeping them up and "alive" will become a major chore. Too much of the economy and too many people's lives will depend on billion-line programs to let them go down for even an instant. David Ackley thinks that reliability and up-time will become the *primary* chore of the software itself. "I claim that for a really complex program sheer *survival* is going to consume more of its resources." Right now only a small portion of a large program is dedicated to maintenance, error correction, and hygiene. "In the future," predicts Ackley, "99 percent of raw computer cycles are going to be spent on the beast watching itself to keep it going. Only that remaining 1 percent is going to be used for user tasks—telephone switching or whatever. Because the beast can't do the user tasks unless it survives."

As software gets bigger, survival becomes critical yet increasingly difficult. Survival in the everyday world of daily use means flexibility and evolvability. And it demands more work to pull off. A program survives only if it constantly analyzes its status, adjusts its code to new demands, cleanses itself, ceaselessly dissects anomalous circumstances, and always adapts and evolves. Computation must seethe and behave as if it is alive. Ackley calls it "software biology" or "living computation." Engineers, even on 24-hour beepers, can't keep billion-line code alive. Artificial evolution may be the only way to keep software on its toes, looking lively.

Artificial evolution is the end of engineering's hegemony. Evolution will take us beyond our ability to plan. Evolution will craft things we can't. Evolution will make them more flawless than we can. And evolution will maintain them as we can't.

But the price of evolution is the title of this book. Tom Ray explains: "Part of the problem in an evolving system is that we give up some control."

Nobody will understand the evolved aviation software that will fly Danny Hillis. It will be an indecipherable spaghetti of 5 million strands of nonsense—of which perhaps only 2 million are really needed. But it will work flawlessly.

No human will be able to troubleshoot the living software running Ackley's evolved telephone system. The lines of program are buried in an uncharted web of small machines, in an incomprehensible pattern. But, when it falters, it will heal itself.

No one will control the destination of Tom Ray's soup of critters. They are brilliant in devising tricks, but there is no telling them what trick to work on next. Only evolution can handle the complexities we are creating, but evolution escapes our total command.

At Xerox PARC, Ralph Merkle is engineering very small molecules that can replicate. Because these replicators dwell in the microscopic scale of nanometers (smaller than bacteria) their construction techniques are called nanotechnology. At some point in the very near future the engineering skills of nanotechnology and the engineering skills of biotechnology converge; they are both treating molecules as machines. Think of nanotechnology as bioengineering for dry life. Nanotechnology has the same potential for artificial evolution as biological molecules. Merkle told me, "I don't want nanotechnology to evolve. I want to keep it in a vat, constrained by international law. The most dangerous thing that could happen to nanotechnology is sex. Yes, I think there should be international regulations against sex for nanotechnology. As soon as you have sex, you have evolution, and as soon as you have evolution, you have trouble."

The trouble of evolution is not entirely out of our control; surrendering some control is simply a tradeoff we make when we employ it. The things we are proud of in engineering—precision, predictability, exactness, and correctness—are diluted when evolution is introduced.

These have to be diluted because survivability in a world of accidents, unforeseen circumstances, shifting environments—in short, the real world—demands a fuzzier, looser, more adaptable, less precise stance. Life is not controlled. Vivisystems are not predictable. Living creatures are not exact.

"'Correct' will go by the board," Ackley says of complex programs. "'Correct' is a property of small systems. In the presence of great change, 'correct' will be replaced by 'survivability'."

When the phone system is run by adaptable, evolved software, there will be no correct way to run it. Ackley continues: "To say that a system is 'correct' in the future will sound like bureaucratic double-talk. What people are going to judge a system on is the ingenuity of its response, and how well it can respond to the unexpected." We will trade correctness for flexibility and durability. We will trade a clean corpse for messy life. Ackley: "It will be to your advantage to have an out-of-control, but responsive, monster spend 1 percent of itself on your problem, than to have a dedicated little correct ant of a program that hasn't got a clue about what in the world is going on."

A student at one of Stuart Kauffman's lectures once asked him, "How do you evolve for things you don't want? I see how you can get a system to evolve what you want; but how can you be sure it won't create what you don't want?" Good question, kid. We can define what we want narrowly enough to breed for it. But we often don't even know what we don't want. Or if we do, the list of things that are unacceptable is so long as to be impractical. How can we select out disadvantageous side effects?

"You can't." Kauffman replied bluntly.

That's the evolutionary deal. We trade power for control. For control junkies like us, this is a devil's bargain.

Give up control, and we'll artificially evolve new worlds and undreamed-of richness. Let go, and it will blossom.

Have we ever resisted temptation before?

# 16

# THE FUTURE OF CONTROL

THE ABSOLUTELY NEAT THING about the dinosaurs in the movie *Jurassic Park* is that they possess enough artificial life so that they can be reused as cartoon dinos in a Flintstones movie.

They won't be completely the same of course. They'll be tamer, longer, rounder, and more obedient. But inside Dino will beat the digital heart of *T. Rex* and *Velociraptor*—different bodies but the same dinosaurness. Mark Dippe, the wizard at Industrial Light and Magic who invented the virtual dinosaurs, has merely to alter the settings in the creatures' digital genes to transform their shape into lovable pets, while maintaining their convincing screen presence.

Yet the *Jurassic Park* dinosaurs are zombies. They have magnificent simulated bodies, but they lack their own behavior, their own will, their own drive for survival. They are ghostly muppets guided by computer animators. Someday, though, the dinosaurs may become Pinocchios—puppets given their own life.

Before the Jurassic dinosaurs were imported into the photorealistic world of a movie, they dwelt in an empty world consisting solely of three dimensions. In this dreamland—let's think of it as that place where all the flying logos for TV stations live—there is volume, light, and space, but not much else. Wind, gravity, inertia, friction, stiffness, and all the subtle aspects of a material world are absent and have to be faked by imaginative animators.

"In traditional animation all knowledge of physics has to come from the animator's head," says Michael Kass, a computer graphics engineer at Apple Computer. For instance, when Walt Disney drew Mickey Mouse bouncing downstairs on his rear end, Disney played out on drawing paper his perception of how the law of gravity works. Mickey obeyed Disney's ideas of physics, whether they were realistic or not. They usually weren't, which has always been their charm. Many animators exaggerated, altered, or ignored the physical laws of the real world for a laugh. But in the current cinematic style, the goal is strict realism. Modern audiences want E.T.'s flying bicycle to behave like a "real" flying bicycle, not like a cartoon version.

Kass is trying to imbue physics into simulated worlds. "We thought about the tradition of having the physics in the animator's head and decided that instead, the computer should have some knowledge of physics."

Say we start with flying logo dreamland. One of the problems with this simple world, Kass says, is that "things look like they don't weigh anything." To increase the realism of the world we could add mass and weight to objects and a gravity law to the environment, so that if a flying logo drops to the floor it falls at the same acceleration as would a solid logo falling to Earth. The equation for gravity is very simple, and implanting it in a small world is not difficult. We could add a bounce formula to the animated logo so that it rebounds from the floor "of its own accord" in a very regular manner. It obeys the rule of gravity and the rules of kinetic energy and friction which slow it down. And it can be given stiffness—say of plastic or metal—so that it reacts to an impact realistically. The final result has the feel of reality, as a chrome logo falls to the floor and bounces in diminishing hops until it clatters to a rest.

We might continue to apply additional formulas of physical rules, such as elasticity, surface tension, and spin effects, and code them into the environment. As we increase the complexity of these artificial environments, they become fertile ground for synthetic life.

This is why the Jurassic dinosaurs were so lifelike. When they lifted their legs, they encountered the virtual weight of meat.

Their muscles flexed and sagged. When the foot came down, gravity pulled it, and the impact of landing reverberated back up the leg.

The talking cat in Disney's summer of '93 movie *Hocus Pocus* was a virtual character similar to the dinosaurs, but in close-up. The animators built a digital cat form and then "texture-mapped" its fur from a photographed cat, which it perfectly resembled except for its remarkable talking. Its mouth behavior was mapped from a human. The thing was a virtual cat-human hybrid.

A movie audience watches autumn leaves blowing down the street. The audience does not realize the scene is computer-generated animation. The event looks real because the video is of something real: individual virtual leaves being blown by a virtual wind down a virtual street. As in Reynolds's flocks of virtual bats, there is a real shower of things really being pushed by a force in a place with physical laws. The virtual leaves have attributes such as weight and shape and surface area. When they are released into a virtual wind they obey a set of laws parallel to the real ones that real leaves obey. The relationship between all the parts is as real as a New England day, although the lack of details in the leaves wouldn't work in close-up. The blowing leaves are not so much drawn as let loose.

Letting animations follow their own physics is the new recipe for realism. When Terminator 2 wells up from a molten pool of chrome, the effect is astoundingly convincing because the chrome is obeying physical constraints of liquids (such as surface tension) in a parallel universe. It *is* a liquid in simulation.

Kass and Apple colleague Gavin Miller came up with computer programs to render the subtle ways in which water trickles down a shallow stream, or falls as rain on a puddle. They transferred the laws of hydrology into a simulated universe by hooking up the formulas to an animating engine. Their video clips show a shallow wave sweeping over a dry sandy shore under a soft light, breaking in the irregular manner of real waves, then receding, leaving wet sand behind. In reality it's all just equations.

To make these digital worlds really work in the future,

everything in creation will have to be reduced to equations. Not just the dinosaurs and water, but eventually the trees the dinos munched on, the jeeps (which *were* digital in some scenes of *Jurassic Park*), buildings, clothes, breakfast tables, and the weather. If this all had to happen just for the movies, it wouldn't. But every manufactured item in the near future will be designed and produced using CAD (computer-assisted design) programs. Already today, automobile parts are simulated on computer screens first, and their equations later transmitted directly to the factory lathes and welders to give the numbers actual form. A new industry called automatic fabrication takes the data from a CAD and instantly generates a 3-D prototype from powered metal or liquid plastic. First an object is just lines on a screen; then it's a solid thing you can hold in your hand or walk around. Instead of printing a picture of a gear, automatic fabrication technology "prints" the actual gear itself. Emergency spare parts for factory machines are now printed out in hi-impact plastic on the factory floor; they'll hold out until the authentic spare part arrives. Someday soon, the printed object will be the authentic part. John Walker, founder of the world's premier CAD program, AutoCAD, told a reporter, "CAD is about building models of real-world objects inside the computer. I believe in the fullness of time, every object in the world, manufactured or not, will be modeled inside a computer. This is a very, very big market. This is everything."

Biology included. Flowers can already be modeled in computers. Przemyslaw Prusinkiewicz, a computer scientist at the University of Calgary, Canada, uses a mathematical model of botanical growth to create 3-D virtual flowers. A few simple laws apparently govern most plant growth. Flowering signals can get complicated. The blossom sequence on a stalk may be determined by several interacting messages. But these interacting signals can be coded into a program quite simply.

The mathematics of growing plants was worked out in 1968 by the theoretical biologist Aristid Lindenmeyer. His equations articulated the distinction between a carnation and a rose; the difference can be reduced to a set of variables in a numerical seed. An entire plant may only take a few kilobytes on a hard disk—a seed. When the seed is decompressed by the computer program, a graphical flower grows on the screen. First a green sprout shoots up, leaves unfurl, a bud takes shape, and then, at the right moment, a flower blossoms. Prusinkiewicz and his students have scoured the botanical literature to discover how multiple heads of flowers bloom, or how a daisy forms, and how an elm or oak fork their distinctive branches. They have also compiled algorithmic laws of growth for hundreds of seashells and butterflies. The graphical results are entirely convincing. A still frame of one of Prusinkiewicz's computer-grown lilac sprays with its myriad florets could pass for a photograph in a seed catalog.

At first this was a fun academic exercise, but Prusinkiewicz is now besieged with calls from horticulturists wanting his software. They'll pay a lot of money if they can get a program that will show their clients what their landscape designs will look like in ten years or even next spring.

The best way to fake a living creature, Prusinkiewicz found, is to grow it. The laws of growth he has extracted from biology and then put into a virtual world are used to grow cinematic trees and flowers. They make a wonderfully apt environment for dinosaurs or other digital characters.

Brøderbund software, a venerable publisher of educational software for personal computers, sells a program that models physical forces as a way of teaching physics. When you boot-up the Physics program on your Macintosh you launch a toy planet that orbits the sun on the computer screen. The virtual planet obeys the forces of gravity, motion, and friction written into the toy universe. By fiddling with the forces of momentum and gravity, a student can get a feel for how the physics of the solar system works.

How far can we press this? If we kept adding other forces that the toy planet had to obey, such as electrostatic attraction, magnetism, friction, thermodynamics, volume, if we kept adding every feature we saw in the real world to this program, what kind of solar system would we eventually have in the computer? If a computer is used to model a bridge—all its forces of steel, wind, and gravity—could we ever get to the point that we could say we have a bridge inside the computer? And can we do this with life?

As fast as physics is encroaching into digital worlds, life is invading faster. To see how far distributed life has infiltrated computational cinema, and to what consequences, I took a tour of the state-of-the-art animation labs.

MICKEY MOUSE is one of the ancestors of artificial life. Mickey, now 67 years old, will soon have to face the digital era. In one of the permanent "temporary" buildings on the backlot of Disney's Glendale studios, his trustees were cautiously planning ways to automate animated characters and backgrounds. I spoke to Bob Lambert, director of new technologies for the Disney animators.

The first thing Bob Lambert made clear to me was that Disney was in no hurry to completely automate animation. Animation was a handcraft, an art. Disney Inc.'s great fortune was sealed in this craft, and their crown jewels—Mickey Mouse and pals—were perceived by their customers as exemplary works of art. If computer animation meant anything like the wooden robots kids see on Saturday morning cartoons then Disney wanted no part of it. Lambert: "We don't need people saying, 'Oh damn, there goes another handcrafted art down the computer hole.'"

Then there was the problem of the artists themselves. Said Lambert, "Look, we have 400 ladies in white smocks who have been painting Mickey for 30 years. We can't change suddenly."

The second thing Lambert wanted to make clear was that Disney had already been using some automated animation in their legendary films since 1990. Gradually they were digitizing their worlds. Their animators had gotten the message that those who didn't transfer their artists' intelligence from their heads into an almost living simulation would soon be dinosaurs of another kind. "To be honest," said Lambert, "by 1992 our animators were clamoring to use computers."

The giant clockwork in Disney's *The Great Mouse Detective* was a computer-generated model of a clock that hand-drawn characters ran over. In *Rescuers Down Under*, Oliver the Albatross dove down through a virtual New York City, a completely computer-generated environment grown from a large database of New York buildings compiled by a large contractor for commercial reasons. And in *The Little Mermaid*, Ariel swam through clusters of fish whose schooling was simulated, seaweed that swayed autonomously, and bubbles that percolated with physics. However, with a nod to the 400 ladies in white, each frame of these computer-generated background scenes was printed out on fine painting paper and hand-colored to match the rest of the movie.

*Beauty and the Beast* was Disney's first movie to use "paperless animation," at least in one scene. The ballroom dance at the end of the film was composed and rendered digitally, except for the hand-drawn characters of the Beast and Belle. The shift in the movie between the real cartoon and the faked cartoon was just slightly noticeable to my eye. The discontinuity protruded not because it was less graceful than the hand animation, but because it was better—because it looked more photographic than the cartoon.

The first Disney character to be completely paperless was the flying (walking, pointing, jumping) carpet in *Aladdin.* To make it, the form of a Persian carpet was rendered on a computer screen. The animator bent it into its poses by moving a cursor, and then the computer filled out the "between" frames. The digitized carpet action was then added into the digitized version of the rest of the hand-drawn movie. *Lion King*, Disney's latest animation, has several animals that are computer-generated in the manner of the Jurassic dinosaurs, including some animals with semi-autonomous herding and flocking behaviors. Disney is now working on their first completely digital animation, to be released in late 1994. It will feature the work of an ex-Disney animator, John Lassiter. Almost the entire computer animation will be done at Pixar, a small innovative studio located in a remodeled business park in Richmond Point, California.

I stopped by Pixar to see what kind of artificial life they were hatching. Pixar has made four award-winning short computer animations done by Lassiter. Lassiter likes to animate normally inanimate objects—a bicycle, a toy, a lamp, or knick-knacks on a shelf. Although Pixar films are considered state-of-the-art computer animations in computer graphic circles, the animation part is mostly handcrafted. Instead of drawing with a pencil, Lassiter uses a cursor to modify his computer-rendered 3-D objects. If he wants his toy soldier character to be depressed he goes into his figure's happy face on the computer screen and drags the toon's mouth into a droop. After testing the expression he may decide the toy soldier's eyebrows really shouldn't droop so fast, or maybe its eyes bat too slowly. So by cursor-dragging he alters the computer form. "I don't know how else to tell it what to do, such as making its mouth like this," says Lassiter, forming an O with his mouth in mock surprise, "that would be any faster or better than doing it myself."

I hear more of this communication problem from Ralph Guggenheim, production director at Pixar: "Most hand animators believe that what Pixar does is feed scripts into a computer and out comes a film. That's why we were once barred from animation festivals. But if we were to really do that, we could not create great stories . . . The chief day-to-day problem we have at Pixar is that computer animation reverses the animation process. It asks animators to describe before they animate what it is they want to animate!"

Animators, true artists, are like writers in that they don't know what they want to say until they hear themselves say it. Guggenheim reiterates, "Animators can't know a character until they animate it. They will tell you that it is very slow going in the beginning of a story because they are becoming familiar with their character. Then it starts speeding up as they become more intimate with it. As they get to the halfway point of the film, now they know the character well and they are screaming through the frames."

In the short animation *Tin Toy*, a plume on the toy soldier's hat shakes naturally when he bobs his head. That effect was achieved with virtual physics, or what the animators call "lag, drag, and wiggle." When the base of the plume moved, the rest of the feather acted as if it were a spring pendulum—a fairly standard physics equation. The exact way the plume quivered was unpredicted and quite realistic because the plume was obeying the physics of shaking. But the face of the toy soldier was still manipulated entirely by an experienced human animator. The animator is a surrogate actor. He acts out a character by drawing it. Every animator's desk has a mirror on it that the animator uses to draw his own exaggerated facial expressions.

I asked the artists at Pixar if they can at least *imagine* an autonomous computer character—you feed in a rough script and out comes a digital Daffy Duck doing his mischief. There was uniform grave denial and shaking of heads. "If animating a believable character was as easy as feeding a script into a computer, then there would be no bad actors in the world," said Guggenheim. "But we know that not all actors are great. You see tons of Elvis or Marilyn Monroe impersonators all the time. Why aren't we fooled? Because the impersonator has a complex job knowing when to twitch the right side of his mouth or how to hold a microphone. If a human actor has difficulty doing that, how will a computer script do it?"

The question they are asking is one of control. It turns out that the special effects and animation business is an industry of control freaks. They feel that the subtleties of acting are so minute that only a human overseer can channel the choices of a digital or drawn character. They are right.

But tomorrow, they won't be. If computer power continues to increase as it has, within five years we'll see a character created by releasing synthetic behavior into a synthetic body star in a film.

The *Jurassic Park* dinos made it very clear how nearly perfect synthetic body representations are today. The flesh of the dinos was visually indistinguishable from what we'd expect a filmed dinosaur to be. A number of digital effects laboratories are compiling the components of a believable digital human actor right now. One lab specializes in creating perfect digital human hair, another concentrates on getting the hands right, and another on generating facial expressions. Already, digital characters are inserted into Hollywood films (without anyone noticing) when a synthetic scene demands people moving in the distance. Realistic clothing that drapes and folds naturally is still a challenge; done imperfectly it gives the virtual person a clunky feel. But at the start, digital characters will be used for dangerous stunts, or worked into composite scenes—but only in long shots, or in crowds, rather than in the full attention of a close-up. An entirely convincing virtual human form is tricky, but close at hand.

What is not very close at hand is simulating convincing human action. Especially out of reach is convincing facial behavior. The final frontier, the graphics experts say, is the human expression. A quest for control of a human face is now a minor crusade.

At Colossal Picture Studios in the industrial outskirts of San Francisco, Brad de Graf works on faking human behavior. Colossal is the little-known special effects studio behind some of the most famous animated commercials on TV such as the Pillsbury Doughboy. Colossal also did the avant garde animation series for MTV called *Liquid TV*, starring animated stick figures, low-life muppets on motorbikes, animated paper cutouts, and the bad boys *Beavis and Butt-head*.

De Graf works in a cramped studio in a redecorated warehouse. In several large rooms under dimmed lights about two dozen large computer monitors glow. This is an animation studio of the '90s. The computers—heavy-duty graphic workstations from Silicon Graphics—are lit with projects in various stages, including a completely computerized bust of rock star Peter Gabriel. Gabriel's head shape and face were scanned, digitized, and reassembled into a virtual Gabriel that can substitute for his live body in his music videos. Why waste time dancing in front of cameras when you could be in a recording studio or in the pool? I watched an animator fiddle with the virtual star. She was trying to close Gabriel's mouth by dragging a cursor to lift his jaw. "Ooops" she said, as she went too far and Gabriel's lower lip sailed up and penetrated his nose, making a disgusting grimace.

I was at de Graf's workshop to see Moxy, the first completely computer-*animated* character. On the screen Moxy looks like a cartoon dog. He's got a big nose, a chewed ear, two white gloves for hands, and "rubber hose" arms. He's also got a great comic voice. His actions are not drawn. They are lifted from a human actor. There's a homemade virtual reality "waldo" in one corner of the room. A waldo (named from a character in an old science-fiction story) is a device that lets a person drive a puppet from a distance. The first waldo-driven computer animation was an experimental Kermit the Frog animated by a hand-size muppet waldo. Moxy is a full-bodied virtual character, a virtual puppet.

When an animator wants to have Moxy dance, the animator puts on a yellow hardhat with a stick taped to the peak. At the end of the stick is a location sensor. The animator straps on shoulder and hip sensors, and then picks up two foam-board pieces cut out in the shape of very large cartoon hand-gloves. He waves these around—they also have location sensors on them—as he dances. On the screen Moxy the cartoon dog in his funky toon room dances in unison.

Moxy's best trick is that he can lip sync automatically. A recorded human voice pours into an algorithm which figures out how Moxy's lips should move, and then moves them. The studio hackers like to have Moxy saying all kinds of outrageous things in other people's voices. In fact, Moxy can be moved in many ways. He can be moved by twirling dials, typing commands, moving a cursor, or even by autonomous behavior generated by algorithms.

That's the next step for de Graf and other animators: to imbue characters like Moxy with elementary moves—standing up, bending over, lifting a heavy object—which can be recombined into smooth believable action. And then to apply that to a complex human figure.

To calculate the move of a human figure is marginally possible for today's computers given enough time. But done on the fly, as your body does in real life, in a world that shifts while you are figuring where to put your foot, this calculation becomes nearly impossible to simulate well. The human figure has about 200 moving joints. The total number of possible positions a human figure can assume from 200 moving parts is astronomical. To simply pick your nose in real time demands more computational power than we have in large computers.

But the complexity doesn't stop there because each pose of the body can be reached by a multitude of pathways. When I raise my foot to slip into a pair of shoes, I steer my leg through that exact pose by hundreds of combinations of thigh, leg, foot, and toe actions. In fact, the sequences that my limbs take while walking are so complex that there is enough room for a million differences in doing so. Others can identify me—often from a hundred feet away and not seeing my face—entirely by my unconscious choice of which feet muscles I engage when I walk. Faking someone else's combination is hard.

Researchers who try to simulate human movement in artificial figures quickly discover what animators of Bugs Bunny and Porky Pig have known all along: that some linkage sequences are more "natural" than others. When Bugs reaches for a carrot, some arm routes to the vegetable appear more human than other routes. (Bugs's behavior, of course, does not simulate a rabbit but a person.) And much depends on the sequential timing of parts. An animated figure following a legitimate sequence of human movements can still appear robotic if the relative speeds of, say, swinging upper arm to striding leg are off. The human brain detects such counterfeits easily. Timing, therefore, is yet another complexifying aspect of motion.

Early attempts to create artificial movement forced engineers far afield into the study of animal behavior. To construct legged vehicles that could roam Mars, researchers studied insects, not to learn how to build legs, but to figure out how insects coordinated six legs in real time.

At the corporate labs of Apple Computer, I watched a computer graphic specialist endlessly replay a video of a walking cat to deconstruct its movements. The video tape, together with a pile of scientific papers on the reflexes of cat limbs, was helping him extract the architecture of cat walking. Eventually he planned to transplant that architecture into a computerized virtual cat. Ultimately he hoped to extract a generic four-footed locomotion pattern that could be adjusted for a dog, cheetah, lion, or whatever. He was not concerned at all with the look of the animal; his model was a stick figure. He was concerned with organization of the complicated leg, ankle, and foot actions.

In David Zeltzer's lab at MIT's Media Lab, graduate students developed simple stick figures which could walk across an uneven landscape "on their own." The animals were nothing more than four legs on a stick backbone, each leg hinged in the middle. The students would aim the "animat" in a certain direction, then it would move its legs upon figuring out where the low or high spots were, adjusting its stride to compensate. The effect was a remarkably convincing portrait of a critter walking across rugged terrain. But unlike an ordinary Road Runner animation, no human decided where each leg had to go at every moment of the picture. The character itself, in a sense, decided. Zeltzer's group eventually populated their world with autonomous six-legged animats, and even got a two-legged thing to ramble down a valley and back.

Zeltzer's students put together Lemonhead, a cartoony figure that could walk on his own. His walking was more realistic and more complicated than the sticks because he relied on more body parts and joints. He could skirt around obstacles such as fallen tree trunks with realistic motion. Lemonhead inspired Steve Strassman, another student in Zeltzer's lab, to see how far he could get in devising a library of behavior. The idea was to make a generic character like Lemonhead and give him access to a "clip book" of behaviors and gestures. Need a sneeze? Here's a disk-full.

Strassman wanted to instruct a character in plain English. You simply tell it what to do, and the figure retrieves the appropriate behaviors from the "four food groups of behavior" and combines them in the right sequence for sensible action. If you tell it to stand up, it knows it has to move its feet from under the chair first. "Look," Strassman warns me before his demo begins, "this guy won't compose any sonatas, but he will sit in a chair."

Strassman fired up two characters, John and Mary. Everything happened in a simple room viewed from an oblique angle above the ceiling—a sort of god's-eye view. "Desktop theater," Strassman called it. The setting, he said, was that the couple occasionally had arguments. Strassman worked on their goodbye scene. He typed: "In this scene, John gets angry. He offers the book to Mary rudely, but she refuses it. He slams it down on the table. Mary rises while John glares." Then he hits the PLAY key.

The computer thinks about it for a second, and then the

characters on the screen act out the play. John frowns; his actions with the book are curt; he clenches his fists. Mary stands up suddenly. The end. There's no grace, nothing very human about their movements. And it's hard to catch the fleeting gestures because they don't call attention to their motions. One does not feel involved, but there, in that tiny artificial room, are characters interacting according to a god's script.

"I'm a couch-potato director," Strassman says. "If I don't like the way the scene went I'll have them redo it." So he types in an alternative: "In this scene, John gets sad. He's holding the book in his left hand. He offers it to Mary kindly, but she refuses it politely." Again, the characters play out the scene.

Subtlety is the difficult part. "We pick up a phone differently than a dead rat," Strassman said. "I can stock up on different hand motions, but the tricky thing is what manages them? Where does the bureaucracy that controls these choices get invented?"

Taking what they learned from the stick figures and Lemonhead, Zeltzer and colleague Michael McKenna fleshed out the skeleton of one six-legged animat into a villainous chrome cockroach and made the insect a star in one of the strangest computer animations ever made. Facetiously entitled "Grinning Evil Death," the token plot of the five-minute video was the story of how a giant metallic bug from outer space invaded Earth and destroyed a city. While the story was a yawner, the star, a six-legged menace, was the first animat—an internally driven artificial animal.

When the humongous chrome cockroach crawled down the street, its behavior was "free." The programmers told it, "walk over those buildings," and the virtual cockroach in the computer figured out how its legs should go and what angle its torso should be and then it painted a plausible video portrait of itself wriggling up and over five-story brick buildings. The programmers aimed its movements rather than dictated them. Coming down off the buildings, an artificial gravity pulled the giant robotic cockroach to the ground. As it fell, the simulated gravity and simulated surface friction made its legs bounce and slip realistically. The cockroach acted out the scene without its directors being drowned in the minutiae of its foot movements.

The next step toward birthing an autonomous virtual character is now in trial: Take the bottom-up behavioral engine of the giant cockroach and surround it with the glamorous carcass of a Jurassic dino to get a digital film actor. Wind the actor up, feed it lots of computer cycles, and then direct it as you would a real actor. Give it general instructions—"Go find food"—and it will, on its own, figure out how to coordinate its limbs to do so.

Building the dream, of course, is not that easy. Locomotion is merely one facet of action. Simulated creatures must not only move, they must navigate, express emotion, react. In order to invent a creature that could do more than walk, animators (and roboticists) need some way to cultivate indigenous behaviors of all types.

In the 1940s, a trio of legendary animal watchers in Europe—Konrad Lorenz, Karl von Frisch, and Niko Tinbergen—began describing the logical underpinnings of animal behavior. Lorenz shared his house with geese, von Frisch lived among honeybee hives, and Tinbergen spent his days with stickleback perch and sea gulls. By rigorous and clever experiments the three ethologists refined the lore of animal antics into a respectable science called ethology (roughly, the study of character). In 1973, they shared a Nobel prize for their pioneering achievements. When cartoonists, engineers, and computer scientists later delved into the literature of ethology, they found, much to their surprise, a remarkable behavioral framework already worked out by the three ethologists, ready to be ported over to computers.

At the core of ethological architecture dwells the crucial idea of decentralization. As formalized in 1951 by Tinbergen in his book *The Study of Instinct*, the behavior of an animal is a decentralized coordination of independent action (drive) centers which are combined like behavioral building blocks. Some behavioral modules consist of a reflex; they invoke a simple function, such as: pull away when hot, or blink when touched. The reflex knows nothing of where it is, what else is going on, or even of the current goal of its host body. It can be triggered anytime the right stimulus appears.

A male trout instinctually responds to the following stimuli: a female trout ripe for copulation, a nearby worm, a predator approaching from behind. But when all three stimuli are presented simultaneously, the predator module always wins out by suppressing feeding or mating instincts. Sometimes, when there is a conflict between action modules, or several simultaneous stimuli, management modules are triggered to decide. For instance, you are in the kitchen with messy hands when the phone rings at the same time someone knocks on the front door. The conflicting drives—jump to the phone! no, wipe hands first! no, dash to the door!—could lead to paralysis unless arbitrated by a third module of learned behavior, perhaps one that invokes the holler, "Please wait!"

A less passive way to view a Tinbergen drive center is as an "agent." An agent (whatever physical form it takes) detects a stimulus, then reacts. Its reaction, or "output" in computer talk, may be considered input by other modules, drive centers, or agents. Output from one agent may *enable* other modules (cocking a gun's hammer) or it may *activate* other modules already enabled (pulling the trigger). Or the signal may *disable* (uncock) a neighboring module. Rubbing your tummy and patting your head at the same time is tricky because, for some unknown reason, one action suppresses the other. Commonly an output may both enable some centers and suppress others. This is, of course, the layout of a network swamped with circular causality and primed to loop into self-creation.

Outward behavior thus emerges from the thicket of these blind reflexes. Because of behavior's distributed origin, very simple agents at the bottom can produce unexpectedly complex behavior at the top. No central module in the cat decides whether the cat should scratch its ear or lick its paw. Instead, the cat's conduct is determined by a tangled web of independent "behavioral agents"—cat reflexes—cross-activating each other, forming a gross pattern (called licking or scratching) that wells up from the distributed net.

This sounds a lot like Brooks's subsumption architecture because it is. Animals are robots that work. The decentralized, distributed control that governs animals is also what works in robots and what works for digital creatures.

Web-strewn diagrams of interlinked behavior modules in

ethology textbooks appear to computer scientists as computer logic flow charts. The message is: Behavior is computerizable. By arranging a circuit of subbehaviors, any kind of personality can be programmed. It is theoretically feasible to generate in a computer any mood, any sophisticated emotional response that an animal has. Film creatures will be driven by the same bottom-up governance of behavior running Robbie the Robot—and the very same scheme borrowed from living songbirds and stickleback fish. But instead of causing pneumatic hoses to pressurize, or fishtails to flick, the distributed system pumps bits of data which move a leg on a computer screen. In this way, autonomous animated characters in film behave according to the same general organizational rules as real animals. Their behavior, although synthetic, is *real* behavior (or at least hyperreal behavior). Thus, toons are simply robots without hard bodies.

More than just movement can be programmed. Character—in the oldfashioned sense of the word—can be encapsulated into bit code. Depression, elation, and rage will all be add-on modules for a creature's operating system. Some software companies will sell better versions of the fear emotion than others. Maybe they'll sell "relational fear"—fear that not only registers on a creature's body but trickles into successive emotion modules and only gradually dissipates over time.

Behavior wants to be free, but to be of any use to humans, artificially generated behavior needs to be supervised or controlled. We want Robbie the Robot, or Bugs Bunny, to accomplish things on his own without our oversight. At the same time, not everything Robbie or Bugs could do would be productive. How can we give a robot, or a robot without a hard body, or any artificial life, the license to behave, while still directing them to be useful to us?

Some answers are unexpectedly being uncovered in a research project on interactive literature begun at Carnegie Mellon University. There researcher Joseph Bates fabricated a world called "Oz," somewhat similar to the tiny room of John and Mary that Steve Strassman created. In Oz there are characters, a physical environment, and a narrative—the same trio of ingredients for classical drama. In traditional drama, the narrative dictates both characters and environment. In Oz, however, the control is inverted somewhat; characters and environment influence the narrative.

Oz is made for humans to enjoy. It is a fanciful virtual world populated with automatons as well as human-directed characters. The goal is to create an environment, a narrative structure, and automatons in such a way that a human can participate in the story without either crashing the story line, or feeling left out as a mere observer in the audience. David Zeltzer, who lent some ideas to the project, gives a wonderful example: "If we provided you with a digitized version of Moby Dick, there's no reason why you couldn't have your own cabin on the *Pequod*. You could talk to Starbuck as he went after the White Whale. There is enough room in the narrative for you to be involved, without changing the plot."

There are three frontiers of control research involved in Oz:

- How do you organize a narrative to allow deviations yet keep it centered on its intended destination?
- How do you construct an environment that can generate surprise events?
- How do you create creatures that have autonomy, but not too much?

From Strassman's "desktop theater" we go to Joseph Bates's "computational drama." Bates envisions a drama of distributed control. A story becomes a type of coevolution, with perhaps only its outer boundaries predestined. You could be in an episode of *Star Trek* attempting to influence alternative storylines, or you could be on a journey with a synthetic Don Quixote confronting new fantasies. Bates, who is chiefly concerned about the experience of the human user of Oz, puts his quest this way. "The question I'm working on is: How do you impose a destiny upon a user without removing their freedom?"

In my search for the future of control from the perspective of the created rather than creator, I will rephrase his question as: How do you impose a destiny upon a character of artificial life without removing its freedom?

Brad de Graf believes this shift in control is shifting the goal of authors. "It's a different medium we are making. Instead of creating a story, I'm creating a world. Instead of creating a character's dialogue and action, I'm creating a personality."

When I had a chance to play with some artificial characters Bates developed, I got a sense of how much fun such personality petlike creatures could be. Bates calls his pets "woggles." Woggles come in three varieties: a blue blob, a red blob, a yellow blob. The blobs are stretchy spheres with two eyes. They hop around in a simple world of stepping-stones and some caves. Each color of woggle is coded with a different suite of behaviors. One is shy, one is aggressive, one is a follower. When a woggle frightens another woggle, the aggressive one stretches tall to scare away the threat. The shy one shrinks and flees.

Ordinarily the woggles hop around doing their woggly thing among themselves. But when a human enters their world by inserting a cursor into their space, they interact with the visitor. They may follow you around, or avoid you, or wait until you aren't around to harass another woggle. You are in the picture, but you are not controlling the show.

I got a better sense of the future of pet control from a prototype world that is somewhat an extension of Bates's woggle world. A virtual reality (VR) group at Fujitsu Laboratories in Japan took wogglelike characters and fleshed them out in virtual three dimensions. I watched a guy wearing a clunky VR helmet on his head and data gloves on his hands give a demonstration.

He was in a fantasy underwater world. A faint impression of a submerged castle shimmered in the distant background. A few old Greek columns and chest-high seaweed furnished the immediate play area. Three "jellyfish" hopped around, and one small sharkish fish circled the area. The jellyfish, in the shape of mushrooms and about the size of dogs, changed color depending on their mood or behavior state. Playing by themselves the three were blue. They would hop around on their fat monopod tirelessly. If the VR-guy beckoned them to come, by waving with his hand, they would excitedly bounce over, turn orange, and jump up and down like friendly dogs waiting to chase a stick. When he showed them attention their eyes would close in a happy expression. The guy could call in the less friendly fish by emitting a blue laser line from his forefinger and touch the fish from afar. This would change the fish's color and interest in humans, so it circled in much closer, and swam nearby—but like a cat, not too close—as long as it was occasionally touched by the blue line.

Even watching from the outside, it was evident that artificial characters with the mildest autonomous behavior and some three-dimensional form in a shared three-dimensional space had a distinct presence of their own. I could imagine having an adventure with them. I could imagine them as Jurassic dinosaurs and me really being scared. Even the Fujitsu guy ducked once when the virtual fish swam too close to his head. "Virtual reality," says de Graf, "is not going to be interesting unless it is populated with interesting characters."

Pattie Maes, an artificial life researcher at the MIT Media Lab, abhors goggle-and-gloves virtual reality. She finds such clothing "too artificial" and confining. She and colleague Sandy Pentland came up with an alternative way to interact with virtual creatures. Her system, called ALIVE, lets a human play with animated creatures via a computer screen and video camera. The camera points back at the human participant, inserting the observer into the virtual world that he or she is watching on the screen.

This neat trick gives a real sense of intimacy. By moving my arms I can interact with little "hamsters" on the screen. The hamsters look like tiny toasters on wheels, but they are autonomous goal-seeking animats that contain a rich repertoire of motivations, sensors, and responses. The hamsters roam the enclosed pen looking for "food" when they haven't eaten in a while. They seek each other's company; sometimes they chase each other. They run from my hand if I move it too fast. If I move it slowly, they try to follow it out of curiosity. A hamster will sit up and beg for food. When they get tired, they fall over and sleep. They are halfway between robots and animated animals, and only several steps away from authentic virtual characters.

Pattie Maes is trying to teach creatures "how to do the right thing." She wants her creatures to learn from their experiences in the environment, without much human supervision. The

Jurassic dinosaurs won't be real characters until they can learn. It will be hardly worth creating a humanist virtual actor unless he or she could learn. Following the subsumption architecture model, Maes is structuring a hierarchy of algorithms that let her creatures not only adapt, but also bootstrap themselves to increasing complex behaviors and—as an essential part of the package—also *let their own goals* emerge from those behaviors.

The animators at Disney and Pixar nearly croak at the thought, but someday Mickey Mouse will have his own agenda.

IT'S THE WINTER OF 2001, in a corner of the Disney studio lot; a trailer is set up as a top-secret research lab. Reels of old Disney cartoons, stacks of gigabyte computer hard drives, and three 24-year-old-computer graphic artists hole up inside. In about three months they deconstruct Mickey Mouse. He is reanimated as a potentially 3-D being who only appears in two dimensions. He knows how to walk, leap, dance, show surprise and wave goodbye on his own. He can lip sync but can't talk. The entire overhauled Mickey fits onto one Syquest 2-gig portable disk.

The disk is walked over to the old animation studio, past its rows of empty and dusty animation stands, to the cubicles where the Silicon Graphics workstations are glowing. Mickey is popped into a computer. The animators have already created a fully detailed artificial world for the Mouse. He's cued up to the scene and the tape turned on. Roll! When Mickey trips on the stairs of his house, gravity hauls him down. The simulated physics of his rubbery rear end bouncing against the wooden stairs generates realistic hops. His cap is blown away by a virtual wind from the open front door, and when the carpet slides out from under him as he attempts to run after his hat, it bunches up in accordance with the physics of fabric, just as Mickey collapses under his own simulated weight. The only instruction Mickey got was to enter the room and be sure to chase his hat. The rest came naturally.

After 1997, nobody ever draws Mickey again. There's no need to. Oh, sometimes the animators butt in and touch up a critical facial expression here or there—mere makeup artists the handlers call them—but by and large Mickey is given a script and he obeys. And he—or one of his clones—works all year round on more than one film at once. Never complains, of course.

The graphic jocks aren't satisfied. They hook up a Maes learning module into Mickey's code. With this on, Mickey matures as an actor. He responds to the emotions and actions of the other great actors in his scenes—Donald Duck and Goofy. Every time a scene is rerun, he remembers what he did on the keeper take and that gesture is emphasized next time. He evolves from the outside as well. The programmers tune up his code, give him improved smoothness, increase the range of his expressions, and beef up the depth of his emotions. He can play the "sensitive guy" now if needed.

But, over five years of learning, Mickey begins to get his own ideas. He somehow reacts hostilely to Donald, and becomes furious when he gets clunked on the head with a mallet. And when he is angry, he becomes obstinate. He balks when the director instructs him to walk off the edge of a cliff, having learned over the years to avoid obstacles and edges. Mickey's programmers complain that they can't code around these idiosyncrasies without disrupting all the other finely tuned traits and skills Mickey has acquired. "It's like an ecology," they say. "You can't remove one thing without disturbing them all." One graphic jock puts it best: "Actually, it's like a psychology. The Mouse has a real personality. You can't separate it. You've just got to work around it."

So by 2007, Mickey Mouse is quite an actor. He is a hot "property" as the agents say. He can speak. He can handle any kind of slapstick situation you can imagine. Does his own stunts. He has a great sense of humor, and the fabulous timing of a comedian. The only problem is that he is an SOB to work with. He'll suddenly fly off the handle and go berserk. Directors hate him. But they put up with him—they've seen worse—because, well, because he's Mickey Mouse.

Best of all, he'll never die, never age.

Disney foreshadowed this liberation of toons in its own film *Roger Rabbit*. Toons in this movie have their own independent life and dreams, but they have to stay in Toon Town, their own virtual world, except when we need them to work in our films. On the set, toons may or may not be cooperative and pleasant. They have the same whims and tantrums that human actors have. Roger Rabbit is just fiction, but someday Disney will have to deal with an autonomous out-of-control Roger Rabbit.

Control is the issue. In his first film, *Steamboat Willie*, Mickey was under the full control of Walt Disney. Disney and the Mouse were one. As more lifelike behaviors are implanted into Mickey, he is less at one with his creators and more out of their control. This is old news to anyone with kids or pets. But it is new news to anyone with a cartoon character, or machines that get smarter. Of course, neither kids nor pets are completely out of our control. There is the direct authority we have in their obedience, and the larger indirect control we have in their training and formation.

The fairest way to state this is that control is a spectrum. At one end there is the total domination of "as one" control. At the other is "out of control." In between are varieties of control we don't have words for.

Until recently, all our artifacts, all our own handmade creations have been under our authority. But as we cultivate synthetic life in our artifacts, we cultivate the loss of our command. "Out of control," to be honest, is a great exaggeration of the state that our enlivened machines will take. They will remain indirectly under our influence and guidance but free of our domination.

Though I have searched everywhere, I could not find the word that describes this type of clout. We simply have no name for the loose relationship between an influential creator and a creation with a mind of its own—a thing we shall see more of. The realm of parent and child should have such a word, but sadly doesn't. We do better with sheep where we have the notion of "shepherding." When we herd a flock of sheep, we know we are not in complete authority, yet neither are we without control. Perhaps we will shepherd artificial lives.

We also "husband" plants, as we assist them in their natural goals, or deflect them slightly for our own. "Manage" is probably the closest in meaning to the general type of control we will need for artificial lives, such as a virtual Mickey Mouse. A woman can "manage" her difficult child, or a barking dog, or the 300-strong sales department under her authority. Disney can manage Mickey in films.

"Manage" is close, but not perfect. Although we manage wilderness areas like the Everglades, we actually have little say in what goes on among the seaweed, snakes and marsh grass. Although we manage the national economy, it does what it wants. And although we manage a telephone network, we have no supervision on how a particular call is completed. The word "management" may imply more oversight than we really have in the examples above, and more than we will have in future very complex systems.

THE WORD I'M LOOKING FOR is more like "co-control." It's seen in some mechanical settings already. Keeping a 747 Jumbo Jet aloft and landing it in bad weather is a very complex task. Because of the hundreds of systems running simultaneously, the immediate reaction time required by the speed of the plane, and disorienting effects of sleepless long trips and hazardous weather, a computer can fly a jet better a human pilot. The sheer number of human lives at stake permits no room for errors or second best. Why not have a very smart machine control the jet?

So engineers wired together an autopilot, and it turns out be very capable. It flies and lands a Jumbo Jet oh so nicely. Flying-by-wire also fits very handily into the craving for order by the air traffic controllers—everything is under digital control. The original idea was that human pilots would monitor the computer in case anything went wrong. The only problem is that humans are terrible at passive monitoring. They get bored. They daydream. Then they start missing critical details. Then an emergency pops up which they have to tackle cold.

So instead of having the pilot watch the computer, the new idea was to invert the relationship and have the computer watch the pilot. This approach was taken in the European Airbus A320, one of the most highly automated planes built to date. Introduced in 1988, the onboard computer supervises the pilot. When he pushes the control stick to turn the plane, the computer figures out how far to bank left or right, but it won't let the plane bank more than 67 degrees or nose up or down more than 30 degrees. This means, in the words of *Scientific American*, "the software spins an electronic cocoon that stops the aircraft from exceeding its structural limitations." It also means, pilots complain, that the pilot surrenders control. In 1989 British Airways pilots flying 747s experienced six different incidents where they had to override a computer-initiated power reduction. Had they not been able to override the erroneous automatic pilot—which Boeing blamed on a software bug—the error could have been fatal. The Airbus A320, however, provides no override of its autosystem.

Human pilots felt they were fighting for control of the plane. Should the computer be a pilot or navigator? The pilots joked that the computer was like putting a dog into the cockpit. The dog's job was to bite the pilot if he tries to touch the controls; and the pilot's only job was to feed the dog. In fact, in the emerging lingo of automated flying, pilots are called "system managers."

I'm pretty sure the computer will end up as co-pilot. There will be much that it does completely out of the reach of the pilot. But the pilot will manage, or shepherd, the computer's behavior. And the two—machine and human—will be in a constant tussle, as are all autonomous things. Planes will fly by co-control.

A graphic jock at Apple, Peter Litwinowicz, fabricated a great hack. He extracted the body and facial movements from a live human actor and applied them to digital actors. He had a human performer ask, in a sort of theatrical way, for a dry martini. He took those gestures—the raised eyebrow, the smirk on the lips, the lilt of the head—to control the face of a cat. The cat delivered the line in exactly the same manner as the actor would. As an encore Litwinowicz then mapped the actor's expressions onto a cartoon, and then onto an inert classical mask, and finally, he animated a tree trunk with the actor's facial controls.

Human actors will not be out of jobs. While some characters will be wholly autonomous, most will be of a cyborgian nature. An actor will animate a cat, while the artificial cat pushes back and helps the actor be a better cat. An actor can "ride" a cartoon, in the same type of co-control that a cowboy rides a horse, or a pilot rides a computer-steered airplane. The green figure of a digital Ninja Turtle may dart about the world on its own, but the human actor sharing control supplies the appropriate nuance every now and then in a smile, or finishes a just-perfect growl with a jeer.

James Cameron, the director of *Terminator 2*, recently told an audience of computer graphic specialists, "Actors love masks. They're willing to sit in makeup chairs for eight hours to put them on. We must make them partners in synthetic character creation. They will be given new bodies and new faces with which to expand their art."

The future of control: Partnership, Co-control, Cyborgian control. What it all means is that the creator must share control, and his destiny, with his creations.

# 17

# AN OPEN UNIVERSE

A SWARM OF HONEYBEES absconds from the hive and then dangles in a cluster from a tree branch. If a nearby beekeeper is lucky, the swarm settles on a branch that is easy to reach. The bees, gorged with honey and no longer protecting their brood, are as docile as ladybugs.

I've found a swarm or two in my time hung no higher than my head, and I've moved them into an empty hive box for my own. The way you move 10,000 bees from a tree branch into a box is one of life's magic shows.

If there are neighbors watching you can impress them. You lay a white sheet or large piece of cardboard on the ground directly under the buzzing cluster of bees. You then slide the bottom entrance lip of an empty hive under one edge of the sheet so that the cloth or cardboard forms a gigantic ramp into the hive's opening. You pause dramatically, and then you give the branch a single vigorous shake.

The bees fall out of the tree in a single clump and spill onto the sheet like churning black molasses. Thousands of bees crawl over each other in a chaotic buzzing mass. Then slowly, you begin to notice something. The bees align themselves toward the hive opening and march into the entrance as if they were tiny robots under one command. And they are. If you bend down to the sheet and put your nose near the pool of crawling bees, you can smell a perfume like roses. You can see that the bees are hunched over and fanning their wings furiously as they walk. They are emitting the rose smell from a gland in their rear ends and fanning the scent back to the troops behind them. The scent says, "The queen is here. Follow me." The second follows the first and the third the second and five minutes later the sheet is almost empty as the last of the swarm sucks itself into the box.

The first life on Earth could not put on that show. It was not a matter of lacking the right variation. There simply was no room in all of the possibilities accorded by its initial genes for such a wild act. To use the smell of a rose to coordinate 10,000 flying beings into a purposeful crawling beast was beyond early life's reach. Not only had early life not yet created the space—worker bee, queen relationship, honey from flowers, tree, hive, pheromones—in which to stage the show, it had not created the tools to make the space.

Nature dispenses breathtaking diversity because its charter is open ended. Life did not confine itself to producing its dazzling variety within the limited space of the few genes it first made. On the contrary, one of the first things life discovered was how to create new genes, more genes, variable genes, and a bigger genetic library.

A book in Borges's Library spans a million genes; a hi-resolution Hollywood movie frame, 30 million. Yet as immense as the libraries built out of these are, they are only a dust mote in the meta-library of all possible libraries.

It is one of the hallmarks of life that it continues to enlarge the space of its own being. Nature is an ever-expanding library of possibilities. It is an open universe. At the same time that life turns up the most improbable books from the Library shelves, it is adding new wings to the collection, making room for more of its improbable texts.

We don't know how life crossed the threshold from fixed gene space to variable gene space. Perhaps it was one particular gene's duty to determine the total number of genes in the chromosome. Then by mutating that one gene, the sum of genes in the string would increase or decrease. Or the size of the genome might have been indirectly determined by more than one gene. Or, more likely, genome size is determined by the structure of the genetic system itself.

Tom Ray showed that in his world of self-replicators, variable genome length emerged instantaneously. His creatures determined their own genome (and thus the size of their possible libraries) in a range from his unexpectedly short "22" to one creature that was 23,000 bytes long.

The consequence of an open genome is open evolution. A system which predetermines what each gene must do or how many genes there are can only evolve to predetermined boundaries. The first systems of Dawkins, Latham, Sims and the Russian El-Fish programmers were grounded by this limitation. They may generate all possible pictures of a given size and depth, but not all possible art. A system that does not predetermine the role or number of genes can shoot the moon. This is why Tom Ray's critters stir such excitement. In theory, his world, run long enough, could evolve anything in the ultimate Library.

There is more than one way to organize an open genome. In 1990, Karl Sims took advantage of the supercomputing power of the CM2 to devise a new type of artificial world formed by genes of unfixed length, a world much improved over his botanical-picture world. Sims accomplished this trick by creating a genome composed of small *equations* rather than of long strings of digits. His original library of fixed genes each controlled one visual parameter of a plant; his second library held equations of variable and open-ended length which drew curves, colors, forms and shapes.

Sims's equation-genes were small self-contained logical units of a computer language (LISP). Each module was an arithmetical command such as *add, subtract, multiply, cosine, sine*. Sims called these units "primitives"—a logical alphabet. If you have a suitable primitive alphabet you can build all possible equations, just as with the appropriately diverse alphabet of sounds you could build all spoken sentences. *Add, multiply, cosine*, etc., can be combined to generate any mathematical equation we can think of. Since any shape can be described by an equation, this primitive alphabet can make any picture. Adding to the complexity of the equation will subtly enlarge the complexity of the resulting image.

There was a serendipitous second advantage to working with a library of equations. In Sims's original world (and in Tom Ray's Tierra and Danny Hillis's coevolutionary parasites), organisms were strings of digits that randomly flipped a digit, just as books in the Borgian Library altered by one letter at a time. In Sims's improved universe, organisms were strings of *logical units* that randomly flipped a unit. This would be like a Borgian Library where words, not letters, were flipped. Every word in every book was correctly spelled, so every page in every book had a more sensible pattern. But whereas the soup for a Borgian Library based on words would necessitate tens of thousands of words in the pot to begin with, Sims could make all possible equations starting with a soup of only a dozen or so mathematical primitives.

Yet, the most revolutionary advantage to evolving logic units rather than digital bits was that it immediately moved the system onto the road toward an open-ended universe. Logic units are functions themselves and not mere values for functions, as digital bits are. By adding or swapping a logical primitive here or there, the entire functionality of the program shifts or enlarges. New *kinds* of functions and new kinds of things will emerge in such a system.

That's what Sims found. Entirely new kinds of pictures evolved by his equations and painted themselves onto the computer monitor. The first thing that struck him was how rich the space was. By restricting the primitives to logical parts, Sims's LISP alphabet ensured that most equations drew *some pattern*. Instead of being full of muddy gray patterns, there were astounding sights almost wherever he went. Just dipping in at random landed him in the middle of "art." The first screen was full of wild red and blue zigzags. The next screen pulsated with yellow hovering orbs. The next generation yielded yellow orbs with a misty horizon, the next, sharpened waves with a horizon of blue. And the next, circular smudges of pastel yellow color reminiscent of buttercups. Almost every turn reeled in a marvelously inventive scene. In an hour, thousands of stunning pictures were roused out of their hiding places and displayed to the living for the first and last time. It was like watching over the shoulder of the world's greatest painter as he sketched without ever repeating a theme or pattern.

While Sims selected one picture, bred variations of it, and then selected another, he was not only evolving pictures. Underneath it all, Sims was evolving logic. A relatively small logic equation drew an eye-boggling complex picture. At one point Sims's system evolved the following eight lines of logic code:

```
(cos (round (atan (log (invert y) (+ (bump (+ (round x y)
y) # (0.46 0.82 0.65) 0.02 # (0.1 0.06 0.1) # (0.99 0.06
0.41) 1.47 8.7 3.7) (color-grad (round (+ y y) (log (invert
x) (+ (invert y) (round (+ y x) (bump (warped-ifs (round
y y) y 0.08 0.06 7.4 1.65 6.1 0.54 3.1 0.26 0.73 15.8 5.7 8.9
0.49 7.2 15.6 0.98) # (0.46 0.82 0.65) 0.02 # (0.1 0.06 0.1)
# (0.99 0.06 0.41) 0.83 8.7 2.6))))) 3.1 6.8 # (0.95 0.7
0.59) 0.57))) # (0.17 0.08 0.75) 0.37) (vector y 0.09 (cos
(round y y)))))
```

When fleshed out on Sims's color monitor, the equation painted what seems to be two sheets of icicles backlit by an arctic sunset. It's an arresting image. The ice is molded in great detail and translucent, the horizon in the background abstract and serene. It could have been painted by a weekend artist. As Sims points out, "This equation was evolved from scratch in only a few minutes—probably much faster than it could be designed."

But Sims is at a total loss to explain the logic of the equation and why it produces a picture of ice. It looks as cryptic and muddled to him as to you. The equation's convoluted reason is beyond quick mathematical understanding.

The bombastic notion of evolving logic programs has been taken up in earnest by John Koza, a professor of computer science at Stanford. Koza was one of John Holland's students who brought knowledge of Holland's genetic algorithms out of the dark ages of the '60s and '70s into the renaissance of parallelism of the late '80s.

Rather than merely explore the space of possible equations, as Sims the artist did, Koza wanted to breed the best equation to solve a particular problem. One could imagine (as a somewhat silly example) that in the space of possible pictures there might be one that would induce cows gazing at it to produce more milk. Koza's method can evolve the equations that would draw that particular picture. In this farfetched idea, Koza would keep rewarding the equations which drew a picture that even minutely increased milk production until there was no further increase. For his actual experiments, though, Koza choose more practical tests, such as finding an equation that could steer a moving robot.

But in a sense his searches were similar to those of Sims and the others. He hunted in the Borgian Library of possible computer programs—not on an aimless mission to see what was there, but to find the best equation for a particular *practical* problem. Koza wrote in *Genetic Programming*, "I claim that the process of solving these problems can be reformulated as a search for a highly fit individual computer program in the space of possible computer programs."

For the same reason computer experts said Ray's scheme of computer evolution couldn't work, Koza's desire to "find" equations by breeding them bucked convention. Everyone "knew" that logic programs were brittle and unforgiving of the slightest alteration. In computer science theory, programs had two pure states: (1) flawlessly working; or (2) modified and bombed. The third state—modified at random yet working—was not in the cards. Slight modifications were known as bugs, and people paid a lot of money to keep them out. If progressive modification and improvement (evolution) of computer equations was at all possible, the experts thought, it must be so only in a few precious areas or specialized types of programs.

The surprise of artificial evolution has been that conventional wisdom was so wrong. Sims, Ray, and Koza have wonderful evidence that logical programs can evolve by progressive modifications.

Koza's method was based on the intuitive hunch that if two mathematical equations are somewhat effective in solving a problem, then some parts of them are valuable. And if the valuable parts from both are recombined into a new program, the result might be more effective than either parent. Koza randomly recombined, in thousands of combinations, parts of two parents, banking on the probabilistic likelihood that one of those random recombinations would include the optimal arrangement of valuable parts to better solve the problem.

There are many similarities between Koza's method and Sims's. Koza's soup, too, was a mixture of about a dozen arithmetical primitives, such as *add, multiply, cosine*, rendered in the computer language LISP. The units were strung together at random to form logical "trees," a hierarchical organization somewhat like a computer flow chart. Koza's system created 500 to 10,000 different individual logic trees as the breeding population. The soup usually converged upon a decent offspring in about 50 generations.

Variety was forced by sexually swapping branches from one tree to the next. Sometimes a long branch was grafted, other times a mere twig or terminal "leaf." Each branch could be thought of as an intact subroutine of logic made of smaller branches. In this way, bits of equation (a branch), or a little routine that worked and was valuable, had a chance of being preserved or even passed around.

All manner of squirrely problems can be solved by evolving equations. A typical riddle which Koza subjected to this cure was how to balance a broom on a skateboard. The skateboard must be moved back and forth by a motor to keep the inverted broom pivoted upright in the board's center. The motor-control calculations are horrendous, but not very different from the control circuits needed for maneuvering robot arms. Koza found he could evolve a program to achieve this control.

Other problems he tested evolutionary equations against included: strategies for navigating a maze; rules for solving quadratic equations; methods to optimize the shortest route connecting many cities (also known as traveling salesman problem); strategies for winning a simple game like tic-tac-toe. In each case, Koza's system sought a *formula* for the test problem rather than a specific answer for a specific instance of the test. The more varied instances a sound formula was tested against, the better the formula became with each generation.

While equation breeding yields solutions that work, they are usually the ugliest ones you could imagine. When Koza began to inspect the insides of his highly evolved prizes, he had the same shock that Sims and Ray did: the solutions were a mess! Evolution went the long way around. Or it burrowed through the problem by some circuitous loophole of logic. Evolution was chock-full of redundancy. It was inelegant. Rather than remove an erroneous section, evolution would just add a countercorrecting section, or reroute the main event around the bad sector. The final formula had the appearance of being some miraculous Rube Goldberg collection of items that by some happy accident worked. And that's exactly what it was, of course.

Take as an example a problem Koza once threw at his evolution machine. It was a graph of two intertwining spirals. A rough approximation would be the dual spirals in a pinwheel. Koza's evolutionary equation machine had to evolve the best equation capable of determining on which of the two intertwined spiral lines each of about 200 data points lay.

Koza loaded his soup with 10,000 randomly generated computer formulas. He let them breed, as his machine selected the equations that came closest to getting the right formula. While Koza slept, the program trees swapped branches, occasionally birthing a program that worked better. He ran the machine while he was on vacation. When he returned, the system had evolved an answer that perfectly categorized the twin spirals.

This was the future of software programming! Define a problem and the machine will find a solution while the engineers play golf. But the solution Koza's machine found tells us a lot about the handiwork of evolution. Here's the equation it came up with:

```
(SIN (IFLTE (IFLTE (+ Y Y) (+ X Y) (- X Y) (+ Y Y)) (*
X X) (SIN (IFLTE (% Y Y) (% (SIN (SIN (% Y
0.30400002))) X) (% Y 0.30400002) (IFLTE (IFLTE (%
(SIN (% (% Y (+ X Y)) 0.30400002)) (+ X Y)) (% X
0.10399997) (- X Y) (* (+ -0.12499994 -0.15999997)
(- X Y))) 0.30400002 (SIN (SIN (IFLTE (% (SIN (% (% Y
0.30400002) 0.30400002)) (+ X Y)) (% (SIN Y) Y) (SIN
(SIN (SIN (% (SIN X) (+ - 0.12499994 -0.15999997)))))
(% (+ (+ X Y) (+ Y Y)) 0.30400002)))) (+ (+ X Y) (+ Y
```

```
Y))))) (SIN (IFLTE (IFLTE Y (+ X Y) (- X Y) (+ Y Y)) (*
X X) (SIN (IFLTE (% Y Y) (% (SIN (SIN (% Y
0.30400002))) X) (% Y 0.30400002) (SIN (SIN (IFLTE
(IFLTE (SIN (% (SIN X) (+ -0.12499994 -0.15999997)))
(% X -0.10399997) (- X Y) (+ X Y)) (SIN (% (SIN X) (+
-0.12499994 -0.15999997))) (SIN (SIN (% (SIN X) (+
-0.12499994 -0.15999997)))) (+ (+ X Y) (+ Y Y)))))))
(% Y 0.30400002))))).
```

Not only is it ugly, it's incomprehensible. Even for a mathematician or computer programmer, this evolved formula is a tar baby in the briar patch. Tom Ray says evolution writes code that only an intoxicated human programmer would write, but it may be more accurate to say evolution generates code that only an alien would write; it is decidedly inhuman. Backtracking through the evolving ancestors of the equation, Koza eventually traced the manner in which the program tackled the problem. By sheer persistence and by hook and crook it found a laborious roundabout way to its own answer. But it worked.

The answer evolution discovered seems strange because almost any high school algebra student could write a very elegant equation in a single line that described the two spirals.

There was no evolutionary pressure in Koza's world toward simple solutions. His experiment could not have found that distilled equation because it wasn't structured to do so. Koza tried applying parsimony in other runs but found that parsimony added to the beginning of a run dampened the efficiency of the solutions. He'd find simple but mediocre to poor solutions. He has some evidence that adding parsimony at the end of evolutionary procedure—that is, first let the system find a solution that kind of works and then start paring it down—is a better way to evolve succinct equations.

But Koza passionately believes parsimony is highly overrated. It is, he says, a mere "human esthetic." Nature isn't particularly parsimonious. For instance, David Stork, then a scientist at Stanford, analyzed the neural circuits in the muscles of a crayfish tail. The network triggers a curious backflip when the crayfish wants to escape. To humans the circuit looks baroquely complex and could be simplified easily with the quick removal of a couple of superfluous loops. But the mess works. Nature does not simplify simply to be elegant.

Humans seek a simple formula such as Newton's *f=ma*, Koza suggests, because it reflects our innate faith that at bottom there is elegant order in the universe. More importantly, simplicity is a human convenience. The heartwarming beauty we perceive in *f=ma* is reinforced by the cold fact that it is a much easier formula to use than Koza's spiral monster. In the days before computers and calculators, a simple equation was *more useful* because it was easier to compute without errors. Complicated formulas were a grind and treacherous. But, within a certain range, neither nature nor parallel computers are troubled by convoluted logic. The extra steps we find ugly and stupefying, they do perfectly in tedious exactitude.

The great irony puzzling cognitive scientists is why human consciousness is so unable to think in parallel, despite the fact that the brain runs as a parallel machine. We have an almost uncanny blind spot in our intellect. We cannot innately grasp concepts in probability, horizontal causality, and simultaneous logic. We simply don't think like that. Instead our minds retreat to the serial narrative—the linear story. That's why the first computers were programmed in von Neumann's serial design: because that's how humans think.

And this, again, is why parallel computers must be evolved rather than designed: because we are simpletons when it comes to thinking in parallel. Computers and evolution do parallel; consciousness does serial. In a very provocative essay in the Winter 1992 *Daedalus*, James Bailey, director of marketing at Thinking Machines, wrote of the wonderful boomeranging influence that parallel computers have on our thinking. Entitled "First We Reshape Our Computers. Then Our Computers Reshape Us," Bailey argues that parallel computers are opening up new territories in our intellectual landscape. New styles of computer logic in turn force new questions and new perspectives from us. "Perhaps," Bailey suggests, "whole new forms of reckoning exist, forms that only make sense in parallel." Thinking like evolution may open up new doors in the universe.

John Koza sees the ability of evolution to work on both

ill-defined and parallel problems as another of its inimitable advantages. The problem with teaching computers how to learn to solve problems is that so far we have wound up explicitly reprogramming them for every new problem we come across. How can computers be designed to do what needs to be done, without being told in every instance what to do and how to do it?

Evolution, says Koza, is the answer. Evolution allows a computer's software to solve a problem to which the scope, kind, or range of the answer(s) may not be evident at all, as is usually the case in the real world. Problem: A banana hangs in a tree; what is the routine to get it? Most computer learning to date cannot solve that problem unless we explicitly clue the program in to certain narrow parameters such as: how many ladders are nearby? Any long poles?

Having defined the boundaries of the answer, we are half answering the question. If we don't tell it what rocks are near, we know we won't get the answer "throw a rock at it." Whereas in evolution, we might. More probably, evolution would hand us answers we could have never expected: use stilts; learn to jump high; employ birds to help you; wait until after storms; make children and have them stand on your head. Evolution did not narrowly require that insects fly or swim, only that they somehow move quick enough to escape predators or catch prey. The open problem of escape led to the narrow answers of water striders tiptoeing on water or grasshoppers springing in leaps.

Every worker dabbling in artificial evolution has been struck by the ease with which evolution produces the improbable. "Evolution doesn't care about what makes sense; it cares about what works," says Tom Ray.

The nature of life is to delight in all possible loopholes. It will break any rule it comes up with. Take these biological jaw-droppers: a female fish that is fertilized by her male mate who lives inside her, organisms that shrink as they grow, plants that never die. Biological life is a curiosity shop whose shelves never empty. Indeed the catalog of natural oddities is almost as long as the list of all creatures; *every* creature is in some way hacking a living by reinterpreting the rules.

The catalog of human inventions is far less diverse. Most

machines are cut to fit a specific task. They, by our old definition, follow our rules. Yet if we imagine an ideal machine, a machine of our dreams, it would adapt, and—better yet—evolve.

Adaptation is the act of bending a structure to fit a new hole. Evolution, on the other hand, is a deeper change that reshapes the architecture of the structure itself—how it *can* bend—often producing new holes for others. If we predefine the organizational structure of a machine, we predefine what problems it can solve. The ideal machine is a general problem solver, one that has an open-ended list of things it can do. That means it must have an open-ended structure, too. Koza writes, "The size, shape, and structural complexity [of a solution] should be part of the answer produced by a problem solving technique—not part of the question." In recognizing that a system itself sets the answers the system can make, what we ultimately want, then, is a way to generate machines that do not possess a predefined architecture. We want a machine that is constantly remaking itself.

Those interested in kindling artificial intelligence, of course, say "amen." Being able to come up with a solution without being unduly prompted to where the solution might exist—lateral thinking it's called in humans—is almost the definition of human intelligence.

The only machine we know of that can reshape its internal connections is the living gray tissue we call the brain. The only machine that would generate its own structure that we can presently even imagine manufacturing would be a software program that could reprogram itself. The evolving equations of Sims and Koza are the first step toward a self-reprogramming machine. An equation that can breed other equations is the basic soil for this kind of life. Equations that breed other equations are an open-ended universe. Any possible equation could arise, including self-replicating equations and formulas that loop back in a Uroborus bite to support themselves. This kind of recursive program, which reaches into itself and rewrites its own rules, unleashes the most magnificent power of all: the creation of perpetual novelty.

"Perpetual novelty" is John Holland's phrase. He has been crafting means of artificial evolution for years. What he is really working on, he says, is a new mathematics of perpetual novelty. Tools to create neverending newness.

Karl Sims told me, "Evolution is a very practical tool. It's a way of exploring new things you wouldn't have thought about. It's a way of refining things. And it's a way of exploring procedures without having to understand them. If computers are fast enough they can do all these things."

Exploring beyond the reach of our own understanding and refining what we have are gifts that directed, supervised, optimizing evolution can bring us. "But evolution," says Tom Ray, "is not just about optimization. We know that evolution can go beyond optimization and create new things to optimize." When a system can create new things to optimize we have a perpetual novelty tool and open-ended evolution.

Both Sims's selection of images and Koza's selection of software via the breeding of logic are examples of what biologists call breeding or artificial selection. The criteria for "fit"—for what is selected—is chosen by the breeder and is thus an artifact, or artificial. To get perpetual novelty—to find things we don't anticipate—we must let the system itself define the criteria for what it selects. This is what Darwin meant by "natural selection." The selection criteria was done by nature of the system; it arose naturally. Open-ended artificial evolution also requires natural selection, or if you will, artificial natural selection. The traits of selection should emerge naturally from the artificial world itself.

Tom Ray has installed the tool of artificial natural selection by letting his world determine its own fitness selection. Therefore his world is theoretically capable of evolving completely new things. But Ray did "cheat" a little to get going. He could not wait for his world to evolve self-replication on its own. So he introduced a self-replicating organism from the beginning, and once introduced, replication never vanished. In Ray's metaphor, he jump-started life as a single-celled organism, and then watched a "Cambrian explosion" of new organisms. But he isn't apologetic. "I'm just trying to get evolution and I don't really care how I get it. If I need to tweak my world's physics and chemistry to the point where they can support rich, open-ended evolution, I'm going to be happy. It doesn't make me feel guilty that I had to manipulate them to get it there. If I can engineer a world to the threshold of the Cambrian explosion and let it boil over the edge on its own, that will be truly impressive. The fact that I had to engineer it to get there will be trivial compared to what comes out of it."

Ray decided that getting artificial open-ended evolution up and running was enough of a challenge that he didn't need to evolve it to that stage. He would engineer his system until it could evolve on its own. As Karl Sims said, evolution is a tool. It can be combined with engineering. Ray used artificial natural selection after months of engineering. But it can go both ways. Other workers will engineer a result after months of evolution.

As a tool, evolution is good for three things:

- How to get somewhere you want but can't find the route to.
- How to get to somewhere you can't imagine.
- How to open up entirely new places to get to.

The third use is the door to an open universe. It is unsupervised, undirected evolution. It is Holland's ever-expanding perpetual novelty machine, the thing that creates itself.

Amateur gods such as Ray, Sims, and Dawkins have all expressed their astonishment at the way evolution seems to amplify the fixed space they thought they had launched. "It's a lot bigger than I thought" is the common refrain. I had a similar overwhelming impression when I stepped and jumped (literally) through the picture space of Karl Sims's evolutionary exhibit. Each new picture I found (or it found for me) was gloriously colored, unexpectedly complex, and stunningly different from anything I had ever seen before. Each new image seemed to enlarge the universe of possible pictures. I realized that my idea of a picture had previously been defined by pictures made by humans, or perhaps by biological nature. But in Sims's world an equally vast number of breathtaking vistas that were neither human-made nor biologically made—but equally rich—were waiting to be unwrapped.

Evolution was expanding my notions of possibilities. Life's biological system is very much like this. Bits of DNA are functional units—logical evolvers that expand the space of possibilities. DNA directly parallels the operation of Sims's and Koza's logical units. (Or should we say their logical units parallel DNA?) A handful of units can be mixed and matched to code for any one of an astronomical number of possible proteins. The proteins produced by this small functional alphabet serve as tissue, disease, medicines, flavors, signals, and the bulk infrastructure of life.

Biological evolution is the open-ended evolution of DNA units breeding new DNA units in a library that is ever-expanding and without known boundaries.

Gerald Joyce, the molecular breeder, says he is happily into "evolving molecules for fun and profit." But his real dream is to hatch an alternative open-ended evolution scheme. He told me, "My interest is to see if we can set in motion, under our own control, the process of self-organization." The test case Joyce and colleagues are working on is to try to get a simple ribozyme to evolve the ability to replicate itself—that very crucial step that Tom Ray skipped over. "The explicit goal is to set an evolving system in motion. We want molecules to learn how to make copies of themselves by themselves. Then it would be autonomous evolution instead of directed evolution."

Right now autonomous and self-sustained evolution is a mere dream for biochemists. No one has yet coerced an evolutionary system to take an "evolutionary step," one that develops a chemical process that heretofore didn't exist. To date, biochemists have only evolved new molecules which *resolve* problems they already knew how to solve. "True evolution is about going somewhere novel, not just reeling in interesting variants," says Joyce.

A working, autonomous, evolving, molecular system would be an incredibly powerful tool. It would be an open-ended system *that could create all possible biologies*. "It would be biology's triumph!" Joyce exclaims, equivalent, he believes, to the impact of "finding another life form in the universe that was happy to share samples with us."

But Joyce is a scientist and does not want to let his enthusiasm run over the edge: "We're not saying we are going to make life and it's going to develop its own civilization. That's goofy. We're saying we are going to make an artificial life form that is going to do slightly different chemistry than it does now. That's not goofy. That's realistic."

But Chris Langton doesn't find the prospect of artificial life creating its own civilization so goofy. Langton has gotten a lot of press for being the maverick who launched the fashionable field of artificial life. He has a good story, worth retelling very briefly because his own journey recapitulates the awakening of human-made, open-ended evolution.

Several years ago Langton and I attended a week-long science conference in Tucson, and to clear our heads, we played hooky for an afternoon. I had an invitation to visit the unfinished Biosphere 2 project an hour away, and so as we cruised the black ribbon of asphalt that winds through the basins of southern Arizona, Langton told me his life story.

At the time, Langton worked at the Los Alamos National Laboratory as a computer scientist. The entire town and lab of Los Alamos were originally built to invent the ultimate weapon. So I was surprised to hear Langton begin his story by saying he was a conscientious objector during the Vietnam War.

As a CO, Langton scored a chance to do alternative service as a hospital orderly at Boston's Massachusetts General Hospital. He was assigned the undesirable chore of transporting corpses from the hospital basement to the morgue basement. On the first week of the job, Langton and his partner loaded a corpse onto a gurney and pushed it through the dank, underground corridor connecting the two buildings. They needed to push it over a small concrete bridge under the only light in the tunnel, and as the gurney hit the bump, the corpse belched, sat upright, and started to slide off its perch! Chris spun around to grab his partner, but he saw only the distant doors flapping behind his coworker. Dead things could behave as if they were alive! Life was behavior; that was the first lesson.

Langton told his boss he couldn't go back to that job. Could he do something else? "Can you program computers?" he was asked. "Sure."

He got a job programming early-model computers. Sometimes he would let a silly game run on the unused computers at night. The game was called Life, devised by John Conway, and written for the mainframe by an early hacker named Bill Gosper. The game was a very simple code that would generate an infinite variety of forms, in patterns reminiscent of biological cells growing, replicating, and propagating on an agar plate. Langton remembered working alone late one night and suddenly feeling the presence of someone, something alive in the room, staring at him. He looked up and on the screen of Life he saw an amazing pattern of self-replicating cells. A few minutes later he felt the presence again. He looked up again and saw that the pattern had died. He suddenly felt that the pattern had been alive—alive and as real as mold on an agar plate—but on a computer screen instead. The bombastic idea that perhaps a computer program could capture life sprouted in Langton's mind.

He started fooling around with the game, probing it, wondering if it was possible to design a game like Life that would be open ended—so that things would start to evolve on their own. He honed his programmer skills. On the job Langton was given the task of transferring a program from an out-of-date mainframe computer to a very different newer one. In order to do this, the trick was to abstract the operation of the *hardware* of the old computer and put it into the *software* of the newer one—to extract the essential behavior of the hardware and cast it in intangible symbols. This way, old programs running on the new machine would be running in a virtual old computer emulated in software in the new computer. Langton said, "This was a first-hand experience of moving a process from one medium to another. The hardware didn't matter. You could run it on any hardware. What mattered was capturing the essential processes." It made him wonder if life could be taken from carbon and put into silicon.

After his service stint Langton spent his summers hang-gliding. He and a friend got a job hang-gliding over Grandfather Mountain in North Carolina for $25 per day as an airborne tourist attraction. They stayed aloft for hours at a time in 40-mile-per-hour winds. Swiped by a freak gust one day, Langton crashed from the sky. He hit the ground in a fetus position and broke 35 bones, including all the bones in his head except his skull. Although he smashed his knees through his face, he was alive. He spent the next six months on his back, half-conscious.

As he recovered from his massive concussions, Langton felt he was watching his brain "reboot," just as computers that are turned off have to rebuild their operating system when turned back on. One by one certain deep functions of his mind reappeared. In an epiphany of sorts, Langton remembers the moment when his sense of proprioception—the sense of being centered in a body—returned. He was suddenly struck with a "deep emotional gut feeling" of his own self becoming integrated, as if his machine had completed its reboot and was now waiting for an application. "I had a personal experience of what growing a mind feels like," he told me. Just as he had seen life in a computer, he now had a visceral appreciation of his own life being in a machine. Surely, life must be independent of its matrix? Couldn't life in both his body and his computer be the same?

Wouldn't it be great, he thought, if he could get something alive with evolution going in a computer! He thought he would start with human culture. That seemed an easier simulation to start with than simulated cells and DNA. As a senior at the University of Arizona, Langton wrote a paper on "The Evolution of Culture." He wanted his anthropology, physics, and computer science professors to let him design a degree around building a computer to run artificial evolution, but they discouraged him. On his own he bought an Apple II and wrote his first artificial world. He couldn't get self-reproduction or natural selection, but he did discover the literature of cellular automata—of which the Game of Life, it turned out, was only one example.

And he came across John von Neumann's proofs of artificial self-replication from the 1940s. Von Neumann had come up with a landmark formula that would self-replicate. But the program was unwieldy, inelegantly large and clumsy. Langton spent months of long nights coding his Apple II (a handy advantage that von Neumann didn't have; he did his with pencil on paper). Eventually guided only by his dream to create life in silicon, Langton came up with the smallest self-replicating machine then known to anyone. On the computer screen the self-replicator looked like a small blue Q. Langton was able to pack into its loop of only 94 symbols a complete representation of the loop, instructions on how to reproduce, and the trick of throwing off another just like itself. He was delirious. If he could engineer such a simple replicator, how many of life's other essential processes could he also mimic? Indeed, what *were* life's other essential processes?

A thorough search of the existing literature showed that very little science had been written on such a simple question, and what little there was, was scattered here and there in hundreds of tiny corners. Emboldened by his new research position at the Los Alamos Labs, in 1987 Langton staked his career on gathering an "Interdisciplinary Workshop on the Synthesis and Simulation of Living Systems"—the first conference on what Langton was now calling Artificial Life. In his search for any and all systems that exhibit the behavior of living systems, Langton opened the workshop to chemists, biologists, computer scientists, mathematicians, material scientists, philosophers, roboticists, and computer animators. I was one of the few journalists attending.

At the workshop Langton began with his quest for a definition of life. Existing ones seemed inadequate. As more research was started over the years following the first conference, physicist Doyne Farmer proposed a list of traits that defined life. Life, he said, has:

- Patterns in space and time
- Self-reproduction
- Information storage of its self-representation (genes)
- Metabolism, to keep the pattern persisting
- Functional interactions—it does stuff
- Interdependence of parts, or the ability to die
- Stability under perturbations
- Ability to evolve.

The list provokes. For although we do not consider computer viruses alive, computer viruses satisfy most of the qualifications above. They are a pattern that reproduce; they include a copy of their own representation; they capture computer metabolistic (CPU) cycles; they can die; and they can evolve. We could say that computer viruses are the first examples of emergent artificial life.

On the other hand, we all know of a few things whose aliveness we don't doubt yet are exceptions to this list. A mule can not self-reproduce, and a herpes virus has no metabolism. Langton's success in creating a self-reproducing entity made him skeptical of arriving at a consensus: "Every time we succeed in synthetically satisfying the definition of life, the definition is lengthened or changed. For instance if we take Gerald Joyce's definition of life—a self-sustaining chemical system capable of undergoing Darwinian evolution—I believe that by the year 2000 one lab somewhere in the world will make a system satisfying this definition. But then biologists will merely redefine life."

Langton had better luck defining artificial life. Artificial life, or "a-life" in short hand, is, he said, "the attempt to abstract the logic of life in different material forms." His thesis was that life is a *process*—a behavior that is not bound to a specific material manifestation. What counts about life is not the stuff it is made of, but what it does. Life is a verb not a noun. Farmer's list of qualifications for life represents actions and behaviors. It is not hard for computer scientists to think of the list of life's qualities as varieties of processing. Steen Rasmussen, a colleague of Langton who was also interested in artificial life, once dropped a pencil onto the desk and sighed, "In the West we think a pencil is more real than its motion."

If the pencil's motion is the essence—the real part—then "artificial" is a deceptive word. At the first Artificial Life Conference, when Craig Reynolds showed how he was able to use three simple rules to get dozens of computer-animated birds to flock in the computer autonomously, everyone could see that the flocking was real. Here were *artificial birds really flocking*. Langton summarized the lesson: "The most important thing to remember about a-life is that the part that is artificial is not the life, but the materials. Real things happen. We observe real phenomena. It is real life in an artificial medium."

Biology—the study of life's general principles—is undergoing an upheaval. Langton says biology faces "the fundamental obstacle that it is impossible to derive general principles from single examples." Since we have only a single collective example of life on Earth, it is pointless to try to distinguish its essential and universal properties from those incidental properties due to life's common descent on the planet. For instance, how much of what we think life is, is due to its being based on carbon chains? We can't know without at least a second example of life not based on carbon chains. To derive general principles and theories of life—that is, to identify properties that would be shared by any vivisystem or any life—Langton argues that "we need an ensemble of instances to generalize over. Since it is quite unlikely that alien life-forms will present themselves to us for study in the near future, our only option is to try to create alternative life-forms ourselves." This is Langton's mission—to create an alternative life, or maybe even several alternative "lifes," as a basis for a true biology, a true logic of Bios. Since these other lifes are artifacts of humans rather than nature, we call them artificial life; but they are as real as we are.

The nature of this ambitious challenge initially sets the science of artificial life apart from the science of biology. Biology seeks to understand the living by taking it apart and reducing it to it pieces. Artificial life, on the other hand, has nothing to dissect, so it can only make progress by putting the living together and assembling it from pieces. Rather than analyze life, synthesize it. For this reason, Langton says, "Artificial life amounts to the practice of synthetic biology."

Artificial life acknowledges new lifes and a new definition of life. "New" life is an old force that organizes matter and energy in new ways. Our ancient ancestors were often generous in deeming things alive. But in the age of science, we make a careful distinction. We call creatures and green plants alive, but when we call an institution such as the post office an "organism," we say it is lifelike or "as if it were alive."

We (and by this I mean scientists first) are beginning to see that those organizations once called metaphorically alive are truly alive, but animated by a life of a larger scope and wider definition. I call this greater life "hyperlife." Hyperlife is a particular type of vivisystem endowed with integrity, robustness, and cohesiveness—a strong vivisystem rather than a lax one. A rain forest and a periwinkle, an electronic network and a servomechanism, SimCity and New York City, all possess degrees of hyperlife. Hyperlife is my word for that class of life that includes both the AIDS virus and the Michelangelo computer virus.

Biological life is only one species of hyperlife. A telephone network is another species. A bullfrog is chock-full of hyperlife. The Biosphere 2 project in Arizona swarms with hyperlife, as do Tierra, and Terminator 2. Someday hyperlife will blossom in automobiles, buildings, TVs, and test tubes.

This is not to say that organic life and machine life are identical; they are not. Water striders will forever retain certain characteristics unique to carbon-based life. But organic and artificial life share a set of characteristics that we have only begun to discern. And of course there easily may be other types of hyperlife to come that we can't describe yet. One can imagine various possibilities of life—weird hybrids bred from both biological and synthetic lines, the half-animal/half-machine cyborgs of old science fiction—that may have emergent properties of hyperlife not found in either parent.

Man's every attempt to create life is a probe into the space of possible hyperlifes. This space includes all endeavors to re-create the origins of life on Earth. But the challenge goes way beyond that. The goal of artificial life is not to merely describe the space of "life-as-we-know-it." The quest that fires up Langton is the hope of mapping the space of *all* possible lifes, a quest that moves us into the far, far vaster realm of "life-as-it-could-be." Hyperlife is that library which contains all things alive, all vivisystems, all slivers of life, anything bucking the second law of thermodynamics, all future and all past arrangements of matter capable of open-ended evolution, and all examples of a type of something marvelous we can't really define yet.

The only way to explore this *terra incognita* is to build many examples and see if they fit in the space. As Langton wrote in his introduction to the proceedings of the Second Artificial Life conference, "If biologists could 'rewind the tape' of evolution and start it over, again and again, from different initial conditions, or under different regimes of external perturbations along the way, they would have a full *ensemble* of evolutionary pathways to generalize over." Keep starting from zero, alter the rules a bit and then build an example of artificial life. Do it dozens of times. Each instance of synthetic life is added to the example of Earth-bound organic life to form the complete ensemble of hyperlife.

Since life is a property of form, and not matter, the more materials we can transplant living behaviors into, the more examples of "life-as-it-could-be" we can accumulate. Therefore the field of artificial life is broad and eclectic in considering all avenues to complexity. A typical gathering of a-life researchers includes biochemists, computer wizards, game designers, animators, physicists, math nerds, and robot hobbyists. The hidden agenda is to hack the definition of life.

One evening after a late-night lecture session at the First Artificial Life Conference, while some of us watched the stars in the desert night sky, mathematician Rudy Rucker came up with the most expansive motivation for artificial life I've heard: "Right now an ordinary computer program may be a thousand lines and take a few minutes to run. Artificial life is about finding a computer code that is only a few lines long and that takes a thousand years to run."

That seems about right. We want the same in our robots: Design them for a few years and then have them run for centuries, perhaps even manufacturing their replacements. That's what an acorn is too—a few lines of code that run out as a 180-year-old tree.

The conference-goers felt the important thing about artificial life was that it not only was redefining biology and life, but it was also redefining the concept of both artificial and real. It was radically enlarging the realm of what seemed important—that is, the realm of life and reality. Unlike the "publish or perish" mode of academic professionalism of yesteryear, most of the artificial life experimenters—even the mathematicians—espoused the emerging new academic creed of "demo or die." The only way to make a dent in artificial and hyperlife was to get a working example up and running. Explaining how he got started in life-as-it-could-be, Ken Karakotsios, a former Apple employee, recalled, "Every time I met a computer I tried to program the Game of Life into it." This eventually led to a remarkable Macintosh a-life program called SimLife. In SimLife you create a hyperlife world and set loose little creatures into it to coevolve into a complexifying artificial ecology. Now Karakotsios seeks to write the biggest and best game of life, an ultimate living program: "You know, the universe is the only thing big enough to run the ultimate game of life. The only problem with the universe as a platform, though, is that it is currently running someone else's program."

Larry Yaeger, a current Apple employee, once handed me his business card. It ran: "Larry Yaeger, Microcosmic God." Yaeger created Polyworld, a sophisticated computer world with organisms in the shape of polygons. The polys fly around by the hundreds, mating, breeding, consuming resources, learning (a power God Yaeger gave them), adapting, and evolving. Yaeger was exploring the space of possible life. What would appear? "At first," said Yaeger, "I did not charge the parents an energy cost when offspring was born. They could have offspring for free. But I kept getting this particular species, these indolent cannibals, who liked to hang around the corner in the vicinity of their parents and children and do nothing, never leave. All they would do was mate with each other, fight with each other, and eat each other. Hey, why work when you can eat your kids!" Life of some hyper-type had appeared.

"A central motivation for the study of artificial life is to extend biology to a broader class of life forms than those currently present on the earth," writes Doyne Farmer, understating the sheer, great fun artificial life gods are having.

But Farmer is onto something. Artificial life is unique among other human endeavors for yet another reason. Gods such as Yaeger are extending the class of life because life-as-it-could-be is a territory we can only study by first creating it. We must manufacture hyperlife to explore it, and to explore it we must manufacture it.

But as we busily create ensembles of new forms of hyperlife, an uneasy thought creeps into our minds. Life is using us. Organic carbon-based life is merely the first, earliest form of hyperlife to evolve into matter. Life has conquered carbon. But now under the guise of pond weed and kingfisher, life seethes to break out into crystal, into wires, into biochemical gels, and into hybrid patches of nerve and silicon. If we look at where life is headed, we have to agree with developmental biologist Lewis Held when he said, "Embryonic cells are just robots in disguise." In his report for the proceedings of the Second Artificial Life Conference Tom Ray wrote, "Virtual life is out there, waiting for us to create environments for it to evolve into." Langton told Steven Levy, reporting in *Artificial Life*, "There are these other forms of life, artificial ones, that want to come into existence. And they are using me as a vehicle for its reproduction and its implementation."

Life—the hyperlife—wants to explore all possible biologies and all possible evaluations, but it uses us to create them because to create them is the only way to explore or complete them. Humanity is thus, depending on how you look at it, a mere passing station on hyperlife's gallop through space, or the critical gateway to the open-ended universe.

"With the advent of artificial life, we may be the first species to create its own successors," Doyne Farmer wrote in his manifesto, *Artificial Life: The Coming Evolution*. "What will these successors be like? If we fail in our task as creators, they may indeed be cold and malevolent. However, if we succeed, they may be glorious, enlightened creatures that far surpass us in their intelligence and wisdom." Their intelligence might be "inconceivable to lower forms of life such as us." We have always been anxious about being gods. If through us, hyperlife should find spaces where it evolves creatures that amuse and help us, we feel proud. But if superior successors should ascend through our efforts, we feel fear.

Chris Langton's office sat catty-corner to the atomic museum in Los Alamos, a reminder of the power we have to destroy. That power stirred Langton. "By the middle of this century, mankind had acquired the power to extinguish life," he wrote in one of his academic papers. "By the end of the century, he will be able to create it. Of the two, it is hard to say which places the larger burden of responsibilities on our shoulders."

Here and there we create space for other varieties of life to emerge. Juvenile delinquent hackers launch potent computer viruses. Japanese industrialists weld together smart painting robots. Hollywood directors create virtual dinosaurs. Biochemists squeeze self-evolving molecules into tiny plastic test tubes. Someday, we will create an open-ended world that can keep going, and keep creating perpetual novelty. When we do we will have created another living vector in the life space.

When Danny Hillis says he wants to make a computer that would be proud of him, he isn't kidding. What could be more human than to give life? I think I know: to give life and freedom. To give *open-ended* life. To say, here's your life *and* the car keys. Then you let it do what we are doing—making it all up as we go along. Tom Ray once told me, "I don't want to download life into computers. I want to upload computers into life."

# 18

# The Structure of Organized Change

Open any book on evolution, and the pages flow with stories of change. The terms *adaptation, speciation, mutation* are all the jargon of transformation—of differences over time. Through the language of change, which evolution science has given us, we tell our history as one of alterations, metamorphosis, and novelty. "New" is our favorite word.

But rare is the book on evolution theory that tells the story of steadfastness. The index will not list *stasis*, or *fixity*, or *stability*, or any of the jargon of permanence. Despite the overwhelming fact that evolution spends almost all of its time not changing very much, teachers and textbooks are silent on the ways of constancy.

The dinosaur is the undeserved emblem of unwillingness to change. We see the towering beast in our mind: with slack-jaw stupidity it gawks at the birdy things flittering around its sluggish feet. Don't be a dinosaur! we admonish the timid. Don't be steamrolled by progress! we tell the slow. Adapt or flatten.

When I type the word "evolution" into my library's online card catalog I get a list of book titles such as these:

*The Evolution of Language in China*
*The Evolution of Music*

*The Evolution of Political Parties in Early United States*
*The Evolution of Technology*
*The Evolution of The Solar System*

It is evident that "evolution," as used in these titles, is a common vernacular term meaning incremental change over time. But what in the world doesn't alter gradually? Nearly all change around us is incremental. Catastrophic change is rare, and continual catastrophic change over long periods is almost unknown. Is all long-term change evolutionary?

Some people take it that way. The charter of the Washington Evolutionary Systems Society, a lively national association of 180 members in the science and engineering professions, considers any and all systems as evolutionary, "placing no constraint on the type of system to be explored . . . All that we see about us and experience are the products of ongoing evolutionary processes." A perusal of the topics they consider evolutionary—"evolution of objectivity, evolution of business firms"—prompted me to ask Bob Crosby, the Society's founder, "Are there any systems you don't consider evolutionary?" His reply: "We don't see anywhere where there isn't evolution." I have tried to avoid using this meaning of the word in this book, but I haven't been perfect.

Despite the confusion about the word "evolution," our strongest terms of change are rooted in the organic: *grow, develop, evolve, mutate, learn, metamorphose, adapt*. Nature is the realm of ordered change.

Disordered change is what technology has been about until now. The strong term for disordered change is "revolution"—a type of drastic discontinuous change peculiar to human-made things. There are no revolutions within nature.

Technology introduced the concept of revolution as an ordinary mode of change. Beginning with the Industrial Revolution, and its spillovers the French and American Revolutions, we've seen an uninterrupted series of revolutions brought on by technological advances—the revolutions of electrical appliances, of antibiotics and surgery, of plastic, of highways, of birth control, and so on. These days, revolutions, both social and technological, are announced weekly. Genetic engineering and nanotechnology—technologies which, by definition, mean we can make anything we desire—promise revolutions daily.

But daily revolution, I predict, will be headed off by daily evolution. The last revolution in technology will be to embrace evolutionary change. Science and commerce now seek to capture change—to instill it in a structured way—so that it works steadily, producing a constant tide of microrevolutions instead of dramatic and disruptive macrorevolutions. How can we implant change into the artificial so that it is both ordered and autonomous?

The science of evolution is no longer valuable only to biologists, but to engineers as well. Artificial evolution arises in our environment; but just as important, the study of evolution (both natural and artificial) rises in our esteem. Alvin Toffler was the first futurist to bring to public consciousness the fact that not only are technological and cultural things changing fast, but the rate of change itself seems to be accelerating. We live in a world of constant change, and we need to understand it. We don't understand natural evolution very well. With our recent invention of artificially natural evolution, and its study, we can understand organic evolution better, and we can better manage, inoculate, and anticipate change in our made world. Artificial evolution is the second course in a new biology of creatures, and the first course in a new biology of machines.

The goal is to make, say, a car that adjusts its frame and wheels to fit the kind of road it's on, to make a road aware of its conditions to repair itself, to make a car factory flexible to produce a personalized car to fit each customer, to make a highway system aware of traffic to minimize it, and to make a city learn to balance the traffic it absorbs. Each of these impute to technology the ability to change itself.

But rather than continually pump in bits of change, we'd like to implant the intact heart of change—an adaptive spirit—into the core of the system itself. This magic ghost is artificial evolution. In stronger doses evolution breeds artificial intelligence, and in dilute form it promotes mild adaptation. Either way, evolution is the broad self-guiding force that machines still lack in larger doses.

The postmodern mind accepts on faith the once disturbing notion that evolution is blind towards the future. After all, we humans are incapable of anticipating all our future needs—and we claim to be above average in the looking-ahead department. The irony is that evolution is even more ignorant than we knew: it is blind both coming and going. Blind not only to how things might be, but also to how they are now and were in the past. Nature doesn't know what it did yesterday, doesn't care. It keeps no audited record of successes, of smart moves, of things that helped. We—all organisms—are a historical record of sorts, but our history is not easy to unravel or decipher without great intelligence.

An ordinary organism hasn't the faintest notion of the details operating in its lower levels. A cell is a bimbo in terms of what it can relate about its own genes. Both plants and animals are small pharmaceutical factories, casually churning out biochemicals that would make Genentech drool, but neither a cell, nor an organ, nor an individual, nor a species keeps track of these achievements—what produces what. "It works, why worry?" is life's deepest philosophy.

When we contemplate nature as a system we don't expect consciousness, just bookkeeping. As far as anyone knows, there is one law biology keeps sacrosanct: The Central Dogma. The Central Dogma states that nature does no bookkeeping. More accurately it states that information travels from gene to body, but never sends an account in the opposite way—from the body back to the genes. In this way, nature is blind about its past.

If nature transmitted information in both directions within organisms, it would allow the possibility of Lamarckian evolution, which requires two-way communication between gene and its products. The advantages of Lamarckism are awesome. When an animal needs faster legs to survive, it could use body-to-gene communication to direct the genes to make faster leg muscles, and then pass that innovation on to its offspring. Evolution would accelerate madly.

But Lamarckian evolution requires an organism to have a working index to its genes. If the organism met a harsh environment—say extreme high altitude—it would notify all the genes in its body able to influence respiration and ask them to adjust. The body of an organism can certainly communicate that message to other organs in the body by hardwired hormone and chemical circuits. And it could communicate the same to the genes if it could pinpoint the right ones. But that is the bookkeeping chore that is missing. The body does not keep track of how it solves a problem, so it cannot pinpoint which genes pump up the muscle on the blacksmith's biceps, or which genes regulate respiration and blood pressure. And because there are millions of genes producing billions of features—and one gene can make more than one feature and one feature can be made by more than one gene—the complexity of accounting and indexing could exceed the complexity of the organism itself.

So it isn't so much that information can't be transmitted in the body to gene direction, it's more that communication is blocked because messages have no distinct destination. There is no central gene-authority to direct traffic. The genome is the ultimate decentralized system—rampant redundancy, massive parallelism, no one in charge, no one looking over the shoulder of every transaction.

But what if there is some way around this? Genuine two-way genetic communication would light up an interesting bunch of questions: Would there be any biological advantage if such a mechanism were possible? What else would it take to have a Lamarckian biology? Could there have been a biological route to such a mechanism at one time? If it is possible, why hasn't it happened? Could we outline a working biological Lamarckism as a thought experiment?

In all probability, Lamarckian biology requires a type of deep complexity—an intelligence—that most organisms can't reach. But where complexity *is* rich enough for intelligence, such as in human organisms and organizations, and their robotic offspring, Lamarckian evolution is possible and advantageous. Ackley and Littman showed that computers programmed by humans could run Lamarckian evolution.

But in the last decade, mainstream biologists have acknowledged an observation a few maverick biologists have preached for a century: that when an organism acquires sufficient complexity in its body, it can use its body to teach the genes what they need to know to evolve. Because this mechanism is a hybrid of evolution and learning, it has great potential in artificial realms.

Every animal's body has a built-in but limited power to adjust to different environments. Humans can acclimatize to life at a significantly higher elevation. Our heart rate, blood pressure, and lung capacity must and will compensate for the lower air pressure. The same changes reverse when we migrate to a lower elevation. But there is a limit to the degree to which we can acclimatize. For us, it's around 20,000 feet above sea level. Beyond this altitude, the human body cannot stretch itself for long-term habitation.

Imagine a settlement of people living high in the Andes. They have moved from the plains into a niche where they are not exactly best suited—the air is thin. For the thousands of years they have lived there, their hearts and lungs—their bodies—have had to work overtime to keep up with the altitude. If a "freak" should be born in their village, one whose body has a genetically more proficient way to handle the stress of high altitudes—say, a better hemoglobin variety rather than faster heartbeat—then the freak has an advantage. If the freak has children, then this trait could potentially spread through the village over generations because it is an advantage to lower stress on the heart and lungs. By the usual Darwinian dynamics of natural selection, the mutation of altitude acclimation comes to dominate the village gene pool.

On the surface there appears to be nothing but classical Darwinism at work here. But in order for Darwinian evolution to take place, the organism first had to survive in the niche for many generations without the benefit of genetic change. *Thus it was the flexibility or the body that kept the population surviving long enough for the mutation to arise and fix itself in the gene.* An adaptation spearheaded by the body (a somatic adaptation) is assimilated over time by the genes. Theoretical biologist C. H. Waddington called this transfer "genetic assimilation." Cyberneticist Gregory Bateson called it "somatic adaptation." Bateson likened it to legislative change in society—first a change is made by the people, then it is made law. Writes Bateson, "The wise legislator will only rarely initiate a new rule of behavior; more usually he will confine himself to affirming in law that which has already become the custom of the people." In the technical literature, this genetic affirmation is also known as the Baldwin effect, after J. M. Baldwin, a psychologist who first published the idea as a "New Factor in Evolution" in 1896.

Let's say there is this other village in the mountains, this time in the Himalayas, in a valley called Shangri La, whose residents' bodies are able to acclimatize up to 30,000 feet—10,000 more than the Andes folks—but who are also able to live at sea level. Over generations a mutation spreads to hardwire this talent into the villagers' genes, just as it did in the Andes. Of the two alpine villages, the Himalayan population now has a body type that is more stretchable, more flexible, and therefore, in essence, more evolutionarily adaptable. It may seem like a textbook example of Lamarckism, but giraffes who can evolve the most stretch in their necks can stake out an adaptation with their bodies long enough for their genes to catch up. As long as they keep their hides adjustable to a wide range of stresses, they'll have a competitive advantage in the long run.

The evolutionary moral is that it pays to invest in a flexible phenotype. It makes better sense to keep an adaptable body in service than to have a rigid body wait around for a mutation to pop up anytime an adaptation is needed. But somatic flexibility is "expensive." An organism cannot be equally flexible everywhere, and accommodating one stress will decrease its ability to accommodate another. Hardwiring is more efficient, but it takes time; for hardwiring to work, the stress must remain constant over a long period. In a rapidly changing environment, the tradeoff favors keeping the body flexible. An agile body can foreshadow, or more accurately, try out possible genetic adaptations, and then hold a steady line to them, as a hunting dog holds to a grouse.

But the story is even more radical than it appears because it is *behavior* that moves the body. The giraffe had to first want (for whatever giraffey reasons) higher leaves, and then had to reach for them over and over again. The humans had to choose to move to more alpine villages. By behavior, an organism can scout its options, and explore its space of possible adaptations.

Waddington said genetic assimilation, or the Baldwin effect, was about converting acquired traits into inherited traits. What it really comes down to is the natural selection of traits *controls*. Genetic assimilation bumps up the reach of evolution a notch. Instead of being able to tune the dial to the best trait, somatic and behavioral adaptation gives evolution quicker control over what the dials are and how far and in what direction they turn.

Behavioral adaptation works in other ways, too. Naturalists have verified that animals are constantly roaming out of their adapted environment and taking up homes in areas where they "don't belong." Coyotes creep too far south, or mockingbirds migrate too far north. And then, they stay. Their genes endorse the change by assimilating an adaptation which began, perhaps, as a vague desire.

What begins as vague desire can skate dangerously close to the edge of classical Lamarckism when it reaches individual learning. One species of finch learned to pick up a cactus needle to poke for insects. By this behavior the finch opened up a new niche to itself. By learning—perceived as a deliberate act—it altered its evolution. It is entirely possible, if not probable, that its learning will affect its genes.

Some computerists use the term "learning" in a loose, cybernetic sense. Gregory Bateson described the flexibility of the body as a type of learning. He saw little in its effect to distinguish the kind of search the body performed from the kind of search that either evolution or mind did. By this reckoning, a flexible body *learns* to acclimatize to stresses. "Learn" means adaptation within a lifetime instead of over lifetimes. The computerists make no real distinction between behavioral learning and somatic learning. What matters is that both types of adaptation search the fitness space *within the lifetime of an individual.*

An organism has great room to reshape itself within its lifetime. Robert Reid, at the University of Victoria, Canada, suggests that organisms can respond to environmental change with the following types of plasticity:

- Morphological plasticity
  (An organism can have more than one body form.)
- Physiological adaptability
  (An organism's tissues can modify themselves to accommodate stress.)
- Behavioral flexibility
  (An organism can do something new or move.)
- Intelligent choice
  (An organism can choose, or not, based on past experiences.)
- Guidance from tradition
  (An organism can be influenced or taught by others' experiences.)

Each of these freedoms is a front along which the organism can search for better ways to refit itself in a coevolutionary environment. In the sense that they are adaptations within a lifetime which can later be assimilated, we can call these five options, five varieties of inheritable learning.

Only in the last couple of years has the exhilarating link between learning, behavior, adaptation, and evolution even begun to be investigated. Most of this exciting work has been performed in computer simulations. It has been more or less ignored by biologists—which is not the stigma it once was. A number of researchers such as David Ackley and Michael Littman (in 1990), and Geoffrey Hinton and Steven Nowlan (in 1987) have shown clearly and unequivocally how a population of organisms that are learning—that is, exploring their fitness possibilities by changing behavior—evolve faster than a population that are not learning. In the words of Ackley and Littman, "We found that learning and evolution together were more successful than either alone in producing adaptive populations that survived to the end of our simulation." Their organism's exploratory learning is essentially a random search of a fixed problem. But in December 1991, two researchers, Parisi and Nolfi, presented results at the First European Conference on Artificial Life which showed that self-guided learning—where the problem task is selected by the population themselves—produced optimal rates of learning, which in turn may increase adaptation. They make a bold claim, which will be heard more and more in biology, that behavior and learning are among the *causes* of genetic evolution.

There is a further caveat. Hinton and Nowlan surmise that Baldwinism most likely works only on severely "rugged" problems. They say, "For biologists who believe that evolutionary spaces contain nice hills . . . the Baldwin effect is of little interest, but for biologists who are suspicious of the assertion that the natural search spaces are so nicely structured, the Baldwin effect is an important mechanism that allows adaptive processes within the organism to greatly improve the space in which it evolves." The organism creates its own possibilities.

"The problem with Darwinian evolution," Michael Littman told me, "is that it is great if you have evolutionary time!" But who can wait a million years? In the collective effort to introduce artificial evolution into manufactured systems, one way to accelerate the speed at which things evolve is to add learning to the soup. Artificial evolution will probably require a certain amount of artificial learning and intelligence to make it happen within human time scales.

Learning plus evolution is basically the recipe for culture. It may be that just as learning and behavior can pass off their information to genes, genes can pass their information off onto learning and behavior. The former is called genetic assimilation; the latter, cultural assimilation.

Human history is a story of cultural takeover. As societies develop, their collective skill of learning and teaching steadily expropriates similar memory and skills transmitted by human biology.

In this view—which is a rather old idea—each step of cultural learning won by early humankind (fire, hammer, writing) prepared a "possibility space" that allowed human minds and bodies to shift so that some of what it once did biologically would afterwards be done culturally. Over time the biology of humans became dependent on the culture of humans, and more supportive of further culturalization, since culture assumed some of biology's work. Every additional week a child was reared by culture (grandparent's wisdom) instead of by animal instinct gave human biology another chance to irrevocably transfer that duty to further cultural rearing.

Cultural anthropologist Clifford Geertz sums up this hand-off:

> The slow, steady, almost glacial growth of culture through the Ice Age altered the balance of selection pressures for the evolving Homo in such a way as to play a major directive role in his evolution. The perfection of tools, the adoption of organized hunting and gathering practices, the beginnings of true family organization, the discovery of fire, and most critically, though it is as yet extremely difficult to trace it out in any detail, the increasing reliance upon systems of significant symbols (language, art, myth, ritual) for orientation, communication, and self-control all created for man a new environment to which he was then obliged to adapt . . . We were obliged to abandon the regularity and precision of detailed genetic control over our conduct . . .

But if we consider culture as its own self-organizing system—a system with its own agenda and pressure to survive—then the history of humans gets even more interesting. As Richard Dawkins has shown, systems of self-replicating ideas or memes can quickly accumulate their own agenda and behaviors. I assign no higher motive to a cultural entity than the primitive drive to reproduce itself and modify its environment to aid its spread. One way the self-organizing system of culture can survive is by consuming human biological resources. And human bodies often have legitimate motivation in surrendering certain jobs. Books relieve the human mind of long-term storage rents, freeing it up for other things, while language compresses awkward hand-waving communication into a thrifty, energy conserving voice. Over generations of society, culture would assimilate more of the functions and information of organic tissue. Sociobiologists E. O. Wilson and Charles Lumsden used mathematical models to arrive at what they call the "thousand-year rule." They calculated that cultural evolution can pull along significant genetic change so that it catches up in only a thousand years. They speculate that the vast changes we have seen in our culture over the last millennium could have some foundation in genetic change, even though genetic change might not be visible.

So tightly coupled are genes and culture, Wilson and Lumsden say, that "genes and culture are inseverably linked. Changes in one inevitably force changes in the other." Cultural evolution can shape genomes, but it can also be said that genes must shape culture. Wilson believes that genetic change is a *prerequisite* for cultural change. Unless the genes are flexible enough to assimilate cultural change, he believes it will not take root for the long term.

Culture follows our bodies, while our bodies follow culture. In the absence of culture, humans seem to lose distinctly human talents. (As somewhat unsatisfactory evidence we have the failures of "wolf children" raised by animals to develop into creative adults.) Culture and flesh, then, meld into a symbiotic relationship. In Danny Hillis's terminology, civilized humans are "the world's most successful symbionts"—culture and biology behaving as mutually beneficial parasites for each other—the coolest example of coevolution we have. And as in all cases of coevolution, it implies positive feedback and the law of increasing returns.

Cultural learning rewires biology (to be precise, it allows biology to remodel itself) so that biology becomes susceptible to further culturalization. Thus, culture tends to accelerate itself. In the same way that life begets more life and more kinds of life, culture begets more culture and more kinds of culture. I mean it in a strong way, that culture produces organisms that are biologically more able to produce, learn, adapt in cultural ways, rather than biological ways. This implies that the reason we have brains that can produce culture is that culture produced brains that could. That is, whatever shred of culture resident in prehuman species was instrumental in molding offspring to produce more culture.

To the human body this accelerating evolution towards an information-based system looks like biological atrophy. From the view of books and learning, it looks like self-organization, culture amplifying itself at the expense of biology. Just as life infiltrates matter mercilessly and then hijacks it forever, cultural life hijacks biology. In the strong sense I'm advocating here, culture modifies our genes.

I have absolutely no biological evidence for all this. I've heard casual things from folks like Stephen Jay Gould who says the "morphology of humans hasn't changed in the 25,000 years from Cro-Magnon," but I don't know what that means for this idea, and how true his assertion is. On the other hand, devolution is weirdly quick. Lizards and mice can lose their eyesight in a blink (so to speak) inhabiting lightless caves. Flesh, it seems to me, is ever ready to give up part of its daily grind if given a chance.

My larger point is that the advantages of Lamarckian evolution are so great that nature *has* found ways to make it happen. In Darwin's metaphor I would put its success this way: Evolution daily scrutinizes the world not just to find fitter organisms, but to find ways to increase its own ability. It hourly seeks to gain an edge in adaptation. Its own ceaseless pushing creates an immense pressure—like the weight of an ocean seeking a crack to seep through—to increase its adaptive abilities. Evolution searches the surface of the planet to find ways to speed itself up, to make itself more nimble, more evolvable—not because it is anthropomorphic, but because the speeding up of adaptation is the runaway circuit it rides on. It searches for the advantages of Lamarckian evolution without realizing it because Lamarckism is a crack of less resistance and more evolvability.

When animals with complex behavior evolved, evolution began to break out of its Darwinian straight jacket. Animals could react, choose, migrate, adapt, and give room for the blossoming of pseudo-Lamarckian evolution. As human brains evolved, they created culture, which permitted the birth of a true Lamarckian system of inherited acquisitions.

Darwinian evolution is not just slow learning. In Marvin Minsky's words, "Darwinian evolution is *dumb* learning." What evolution later found in primitive brains is a way to quicken itself by introducing learning into the equation. What evolution eventually found in the human brain was the complexity needed to peer ahead in anticipation and direct evolution's course.

Evolution is a structure of organized change. But it is more. Evolution is a structure of organized change which is itself undergoing change and reorganization.

Evolution on Earth has already undergone structural changes in its four-billion-year lifespan and will probably undergo more. The evolution of evolution can be summed up by the following series of historical evolution types:

1) Auto-genesis of systems
2) Replication

3) Genetic control
4) Somatic plasticity
5) Memetic culture
6) Self-directed evolution.

In the prebiotic conditions of early Earth, before there was any life to evolve, the dynamics of evolution favored the survival of anything stable. (There is a Uroboric tautology lurking here because in the very beginning stability *is* survival.)

Stability permitted evolution to operate longer, and so stability allowed evolution to generate further stability. We know from the work of Walter Fontana and Stuart Kauffman (*see* chapter 20) that a fairly straightforward chemistry of simple compounds which can catalyze their own production results in a kind of chemical self-supporting ring. The first stage of evolution was thus the evolution of a matrix of self-generating complexity, which gave evolution a population of persistent things to work on.

At the next stage, evolution evolved *self-replicating* stabilities. Self-reproduction provided the possibility of errors and variation. Evolution then evolved natural selection and unleashed its remarkable search power.

Next, the mechanics of inheritance split from mechanics of survival, and evolution evolved the dual system of genotype and phenotype. By allowing a compact genotype to describe huge libraries of possible forms, evolution entered into a vast space to operate within.

As evolution evolved more complex body forms and behaviors, it made bodies that reshaped themselves and animals that chose their own niches. These choices opened up the space of bodily "learning" for evolution to evolve further.

Learning hastened the next step which was the evolution of a complex symbolic learning machine—the human brain. Human thinking evolved culture and memetic (idea) evolution. Evolution could now accelerate itself in a self-aware and "smarter" way through a vast new library of possibilities. This is the stage of history we are at now.

God only knows where evolution may evolve next. Will human-made artificial evolution set the stage for another realm of evolution? The obvious course that evolution seems bound to hit sooner or later is self-direction. In *self-direction*, evolution itself chooses where it wants to evolve. This is not discussed by biologists.

I prefer to rephrase this history and say that evolution has been, and will keep on, exploring the space of possible *evolutions*. Just as there is a space of possible pictures, a space of possible biological forms, and a space of possible computations, there is also a space—how large we don't know—of ways to explore spaces. This metaevolution, or hyperevolution, or deep evolution, or perhaps even ultimate evolution, wanders the landscape of all possible evolutionary games looking for the trick that will allow it to complete its search of all possible evolutions.

Organisms, memes, biomes—the whole ball of wax—are only evolution's way to keep evolving. What evolution really wants—that is, where it is headed—is to uncover (or create) a mechanism that will most quickly uncover (or create) possible forms, things, ideas, processes in the universe. Its ultimate goal is not only to create forms, things, and ideas, but to create new ways in which new things are found or created. Hyperevolution does this by bootstrapping itself into a layered strategy that continually increases its reach, continually creates new libraries of possible places to explore, and continually searches for better, more creative ways to create.

That sounds like fiddle-faddle double-talk, but I don't know any less recursive way to say it. Perhaps: Evolution's job is to create all possible possibilities by creating the spaces in which they could be.

THE BALD CONCEPT OF EVOLUTION is so powerful and universal that at times it seems to touch everything. The pioneer geneticist Theodosius Dobzhansky writing in *Mankind Evolving*, says:

> Is evolution a theory, a system, or a hypothesis? It is much more—it is a general postulate to which all theories, all hypotheses, all

systems must henceforth bow and which they must satisfy in order to be thinkable and true. Evolution is a light which illuminates all facts, a trajectory which all lines of thought must follow—this is what evolution is.

Evolution's role to explain everything, however, stains it with a tinge of religiosity. As Bob Crosby of the Washington Evolutionary Systems Society unabashedly says, "Where other people see the hand of God, we see evolution."

Much can be said of viewing evolution as a religion. Evolution theory's framework is encompassing, rich, almost self-evident, inarguable, and it has now spawned local home fellowships that meet monthly, as Crosby's large group does. Author Mary Midgley begins her slim and wonderful monograph *Evolution as a Religion*, with these four sentences: "The theory of evolution is not just an inert piece of theoretical science. It is, and cannot help being, also a powerful folk-tale about human origins. Any narrative must have symbolic force. We are probably the first culture not to make that its main function."

Her arguments are not against the veracity of evolutionary theory in the least, but rather against the idea that we can divorce the logical aspects of evolution from all the other things this powerful notion does to us as humans.

It is the unexamined consequences of evolution—however it comes about, and wherever it is headed—that I believe will shape our future in the long term. I don't doubt that our discoveries about the hidden nature of deep evolution will also touch our souls.

# 19

# POSTDARWINISM

"IT IS TOTALLY WRONG. It's wrong like infectious medicine was wrong before Pasteur. It's wrong like phrenology is wrong. Every major tenet of it is *wrong*," said the outspoken biologist Lynn Margulis about her latest target: the dogma of Darwinian evolution.

Margulis has been right about what is wrong before. She shook up the world of microbiology in 1965 with her outrageous thesis of the symbiotic origin of nucleated cells. To the disbelief of traditionalists, she claimed that free-roaming bacteria cooperated to form cells. Then in 1974, Margulis again rattled the cage of biology by suggesting (jointly with James Lovelock) that atmospheric, geological, and biological processes on Earth are so interconnected that they act as a single living, self-regulating system—Gaia. Margulis was now denouncing the modern framework of the century-old theory of Darwinism, which holds that new species build up from an unbroken line of gradual, independent, random variations.

Margulis is not alone in challenging the stronghold of Darwinian theory, but few have been so blunt. Disagreeing with Darwin resembles creationism to the uninformed; therefore the stigma that any taint of creationism can bring to a scientific reputation, coupled with the intimidating genius of Darwin, have kept all but the boldest iconoclasts from doubting Darwinian theory in public.

What excites Margulis is the remarkable *incompleteness* of general Darwinian theory. Darwinism is wrong by what it omits and by what it incorrectly emphasizes.

A number of microbiologists, geneticists, theoretical biologists, mathematicians, and computer scientists are saying there is more to life than Darwinism. They do not reject Darwin's contribution; they simply want to move beyond it. I call them the "postdarwinians." Neither Lynn Margulis nor any other postdarwinian denies the true ubiquity of natural selection in evolution. Their disagreement is with the very sweeping nature of the Darwinian argument, the fact that in the end it doesn't explain much, and the emerging evidence that Darwinism alone may not be sufficient to explain all we see. The vital questions the postdarwinians raise are: What are the limits to natural selection? What can't evolution make? And if blind natural selection has limits, what else is operating within or beyond evolution as we understand it?

According to the ordinary contemporary Darwinian biologist, there is nothing we see in nature that cannot be explained by the elemental process of natural selection. In academic jargon this stance is called selectionism, and the position is nearly universal among biologists working today. Because this stance is more extreme than what Darwin himself believed, it is sometimes called neodarwinism.

In the pursuit of artificial evolution, the limits (if any) to natural selection, or to evolution in general, take on practical importance. We'd like an artificial evolution that generates neverending diversity, but so far, that isn't so easy to do. We'd like to extend the dynamics of natural selection to very large systems with many levels of scale, but we don't know how far natural selection can be extended. We'd like an artificial evolution that we could control a bit more than we control organic evolution. Is that possible?

Questions like these have prompted the postdarwinians to reconsider alternative theories of evolution—many that existed before Darwin—that were eclipsed by the dominance of Darwinism. In a kind of intellectual survival of the fittest, contemporary biology places very little importance on these "inferior" beaten theories, so they survive only in marginal out-of-print books. But the ideas of these creative theories are suited to a new niche called artificial evolution and are cautiously being resurrected for examination.

The most stellar naturalists, geologists, and biologists of Darwin's time hesitated (despite Darwin's constant badgering) to accept his general theory in full when it was published in 1859. They accepted his transmutation theory—"descent with modification," or the gradual transmutation of new species from preexisting species. But they remained skeptical of his selectionist reasoning—that tiny random improvements were all there was to it—because they felt Darwin's explanation did not accurately fit the facts of nature, facts with which they were intimately familiar in a way that is rare today in this era of specialization and indoor laboratories. But since they could offer neither compelling disproof nor an alternative theory of equal quality, their forceful criticisms were buried in correspondence and scholarly disputes.

Darwin didn't offer a concrete mechanism by which his proposed natural selection would take place, either. He was ignorant about genes, for starters. The first fifty years following the publication of Darwin's tour de force were ripe with supplemental theories of evolution, until Darwin's dominance was clinched by the discovery of genes and later DNA. Almost every radical evolutionary conviction circulating today has as its source some thinker in the years after Darwin but before acceptance of his theory as dogma.

No one was more sensitive to the weaknesses of Darwinian theory than Darwin himself. As an example of trouble, Darwin volunteered the astounding multifaceted sophistication of the human eye. (Every critic of Darwin since has also used his example.) The exquisite design of interacting lens, iris, retina, etc., seems to defy the plausibility of Darwin's "slight, incremental" chance improvements. As Darwin wrote to his American friend Asa Gray, "About the weak points I agree. The eye to this day gives me a cold shudder." The difficulty Gray had was imagining how any portion of an unfinished eye, a retina without lens or vice versa, would be useful to its possessor. Since nature cannot hoard innovations ("Hey, this will come in handy in the Cretaceous!"), every stage in development must be immediately useful and viable. Breakthroughs have to work the first time. Even clever humans can't design in such a consistently demanding manner. Therefore nature appears superhuman in its ability to create.

Imagine, says Darwin, that we extrapolate the tiny microevolutionary changes we see in domesticated breeding—a pea with extra-large pods made larger, or a short horse bred shorter. Imagine if we extend those slight changes caused by selection over millions of years; we add up all the minute differences until we see major change. This is what makes coral reefs and armadillos out of bacteria, Darwin said—accumulated microchange. Darwin asks that we extend the logic of microchange to cover the grand scale of Earth and Time.

The argument that natural selection can be extended to explain everything in life is a *logical* argument. But human imagination and human experience know that what is logical is not always what is so. To be logical is a necessary but insufficient reason to be true. Every swirl on a butterfly wing, every curve of leaf, every species of fish is explained by adaptive selection in neodarwinism. There seems to be absolutely nothing that cannot be explained in some way as an adaptive advantage. But, as Richard Lewontin, a renowned neodarwinist, says, "Natural selection explains nothing, because it explains everything."

Biologists cannot (or at least they have not) ruled out the role of other forces at work in nature producing similar effects in evolution. Therefore, until evolution is duplicated under controlled conditions, in the wild, or in a lab, neodarwinism remains a nice "just-so" story—more like history than science. Philosopher of science Karl Popper said bluntly that neodarwinism is not a scientific theory at all, since it cannot be falsified. "Neither Darwin, nor any Darwinian, has so far given an actual causal explanation of the adaptive evolution of any single organism or any single organ. All that has been shown—and this is very much [sic]—is that such an explanation might exist—that is to say, [these theories] are not logically impossible."

Life has a causality problem. Any coevolved organism seems to be self-created, making causality onerous to pin down. Part of the search for more complete explanations of evolution is a search for a more complete logical understanding of spontaneous complexity and the rules by which entities may emerge from a web of parts. The quest for artificial evolution—so far done primarily in computer simulations—is very much tied into a new way of establishing proof in science. Previous to the advent of ubiquitous computers, science consisted of two facets: theory and experiment. A theory would shape an experiment, and then the experiment would confirm or disprove the theory.

But computers have birthed a third way of doing science: by simulation. A simulation is at once both a theory and an experiment. By running a computer model, such as Tom Ray's artificial evolution, we are trying out a theory and also running something real and accumulating falsifiable data. It may be that the dilemma of ascertaining causality in complex systems will be bypassed by these new methods of understanding, wherein one studies the real by modeling working surrogates.

Artificial evolution is at once a theory and test for natural evolution, and something original in itself.

AROUND THE WORLD, a few naturalists are conducting long-term observations of evolving populations of organisms in the wild: snails in Tahiti, fruitflies in Hawaii, finches in the Galapagos, and lake fish in Africa. Every year that these studies go on, there is a better chance that scientists can unequivocally demonstrate long-term evolution *in action* in the field. Shorter-term studies using bacteria, and recently flour beetles, show short-term evolution of organisms in the lab. So far, these experiments with populations of living creatures have matched the results expected from neodarwinian theory. The beaks of finches in the Galapagos really do thicken over time in response to drought-induced changes in their food supply, just as Darwin predicted.

These careful measurements prove that self-governing adaptation does spontaneously occur in nature. They also unequivocally demonstrate that noticeable change can emerge on its own by summing up the steady unnoticeable work of incremental deletions of the unfit. But the results do not show new levels of diversity, new kinds of creatures, or even new complexity emerging.

Despite a close watch, we have witnessed no new species emerge in the wild in recorded history. Also, most remarkably, we have seen no new animal species emerge in domestic breeding. That includes no new species of fruitflies in hundreds of millions of generations in fruitfly studies, where both soft and harsh pressures have been deliberately applied to the fly populations to induce speciation. And in computer life, where the term "species" does not yet have meaning, we see no cascading emergence of entirely new *kinds* of variety beyond an initial burst. In the wild, in breeding, and in artificial life, we see the emergence of variation. But by the absence of greater change, we also clearly see that the limits of variation appear to be narrowly bounded, and often bounded within species.

The standard explanation is that we are measuring a geological event in real time on a ridiculously infinitesimally small time span, so what do we expect? Life was bacterialike for billions of years before much happened. Patience, please! This is why Darwin and other biologists turned to the fossil record for proof of evolution. And although the fossil record indisputably exhibits Darwin's larger thesis—that over time modification of form is accumulated in descendants—the fossil record has not proved that this change is due solely or even primarily to natural selection.

No one has yet witnessed, in the fossil record, in real life, or in computer life, the exact transitional moments when natural selection pumps its complexity up to the next level. There is a suspicious barrier in the vicinity of species that either holds back this critical change or removes it from our sight.

Stephen Jay Gould believes the exact transformation periods are removed from the sight of the fossil record by their incredibly instantaneous (evolutionarily speaking) mode. Whether his theory is correct or not, the evidence points to a natural limiting factor for extrapolated microchange that must somehow be overcome by evolution.

Synthetically reproduced protolife and artificial evolution in computers have already unearthed a growing body of nontrivial surprises. Yet artificial life suffers from the same malaise that afflicts its cousin, artificial intelligence. No artificial intelligence that I am aware of—be it autonomous robot, learning machine, or massive cognition program—has run more than 24 hours in succession. After a day, artificial intelligence stalls. Likewise, artificial life. Most runs of computational life fizzle out of novelty quickly. While the programs sometimes keep running, churning out minor variation, they ascend to no new levels of complexity or surprise after the first spurt (and that includes Tom Ray's world of Tierra). Perhaps given more time to run, they would. Yet, for whatever reason, computational life based on unadorned natural selection has not seen the miracle of open-ended evolution that its creators, and I, would love to see.

As the French evolutionist Pierre Grasse said, "Variation is one thing, evolution quite another; this cannot be emphasized strongly enough . . . Mutations provide change, but not progress." So while natural selection may be responsible for microchange—a trend in variations—no one can say indisputably that it is responsible for macrochange—the open-ended creation of an unexpected novel form and progress toward increasing complexity.

Many of the promises for artificial evolution foretold in this book will still come about if artificial evolution is merely adaptive microchange. Spontaneously directed variation and selection is an incredibly powerful problem solver. Natural selection indeed works over the immediate short term. We can use it to find what we can't see and fill in what we can't imagine. The question comes down to whether random variation and selection are sufficient alone to produce ever increasing novelty over the very long term. And if "natural selection is not enough" then what else might be at work in wild evolution, and what may we import into artificial evolution that will generate self-organizing complexity?

Most critics of natural selection concede that Darwin got

"survival of the fittest" right. Natural selection primarily means the destruction of the unfit. Once fitness is created, natural selection is peerless for winnowing out the duds.

But creating something useful is the bugaboo. What the Darwinian perspective neglects is a plausible explanation for the origin of fitness. Where does fitness come from before it is selected? In the popular rendition of neodarwinism today, the origin of fitness is credited to random variation. Random variation within chromosomes produces a random variation in the developmental growth of the organism, which every now and then bestows increased fitness on the whole organism. Fitness is generated randomly.

As experiments in wild and artificial evolution have shown, this simple process can steer coordinated change over the short time. But given that natural selection weeds out all the uncountable failures, and that there is uncountable time, can random mutation *generate* the unbroken series of needed winners for selection to choose from? Darwinian theory has the sizable burden of proving that the negative, braking power of selective demise, coupled with the blind chaotic power of randomness, can produce the persistent, creative, positive drive toward more complexity we see sustained in nature over billions of years.

Postdarwinism suggests that other forces are at work in evolution in the long run. These lawful mechanisms of change reorganize life into new fitnesses. These unseen dynamics extend the Library in which natural selection may operate. This deepened evolution need not be any more mystical than natural selection is. Think of each dynamic—symbiosis, directed mutation, saltationism, self-organization—as a mechanism that will foster evolutionary innovation over the long term in complement to Darwin's ruthless selection.

Symbiosis —the merger of two organisms into one—was once thought to occur only in isolated curiosities like lichens. After

Lynn Margulis postulated bacterial symbiosis as a central event in the formation of the ancestral cell, biologists found symbiosis popping up frequently in microbial life. Since microbial life is (and has always been) the bulk of all life on Earth, and the primary Gaian workhorse, widespread microbial symbiosis makes symbiosis fundamental, both in the past and in the present.

In contrast to the traditional picture of a population seething with tiny, random, incremental changes in their routine until they hit upon a stable new configuration, Margulis would have us consider the accidental merging of two working simple systems into one larger, more complex system. As illustration, a proven system for oxygen transport inherited by one cell line might be married to an existing system for air exchange in another cell line. Combined in symbiosis, the two might form a respiratory system unlikely to develop incrementally.

For a historical example, Margulis suggests her own studies on the symbiotic nature of nucleated cells. These emerging cells did not have to reinvent by trial and error over a billion years the clever processes of photosynthesis and respiration worked out by several types of bacteria. Instead, the membraned cells incorporated the bacteria and their informational assets as wholly owned subsidiaries working for the cells. They kidnapped the innovations.

In some cases the genetic strands of two symbiotic partners may fuse. One proposed mechanism for the informational coordination needed for this kind of symbiosis is the known intercell gene transfer, which happens at a terrific rate among bacteria in the wild. The know-how of one system can be shuttled back and forth between separate species. A new bacteriology views all the bacteria of the world as a single genetically interacting superorganism that rapidly absorbs and broadcasts genetic innovations among its members. Interspecies gene transfer also occurs (at an unknown rate) among more complex species, including humans. Species of every sort are constantly swapping genes, often with naked viruses as the messengers. Viruses themselves are sometimes taken in symbiotically. A number of biologists believe that large chunks of human DNA were inserted viruses. A few even think that it's a loop—that many human disease viruses are escaped hunks of human DNA.

If true, the symbiotic nature of a cell provides a couple of lessons. First, it gives an example of a significant evolutionary change that lessens immediate benefits to the individual (since the individual disappears), in contradiction to classical Darwinian dogma. Second, it gives an example of evolutionary change that is not amassed by slight incremental differences, also in contradiction to Darwinian dogma.

Routine symbiosis on a large scale could drive many of the complexities in nature that seem to require multiple simultaneous innovations. It would provide evolution with several other advantages; for instance, it would exploit the power of cooperation, rather than competition, exclusively. At the very least, cooperation nurtures a distinct set of niches and a type of diversity that competition cannot produce—such as lichens. In other words, it unleashes another dimension in evolution by enlarging its library of forms. Also, a small amount of symbiotic coordination at the right time could replace an eon of minor alterations. In one mutual relationship, evolution could jump past a million years of individual trial and error.

Perhaps evolution could have discovered nucleated cells directly, without symbiosis, but it might have taken another billion years, or five, to do so. Lastly, symbiosis recombines widely diverse know-how separated in life's divergent genealogy. The picture to keep in mind is the diagrammatic tree of life, with ever dividing, ever spreading branches. Symbiotic alliances, on the other hand, bring divergent branches of the tree of life together again, to intersect. Evolution, charted with symbiosis included, may resemble a briar patch more than a tree—the Thicket of Life. If the Thicket of Life is sufficiently tangled, it may require a rethinking of our past and future.

NATURAL SELECTION is a very grim natural reaper. Darwin made the bold claim that, at the very heart of evolution, many small deletions in bulk—many small wanton deaths—feeding on the throwaway optimism of minor variation, could, in a counter-intuitive way, add up to something truly new and meaningful. In the drama of traditional selection theory, death plays the star role. It works single-mindedly by attrition. It is an editor that knows only one word: "No." Variation counterbalances the one-note song of death by giving birth to the new in cheap abundance. It too knows only one word: "Maybe." Variation cranks out disposable "maybes" in bulk, which are immediately mowed down by death. Bulk mediocrity is dismissed by wanton death. Occasionally, the theory goes, this duet produces a "Yes!"—a starfish, kidney cells, or Mozart. On the face of it, evolution by natural selection is still a startling hypothesis.

Death gives room for the new, it eliminates the ineffective. But to say that death *causes* wings to be formed, or eyeballs to work, is essentially wrong. Natural selection merely selects away the deformed wing, the unseeing eye. "Natural selection is the editor, not the author," says Lynn Margulis. What, then, authors innovation in flight and sight?

Evolution theory, from Darwin on, has had a dismal record in dealing with the origin of innovation. As his book title made clear, the question of the origin of species was the great riddle Darwin hoped to solve, not the origin of individuality. He asked, Where did new kinds of creatures come from? He did not ask, Where did variation among individuals come from?

Genetics, which began as a distinctly separate field of science, did pay attention to variation and origin of innovation. Early geneticists like Mendel and William Bateson (Gregory Bateson's father and the man who coined the term "genetics") struggled with explanations of how variations arose and were passed on to descending generations. Sir Francis Galton showed that for statistical purposes—the main bent of genetics until bioengineering came along—the propagation of variation within populations could be considered to have a random origin.

Later, when the mechanism for heredity was discovered to be a code of four symbols strung on a long chain of molecules, the random flip of a symbol at a random point on the thread was easy to visualize as a cause of variation and easy to model in mathematics. These molecular flips are generally attributed to cosmic rays or thermodynamic noise. A monstrous mutation, once implying freakish severity, was newly seen as simply a flip, a mere deviation from the average variation. It was not long before all variations in an organism—from freckles to cleft palates—were treated as statistical degrees of mutational error. Variation thus became mutation and "mutation" became inseparably compounded into "random mutation." Today, the term "random mutation" seems redundant. What other kind of mutation could there possibly be?

In computer-intensive artificial evolution, mutations are manufactured by electronic, pseudo-random generators. But the exact nitty-gritty origins of mutations and variations in biology are still uncertain. We do know this: variation is emphatically not due to *random* mutation—at least not always; it has some measure of order. This is an old idea. As early as 1926, theorist Jan Christiaan Smuts gave this genetic semi-order a name: internal selection.

A plausible scenario for internal selection allows cosmic rays to produce supposedly random errors in the DNA code, which are then corrected in cells by a known self-repair apparatus working in a discriminate (but unknown) fashion—correcting some and passing others. There is a high energetic cost to the correction of errors, a cost which must be weighed against the possible benefit of the variations. If the error occurred where it is probably opportune, it stays; if it occurs where it is bothersome, it is corrected. For a hypothetical example, the Krebs cycle is the basic fuel plant in every cell of your body. It has worked fine for hundreds of millions of years. There is simply too little to gain, and far too much to lose, in fiddling with it now. When a variation is detected in the code for the Krebs cycle, it is quickly extinguished. On the other hand, body size and body proportions might be worth tweaking; let's leave that area open to variation. If this were how it worked, differential variation would mean that some randomness is "more equal" than others. One fascinating consequence of this setup is that a mutation in the regulatory apparatus itself could have a large-scale effect far beyond a mutation in the strings it governs. I'll get back to that later.

Because genes interact and regulate each other so extensively, the genome forms a complex whole that resists change. Only certain areas can vary at all because most of the genes are so interdependent upon each other—almost grid-locked—that variation is not a choice. As evolutionist Ernst Mayr puts it, "Free variability is found only in a limited portion of the genotype." The power of this genetic holism can be seen in animal breeding. Breeders commonly encounter undesirable side effects triggered when unknown genes are activated in the process of selecting for one particular trait. However, when pressure for that one trait is let up, organisms in succeeding generations rapidly revert to the original type, much as if the genome has sprung back to its set point. Variation in real genes is quite different than we imagined. The evidence suggests that not only is it nonrandom and parochial, but it is difficult to come by at all.

The impression one gets is of a highly flexible bureaucracy of genes managing the lives of other genes. Most astounding, the same gene bureaucracy is franchised throughout life, from fruitfly to whale. For example, a nearly identical homeobox self-control sequence (a master-switch gene which turns hunks of other genes on) is found in every vertebrate.

So prevailing is the logic of nonrandom variation that I was at first flabbergasted in my failure to find any biologists working today who still believe mutations to be truly random. Their nearly unanimous acknowledgment that mutations are "not truly random" means to them (as far as I can tell) that individual mutations may be less than random—ranging from near-random to plausible; but they still believe that statistically, over the long haul, a mass of mutations behaves randomly. "Oh, randomness is just an excuse for ignorance," quips Lynn Margulis.

This weak version of nonrandom mutation is hardly even an issue anymore, but a stronger version is more of a juicy heresy. It says that variations can be chosen in a deliberate way. Rather than have the gene bureaucracy merely edit random variations, have it produce variations by some agenda. Mutations would be created by the genome for specific purposes. Direct mutations could spur the blind process of natural selection out of its slump and propel it toward increasing complexity. In a sense, the organism would direct mutations of its own making in response to environmental factors. Ironically, there is more hard lab evidence at hand for the strong version of directed mutation than for the weak version.

According to the laws of neodarwinism, the environment, and only the environment, can select mutations; and the environment can never induce or direct mutations. In 1988 Harvard geneticist John Cairns and colleagues published evidence of environmentally induced mutations in the bacterium *E. coli.* Their claim was audacious: that under certain conditions the bacteria spontaneously crafted needed mutations in direct response to stresses in their environment. Cairns also had the gall to end his paper by suggesting that whatever process was responsible for the directed mutations "could, in effect, provide a mechanism for the inheritance of acquired characteristics"—a bald allusion to Darwin's rival-in-theory Jean-Baptiste Lamarck.

Another molecular biologist, Barry Hall, published results which not only confirmed Cairns's claims but laid on the table startling additional evidence of direct mutation in nature. Hall found that his cultures of *E. coli* would produce needed mutations at a rate about 100 million times greater than would be statistically expected if they came by chance. Furthermore, when he dissected the genes of these mutated bacteria by sequencing them, he found mutations in no areas other than the one where there was selection pressure. This means that the successful bugs did not desperately throw off all kinds of mutations to find the one that works; they pinpointed the one alteration that fit the bill. Hall found some directed variations so complex they required the mutation of two genes simultaneously. He called that "the improbable stacked on top of the highly unlikely." These kinds of miraculous change are not the kosher fare of serial random accumulation that natural selection is supposed to run on. They have the smell of some design.

Both Hall and Cairns claim that they have carefully eliminated all other explanations for their results, and stick by their claim that the bacteria are directing their own mutations. However, until they can elucidate a mechanism for the way in which a stupid bacterium can become aware of which mutation is required, few other molecular geneticists are ready to give up strict Darwinism.

THE DIFFERENCE BETWEEN wild evolution in nature and synthetic evolution in computers is that software has no body. The kind of software you load with floppy disks is straightforward. If you alter the code (for the better, you hope), you execute the program and it fulfills its orders. There is nothing between what the code is and what it does, except the wiring of the machine it runs on.

Biology is vastly different. If we take a hypothetical hunk of DNA as software code, and alter it, there is a consequential body that must be grown before the effects of the alteration can manifest itself. The development of an animal from fertilized egg, to egg producer may take years to complete; so the effect of that alteration can be judged differently depending on the stage of the growth. The same initial alteration of code can have one effect on the growing microscopic fetus and another effect on the sexually mature organism, if it survives that long. In every case, between the code alteration and the terminal effect (say, longer fingers), there is a chain of intermediate bodies governed by physics and chemistry—the enzymes, proteins, and tissues of life—which also must be indirectly altered by the software change. This vastly complicates mutational variation. Programming computers is no longer an adequate comparison.

You were once the size of a period. For a brief time you tumbled about as a multicellular sphere, much like pond algae. Currents swept and washed over you. Remember? Then you grew. You became sponge life, tubular, all gut. To eat was life. You grew a spinal cord to feel. You put on gill arches in preparation to breathe and burn food with intensity. You grew a tail to move, to steer, to decide. You were not a fish, but a human embryo role-playing a fish embryo. At every ghost-of-embryonic-animal you slipped into and out of, you replayed the surrender of possibilities needed for your destination. To evolve is to surrender choices. To become something new is to accumulate all the things you can no longer be.

While evolution is inventive, it is also conservative, making do with what is available. Biology rarely starts over. It begins with the past, which is distilled in the development of the organism. By the time an organism arrives at the end of its natal development, the millions of tradeoffs it has incurred forever block the chance to evolve in certain other directions. Evolution without a body is limitless. Evolution with a body, wrapped in development and prevented from retreating by its current success, is bound by endless constraints. But these constraints give it a place to stand. It may be that for artificial evolution to get anywhere, it too may need to wear a body.

When there are bodies in space, there is time. Mutations bloom in a body grown—in time's dimension. (That's something else artificial evolution has little of so far: developmental time.) To alter development early in the embryo is to fiddle with time. The earlier a mutation expresses itself in embryonic development, the more forcefully it will resound through the organism. This also loosens the constraints against failure, so the earlier the mutation is in development, the less likely it will be workable. In other words, the more complex an organism becomes, the less likely a very early change will survive.

Early developmental change has the advantage that a small mutation can affect a suite of things in a single blow. An appropriate early tweak can invoke or erase ten million years of evolution. The famous Antennapedia mutant of the *Drosophila* fruitfly is an example. This single-point mutation engages the leg-making apparatus of the embryo fly to build a leg where its antenna should be. The afflicted fly is born with a fake foot sticking out of its forehead—all triggered by one tiny alteration of code, which in turn triggers a suite of other genes. All kinds of monsters can be hatched this way. Which leads developmental biologists to wonder if the self-regulating genes of an organism might be able to tweak the genes governing these early suites into useful freaks, thus bypassing Darwin's incremental natural selection.

The curious thing about monsters, though, is that they seem to follow internal laws. While a two-headed calf may seem to us to be randomly defective, it isn't. When biologists studied freaks they found that the same type of monstrosities appeared in many species, and that their freakishness could even be categorized. For instance, a cyclops—a relatively common freak in mammals, including humans—born with a single centrally positioned eye, will almost always have its nostrils located above its eye. This is true regardless of the species in which it appears. Similarly, two-headedness is much more common than three-headedness. Since neither mutation is a variation that offers reproductive advantage, since few of these freaks survive, natural selection cannot be selecting one over the other. This mutant order must be internally generated.

In the early and mid-19th century a French father and son team, Etienne and Isidore Geoffroy Saint Hilaire, devised a classification scheme for natural monsters. Their taxonomy of mutants paralleled the Linnean system of natural species: every monstrosity was assigned a class, order, family, genus, and even species. Their work became the foundation of the modern science of monsters—teratology. Orderly form, the Hilaires implied, extended beyond natural selection.

Pere Alberch, at the Museum of Comparative Zoology at Harvard, is the modern spokesman for the importance of teratology in evolutionary biology. He interprets teratologies as overlooked blueprints for strong internal self-organization within living organisms. He states, "Teratologies are a superb document of the potentiality of a given developmental process. In spite of strong negative selection, teratologies are not only generated in an organized and discrete manner but they also exhibit generalized transformational rules. These properties are not exclusive to teratology; rather they are general properties of all developmental systems."

The orderly makeup of monsters—it is after all a well-formed foot which erupts out of a mutant *Drosophila*'s forehead—speaks of a deep underlying internal force which helps guide the outward shape of organisms. This "internalist" approach differs from the orthodox "externalist" approach of most adaptationists who see ubiquitous natural selection as the major shaping force. As a dissenting internalist, Alberch writes:

> The internalist approach assumes, and this is a key assumption, that morphological diversity is generated by perturbations in parameter values (such as rates of diffusion, cell adhesion, etc. . . .) while the structure of the interactions among the components remains constant. Given this assumption, even if the parameters of the system are randomly perturbed, by either genetic mutation, environmental variance or experimental manipulation during development, the system will generate a limited and discrete subset of phenotypes. Thus the realm of possible forms is a property of the internal structure of the system.

Thus we have two-headed freaks for perhaps the same reason we have bilateral arms; most likely neither is due to natural selection. Rather, internal structure, particularly the structure of the genome, and the accumulated morphogenesis of development, may be an equal or greater influence upon the variety of biological organizations possible.

The bodies that genes wear play an incredible role in the gene's evolution. When two chromosomes recombine in sex they do so not in nakedness but clothed inside a gigantic egg cell. The overstuffed egg has a great deal of say in how the genes are implemented. The yolky cell is chock-full of protein factors and hormonelike agents, and controlled by its own nonchromosomal DNA. The egg cell directs the chromosomal genes as they begin to differentiate, guiding them, orienting them, and orchestrating the construction of their baby. It is no exaggeration to say that the final organism reproduced is partly under the control of the egg cell, and out of the control of the genes. The state of the egg cell can be affected by stress, age, nutrition, etc. (There is one claim that Down's Syndrome, common in babies born to older women, happens because the two chromosomes responsible for the birth defect become physically entangled by lying so close to each other for so many years in the mother's egg cell.) Even before you are born—indeed from the moments of conception onward—forces outside of your genetic information form you genetically. Hereditary information does not exist independently of its embodiment. The origin of an organism's inheritable body, or morphogenesis, is due then to a partnership of nongenetic cell material and hereditary genes—body and genes. Evolution theory, and in particular evolutionary genetics, cannot understand evolution in full unless it remembers the complicated morphology of life. Artificial evolution will only take off when it is embodied.

Each biological egg cell, like most nucleated cells, carries several libraries of DNA information outside of the chromosomes. Most disturbing to standard theory, the egg cell may be constantly swapping bits of code within itself, between the files of its in-house DNA and the files of inherited chromosomal DNA. If information in the house DNA could be shaped by the experience of the egg cell, then transmitted to the chromosomal DNA, it would transgress the stern Central Dogma, which states that in biology information can only flow from the genes to the cellular body—not vice versa. That is, there is no direct feedback from the body (phenotype) to the gene (genotype). We should be suspicious of any rule such as the Central Dogma, Darwinian critic Arthur Koestler pointed out, because "it would be the only example found in nature of a biological process devoid of feedback."

There are two lessons in morphogenesis for creators of artificial evolution. The first is that changes in an adult organism are triggered in embryos indirectly through the environment of the mother's egg, as well as directly by genealogy. There is plenty of room in this process for unconventional information flow from the cell (the mother's cell) to the genes via control factors and intracellular DNA swap. As German morphologist Rupert Riedl puts it, "Neolamarckism postulates that there is direct feedback. Neodarwinism postulates that there is no feedback. Both are mistaken. Truth lies in the middle. There is feedback but it is not direct." One major route for indirect feedback to the genes is the very early stages of embryonic growth, the hours of incarnation when the genes become flesh.

During these hours, the embryo is an amplifier. Hence the second lesson: Small changes can be magnified as development unfolds. In this way, morphogenesis skips Darwinian gradualism. This point was made by the Berkeley geneticist Richard Goldschmidt, whose ideas on nongradual evolution were derided and scorned throughout his life. His major work, *The Material Basis of Evolution* (1940), was dismissed as near-crackpot until Stephen Jay Gould began a campaign to resurrect his ideas in the 1970s. Goldschmidt's title mirrors a theme of mine here: that evolution is an intermingling of material and information, and that genetic logic cannot be divorced from the laws of material form in which it dwells. (An extrapolation of this idea would be that artificial evolution will run slightly differently from natural evolution as long as it is embedded on a different substrate.)

Goldschmidt spent an unrewarded lifetime showing that extrapolating the gradual transitions of microevolution (red rose to yellow rose) could not explain macroevolution (worm to snake). Instead, he postulated from his work on developing insects that evolution proceeded by jumps. A small change made early in development would lead to a large change—a monster—at the adult stage. Most radically altered forms would abort, but once in a while, large change would cohere and a hopeful monster would be born. The hopeful monster would have a full wing, say, instead of the half-winged intermediate form Darwinian theory demanded. Organisms could arrive fully formed in niches that a series of partially formed transitional species would never get to. The appearance of hopeful monsters would also explain the real absence of transitional forms in fossil lineages.

Goldschmidt made the intriguing claim that his hopeful monsters could most easily be generated by small shifts in developmental timing. He found "rate genes" that controlled the timing of local growth and differentiation processes. For instance, a tweak in the gene controlling the rates of pigmentation would produce caterpillars of wildly different color patterns. As his champion Gould writes, "Small changes early in embryology accumulate through growth to yield profound differences among adults . . . Indeed, if we do not invoke discontinuous change by small alterations in rates of development, I do not see how most major evolutionary transitions can be accomplished at all."

There is a grave and unmistakable lack of intermediates in the fossil record. The fact that creationists gloat over it should not tempt others to ignore it. The "fossil gaps" were a hole in

Darwin's theory that he promised would go away in the future, when more areas of Earth were searched by professional evolutionists. The gaps did not go away in the least. Once a "trade secret" of paleontologists, the gaps are now acknowledged by every leading authority on evolution. Here are two: "The known fossil record fails to document a single example of phyletic [gradual] evolution accomplishing a major morphologic transition and hence offers no evidence that the gradualistic model can be valid," says Steven Stanley, evolutionary paleontologist. And here's Stephen Jay Gould again, speaking as the expert paleontologist he is:

> All paleontologists know that the fossil record contains precious little in the way of intermediate forms; transitions between major groups are characteristically abrupt . . . The history of most fossil species includes two features particularly inconsistent with gradualism:
>
> 1. Stasis. Most species exhibit no directional change during their tenure on Earth. They appear in the fossil record looking much the same way as when they disappear . . .
>
> 2. Sudden appearance. In any local area, a species does not arise gradually by the steady transformation of its ancestors; it appears all at once and "fully formed."

In the eyes of science historians, Darwin's most consequential claim was that the discontinuous face of life as a whole was an illusion. The separateness of species, the "immutable essence" intrinsic to each type of animal or plant—a principle which the ancient philosophers had taught forever—was, he claimed, false. The Bible spoke of creatures "each made in their kind," and most biologists of the day, including the young Darwin, thought species kept to their breed in an idealized way. It was the type that mattered, while individuals conformed more or less to the type. The enlightened Darwin announced, however, that (1) every individual differed significantly; (2) all life was dynamically plastic, infinitely malleable between individuals, so (3) individuals arranged in populations were all that mattered. The barriers erected by species were porous and illusory. By shifting the discontinuity from species to every individual, Darwin vaporized it. Life was one evenly distributed being.

But intriguing suspicions now accumulating in the study of complex systems, particularly complex systems that adapt, learn, and evolve, suggest Darwin was wrong in his most revolutionary premise. Life is largely clumped into parcels and only mildly plastic. Species either persist or die. They transmute into something else under only the most mysterious and uncertain conditions. By and large, complex things fall into categories and the categories persist. Stasis of the category is the norm: the typical lifespan for a species is between one and ten million years.

Things that resemble organisms—economic firms, thoughts in the brain, ecological communities, nation-states—also naturally differentiate into persistent clumps. Human institution clumps—churches, departments, companies—find it easier to grow than to evolve. Required to adapt too far from their origins, most institutions will die.

"Organic" entities are not infinitely malleable because complex systems cannot easily be gradually modified in a sequence of functional intermediates. A complex system (such as a zebra or a company) is severely limited in the directions and ways it can evolve, because it is a hierarchy composed entirely of subentities, which are also limited in their room for adaptation because they are composed of sub-subentities, and so on down the tower.

It should be no surprise, then, to find that evolution works in quantum steps. The given constituents of an organism can collectively make this or that, but not everything in between this and that. The hierarchical nature of the whole prevents it from reaching all the possible states it might theoretically hit. At the same time, the hierarchical arrangement of the whole gives it power to make some large-scale shifts. So a record of this organism would show it leaping from this to that. In biology, this is called saltationism (from the Latin *saltare*, to jump) and it is totally out of favor among professional biologists. Mild saltationism was rejuvenated with interest in Goldschmidt's genetic hopeful monsters, but a complex saltationism that would significantly leap over transitional forms is pure heresy at the moment. Yet the interdependent coadaptations that constitute a complex being must produce quantum evolution. Artificial evolution has not yet produced an "organism" complex enough to contain hierarchical depth, and so we don't know yet in what way saltationism might appear in synthetic worlds.

The morphogenic development of an egg cell into a living creature is full of inherited baggage that constrains the possible variety of its potential descendants. Overall, materials that constitute bodies impose physical constraints that limit what kind of animals can be formed. There'll be no elephants with legs as thin as an ant's. Genetic constraints—the physical nature of genes—likewise narrow what kind of animals can be formed. Each hunk of genetic information is a protein that must physically move to communicate. As general as DNA is, some messages will be difficult or impossible to code in a complex body because of the physical constraints of the genes.

Because genes have their own dynamics independent of the organism, they dictate what can be birthed from them. Inside the genome, genes are interconnected to the point that the gene can become grid-locked—*A* is waiting on *B*, *B* is waiting on *C*, and *C* is waiting on *A*. This internal linkage raises a conservative force within the genome that pushes on itself to keep the genome unchanged—regardless of what body it makes. Like a complex system, the genetic circuitry tends to resist perturbations by restricting allowable variations. The genome seeks to persist as a cohesive unity.

When artificial or natural selection moves a genotype (say, of a pigeon) out of one stability toward a preferred character (say, white color), the interlinked character of the genome kicks in to produce multiple side effects (say, nearsightedness). Darwin, pigeon breeder that he was, noticed this and called it "the mysterious law of correlation of growth." Ernst Mayr, the grand old man of neodarwinism, states, "I do not know of a single intensive selection [breeding] experiment during the past 50 years during which some such undesirable side effects have not appeared." The single-point mutations that traditional population genetics are built upon are rare. Genes usually work in complexes, and are themselves a complex, adaptive system. The genes harbor their own wisdom and their own inertia. This is why even monsters follow rules.

The genome must stray far enough from its usual arrangement before it can create a substantially different outward form. When the genome is "pulled" by competitive pressures outside its usual orbit, it must materially rearrange its patterns of linkage in order to remain stable. In cybernetic terms, it must settle into a different basin of attraction, one that has its own unity and cohesion, its own homeostasis.

Before an organism takes a stand in the world, before it directly meets the natural selection of competition and survival, it has already been subjected to two degrees of internal selection—first by the internal constraints of the genome, and secondly by the laws of bodily form. There is yet a third degree of internal selection that affects an organism before it can truly deal with natural selection. A change accepted by the genome, and then accepted by the bodily form, must then be accepted by the population at large. A single individual with a brilliant mutation will bury that innovation when it dies unless those genes are spread throughout the population. Populations (or demes) exhibit their own cohesive drive toward unity, contributing to an emergent behavior of the whole, as if they were one large, homeostatically balanced system—the population as an individual.

That anything novel ever surmounts these hurdles to evolve is astounding. Mayr writes in *Toward a New Philosophy of Biology*: "The most difficult feat of evolution is to break out of the straight-jacket of this cohesion. This is the reason why only so relatively few new structural types have arisen in the last 500 million years, and this may well also be the reason why 99.999 percent of all evolutionary lines have become extinct. They did so because the cohesion prevented them from responding quickly to sudden new demands by the environment." Stasis, long a major riddle in a constantly changing, coevolving world, now has an alibi.

I delve into these matters deeply because the constraints on biological evolution are the hope of artificial evolution. Every negative constraint within the kinetics of evolution may be viewed in the positive. The power of constraints that retain the old also assemble the new. The delicate gravity that holds organisms in their places, preventing them from casually drifting off to other forms, is the same gravity that pulls in organisms to certain forms in the first place. The self-reinforcing aspect of a gene's internal genetic selection—which makes leaving its stability so difficult—acts as a valley drawing in random arrangements until they rest in that basin of the possible. Over millions of years, the multiple stabilities of genome and body keep a species centered, overriding the action of natural selection. When a species does break away by a radical jump, the same cohesion—again beyond influence of natural selection—lures it into a new homeostasis. It seems odd at first, but constraints create.

Therefore what is said about extinctions—that constraints caused them—may be equally true about origins. The emergent cohesion at various levels of biology, and not natural selection per se, may well be the reason why 99.999 percent of life forms originated. The role of constraints to assemble life—what some call self-organization—is unmeasured, but probably immense.

---

A FAMOUS IMAGE from Darwin's *Origin of Species*, written over a century before the dawn of the first computer, precisely embodies the task of evolution in computerese. Evolution, Darwin said, "is daily and hourly scrutinizing, throughout the world, every variation, even the slightest; rejecting that which is bad, preserving and adding up all that is good; silently and insensibly working . . ." This is the algorithmic search through the Library of forms. Is the Library of possible biological life forms a vast space with only a few sparse coherent works, or is it filled with many of them? How likely is it that a random evolutionary step will land on a possibility with real life? How closely bunched are functioning organisms in the space of possibilities? How isolated are viable lineages from each other?

If the density of possible life forms is sufficiently crowded with feasible beings, then the space of possibilities can be more easily searched by the chance-driven walk of natural selection. A space thick with prospects and searchable by randomness provides uncountable paths for evolution to follow through time. On the other hand, if functioning life forms are sparse and isolated from each other, natural selection alone will probably be unable to reach new forms of life. The distribution of functional units in life may be so scant that most of the space of possible organisms lies empty of workable cases. In this vast space of failure, viable life forms may be found lumped together in patches, or conglomerated onto a few crooked paths through the space.

If the space of functioning organisms is at all sparse, then it is clear that in order to proceed from one patch of viable creatures to the next, evolution needs something to guide it through empty wastelands. A trial-and-error walk, such as that which underlies natural selection, can only get you nowhere fast.

We know virtually nothing of the real distribution of life in the Library of realities. It may be so sparse and unpregnant with possibilities that there is only one living path through it—the path we are currently on. Or there might be broad highways in the Library that channel a number of paths into a few bottlenecks that all beings must cross—say, the resonant attractor of four legs, a tubular gut, five-digit hands. Or there may be a submerged bias in life's substrate, so that no matter where you start you eventually arrive on the shores of bilateral symmetry, segmented limbs, and intelligence of one kind or another. We just don't know. But with artificial evolution at work, we could know.

These fruitful questions about the constitutional laws of evolution are being asked, not in biological terms, but in the language of a new science, the science of complexity. Biologists find it most grating that the impetus for this postdarwinian convergence comes chiefly from mathematicians, physicists, computer scientists, and whole systems theorists—people who couldn't tell the difference between *Cantharellus cibarius* and *Amanita muscaria* (one of them a deadly mushroom) if their lives depended on it. Naturalists have had nothing but scorn for those so willing to simplify nature's complexity into computer models, and to disregard the conclusions of that most awesome observer of nature, Charles Darwin.

Of Darwin's insights, Darwin himself reminded readers in his update to the third edition of *Origin of Species*:

> As my conclusions have lately been much misrepresented, and it has been stated that I attribute the modification of species exclusively to natural selection, I may be permitted to remark that in the first edition of this work, and subsequently, I place in a most conspicuous position—namely at the close of the Introduction—the following words: "I am convinced that natural selection has been the main, but not the exclusive means of modification." This has been of no avail. Great is the power of steady misrepresentation.

Neodarwinism presented a wonderful story of evolution through natural selection, a just-so story whose logic was impossible to argue with: since natural selection could logically create all things, all things were created via natural selection. As long as the argument was over the history of our one life on Earth, one had to settle for this broad interpretation unless inarguable evidence would come along to prove otherwise.

It has not yet come. The clues I present here of symbiosis, directed mutation, saltationism, and self-organization, are far from conclusive. But they are of a pattern: that evolution has multiple components in addition to natural selection. And furthermore, these bits and questions are being stirred up by a bold and daring vision: to synthesize evolution outside of biology.

The moment we tried to transfer the dynamics of evolution out of history and into a manufactured medium, the inner nature of evolution was exposed to scrutiny. Evolution pressed into artificial evolution within computers has passed the first neodarwinist test. It demonstrates spontaneous self-selection as a means of adaptation, and as a means of generating some initial novelty.

But if artificial evolution is to become a powerhouse of creativity on par with natural evolution, we must either grant it immense time periods we don't have, or enhance it with further creative aspects of natural evolution, if they are indeed there. At the very least, messing with artificial evolution will illuminate the true character of historical evolution of life on Earth in a way that neither current observations nor past fossils can hope to do.

I do not find it alarming at all that evolution theory may be taken over by postdarwinians without biology degrees. The great lesson which artificial evolution has *already* imparted is that evolution is not a biological process. It is a technological, mathematical, informational, and biological process rolled into one. It could almost be said it is a law of physics, a principle that reigns over all created multitudes, whether they have genes or not.

The least-appreciated aspect of Darwin's natural selection is how unavoidable it is. The conditions for natural selection are very specific, but if these conditions are met, natural selection is inevitable!

Natural selection can only occur in populations and swarms of things. It's a phenomenon of mobs distributed in space and time. The process must involve a population having (1) variation among individuals in some trait, (2) where that trait makes some difference in fertility, fecundity, or survival ability, and (3) where that trait is transmitted in some fashion from parents to their offspring. If those conditions exist, natural selection will happen as inevitably as seven follows six, or heads and tails split. As evolution theorist John Endler says, "Natural selection probably should not be called a biological law. It proceeds not for biological reasons, but from the laws of probability."

But natural selection is not evolution, nor can evolution be equated with natural selection. In the same way, arithmetic is not mathematics nor can mathematics be equated with arithmetic. One can claim that all of mathematics is just addition compounded. Subtraction is addition in reverse, multiplication addition in sequence, and all complex functions built upon those mere extrapolation of addition. This is somewhat the same argument of the neodarwinists: all evolution is the extrapolation of natural selection compounded. While there is a grain of truth in this perspective, it shuts off understanding and appreciation of more complex things. While multiplication is precisely a form of serial additions, wholly new powers emerge from this shortcut that would not be understood if multiplication was only thought of as addition repeated. Dwelling on addition will not get you to $E=mc^2$.

I believe there is a mathematics of life. Natural selection may be its additive function. But to fully explain the origin of life, the remarkable trend toward complexity, and the invention of intelligence requires more than addition. It needs a rich mathematics of complex functions built upon each other; it needs deeper evolution. Natural selection alone is not enough, not by miles. It must be alloyed with more creative, generative processes to accomplish much. It must have more to naturally select from.

What the postdarwinians have shown is that there is no such thing as monolithic evolution run by one-dimensional natural selection. It would be more fitting to say that evolution is plural and deep. Deep evolution is an aggregate of many kinds of evolutions; it is a multifaced god, a creator with many arms, working by many methods, of which natural selection of variation is perhaps the most universal factor. An uncharted variety of evolutions make up deep evolution, just as our minds comprise a society of dimwitted agents and a variety of types of thinking. Various evolutions proceed at different scales, at different tempos, in different styles. Furthermore, this blend of evolutions changes over time. Certain types of evolution were important in early protolife; some are more emphasized now, four billion years later. One variety (natural selection) will be ubiquitous throughout the plurality, while others will be rare and specialized in their roles. Deep, pluralistic evolution, like intelligence, is an emergent property of a community of dynamics.

As we construct an artificial evolution to breed machines and software, we will also need to allow for this homogenous character of evolution. In a functioning artificial evolution capable of open-ended, sustainable creativity, I would expect to see the following dynamics (which I believe reside to some degree in biological evolution but which may appear artificially in a stronger form than we find in biology):

- *Symbiosis*—Easy informational swaps that permit convergence of distinct lines

- *Directed Mutations*—Nonrandom mutation and crossover mechanisms with direct communication from the environment

- *Saltationism*—Clustering of functions, hierarchical levels of control, modularization of components, and adaptive processes that modify a cluster all at once
- *Self-organization*—Development biased toward certain forms (like four wheels), which become pervasive standards

Artificial evolution will not be able to make everything. There will be many things that we can imagine in full detail—and that by the laws of both physics and logic should work—that synthetic evolution will not be able to reach because of its constraints. In an unconscious way the computer-toting postdarwinians are asking the question: What are the limits of evolution? What can evolution not do? The limits to organic evolution may not be ultimate, but its biases and inabilities may hold answers to evolution's creative talents. Where are the vacant black holes in the landscape of possible creatures? I can only echo Alberch, the monster guy, who said, "I am more concerned about the empty spaces, about the morphologies that, although conceivable, are not realized." To paraphrase Lewontin, "An evolution that cannot make all things, explains some things."

# 20

# THE BUTTERFLY SLEEPS

SOME IDEAS ARE REELED into our mind wrapped up in facts; and some ideas burst upon us naked without the slightest evidence they could be true but with all the conviction they are. The ideas of the latter sort are the more difficult to displace.

The idea of antichaos—order for free—came in a vision of the unverifiable sort.

The idea was dealt to Stuart Kauffman, an undergraduate medical student at Dartmouth College some thirty years ago. As Kauffman remembers it, he was standing in front of a bookstore window daydreaming about the design of a chromosome. Kauffman was a sturdy guy with curly hair, easy smile, and no time to read. As he stared in the window, he imagined a book, a book with his name on it in the author's slot, a book that he would write in the future.

In his vision the pages of the book were filled with a web of arrows connecting other arrows, weaving in and out of a living tangle. It was the icon of the Net. But the mess was not without order. The tangle sparked mysterious, almost cabalistic, "currents of meanings" along the threads. Kauffman discerned an image emerging out of the links in a "subterranean way," just as recognition of a face springs from the crazy disjointed surfaces in a cubist painting.

As a medical student studying cell development, Kauffman

saw the intertwined lines in his fantasy as the interconnections between genes. Out of that random mess, Kauffman suddenly felt sure, would come inadvertent order—the architecture of an organism. Out of chaos would come order for no reason: order for free. The complexity of points and arrows seemed to be generating a *spontaneous order*. To Kauffman the depiction was intimately familiar; it felt like home. His task would be to explain and prove it. "I don't know why this question, this ill-lit path," he says, but it has become a "deeply felt, deeply held image."

Kauffman pursued his vision by taking up academic research in cell development. As many other developmental biologists had, he studied *Drosophila*, the famous fruitfly, as it progressed from fertilized egg to adult. How did the original lone egg cell of any creature manage to divide and specialize first into two, then four, then eight new kinds of cells? In a mammal the original egg cell would propagate an intestinal cell line, a brain cell line, a hair cell line; yet each substantially specialized line of cells presumably ran the same operating software. After a relatively few generations of division, one cell type could split into all the variety and bulk of an elephant or oak. A human embryo egg needed to divide only 50 times to produce the trillions of cells that form a baby.

What invisible hand controlled the fate of each cell, as it traveled along a career path forking 50 times, guiding it from general egg to hundreds of kinds of specialized cells? Since each cell was supposedly driven by identical genes (or were they actually different?), how could cells possibly become different? What controlled the genes?

Françoise Jacob and Jacques Monod discovered a major clue in 1961 when they encountered and described the regulatory gene. The regulatory gene's function was stunning: to turn other genes on. In one breath it blew away all hopes of immediately understanding DNA and life. The regulatory gene set into motion the quintessential cybernetic dialogue: What controls genes? Other genes! And what controls those genes? Other genes! And what . . .

That spiraling, darkly modern duet reminded Kauffman of his home image. Some genes controlling other genes which in turn might control still others was the same tangled web of arrows of influence pointing in every direction in his vision book.

Jacob and Monod's regulatory genes reflected a spaghetti-like vision of governance—a decentralized network of genes steering the cellular network to its own destiny. Kauffman was excited. His picture of "order for free" suggested to him a fairly far-out idea: that some of the differentiation (order) each egg underwent was inevitable, no matter what genes you started out with!

He could think of a test for this notion. Replace all the genes in the fruitfly with random genes. His bet: you would not get *Drosophila*, but you would get the same order of monsters and freak mutations *Drosophila* produced in the natural course of things. "The question I asked myself," Kauffman recalls, "was the following. If you just hooked up genes at random, would you get anything that looked useful?" His intuitive hunch was that simply because of distributed bottom-up control and everything-is-connected-to-everything type of cell management, certain classes of patterns would be inevitable. *Inevitable!* Now here was a germ of heresy. Something to devote one's years to!

"I had a hard time in medical school," he continues, "because instead of studying anatomy I was scribbling all these notebooks with little model genomes." The way to prove this heresy, Kauffman cleverly decided, was not to fight nature in the lab, but to model it mathematically. Use computers as they became accessible. Unfortunately there was no body of math with the ability to track the horizontal causality of massive swarms. Kauffman began to invent his own. At the same time (about 1970) in about a half-dozen other fields of research, the mathematically inclined (such as John Holland) were coming up with procedures that allowed them to simulate the effects of a mob of interdependent nodes whose values simultaneously depend on each other.

This set of math techniques that Kauffman, Holland and others devised is still without a proper name, but I'll call it here "net math." Some of the techniques are known informally as parallel distributed processing, Boolean nets, neural nets, spin glasses, cellular automata, classifier systems, genetic algorithms, and swarm computation. Each flavor of net math incorporates the lateral causality of thousands of simultaneous interacting functions. And each type of net math attempts to coordinate massively concurrent events—the kind of nonlinear happenings ubiquitous in the real world of living beings. Net math is in contradistinction to Newtonian math, a classical math so well suited to most physics problems that it had been seen as the only kind of math a careful scientist needed. Net math is almost impossible to use practically without computers.

The wide variety of swarm systems and net maths got Kauffman to wondering if this kind of weird swarm logic—and the inevitable order he was sure it birthed—were more universal than special. For instance, physicists working with magnetic material confronted a vexing problem. Ordinary ferromagnets—the kind clinging to refrigerator doors and pivoting in compasses—have particles that orient themselves with cultlike uniformity in the same direction, providing a strong magnetic field. Mildly magnetic "spin glasses," on the other hand, have wishy-washy particles that will magnetically "spin" in a direction that *depends in part on which direction their neighbors spin*. Their "choice" places more clout on the influence of nearby ones, but pays some attention to distant particles. Tracing the looping interdependent fields of this web produces the familiar tangle of circuits in Kauffman's home image. Spin glasses used a variety of net math to model the material's nonlinear behavior that was later found to work in other swarm models. Kauffman was certain genetic circuitry was similar in its architecture.

Unlike classical mathematics, net math exhibits nonintuitive traits. In general, small variations in input in an interacting swarm can produce huge variations in output. Effects are disproportional to causes—the butterfly effect.

Even the simplest equations in which intermediate results flow back into them can produce such varied and unexpected turns that little can be deduced about the equations' character merely by studying them. The convoluted connections between parts are so hopelessly tangled, and the calculus describing them so awkward, that the only way to even guess what they might produce is to run the equations out, or in the parlance of computers, to "execute" the equations. The seed of a flower is similarly compressed. So tangled are the chemical pathways stored in it, that inspection of an unknown seed—no matter how intelligent—cannot predict the final form of the unpacked plant. The quickest route to describing a seed's output is therefore to sprout it.

Equations are sprouted on computers. Kauffman devised a mathematical model of a genetic system that could sprout on a modest computer. Each of the 10,000 genes in his simulated DNA is a teeny-weenie bit of code that can turn other genes either on or off. What the genes produced and how they were connected were assigned at random.

This was Kauffman's point: that the very topology of such complicated networks would produce order—spontaneous order!—no matter what the tasks of the genes.

While he worked on his simulated gene, Kauffman realized that he was constructing a generic model for any kind of swarm system. His program could model any bunch of agents that interact in a massive simultaneous field. They could be cells, genes, business firms, black boxes, or simple rules—anything that registers input and generates output interpreted as input by a neighbor.

He took this swarm of actors and randomly hooked them up into an interacting network. Once they were connected he let them bounce off one another and recorded their behavior. He imagined each node in the network as a switch able to turn certain neighboring nodes off or on. The state of the neighbor nodes looped back to regulate the initial node. Eventually this gyrating mess of he-turns-her-who-turns-him-on settled down into a stable and measurable state. Kauffman again randomly rearranged the entire net's connections and let the nodes interact until they all settled down. He did that many times, until he had "explored" the space of possible random connections. This told him what the generic behavior of a net was, independent of its contents. An oversimplified analogous experiment would be to take ten thousand corporations and randomly link up the employees in each by telephone networks, and then measure the average effects of these networks, independent of what people said over them.

By running these generic interacting networks tens of thousands of times, Kauffman learned enough about them to paint a rough portrait of how such swarm systems behaved under specific circumstances. In particular, he wanted to know what *kind of* behavior a generic genome would create. He programmed thousands of randomly assembled genetic systems and then ran these ensembles on a computer—genes turning off and on and influencing each other. He found they fell into "basins" of a few types of behaviors.

At a slow speed water trickles out of a garden hose in one uneven but consistent pattern. Turn up the tap, and it abruptly sprays out in a chaotic (but describable) torrent. Turn it up full blast, and it gushes out in a third way like a river. Carefully screw the tap to the precise line between one speed and a slower one, and the pattern refuses to stay on the edge but reverts to one state or the other, as if it were attracted to a side, any side. Just as a drop of rain falling on the ridge of a continental divide must eventually find its way down to either the Pacific Basin or the Atlantic Basin, roll down one side or the other it must.

Sooner or later the dynamics of the system would find its way to at least one "basin" that entrapped the shifting motions into a persistent pattern. In Kauffman's view a randomly assembled system would find its way to a stock pattern (a basin); thus, out of chaos, order for free emerges.

As he ran uncounted genetic simulations, Kauffman discovered a rough ratio (the square root) between the number of genes and the number of basins the genes in the system settled into. This proportion was the same as the number of genes in biological cells and the number of cell types (liver cells, blood cells, brain cells) those genes created, a ratio that is roughly constant in all living things.

Kauffman claims this universal ratio across many species suggests that the number of cell types in nature may derive from cellular architecture itself. The number of types of cells in your body, then, may have little to do with natural selection and more to do with thc mathematics of complex gene interactions. How many other biological forms, Kauffman gleefully wonders, might also owe little to selection?

He had a hunch about a way to ask the question experimentally. But first he needed a method to cook up random ensembles of life. He decided to simulate the origin of life by generating all possible pools of prelife parts—at least in simulation. He would let the virtual pool of parts interact randomly. If he could then show that out of this soup order inevitably emerged, he would have a case. The trick would be to allow molecules to converge into a lap game.

The lap game peaked in popularity a decade ago. It is a spectacular outdoor game that advertises the power of cooperation. The facilitator of the lap game takes a group of 25 or more people and has them stand fairly close together in a circle, so that each participant is staring at the back of the head of the person in front of him. Just picture a queue of people waiting in line for a movie and connect them in a tidy circle.

At the facilitator's command this circle of people bend their knees and sit on the spontaneously generated knee-lap of the person behind them. If done in unison, the ring of people lowering to sit are suddenly propped up on a self-supporting collective chair. If one person misses the lap behind him, the whole circling line crashes. The world's record for a stable lap game is several hundred people.

Auto-catalytic sets and the selfish Uroborus snake circle are much like lap games. Compound (or function) *A* makes compound (or function) *B* with the aid of compound (or function) *C*. But *C* itself is produced by *A* and *D*. And *D* is generated by *E* and *C*, and so on. Without the others none can be. Another way of saying this is to state that the only way for a particular compound or function to survive in the long run is for it to be a product of another compound or function. In this circular world all causes are results, just as all knees are laps. Contrary to common sense, all existences depend on the consensual existence of all others.

As the reality of the lap game proves, however, circular causality is not impossible. Tautology can hold up 200 pounds of flesh. It's real. Tautology is, in fact, an essential ingredient of stable systems.

Cognitive philosopher Douglas Hofstadter calls these paradoxical circuits "Strange Loops." As examples, Hofstadter points to the seemingly ever rising notes in a Bach canon, or the endlessly rising steps in an Escher staircase. He also includes as Strange Loops the famous paradox about Cretan liars who say they never lie, and Gödel's proof of unprovable mathematical axioms. Hofstadter writes in *Gödel, Escher, Bach*: "The 'Strange Loop' phenomenon occurs whenever, by moving upwards (or downwards) through the levels of some hierarchical system, we unexpectedly find ourselves right back where we started."

Life and evolution entail the necessary strange loop of circular causality—of being tautological at a fundamental level. You can't get life and open-ended evolution unless you have a system that contains that essential logical inconsistency of circling causes. In complex adapting processes such as life, evolution, and consciousness, prime causes seem to shift, as if they were an optical illusion drawn by Escher. Part of the problem humans have in trying to build systems as complicated as our own human biology is that in the past we have insisted on a degree of logical consistency, a sort of clockwork logic, that blocks the emergence of autonomous events. But as the mathematician Gödel showed, inconsistency is an inevitable trait of any self-sustaining system built up out of consistent parts.

Godel's 1931 theorem demonstrates, among other things, that attempts to banish self-swallowing loopiness are fruitless, because, in Hofstadter's words, "it can be hard to figure out just where self-referencing is occurring." When examined at a "local" level every part seems legitimate; it is only when the lawful parts form a whole that the contradiction arises.

In 1991, a young Italian scientist, Walter Fontana, showed mathematically that a linear sequence of function *A* producing function *B* producing function *C* could be very easily circled around and closed in a cybernetic way into a self-generating loop, so that the last function was coproducer of the initial function. When Kauffman first encountered Fontana's work he was ecstatic with the beauty of it. "You have to fall in love with it! Functions mutually making one another. Out of all function space, they come gripping one another's arms in an embrace of creating!" Kauffman called such an autocatalytic set an "egg."

He said, "An egg would be a set of rules having the property that the rules they pose are precisely the ones that create them. That's really not crazy at all."

To get an egg you start with a huge pool of different agents. They could be varieties of protein pieces or fragments of computer code. If you let them interact upon each other long enough, they will produce small loops of thing-producing-other things. Eventually, if given time and elbowroom the spreading network of these local loops in the system will crowd upon itself, until *every* producer in the circuit is a product of another, until every loop is incorporated into all the other loops in massively parallel interdependence. At this moment of "catalytic closure" the web of parts suddenly snaps into a stable game—the system sits in its own lap, with its beginning resting on its end, and vice versa.

Life began in such a soup of "polymers acting on polymers to form new polymers," Kauffman claims. He demonstrated the theoretical feasibility of such a logic by running experiments of "symbol strings acting on symbol strings to form new symbol strings." His assumption was that he could equate protein fragments and computer code fragments as logical equivalents. When he ran networks of bits of code-which-produce-code as a model for proteins, he got autocatalytic systems that are circular in the sense of the lap game: they have no beginning, no center, and no end.

Life popped into existence as a complete whole much as a crystal suddenly appears in its final (though miniature) form in a supersaturated solution: not beginning as a vague half-crystal, not appearing as a half-materialized ghost, but wham, being all at once, just as a lap game circle suddenly emerges from a curving line of 200 people. "Life began whole and integrated, not disconnected and disorganized," writes Stuart Kauffman. "Life, in a deep sense, *crystallized*."

He goes on to say, "I hope to show that self-reproduction and homeostasis, basic features of organisms, are natural collective expressions of polymer chemistry. We can expect any sufficiently complex set of catalytic polymers to be collectively autocatalytic." Kauffman was creeping up on that notion of inevitability again. "If my model is correct then the routes to life in the universe are boulevards, rather than twisted back alleyways." In other words, given the chemistry we have, "life is inevitable."

"We've got to get used to dealing in billions of things!" Kauffman once told an audience of scientists. Huge multitudes of anything are different: the more polymers, the exponentially more possible interactions where one polymer can trigger the manufacture of yet another polymer. Therefore, at some point, a droplet loaded up with increasing diversity and numbers of polymers will reach a threshold where a certain number of polymers in the set will suddenly fall out into a spontaneous lap circle. They will form an auto-generated, self-sustaining, self-transforming network of chemical pathways. As long as energy flows in, the network hums, and the loop stands.

Codes, chemicals, or inventions can in the right circumstances produce new codes, chemicals, or inventions. It is clear this is the model of life. An organism produces new organisms which in turn create newer organisms. One small invention (the transistor) produces other inventions (the computer) which in turn permit yet other inventions (virtual reality). Kauffman wants to generalize this process mathematically to say that functions in general spawn newer functions which in turn birth yet other functions.

"Five years ago," recalls Kauffman, "Brian Goodwin [an evolutionary biologist] and I were sitting in some World War I bunker in northern Italy during a rainstorm talking about autocatalytic sets. I had this profound sense then that there's a deep similarity between natural selection—what Darwin told us—and the wealth of nations—what Adam Smith told us. Both have an invisible hand. But I didn't know how to proceed any further until I saw Walter Fontana's work with autocatalytic sets, which is gorgeous."

I mentioned to Kauffman the controversial idea that in any society with the proper strength of communication and information connection, democracy becomes inevitable. Where ideas are free to flow and generate new ideas, the political organization will eventually head toward democracy as an unavoidable self-organizing strong attractor. Kauffman agreed with the parallel: "When I was a sophomore in '58 or '59 I wrote a paper in philosophy that I labored over with much passion. I was trying to figure out why democracy worked. It's obvious that democracy doesn't work because it's the rule of the majority. Now, 33 years later, I see that democracy is a device that allows conflicting minorities to reach relative fluid compromises. It keeps subgroups from getting stuck on some locally good but globally inferior solution."

It is not difficult to imagine Kauffman's networks of Boolean logic and random genomes mirroring the workings of town halls and state capitals. By structuring miniconflicts and micro-revolutions as a continuous process at the local level, large scale macro- and mega-revolutions are avoided, and the whole system is neither chaotic nor stagnant. Perpetual change is fought out in small towns, while the nation remains admirably stable—thus creating a climate to keep the small towns in ceaseless compromise-seeking modes. That circular support is another lap game, and an indication that such systems are similar in dynamics to the self-supporting vivisystems.

"This is just intuitive," Kauffman cautions me, "but you can *feel* your way from Fontana's 'string-begets-string-begets-string' to 'invention-begets-invention-begets-invention' to cultural evolution and then to the wealth of nations." Kauffman makes no bones about the scale of his ambition: "I am looking for the self-consistent big picture that ties everything together, from the origin of life, as a self-organized system, to the emergence of spontaneous order in genomic regulatory systems, to the emergence of systems that are able to adapt, to nonequilibrium price formation which optimizes trade among organisms, to this unknown analog of the second law of thermodynamics. It is all one picture. I really feel it is. But the image I'm pushing on is this: Can we prove that a finite set of functions generates this infinite set of possibilities?"

*Whew*. I call that a "Kauffman machine." A small but well-chosen set of functions that connect into an auto-generating ring and produce an infinite jet of more complex functions. Nature is full of Kauffman machines. An egg cell producing the body of a whale is one. An evolution machine generating a flamingo over a billion years from a bacterial blob is another. Can we make an artificial Kauffman machine? This may more properly be called a von Neumann machine because von Neumann asked the same question in the early 1940s. He wondered, Can a machine make another machine more complex than itself? Whatever it is called, the question is the same: How does complexity build itself up?

"You can't ask the experimental question until, roughly speaking, the intellectual framework is in place. So the critical thing is asking important questions," Kauffman warned me. Often during our conversations, I'd catch Kauffman thinking aloud. He'd spin off wild speculations and then seize one and twirl it around to examine it from various directions. "How do you ask that question?" he asked himself rhetorically. His quest was for the Question of All Questions rather than the Answer of All Answers. "Once you've asked the question," he said, "there's a good chance of finding some sort of answer."

A Question Worth Asking. That's what Kauffman thought of his notion of self-organized order in evolutionary systems. Kauffman confided to me: "Somehow, each of us in our own heart is able to ask questions that we think are profound in the sense that the answer would be truly important. The enormous puzzle is why in the world any of us ask the questions that we do."

There were many times when I felt that Stuart Kauffman, medical doctor, philosopher, mathematician, theoretical biologist, and MacArthur Award recipient, was embarrassed by the wild question he had been dealt. "Order for free" flies in the face of a conservative science that has rejected every past theory of creative order hidden in the universe. It would probably reject his. While the rest of the contemporary scientific world sees butterflies of random chance sowing out-of-control, nonlinear effects in every facet of the universe, Kauffman asks if perhaps the butterflies of chaos sleep. He wakes the possibility of an overarching design dwelling within creation, quieting disorder and birthing an ordered stillness. It's a notion that for many sounds like mysticism. At the same time, the pursuit and framing of this single huge question is the quasar source of Kauffman's considerable pride and energy: "I would be lying if I didn't tell you that when I was 23 and started wondering how in the world a genome with 100,000 genes controls the emergence of different cell types, I felt that I had found something profound, I had found a profound question. And I still feel that way. I think God was very nice to me.

"If you write something about this," Kauffman says softly, "make sure you say that this is only something crazy that people are thinking about. But wouldn't it be wonderful if somehow there are laws that make laws that make laws, so that the universe is, in John Wheeler's words, something that is looking in at itself!? The universe posts its own rules and emerges out of a self-consistent thing. Maybe that's not impossible, this notion that quarks and gluons and atoms and elementary particles have invented the laws by which they transform one another."

Deep down Kauffman felt that his systems built themselves. In some way he hoped to discover, evolutionary systems controlled their own structure. From the first glimpse of his visionary network image, he had a hunch that in those connections lay the answer to evolution's self-governance. He was not content to show that order emerged spontaneously and inevitably. He also felt that *control* of that order also emerged spontaneously. To that end he charted thousands of runs of random ensembles in computer simulation to see which type of connections permitted a swarm to be most adaptable. "Adaptable" means the ability of a system to adjust its internal links so that it fits its environment over time. Kauffman views an organism, a fruitfly say, as adjusting the network of its genes over time so that the result of the genetic network—a fly body—best fits its changing surroundings of food, shelter, and predators. The Question Worth Asking was: what controlled the evolvability of the system? Could the organism itself control its evolvability?

The prime variable Kauffman played with was the connectivity of the network. In a sparsely connected network, each node would on average only connect to one other node, or less. In a richly connected network, each node would link to ten or a hundred or a thousand or a million other nodes. In theory the limit to the number of connections per node is simply the total number of nodes, minus one. A million-headed network could have a million-minus-one connections at each node; every node is connected to every other node. To continue our rough analogy, every employee of GM could be directly linked to all 749,999 other employees of GM.

As Kauffman varied this connectivity parameter in his generic networks, he discovered something that would not surprise the CEO of GM. A system where few agents influenced other agents was not very adaptable. The soup of connections was too thin to transmit an innovation. The system would fail to evolve. As Kauffman increased the average number of links between nodes, the system became more resilient, "bouncing back" when perturbed. The system could maintain stability while the environment changed. It would evolve. The completely unexpected finding was that beyond a certain level of linking density, continued connectivity would only *decrease* the adaptability of the system as a whole.

Kauffman graphed this effect as a hill. The top of the hill was optimal flexibility to change. One low side of the hill was a sparsely connected system: flat-footed and stagnant. The other low side was an overly connected system: a frozen grid-lock of a thousand mutual pulls. So many conflicting influences came to bear on one node that whole sections of the system sank into rigid paralysis. Kauffman called this second extreme a "complexity catastrophe." Much to everyone's surprise, you could have too much connectivity. In the long run, an overly linked system was as debilitating as a mob of uncoordinated loners.

Somewhere in the middle was a peak of just-right connectivity that gave the network its maximal nimbleness. Kauffman found this measurable "Goldilocks" point in his model networks. His colleagues had trouble believing his maximal value at first because it seemed counterintuitive at the time. The optimal connectivity for the distilled systems Kauffman studied was very low, "somewhere in the single digits." Large networks with thousands of members adapted best with less than ten connections per member. Some nets peaked at less than two connections on average per node! A massively parallel system did not need to be heavily connected in order to adapt. Minimal average connection, done widely, was enough.

Kauffman's second unexpected finding was that this low optimal value didn't seem to fluctuate much, no matter how many members comprised a specific network. In other words, as more members were added to the network, it didn't pay (in terms of systemwide adaptability) to increase the number of links to each node. To evolve most rapidly, add members but don't increase average link rates. This result confirmed what Craig Reynolds had found in his synthetic flocks: you could load a flock up with more and more members without having to reconfigure its structure.

Kauffman found that at the low end, with less than two connections per agent or organism, the whole system wasn't nimble enough to keep up with change. If the community of agents lacked sufficient internal communication, it could not solve a problem as a group. More exactly, they fell into isolated patches of cooperative feedback but didn't interact with each other.

At the ideal number of connections, the ideal amount of information flowed between agents, and the system as a whole found the optimal solutions consistently. If their environment was changing rapidly, this meant that the network remained stable—persisting as a whole over time.

Kauffman's Law states that above a certain point, increasing the richness of connections between agents freezes adaptation. Nothing gets done because too many actions hinge on too many other contradictory actions. In the landscape metaphor, ultra-connectance produces ultra-ruggedness, making any move a likely fall off a peak of adaptation into a valley of nonadaptation. Another way of putting it, too many agents have a say in each other's work, and bureaucratic rigor mortis sets in. Adaptability conks out into grid-lock. For a contemporary culture primed to the virtues of connecting up, this low ceiling of connectivity comes as unexpected news.

We postmodern communication addicts might want to pay attention to this. In our networked society we are pumping up both the total number of people connected (in 1993, the global network of networks was expanding at the rate of 15 percent additional users per month!), and the number of people and places to whom each member is connected. Faxes, phones, direct junk mail, and large cross-referenced data bases in business and government in effect increase the number of links between each person. Neither expansion particularly increases the adaptability of our system (society) as a whole.

Stuart Kauffman's simulations are as rigorous, original, and well-respected among scientists as any mathematical model can be. Maybe more so, because he is using a real (computer) network to model a hypothetical network, rather than the usual reverse of using a hypothetical to model the real. I grant, though, it is a bit of a stretch to apply the results of a pure mathematical abstraction to irregular arrangements of reality. Nothing could be more irregular than online networks, biological genetic networks, or international economic networks. But Stuart Kauffman is himself eager to extrapolate the behavior of his generic testbed to real life. The grand comparison between complex real-world networks and his own mathematical simulations running in the heart of silicon is nothing less than Kauffman's holy grail. He says his models "smell like they are true." Swarmlike networks, he bets, all behave similarly on one level. Kauffman is fond of speculating that "IBM and *E. coli* both see the world in the same way."

I'm inclined to bet in his favor. We own the technology to connect everyone to everyone, but those of us who have tried living that way are finding that we are disconnecting to get anything done. We live in an age of accelerating connectivity; in essence we are steadily climbing Kauffman's hill. But we have little to stop us from going over the top and sliding into a descent of increasing connectivity but diminishing adaptability. Disconnection is a brake to hold the system from overconnection, to keep our cultural system poised on the edge of maximal evolvability.

The art of evolution is the art of managing dynamic complexity. Connecting things is not difficult; the art is finding ways for them to connect in an organized, indirect, and limited way.

From his experiments in artificial life in swarm models, Chris Langton, Kauffman's Santa Fe Institute colleague, derived an abstract quality (called the lambda parameter) that predicts the likelihood that a particular set of rules for a swarm will produce a "sweet spot" of interesting behavior. Systems built upon values outside this sweet spot tend to stall in two ways. They either repeat patterns in a crystalline fashion, or else space out into white noise. Those values within the range of the lambda sweet spot generate the longest runs of interesting behavior.

By tuning the lambda parameter Langton can tune a world so that evolution or learning can unroll most easily. Langton describes the threshold between a frozen repetitious state and a gaseous noise state as a "phase transition"—the same term physicists use to describe the transition from liquid to gas or liquid to solid. The most startling result, though, is Langton's contention that as the lambda parameter approaches that phase transition—the sweet spot of maximum adaptability—it slows down. That is, the system tends to *dwell* on the edge instead of zooming through it. As it nears the place it can evolve the most from, it lingers. The image Langton likes to raise is that of a system surfing on an endless perfect wave in slow motion; the more perfect the ride, the slower time goes.

This critical slowing down at the "edge" could help explain why a precarious embryonic vivisystem could keep evolving. As a random system neared the phase transition, it would be "pulled in" to rest at that sweet spot where it would undergo evolution and would then seek to maintain that spot. This is the homeostatic feedback loop making a lap for itself. Except that since there is little "static" about the spot, the feedback loop might be better named "homeodynamic."

Stuart Kauffman also speaks of "tuning" the parameters of his simulated genetic networks to the "sweet spot." Out of all the uncountable ways to connect a million genes, or a million neurons, some relatively few setups are far more likely to encourage learning and adaptation throughout the network. Systems balanced to this evolutionary sweet spot learn fastest, adapt more readily, or evolve the easiest. If Langton and Kauffman are right, an evolving system will find that spot on its own.

Langton discovered a clue as to how that may happen. He found that this spot teeters right on the edge of chaotic behavior. He says that systems that are most adaptive are so loose they are a hairsbreadth away from being out of control. Life, then, is a system that is neither stagnant with noncommunication nor grid-locked with too much communication. Rather life is a vivisystem tuned "to the edge of chaos"—that lambda point where there is just enough information flow to make everything dangerous.

Rigid systems can always do better by loosening up a bit, and turbulent systems can always improve by getting themselves a little more organized. Mitch Waldrop explains Langton's notion in his book *Complexity*, thusly: if an adaptive system is not riding on the happy middle road, you would expect brute efficiency to push it toward that sweet spot. And if a system rests on the crest balanced between rigidity and chaos, then you'd expect its adaptive nature to pull it back onto the edge if it starts to drift away. "In other words," writes Waldrop, "you'd expect learning and evolution to make the edge of chaos stable." A self-reinforcing sweet spot. We might call it dynamically stable, since its home migrates. Lynn Margulis calls this fluxing, dynamically persistent state "homeorhesis"—the homing in on a moving point. It is the same forever almost-falling that poises the chemical pathways of the Earth's biosphere in purposeful disequilibrium.

Kauffman takes up the theme by calling systems set up in the lambda value range "poised systems." They are poised on the edge between chaos and rigid order. Once you begin to look around, poised systems can be found throughout the universe, even outside of biology. Many cosmologists, such as John Barrow, believe the universe itself to be a poised system, precariously balanced on a string of remarkably delicate values (such as the strength of gravity, or the mass of an electron) that if varied by a fraction as insignificant as 0.000001 percent would have collapsed in its early genesis, or failed to condense matter. The list of these "coincidences" is so long they fill books. According to mathematical physicist Paul Davies, the coincidences "taken together . . . provide impressive evidence that life as we know it depends very sensitively on the form of the laws of physics, and on some seemingly fortuitous accidents in the actual values that nature has chosen for various particle masses, force strengths, and so on." In brief, the universe and life as we know it are poised on the edge of chaos.

What if poised systems could tune themselves, instead of being tuned by creators? There would be tremendous evolutionary advantage in biology for a complex system that was auto-poised. It could evolve faster, learn more quickly, and adapt more readily. If evolution selects for a self-tuning function, Kauffman says, then "the capacity to evolve and adapt may itself be an achievement of evolution." Indeed, a self-tuning function would inevitably be selected for at higher levels of evolution. Kauffman proposes that gene systems do indeed tune themselves by regulating the number of links, size of genome, and so on, in their own systems for optimal flexibility.

Self-tuning may be the mysterious key to evolution that doesn't stop—the holy grail of open-ended evolution. Chris Langton formally describes open-ended evolution as a system that succeeds in ceaselessly self-tuning itself to higher and higher levels of complexity, or in his imagery, a system that succeeds in *gaining control over more and more parameters affecting its evolvability* and staying balanced on the edge.

In Langton's and Kauffman's framework, nature begins as a pool of interacting polymers that catalyze themselves into new sets of interacting polymers in such a networked way that maximal evolution can occur. This evolution-rich environment produces cells that also learn to tune their internal connectivity to keep the system at optimal evolvability. Each step extends the stance at the edge of chaos, poised on the thin path of optimal flexibility, which pumps up its complexity. As long as the system rides this upwelling crest of evolvability, it surfs along.

What you want in artificial systems, Langton says, is something similar. The primary goal that any system seeks is survival. The secondary search is for the ideal parameters to keep the system tuned for maximal flexibility. But it is the third order search that is most exciting: the search for strategies and feedback mechanisms that will increasingly self-tune the system each step on the way. Kauffman's hypothesis is that if systems constructed to self-tune "can adapt most readily, then they may be the inevitable target of natural selection. The ability to take advantage of natural selection would be one of the first traits selected."

As Langton and colleagues explore the space of possible worlds searching for that sweet spot where life seems poised on the edge,

I've heard them call themselves surfers on an endless summer, scouting for that slo-mo wave.

Rich Bageley, another Santa Fe Institute fellow, told me, "What I'm looking for are things that I can almost predict, but not quite." He explained further that it was not regular but not chaotic either. Some almost-out-of-control and dangerous edge in between.

"Yeah," replied Langton who overheard our conversation. "Exactly. Just like ocean waves in the surf. They go thump, thump, thump, steady as a heartbeat. Then suddenly, WHUUUMP, an unexpected big one. That's what we are all looking for. That's the place we want to find."

21

# Rising Flow

Heat was a profound puzzle in the early 19th century. Everyone intuitively knew that a hot object cooled to its surroundings and a cool object likewise warmed up. But a comprehensive theory of how heat really worked eluded scientists.

A real theory of heat had to explain some weird happenings. Yes, a very hot object and a very cold object in a room would converge to the same warmth over time. But some objects, like a basin of ice and water mixture, would not warm up equally fast as the same basin of all ice or all water. Hot things expanded; cold things contracted. Motion could disappear into heat. Heat could spark motion. And when certain metals were heated, they gained weight, so therefore, heat had weight.

The early explorers into heat had no idea that they were investigating temperature, calories, friction, work, efficiency, energy and entropy—all terms they were to invent later. For many decades no one was sure what it was they were actually studying. The most accepted theory among them was that heat was an all-pervading elastic fluid—a material ether.

In 1824, the French military engineer Carnot (rhymes with Godot, the tardy lead in Samuel Beckett's play) derived a principle that later became known as the Second Law of Thermodynamics. Roughly paraphrased it goes thus: all systems everywhere run down over time. Together with the First Law (that energy is conserved overall), Carnot's Second Law was the key framework in the following century for understanding not only heat but most of physics, chemistry, and quantum mechanics. In short, the theory of heat undergirds all of modern physical science.

Biology, however, has no grand theory. The joke currently making the rounds of complexity researchers is that biological science today is "Waiting for Carnot." Theoretical biologists feel equivalent to the 19th-century thermalists just before the advent of thermal dynamics. Biologists talk about complexity without having a measure for complexity; they hypothesize about evolution without having a second instance of it. That reminds them of discussing heat without having the concepts of calories, friction, work, or even energy. Just as Carnot framed physics by his overarching law of heat death and plunge to disorder, some theoretical biologists hope for a Second Law of Biology, which would frame the overarching tendency of life to find order amid disorder. There is a touch of satire within the joke, because in Beckett's notorious play, Godot is a mysterious figure who never shows up!

The search for a Second Law of Biology, a law of rising order, is unconsciously behind much of the search for deeper evolutions and the quest for hyperlife. Many postdarwinians doubt that natural selection alone is powerful enough to offset Carnot's Second Law of Thermodynamics. Yet, we are here, so something has. They are not sure what they are looking for, but they intuitively feel that it can be stated as a complementary force to entropy. Some call it anti-entropy, some call it negentropy, and a few call it extropy. Gregory Bateson once asked: "Is there a biological species of entropy?"

This quest for the secret of life is not usually made explicit in scientists' formal papers. Yet in conversations with them late at night, this is what many of them feel. They allude to a vision only half-glimpsed. Each sees a different part, like the blind men patting an elephant. They hunt for cautious scientific words to cover their beliefs and hunches. The vision they hint at, I synthesize thus:

From the crack of the big bang a hot universe runs down for

ten billion years or so. About two-thirds along into its history something clicks, and an insatiable force begins hijacking the slipping heat and order into local areas of higher order. The remarkable thing about this hijacker is that (a) it is self-sustaining, and (b) it is self-reinforcing: the more of it around, the more it makes of itself.

Two currents were thus born out of the white flash. One current runs downhill all the way. This force begins as a wild hot party and fizzes out into silent coldness. This dive is Carnot's depressing Second Law, a ghoulish rule if there ever was one: all order will eventually succumb to chaos, all fire will die, all variety goes bland, all structure will eventually extinguish itself.

The second current runs in parallel, but with opposite effect. It diverts the heat before the heat disperses (since disperse it must) and extracts order out of disorder. It borrows the failing energy and raises the ante into a rising flow.

The rising flow uses its short moment of order to snatch whatever dissipating power it can to build a platform upon which to extract the next round of order. It saves nothing and spends all. It invests all the order it has to amplify the next round of complexity, growth, and order. In this way it taps chaos to breed antichaos. We call it life.

The rising flow is a wave: a slight rise amid a degrading sea of entropy; a sustainable crest always falling upon itself, forever in the state of almost-toppled.

The wave is a moving edge throughout the universe, a thin line between the plunging sides of chaos. One side slopes away to frozen gray solidness, the other slips into overexcited black gaseousness. The wave is the eternally moving moment between the two—the eternal liquid. The gravity of entropy cannot be defied; but as the crest forever falls, biological order rides it down like a surfer.

The order accumulated by the rising wave serves as a plank to extend itself, using energy from outside, into the next realm of further order. As long as Carnot's force flows downhill and cools the universe, the rising flow can steal heat to flow uphill in places, building itself high by pulling on its bootstraps.

Like a pyramid scheme, or building a castle in the air, the

game of leveraging order as a means to buy more order is a game that's got to keep expanding or collapse. Our collective history as living beings is the story of a trickster who has found a foolproof gimmick and is pulling a fast one—and getting away with it so far. "Life might be defined as the art of getting away with it," said the theoretical biologist C. H. Waddington.

Perhaps this rather broadly poetic vision is mine alone, a vision which I have mistakenly read into the comments of others. But I don't think so. I have heard strands of it from too many scientists. Nor do I think it is pure mysticism any more than one would call Carnot's Law mysticism. Sure, the story is couched in human hope, but the hope I share is to find a falsifiable scientific law. Although there have been theories akin to the rising flow that were outright vehicles for vitalism, a second force doesn't have to be any less scientific than the laws of probability or Darwin's force of natural selection.

Still, an air of hesitancy blocks the vision of the rising flow. It stirs up larger concerns, chiefly that a Rising Flow implies a directional charge within the universe. While the rest of the universe runs down, hyperlife steadily proceeds in the contrary direction up the universe. Life progresses toward more life, more kinds of life, more complexity of life, more something. At this point skepticism sets in. A modern intellectual detects the scent of progress.

Progress smells of human-centeredness. To some it stinks of religiosity. Among the earliest and most fervent supporters of Darwin's scandalous theories were Protestant theologians and seminarians. Here was scientific proof of the dominant status of mankind. Darwinism offered a beautiful model for the orderly march of insentient life toward the peak of known perfection: the human male.

The continuing abuse of Darwin's theories to bolster racism didn't help the notion of evolutionary "progress" either. More important in the story of progress's demise has been the wholesale downshift of human position from the center of the cosmos to an insignificant wisp on the edge of an insignificant spiral in a dusty corner of the universe. If we are marginal, then what progress can evolution have?

Progress is dead, and there is nothing to replace it. The death of progress is nearly official in the study of evolution, as well as in postmodern history, economics, and sociology. Change without progress is how we moderns see our destiny.

A theory of a second force rekindles the possibility of progress and raises troublesome questions: If there is a second law of life—a rising flow—what is it flowing toward? What direction could evolution have if indeed it has a direction? Does life progress, or just wander? Perhaps evolution has a mere slope, which shapes its possibilities and makes it partially predictable? Does the evolution of life (both organic and artificial) follow even small trends? Do human culture and other vivisystems mirror organic life, or can one variety progress without the others? Would an artificial evolution have its own agenda and goals completely outside the desires of its creators?

Our first answer would have to be that all progress seen in life and society is a human-induced illusion. The prevalent notion of a "ladder of progress" or a "great chain of being" in biology doesn't hold up under the facts of geological history.

Start with the first instance of life as the initial point. In a visual metaphor, imagine all descendants of that first life forming a slowly inflating sphere. The radius is time. Each creature alive at a given time becomes a spot on the surface of the sphere at that time.

At the 4-billion-year mark (today's date), the globe of life on Earth shows some 30 million species cramming its circumference. One dot, for example, represents humans; another dot on the far side of the sphere, the bacterium *E. coli*. All points on the sphere are equidistant from the first life; therefore none is superior to the other. All creatures on the globe at any one time are *equally evolved*, having engaged in evolution for an equal amount of time. To put it bluntly, humans are no more evolved than most bacteria.

Gazing at this spherical graph, it is hard to imagine how one spot, the humans, could somehow be the apex of the entire globe. Perhaps any of the other 30 million coevolved spots—say, the flamingo, or poison oak—are the whole point of evolution. As life explores new niches, the whole globe expands, increasing the number of coevolved positions.

The globe graph of life quietly undermines the recurring image of progressive evolution: that of life beginning as a blob and climbing the ladder of success to the pinnacle of humanness. That image leaves out a billion other ladders that should be in the picture, including the all-too-common story of life as a blob climbing a ladder-going-nowhere to the pinnacle of a slightly different blob. In nature, there is no pinnacle, just a billion-spotted sphere. It doesn't matter what you do as long as you make it.

Hanging out and staying the same works too. There are many more cases of species who spent their evolutionary time treading water than who spent it transforming radically. The rewards are identical, however. Both *Homo sapiens* and *E. coli* are elite cosurvivors. And neither particularly has an advantage over the other in surviving the next million years. (Actually, some pessimists give *E. coli* 100-to-1 odds on outliving humans, even though *E. coli* can currently live only in our guts.)

While we can agree that evolutionary life exhibits no progress, perhaps it has a general direction?

In a quick survey of textbooks on evolution, I couldn't find a single one with the word "trends" or "direction" in the index. In the heated zeal to eradicate the notion of progress in evolution, many neodarwinians have banned any notions of trends or direction in evolution whatsoever. Stephen Jay Gould, one of the most outspoken naysayers about evolutionary trends, is actually one of the few biologists who even discusses the idea.

The central metaphor in *Wonderful Life*, Gould's entertaining book about the reinterpretation of the Burgess Shale fossils, is that the history of life can be thought of as a video tape. One can imagine rewinding life, and by some divine miracle, changing a pivotal scene at the beginning, and then rerunning life again from that point. This time-honored literary technique reached its apex in the all-American classic Christmas movie *It's a Wonderful Life*, from which Gould adapted his title. In this nearly archetypal drama, Jimmy Stewart's guardian angel replays Stewart's life without him.

If we could replay the epic story of biological life unfolding on Earth, would it progress in a similar story as the one we know?

Would life recapitulate any of its familiar stages, or would it stun us with contrary alternatives? Gould spins a masterful narrative of why he thinks we would not recognize life on Earth if evolution could be run again.

But since we have this magical tape of life mounted in our machine, there are further, and perhaps more interesting, things to do with it. If we turned out the lights, flipped the cassette at random, and then played it, would a visitor from another universe be able to tell if the tape was running properly forward or unconventionally backward?

What would the screen show if we played the epic Wonderful Life in reverse? Let's dim the lights and see. The story opens with a glorious, bluish Earth wrapped in a very thin film of living things, some mobile, some rooted. The cast of character types totals in the millions, half of them insects. In the opening scenes, not much happens. Plants morph into endless shapes. Some larger, very agile mammal things dissolve into similar, but smaller mammal things. Lots of insects melt into other insects, while some wholly new insect creatures appear. They too gradually merge into others. If we inspect any single character and follow it in slow motion, it's difficult to discern much sensible change going either forward or reverse. To speed the show, we fast-forward (fast-backward to us).

The screen shows life becoming sparse on the planet. Many, but not all, of the animal creatures begin to shrink in size. The total number of kinds of things decreases. The plot slows down. Living creatures inhabit fewer roles, and the roles change less and less as the tape proceeds. Life steadily collapses in scope and size until it becomes small, bland, and naked to the elements. In a very boring ending, the last variety of animated things disappear as they melt into a single tiny amorphous blob.

To review: a wide, complex, convoluted web of diverse forms just collapsed into a relatively simple, unitary speck of protein that mostly just copies itself.

What do you think, friend from Thor? Is the speck the alpha or the omega?

Life surely has a direction of time, but beyond that, neodarwinists would argue, nothing is sure. Since there are no directional trends in organic evolution, nothing about life's future can be forecast. Therefore the unpredictable nature of evolution is one of the few predictions we can make about it. Neodarwinists count on evolution being unpredictable. Who could have guessed while the fishes leaped in the oceans—the "pinnacle" of life and complexity at the time—that the really momentous long-range work was being done by some ugly freaks in dried up mud pools near land? Land, what's that?

The postdarwinists on the other hand keep bringing up the word "inevitable." In 1952, engineer Ross Ashby wrote in his influential book *Design for a Brain*, "The development of life on earth must *not* be seen as something remarkable. On the contrary, it was inevitable. It was inevitable in the sense that if a system as large as the surface of the earth, basically polystable, is kept gently simmering dynamically for five thousand million years, then nothing short of a miracle could keep the system away from those states in which the variables are aggregated into intensely self-preserved forms."

Real biologists cringe when "inevitable" is used in the same sentence as evolution. I believe the reflex is a vestigial response from the time when inevitable meant "God." But one of the few legitimate uses for artificial evolution—that even orthodox biologists will grant it—is as a test-bed for directional trends in evolution.

Might there be some fundamental constraints in the physical universe that channel life along a certain grain? Gould addresses this concern by comparing the possibility-space of life to the metaphor of "a very broad, low and uniform slope." Water dropped randomly onto this slope trickles down, eroding a chaotic path of microcanals. Newly hatched channels are reinforced as more water flows down, quickly carving out small valleys and permanently setting the location of succeeding larger canyons.

In Gould's metaphor, each tiny groove represents the historical timeline of a species. The initial groove sets the course for succeeding forms of genus, family and taxa. In the beginning, where the groove meanders is totally random, but once established, the course of the following canyons are fixed. Even though he admits his metaphor has an initial slope that "does impart a preferred direction to the water dropping on top," Gould insists that nothing disrupts the sure *uncertain* course of evolution. In his favorite refrain, if you replay this experiment over and over again, starting with a blank slope each time, you would get a vastly different landscape of valleys and peaks each run.

The curious thing is that if you actually set up Gould's thought experiment as a real test in a sand box, the results suggest an alternative view. First thing you notice as you repeat the experiment over and over again, as I have, is that the landscape formations are a very limited subset of all possible forms. Many landforms we are familiar with—rolling hills, volcano cones, arches, hanging valleys—will never appear. Thus one can safely predict what general structure the valleys and subsequent canyons will take: gentle gullies.

Second, while the starting groove begins at random in response to a random falling drop, the shape of further channel erosion follows a very homogeneous course. The canyon unfolds in an inevitable sequence. Continuing Gould's analogy, the initial drop is the first species on the scene; it might be any unexpected organism. Although its traits cannot be predicted, the sand-box analogy says that its descendants unfold somewhat predictably, according to trends inherent in the makeup of sand. So while there are points in evolution where results are sensitive to initial conditions (the birth of the Cambrian explosion could be one) this by no means rules out the influences of large trends.

Evolutionary trends were once promoted by prestigious biologists at the turn of the last century. One version is known as orthogenesis. Orthogenic (straight) life advanced in a direct line, from organism *A* through the alphabet of life to organism *Z*. A few orthogenesists in the past really thought evolution proceeded without branching: imagine a ladder climbing upwards, each species stationed on a rung, and every rung closer to heavenly perfection.

But even those orthogenesists who weren't so linear were often supernaturalists. They felt that evolution had direction because it was directed. The directing forces were supernatural purpose or some mysterious vital force that infused living things, or God himself. These notions were clearly outside the ken of science, so what little attraction the idea had to scientists was poisoned by its attraction to the mystical and new-agey.

But in the last several decades, godless engineers have made machines that set their own goals and seem to have their own purpose. One of the first to discover self-direction within machines was Norbert Wiener, the original cybernetic man. Wiener writes in 1950: "Not only can we build purpose into machines, but in an overwhelming majority of cases a machine designed to avoid certain pitfalls of breakdown will look for purposes which it can fulfill." Wiener implied that at a certain threshold of complexity of mechanical design, emergent purpose was inevitable.

Our own minds are a society of mindless agents; purpose emerges from that mix in exactly the same way purpose emerges from other nonintentional vivisystems. In a very real sense, a lowly thermostat has a purpose and a direction—to find the set temperature and hold it there. Astoundingly purposeful behavior can emerge from purposeless subbehaviors cultivated in software. Rod Brooks's MIT mobots built with bottom-up designs perform complicated tasks based on decisions and goals which percolate up from simple goal-less circuits. Genghis the robot insectoid *wants* to climb over phonebooks.

When evolutionists shook off God from evolution, they believed they had shaken off any trace of purpose and direction. Evolution was a machine without a designer, a watch made by a blind watchmaker.

Yet when we actually construct very complex machines, and when we dabble with synthetic evolution, we find that both run by themselves and acquire a sliver of their own agenda. Is the self-organizing order-for-free that Stuart Kauffman sees in adaptive systems, and the teleological goals that Rod Brooks can grow in machines, enough to suggest that evolution—however it came about—might have also evolved some goals and directions of its own?

If we look we may find that direction and goals can emerge in biological evolution from a mob of directionless and goal-less parts, without invoking vitalistic or supernatural explanations.

Experiments in computational evolution confirm this inherent teleogism, this self-produced "trend." Two complexity theorists, Mark Bedau and Norman Packard, have measured a number of evolutionary systems and concluded, "Just as recent studies of chaos have shown that deterministic systems could be unpredictable, we claim that deterministic systems may be teleological." For those with an ear that burns at the combined sound of "goal and evolution," it helps to consider this trait less as a conscious goal, plan, or willful purpose, and more as an "urge" or "tendency."

In the following list I suggest possible large-scale, self-generated tendencies in evolution. Tendencies, as I'm using the word here, are general and provide for exceptions. Not every lineage in a category will follow that trend.

As an example, take Cope's Law, a principle often found in textbooks. Cope was a swashbuckling bone collector in the 1920s who put dinosaurs on the map in more ways than one. He was a pioneer dinosaur surveyor and a tireless promoter of these exotic creatures. Cope noticed that, overall, mammals and dinosaurs seemed to increase in size over time. When studied carefully by later paleontologists, though, his observation applies to only about two-thirds of the cases on record; one can find plenty of exceptions to his rule even in the species lines he had in mind. If Cope's law was without exceptions then the largest living things on Earth would not be "primitive" fungi as large as city blocks hiding under the forest floors. Still, there *is* definitely a long-term trend in evolution that small things such as bacteria have preceded big ones such as whales.

Caveats aside, I discern about seven large trends or directions emerging from the ceaseless, hourly toil of organic evolution. These trends, as far as anyone can tell, are also the seven trends that will bias artificial evolution when it goes marathon; they may be said to be the Trends of Hyperevolution: *Irreversibility, Increasing Complexity, Increasing Diversity, Increasing Numbers of Individuals, Increasing Specialization, Increasing Codependency, Increasing Evolvability.*

***Irreversibility.*** Evolution doesn't back up. (Also known as Dollo's Law.) There are exceptions to the no-backup principle.

A whale in one sense backed up to be a fish again. But it is the exception that proves the rule. In general, current manifestations of life do not work on invading past niches.

Nor are hard-won attributes easily given up. It is an axiom in cultural evolution that technologies once invented are never uninvented. Once a vivisystem discovers language or memory it does not retreat from it.

The presence of life also does not retreat. I am aware of no geological domain that organic life has infiltrated and then retreated from. Once life settles in an environment (hot springs, alpine rock, robots) it will tenaciously maintain some presence there. Life exploits the inorganic world, recklessly transforming it into the organic. "Atoms, once drawn into the torrent of living matter, do not readily leave it," writes Vernadsky.

Prelife Earth was, by definition, a sterile planet. It is commonly accepted that although sterile, the Earth was simmering with the ingredients life needed. In essence it was a global agar plate waiting to be inoculated. Think of an immense 8,000-mile-wide bowl of pasteurized chicken broth. One day you drop a cell into it, and the next day, by the power of exponential growth, the oceanic bowl is thick with cells. In a few decades, all varieties of cells have wormed their way into every nook. Even if it took a hundred years, that is but a nano-blink in geological time. Life is born. Blink. Life is irrepressible.

Having infiltrated computers, artificial life will henceforth never retreat from being in some computer, somewhere.

***Increasing Complexity.*** When I ask friends if evolution has a direction the common answer I get (if I get any at all) is "towards more complexity."

While it seems obvious to almost everyone that evolution moves toward greater complexity, we have few definitions of complexity that really mean anything. Modern biologists question the notion that life heads toward complexity. Stephen Jay Gould has told me flatly, "The illusion of a move toward increasing complexity is an artifact. You need to build simple things first, so naturally complex things come later."

But there are plenty of simple things nature has never made.

If there was not a drive toward complexity, why not stop at bacteria and invent millions of more one-celled varieties. Or why not stop at fish and fill in all possible fish forms? Why make things more complicated? For that matter, why did life start out simple? There is no law we know of that says things *have* to get more complex.

If there is a true trend toward complexity, there must be something pushing it. In the last hundred years a number of theories have been proposed as to what drives apparent complexity. They could be listed by the following overlapping summaries (and the year they were first postulated):

- Runaway replication and duplication of parts make complexity (1871).
- The ruggedness of real environments causes differentiation of parts, which aggregate into complexity (1890).
- Complexity is more thermodynamically efficient (1960).
- Complexity is an inadvertent by-product of selection for other characteristics (1960).
- A complex organism creates a niche for more complexity around it; thus complexity is a positive feedback loop amplifying itself (1969).
- Since it is easier for a system to add a part than to remove a part, complexity accumulates (1976).
- Nonequilibrial systems accumulate complexity when they dissipate entropy, or wasted heat ( 1972).
- Chance alone produces complexity (1986).
- Endless arms races escalate complexity (1986).

Because the term complexity is vague and unscientific at present, no one has done a systematic study of the fossil record to determine whether or not quantitative complexity increases over time. A few studies of particular short lineages of organisms have been done (using differing measures of complexity) and they have shown that sometimes some aspects of these creatures increase in complexity and sometimes they don't. In brief, we don't know for sure what happens as organisms apparently complexify.

***Increasing Diversity.*** This one needs some careful clarification. One famous bed of fossils, the soft-bodied animals in the Burgess

Shale, is currently forcing a rethinking of what we mean by diversity. As Gould tells in *Wonderful Life*, the Burgess Shale show a remarkable range of alien organisms thriving during the innovation boom of the Cambrian. These fantastic creatures are far more diverse in their basic plan than the creatures we descended from. What we see since the Burgess Shale, Gould argues, is decreasing diversity of basic plans, with vastly increasing quantities of minor gingerbreading.

For instance, life churns out millions more kinds of insects, in ever more glorious modifications, but no more new kinds of things such as insects. Endless variations of trilobites, but no new classes such as trilobites. And since the Burgess Shale displays a smorgasbord of structural variety that beats the paltry choice of basic plans which life now offers in the same area, one could argue that the conventional view of diversity beginning small and ballooning over time is inverted.

If you count diversity as significant variety, then diversity is shrinking. Some paleontologists are calling this more fundamental diversity of ground plan "disparity" to distinguish it from the ordinary diversity of species. There is more significant difference (fundamental disparity) between a hammer and a saw, than there is between an electric table saw and a power circular saw or all the thousands of baroque electrical appliances manufactured today. Gould puts it this way, "Three blind mice of differing species do not make a diverse fauna, but an elephant, a tree, and an ant do—even though each assemblage contains just three species." We give more weight to fundamentals of clearly different logic in recognition that it's hard to come up with really innovative basic plans (try to imagine a universal alternative to the tubular gut!).

Because versatile basic plans are rare, when the majority of them go belly up, as they did after the Cambrian, never to be replaced, it's big news. This leads Gould to the "surprising fact of life's history—marked decrease in disparity followed by an outstanding increase in diversity within the few surviving designs." Take ten designs, throw away nine, and do the tenth one up in a bazillion variations, like beetles. The "cone of increasing diversity" we associate with evolution since the

Cambrian, then, is more appropriately figured within the level of species diversity, because more species types are alive today than ever before.

***Increasing Numbers of Individuals.*** There are also more individual organisms in total living now than a billion years ago, or perhaps even a million years ago. Presumably life originated only once, so there was once only the first living organism of Adamlike oneness. Now there are uncounted legions.

There is another important way the sheer number of living entities increases. In a hierarchical manner, supergroups and subgroups create individuals. Bees band together to form a colony, so now the number of individuals totals the number of bees plus one superorganism. A person is an individual made up of millions of individual cells which may also be counted and added to the increasing total of individual lives. Each of these cells may have a parasite, thus more individuals. In many overlapping ways, notions of individuals can be nested within each other in the same limited space. So within one cubic volume, a hive of bees with cells and mites and viral infections may have more individuals than the same volume full of bacteria. As Stanley Salthe writes in *Evolving Hierarchical Systems*, "An indefinite number of unique individuals can exist in a finite material world if they are nested within each other and that world is expanding."

***Increasing Specialization.*** Life starts as a process accomplishing many things in general. Over time a single life is differentiated into many individuals doing more specialized things. Just as a general egg cell differentiates through epigenesis to become a legion of specialized cells, so in evolution animals and plants split up into varieties more dependent on narrower niches. The word "evolution," in fact, originally meant the unrolling development of an egg cell into an embryonic creature. The term was only later applied to organic change over time for the first time by Herbert Spencer, who defined evolution (in 1862) as "a change from an indefinite, incoherent homogeneity, to a definite, coherent heterogeneity; through continuous differentiations and integrations."

The trends listed above can be gathered together with increasing specialization to create the following broad picture: Life begins as one, simple, vague, unformed creativity which, over time, becomes more and more fixed into a cloud of precise, inflexible, machinelike structures. Once differentiated, cell lines rarely revert to the more general. Once specialized, animal lines rarely revert to the more general. Over time the percentage of specialized organisms increases, the kinds of specialization increase, and the degree of specialization increases. Evolution moves toward more detail.

***Increasing Codependency.*** Biologists have noticed that primitive organisms have a direct dependency on the physical environment. Some bacteria live inside rock; some lichens eat stone. Slight perturbations of these organisms' physical habitat have a strong impact (lichens are miners' canaries for acid-rain pollution for this reason). As life evolves it unbinds from the inorganic and interacts more with the organic. While plants are rooted directly to the earth, animals, which are rooted to the plants, are freer from the earth. Amphibians and reptiles generally fertilize their eggs and abandon them to the elements, while birds and mammals raise their young, and so are bound closer to life from birth. Over time the close intimacy with earth and minerals is replaced by a dependence on other living things. Parasites cuddling in the warm interior of an animal's gut may never touch anything outside of organic life. Likewise social animals: while ants may live in the ground, their individual lives are far more dependent upon the other ants than upon the soil around them. Deepening sociality is yet another form of life's increasing codependence on other life. Humans are an extreme example of increasing dependence on life rather than the abiotic.

Evolution pulls life away from the inert and binds to itself whenever possible, manufacturing a great something out of nothing.

***Increasing evolvability.*** In 1987, Cambridge zoologist Richard Dawkins presented a paper at the First Artificial Life Workshop entitled "The Evolution of Evolvability," wherein he explored the feasibility and advantages of evolution evolving itself. Around the same time Christopher Wills, writing in *Wisdom of the Genes*, also published a scenario of how genes might control their own evolvability.

Dawkins's thinking was inspired by his attempts to create an artificial evolution in Biomorph Land. He realized while playing God that certain rare innovations would not only provide an immediate advantage to an individual but were "evolutionarily pregnant" and loosened up future offspring's ability to vary widely. He used the example of the first segmented animal in real life which he called "a freak . . . [which was] not a dramatically successful individual." But something about animal segmentation was a watershed event that birthed a line of descendants who were champion *evolvers*.

Dawkins proposed a higher-level natural selection "which favors, not just adaptively successful phenotypes, but a tendency to evolve in certain directions, or even just a tendency to evolve at all." In other words, evolution would select not only for survivability, but also for evolvability.

The ability to evolve does not rest in a single trait or function—such as mutation rate—yet a function such as mutation rate will play a role in an organism's evolvability. If a species cannot generate requisite variety, it won't evolve. Its ability to modify its body plays a role in its evolvability, as does its behavioral plasticity. The flexibility of its genome is of critical importance. Ultimately the evolvability of a species is a systems characteristic that does not dwell in any single place, just as an organism's ability to survive does not rest in any single place.

Like all traits selected by evolution, evolvability must be accumulative. A weak innovation once adopted can serve as the platform for the birth of a stronger innovation. In this way, weak evolvability establishes an ongoing base for further evolvability to arise. Over the very long term, evolvability is an essential component of survivability. Thus a line of organisms with genes wired to increase evolvability would accumulate a decided ability (and advantage) to evolve. And so on ad infinitum.

The evolution of evolution is like getting the wish that Aladdin's lamp won't let you have: the wish for three more wishes.

It's the power to change the rules of the game legally. Marvin Minsky noticed a similar power of change-which-changes-its-own-rules in the development of a child's mind. Minsky: "A mind cannot really grow very much by only accumulating more and more new knowledge. It must also develop new and better ways to use what it already knows. That's Papert's Principle: Some of the most crucial steps in mental growth are based not simply on acquiring new skills but on acquiring new administrative ways to use what one already knows."

The process by which change is altered is the larger target of evolution. The evolution of evolution does not mean merely that the mutation rate is evolving, although it could entail this. In fact, the mutation rate is remarkably constant over time throughout not only the organic world but also the world of machines and hyperlife. (It is rare for mutation rates to go above a few percent and rare for them to drop below a hundredth of a percent. Somewhere around a tenth of a percent seems to be ideal. That means that a nonsensical wild idea once in a thousand is all that is needed to keep things evolving. Of course one in a thousand is pretty wild for some places.)

Natural selection tends to maintain a mutation rate for maximal evolvability. But for the same advantage, natural selection will move all parameters of a system to the optimal point where further natural selection can take place. However that point of optimal evolvability is a moving target shifted by the very act of reaching for it. In one sense, an evolutionary system is stable because it continually returns itself to the preferred state of optimal evolvability. But because that point is moving—like a chameleon's colors on a mirror—the system is perpetually in disequilibrium.

The genius of an evolutionary system is that it is a mechanism for generating perpetual change. Perpetual change does not mean recurrent change, as the kaleidoscope of pedestrian action on a street corner may be said to endure perpetual change. That's really perpetual dynamism. Perpetual change means persistent disequilibrium, the permanent almost-fallen state. It means change that undergoes change itself. The result will be a system that is always on the edge of changing itself out of existence.

Or into existence. The capacity to evolve must be evolved itself. Where else did evolution come from in the first place?

If we accept the theory that life evolved from some kind of nonlife, or protolife, then *evolution had to precede life*. Natural selection is an abiological consequence; it could very well work on protoliving populations. Once fundamental varieties of evolution were operating, more complex varieties kicked in as the complexity of forms allowed. What we witness in the fossil record of Earthly life is the gradual accumulation of various types of simpler evolutions into the organic whole we now call evolution. Evolution is a conglomeration of many processes which form a society of evolutions. As evolution has evolved over time, *evolution itself has increased in diversity and complexity and evolvability*. Change changes itself.

A summary of evolution's evolution may be hypothesized as follows. In the beginning, evolution started as varying self-replication that produced enough of a population to induce natural selection. Once populations bubbled up, directed mutation became important. Next symbiosis became a major mover and shaker feeding off the change produced by natural selection. As forms grew larger, the constraints of form set in. As genomes grew in length, internal selection began to rule the genome. With the cohesion of the gene, speciation and species level selection kicked in. With organisms of sufficient complexity, behavioral and somatic evolution emerged. Eventually, when intelligence came on the scene, Lamarckian cultural evolution took over. As we humans introduce genetic engineering and self-programming robots, the makeup of evolution on Earth will continue to evolve.

The history of life, then, is a progression through a variety of evolutions brought about by the expanding complexity of life. As life becomes more hierarchical—genes, cells, organisms, species—evolution shifts its work. Yale University biologist Leo Buss claims that in each stage of evolution's evolution the unit subjected to natural selection shifts the tangled hierarchy to a new level of selection. Buss writes, "The history of life is a history of different units of selection." Natural selection selects individuals; Buss says that what constitutes an individual evolves over time. As an example, billions of years ago cells were the unit of natural selection, but eventually cells banded together and natural selection shifted to selecting their group—a multicellular organism—as the individual to select upon. One way to look at this is to say what constitutes an evolutionary individual evolves. At first an individual was a stable system, then a molecule, then a cell, then an organism. What next? Ever since Darwin, many imaginative evolutionists have proposed "group selection," evolution that works on groups of species as if a species were an individual. Certain *kinds* of species would survive or die not because of the survivability of the organism but because of unknown qualities of its specieshood—perhaps its evolvability.

Group selection is still a controversial idea but no less controversial than Buss's larger conclusion that "the major features of evolution were shaped during periods of transition between units of selection." Thus, he says, "At each transition—at each stage in the history of life in which a new self-replicating unit arose—the rules regarding the operation of natural selection changed utterly." In brief, natural evolution evolves.

Artificial evolution will likewise evolve, both artificially and naturally. We will engineer it to accomplish certain jobs, and we'll breed many species of artificial evolution to do particular jobs better. Many years hence, you'll be able to select a particular brand of artificial evolution out of a catalog to get just that right amount of novelty, or the perfect touch of self-guidance. But artificial evolution will also evolve with a certain bias that it shares with all evolutionary systems. Each variety will, for certain, remain out of our exclusive control and carry its own agenda.

If there truly are varieties of artificial evolution and a mixture of subevolutions themselves evolving within that thing we call evolution, then what are the characteristics of this larger evolution, this change of change? What are the traits of hyperevolution—both the general class of evolutions, and the greater evolution that moves through them—and where is it headed? What does evolution want?

I tally the evidence and say that evolution moves towards itself.

The process of evolution gathers itself up ceaselessly and

remakes itself over and over again in time. With every remaking, evolution becomes a process more able to alter itself. It is thus "source and fruition at once."

The mathematics of evolution is not driving it toward more flamingos, more dandelions, or more of any particular entity. Fecundity is a free by-product of evolution—here, have a few million frogs—rather than a goal. Instead evolution moves in the direction of actualizing itself.

Life is the substrate for evolution. Life provides the raw material of organisms and species which allows evolution to evolve further. Without a parade of complexifying organisms, evolution cannot evolve more evolvability. So evolution generates complexity and diversity and millions of beings and thereby gives itself room to evolve into a more powerful evolver.

Any self-evolver must be a coyote trickster. The trickster is never satisfied in remaking itself. Every time it takes its tail and turns itself inside out, becoming a thing more convoluted, more flexible, more lobed and frilled, more dependent upon itself, it rests less and less before it grabs its tail again.

What does the universe gain by tolerating this relentless evolution accumulating ever more evolvability?

Possibilities, as far as I can see.

And, possibilities suit me fine as a destination.

# 22

# PREDICTION MACHINERY

"TELL ME ABOUT THE FUTURE," I plead.

I'm sitting on a sofa in the guru's office. I've trekked to this high mountain outpost at one of the planet's power points, the national research labs at Los Alamos, New Mexico. The office of the guru is decorated in colorful posters of past hi-tech conferences that trace his almost mythical career: from a maverick physics student who formed an underground band of hippie hackers to break the bank at Las Vegas with a wearable computer, to a principal character in a renegade band of scientists who invented the accelerating science of chaos by studying a dripping faucet, to a founding father of the artificial life movement, to current head of a small lab investigating the new science of complexity in an office kitty-corner to the museum of atomic weapons at Los Alamos.

The guru, Doyne Farmer, looks like Ichabod Crane in a bolo tie. Tall, bony, looking thirty-something, Doyne (pronounced Doan) was embarking on his next remarkable adventure. He was starting a company to beat the odds on Wall Street by predicting stock prices with computer simulations.

"I've been thinking about the future, and I have one question," I begin.

"You want to know if IBM is gonna be up or down!" Farmer suggests with a wry smile.

"No. I want to know why the future is so hard to predict."

"Oh, that's simple."

I was asking about predicting because a prediction is a form of control. It is a type of control particularly suited to distributed systems. By anticipating the future, a vivisystem can shift its stance to preadapt to it, and in this way control its destiny. John Holland says, "Anticipation is what complex adaptive systems do."

Farmer likes to use a favorite example when explaining the anatomy of a prediction. "Here catch this!" he says tossing you a ball. You grab it. "You know how you caught that?" he asks. "By prediction."

Farmer contends you have a model in your head of how baseballs fly. You could predict the trajectory of a high-fly using Newton's classic equation of *f=ma*, but your brain doesn't stock up on elementary physics equations. Rather, it builds a model directly from experiential data. A baseball player watches a thousand baseballs come off a bat, and a thousand times lifts his gloved hand, and a thousand times adjusts his guess with his mitt. Without knowing how, his brain gradually compiles a model of where the ball lands—a model almost as good as *f=ma*, but not as generalized. It's based entirely on a series of hand-eye data from past catches. In the field of logic such a process is known as induction, in contradistinction to the deduction process that leads to *f=ma*.

In the early days of astronomy before the advent of Newton's *f=ma*, planetary events were predicted on Ptolemy's model of nested circular orbits—wheels within wheels. Because the central premise upon which Ptolemy's theory was founded (that all heavenly bodies orbited the Earth) was wrong, his model needed mending every time new astronomical observations delivered more exact data for a planet's motions. But wheels-within-wheels was a model amazingly robust to amendments. Each time better data arrived, another layer of wheels inside wheels inside wheels was added to adjust the model. For all its serious faults, this baroque simulation worked and "learned." Ptolemy's simple-minded scheme served well enough to regulate the calendar and make practical celestial predictions for 1400 years!

An outfielder's empirically based "theory" of missiles is reminiscent of the latter stages of Ptolemic epicyclic models. If we parsed an outfielder's "theory" we would find it to be incoherent, ad-hoc, convoluted, and approximate. But it would also be evolvable. It's a rat's-nest of a theory, but it works and improves. If humans had to wait until each of our minds figured out *f=ma* (and half of *f=ma* is worse than nothing), no one would ever catch anything. Even knowing the equation now doesn't help. "You can do the flying baseball problem with *f=ma*, but you can't do it in the outfield in real-time," says Farmer.

"Now catch this!" Farmer says as he releases an inflated balloon. It ricochets around the room in a wild, drunken zoom. No one ever catches it. It's a classic illustration of chaos—a system with sensitive dependence on initial conditions. Imperceptible changes in the launch can amplify into enormous changes in flight direction. Although the *f=ma* law still holds sway over the balloon, other forces such as propulsion and airlift push and pull, generate an unpredictable trajectory. In its chaotic dance, the careening balloon mirrors the unpredictable waltz of sunspot cycles, Ice Age's temperatures, epidemics, the flow of water down a tube, and, more to the point, the flux of the stock market.

But is the balloon really unpredictable? If you tried to solve the equations for the balloon's crazy flitter, its path would be nonlinear, therefore almost unsolvable, and therefore unforeseeable. Yet, a teenager reared on Nintendo could learn how to catch the balloon. Not infallibly, but better than chance. After a couple dozen tries, the teenage brain begins to mold a theory—an intuition, an induction—based on the data. After a thousand balloon takeoffs, his brain has modeled some aspect of the rubber's flight. It cannot predict precisely where the balloon will land, but it detects a direction the missile favors, say, to the rear of the launch or following a certain pattern of loops. Perhaps over time, the balloon-catcher hits 10 percent more than chance would dictate. For balloon catching, what more do you need? In some games, one doesn't require much information to make a prediction that is *useful*. While running from lions, or investing in stocks, the tiniest edge over raw luck is significant.

Almost by definition, vivisystems—lions, stock markets, evolutionary populations, intelligences—are unpredictable. Their messy, recursive field of causality, of every part being both cause and effect, makes it difficult for any part of the system to make routine linear extrapolations into the future. But the whole system can serve as a distributed apparatus to make approximate guesses about the future.

Farmer was into extracting the dynamics of financial markets so that he could crack the stock market. "The nice thing about markets is that you don't really have to predict very much to do an awful lot," says Farmer.

Plotted on the gray, end-pages of a newspaper, the graphed journey of the stock market as it rises and falls has just two dimensions: time and price. For as long as there has been a stock market, investors have scrutinized that wavering two-dimensional black line in the hopes of discerning some pattern that might predict its course. Even the vaguest, if reliable, hint in direction would lead to a pot of gold. Pricey financial newsletters promoting this or that method for forecasting the chart's future are a perennial fixture in the stock market world. Practitioners are known as chartists.

In the 1970s and 1980s chartists had modest success in predicting currency markets because, one theory says, the strong role of central banks and treasuries in currency markets constrained the variables so that they could be described in relatively simple linear equations. (In a linear equation, a solution can be expressed in a graph as a straight line.) As more and more chartists exploited the easy linear equations and successfully spotted trends, the market became less profitable. Naturally, forecasters began to look at the wild and woolly places where only chaotic *nonlinear* equations ruled. In nonlinear systems, the outcome is not proportional to the input. Most complexity in the world—including all markets—are nonlinear.

With the advent of cheap, industrial-strength computers, forecasters have been able to understand certain aspects of nonlinearity. Money, big money, is made by extracting reliable patterns out of the nonlinearity behind the two-dimensional plot of financial prices. Forecasters can extrapolate the graph's future and then bet on the prediction. On Wall Street the computer nerds who decipher these and other esoteric methods are called "rocket scientists." These geeks in suits, working in the basements of trading companies, are the hackers of the '90s. Doyne Farmer, former mathematical physicist, and colleagues from his earlier mathematical adventures, set up in a small, four-room house which serves as an office in adobe-baked Santa Fe—as far from Wall Street as one can get in America—are currently some of Wall Street's hottest rocket scientists.

In reality, the two-dimensional chart of stocks does not hinge on several factors but on thousands of them. The stock's thousands of vectors are whited-out when plotted as a line, leaving only its price visible. The same goes for charts of sunspot activity and seasonal temperature. You can plot, say, solar activity as a simple thin line over time, but the factors responsible for that level are mind-bogglingly complicated, multiple, intertwined, and recursive. Behind the facade of a two-dimensional line seethes a chaotic mixture of forces driving the line. A true graph of a stock, sunspot, or climate would include an axis for every influence, and would become an unpicturable thousand-armed monster.

Mathematicians struggle with ways to tame these monsters, which they call "high-dimensional" systems. Any living creature, complex robot, ecosystem, or autonomous world is a high-dimensional system. The Library of form is the architecture of a high-dimensional system. A mere 100 variables create a humongous swarm of possibilities. Because each behavior impinges upon the 99 others, it is impossible to examine one parameter without examining the whole interacting swarm at once. Even a simple three-variable model of weather, say, touches back upon itself in strange loops, breeding chaos, and making any kind of linear prediction unlikely. (The failure to predict weather led to the discovery of chaos theory in the first place.)

Pop wisdom says that chaos theory proves that these high-dimensional complex systems—such as the weather, the economy, army ants, and, of course, stock prices—are intrinsically no-way-around-it-unpredictable. So ironclad is the assumption, that in common perception any design for predicting the outcome of a complex system is considered naive or mad.

But chaos theory is vastly misunderstood. It has another face. Doyne Farmer, a boomer born in 1952, illustrates this with a metaphor from the age when music came on vinyl:

Chaos is like a hit record with two sides, he suggests.

- The lyrics to the hit side go: By the laws of chaos, initial order can unravel into raw unpredictability. You can't predict far.
- But the flip side goes: By the laws of chaos, things that look completely disordered may be predictable over the short term. You can predict short.

In other words, the character of chaos carries both good news and bad news. The bad news is that very little, if anything, is predictable far into the future. The good news—the flip side of chaos—is that in the short term, more may be more predictable than it first seems. Both the long-term, unpredictable nature of the high-dimensional systems, and the short-term, predictable nature of low-dimensional systems, derive from the fact that "chaos" is not the same thing as "randomness." "There is order in chaos," Farmer says.

Farmer should know. He was an original pioneer into the dark frontier of chaos before it gelled into a scientific theory and faddish field of study. In the hip California town of Santa Cruz of the 1970s, Doyne Farmer and friend Norm Packard cofounded a commune of nerd hippies who practiced collective science. They shared a house, meals, cooking, and credit on scientific papers. As the "Chaos Cabal," the band investigated the weird physics of dripping faucets and other seemingly random generating devices. Farmer in particular was obsessed with the roulette wheel. He was convinced that there must be hidden order in the apparently random spinning of the wheel. If one could discern secret order among the spinning chaos, then . . . why, one could get rich . . . very rich.

In 1977, long before the birth of commercial microcomputers such as the Apple, the Santa Cruz Chaos Cabal built a set of handcrafted programmable tiny microcomputers into the bottoms of three ordinary leather shoes. The computers were keyboarded with toes; their function was to predict the toss of a roulette ball. The home-brew computers ran code devised by

Farmer based on the group's study of a purchased second-hand Las Vegas roulette wheel set up in one of the commune's crowded bedrooms. Farmer's computer algorithm was based not on the mathematics of roulette but on the physics of the wheel. In essence, the Cabal's code *simulated the entire rotating roulette wheel and bouncing ball inside the chip in the shoe*. And it did this in a minuscule 4K of memory, in an era when computers were behemoths demanding 24-hour air-conditioning and an attendant priesthood.

On more than one occasion the science commune played out the flip side of chaos in the scene like this: Wired-up at the casino, one person (usually Farmer) wore a pair of magic shoes to calibrate the roulette operator's flick of the wheel, the speed of the bouncing ball, and the tilt of the wheel's wobble. Nearby, a Cabal cohort wore the third magic shoe linked by radio signals, and placed the actual bet on the table. Earlier, using his toes, Farmer had tuned his algorithm to the idiosyncrasies of a particular wheel in the casino. Now, in the mere 15 seconds or so between the drop of the ball and its decisive stop, his shoe-computer simulated the full chaotic run of the ball. About a million times faster than it took the real ball to land in a numbered cup, Farmer's prediction machinery buzzed out the ball's future destination on his right big toe. Typing with his left big toe, Farmer transmitted that information to his partner, who "heard" it on the bottom of his feet, and then, with a poker face, pushed the chips onto the predetermined squares before the ball stopped.

When everything worked, the chips won. The system never predicted the exact winning number; the Cabal were realists. Their prediction machinery forecasted a small neighborhood of numbers—one octave section of the wheel—as the bettable destination of the ball. The gambling partner spread the bets over this neighborhood as the ball finished spinning. Out of the bunch, one won. While the companion bets lost, the neighborhood as a whole would win often enough to beat the odds. And make money.

The group sold the system to other gamblers because of unreliability in the hardware. But Farmer learned three important things about predicting the future from this adventure:

- First, you *can* milk underlying patterns inherent in chaotic systems to make good predictions.
- Second, you don't need to look very far ahead to make a *useful* prediction.
- And third, even a *little bit* of information about the future can be valuable.

With these lessons firmly in mind, Farmer together with five other physicists (one of them a former Chaos Cabal member) engineered a start-up company to crack every gambler's dream: Wall Street. They would use high-powered computers. They would stuff them with experimental nonlinear dynamics and other esoteric rocket-scientist tricks. They would think laterally and let the technology do as much as possible without their control. They would create a thing, an organism if you will, that would on its own gamble millions of dollars. They would make it . . . (drum roll, please) . . . predict the future. With a bit of bravado, the old gang hung out their new shingle: the Prediction Company.

The guys in the Prediction Company figure that looking ahead a few days into the financial market future is all that is needed to make big bucks. Indeed, recent research done at the Santa Fe Institute, where Farmer and colleagues hang out, makes it clear that "seeing further is not seeing better." When immersed in real world complexity, where few choices are clear cut and every decision is clouded by incomplete information, evaluating choices too far ahead becomes counterproductive. Although this conclusion seems intuitive for humans, it has not been clear why it should pertain to computers and model worlds. The human brain is easily distracted. But let's say you have unlimited computing power specifically dedicated to the task of seeing ahead. Why wouldn't deeper, farther be better?

The short answer is that tiny errors (caused by limited information) compound into grievous errors when extended very far into the future. And the cost of dealing with exponentially increasing numbers of error-tainted possibilities just isn't worth the immense trouble, even if computation is free (which it never is). Santa Fe Institute investigators, Yale economist John Geanakoplos and Minnesota professor Larry Gray, used chess-playing computer programs as the test-bed for their forecasting work. (The best computer chess programs, such as the top-ranked Deep Thought, can beat all human players except for the very best grandmasters.)

Contrary to the expectations of computer scientists, neither Deep Thought nor human grandmasters need to look very far ahead to play excellent games. This limited look-ahead is called "positive myopia." Generally grandmasters survey the chess board and forecast the pieces only one move ahead. Then they select the most plausible play or two and investigate its consequences deeper. At every move ahead the number of choices to consider explodes exponentially, yet great human players will concentrate only on a few of the most probable countermoves at each rehearsed turn. Occasionally they search far ahead when they spot familiar situations they know from experience to be valuable or dangerous. But in general, grandmasters (and now Deep Thought) work from rules of thumb. For instance: Favor moves that increase options; shy from moves that end well but require cutting off choices; work from strong positions that have many adjoining strong positions. Balance looking ahead to really paying attention to what's happening now on the whole board.

Every day we confront similar tradeoffs. We must anticipate what lies around the corner in business, politics, technology, or life. However, we never have sufficient information to make a fully informed decision. We operate in the dark. To compensate we use rules of thumb or rough guidelines. Chess rules of thumb are actually pretty good rules to live by. (Notes to my daughters: Favor moves that increase options; shy away from moves that end well but require cutting off choices; work from strong positions that have many adjoining strong positions. Balance looking ahead to really paying attention to what's happening now on the whole board.)

Common sense embodies a "positive myopia." Rather than spend years developing a company employee manual that anticipates every situation that might arise—yet be out of date the moment it is printed—how much better to adopt positive myopia and not look so far ahead. Devise some general guidelines for the events that seem sure to arise "on the next move" and treat extreme cases if and when they come up. To navigate through rush-hour traffic in an unfamiliar city we can either plan detailed routes through the town on a map—thinking far ahead—or adopt a heuristic such as "Go west until we hit the river road, then turn left." Usually, we do a bit of both. We refrain from looking too far ahead, but we do look immediately in front. We meander west, or uphill, or downtown, while using the map to evaluate the next immediate turn ahead, wherever we are. We employ *limited* look-ahead guided by rules of thumb.

Prediction machinery need not see like a prophet to be of use. It needs only to detect limited patterns—almost any pattern—out of a background camouflage of randomness and complexity.

According to Farmer, there are two kinds of complexity: inherent and apparent. Inherent complexity is the "true" complexity of chaotic systems. It leads to dark unpredictability. The other kind of complexity is the flip side of chaos—apparent complexity obscuring exploitable order.

Farmer draws a square in the air. Going up the square increases apparent complexity; going across the square increases inherent complexity. "Physics normally works down here," Farmer says, pointing to the bottom corner of low complexity for both sorts, home of the easy problems. "Out there," pointing to the opposite upper corner, "it's all hard. But we are now sliding up to here, where it gets interesting—where the apparent complexity is high, but the true complexity is still low. Up here complex problems have something in them you can predict. And those are exactly the ones we are looking for in the stock market."

With crude computer tools that take advantage of the flip side of chaos, the Prediction Company hopes to knock off the easy problems in financial markets.

"We are using every method we can find," says partner Norman Packard, a former Chaos Cabalist. The idea is to throw proven pattern-finding strategies of any stripe at the data and "keep pounding on them" to optimize the algorithms. Find the merest hint of a pattern, and then exploit the daylights out of it. The mindset here is that of a gambler's: any advantage is an advantage.

Farmer and Packard's motivating faith that chaos possesses a flip side firm enough to bank on is based on their own experience. Nothing overcomes doubts like the tangible money they won from their Las Vegas roulette wheel experiments. It seems dumb not to take advantage of these patterns. As the chronicler of their high-rolling adventure exclaims in the book *The Eudaemonic Pie*, "Why would anyone play roulette *without* wearing a computer in his shoe?"

In addition to experience, Farmer and Packard place a lot of faith in the well-respected theories they invented during their years in chaos research. Now they are testing their wildest, most controversial theory yet. They believe, against the unbelief of most economists, that certain regions of otherwise complicated phenomena can be predicted accurately. Packard calls these areas "pockets of predictability" or "local predictability." In other words, the distribution of unpredictability is not uniform throughout systems. Most of the time, most of a complex system may not be forecastable, *but some small part of it may be for short times*. In hindsight, Packard believes local predictability is what allowed the Santa Cruz Cabal to make money forecasting the approximate path of a roulette ball.

If there are pockets of predictability, they will surely be buried under a haystack of gross unpredictability. The signal of local predictability can be masked by a swirling mess of noise from a thousand other variables. The Prediction Company's six rocket scientists use a mixture of old and new, hi-tech and low-tech search techniques to scan this combinatorial haystack. Their software examines the mathematically high-dimensional space of financial data and searches for local regions—any local region—that might match low-dimensional patterns they can predict. They search the financial cosmos for hints of order, any order.

They do this in real time, or what might be called hyperreal time. Just as the simulated bouncing roulette ball in the shoe-computer comes to rest before the real ball does, the Prediction Company's simulated financial patterns are played out faster than they happen on Wall Street. They reenact a simplified portion of the stock market in a computer. When they detect the beginnings of a wave of unfolding local order, they simulate it faster than real life and then bet on where they think the wave will approximately end.

David Berreby, writing in the March 1993 *Discover*, puts the search for pockets of predictability in terms of a lovely metaphor: "Looking at market chaos is like looking at a raging white-water river filled with wildly tossing waves and unpredictably swirling eddies. But suddenly, in one part of the river, you spot a familiar swirl of current, and for the next five or ten seconds you know the direction the water will move in that section of the river."

Sure, you can't predict where the water will go a half-mile downstream, but for five seconds—or five hours on Wall Street—you can predict the unfolding show. That's all you really need to be useful (or rich). Find *any* pattern and exploit it. The Prediction Company's algorithms grab a fleeting bit of order and exploit this ephemeral archetype to make money. Farmer and Packard emphasize that while economists are obliged by their profession to unearth the cause of such patterns, gamblers are not bound so. The exact reason why a pattern forms is not important for the Prediction Company's purposes. In inductive models—the kind the Prediction Company constructs—the abstracted causes of events are not needed, just as they aren't needed for an outfielder's internalized ballistic notions, or for a dog to catch a tossed stick.

Rather than worry about the dim relationships between causes and effects in these massively swarmy systems crowded with circular causality, Farmer says, "The key question to ask in beating the stock market is, what patterns should you pay attention to?" Which ones disguise order? Learning to recognize order, not causes, is the key.

Before a model is used to bet with, Farmer and Packard test it with backcasting. In backcasting techniques (commonly used by professional futurists) a model is built withholding the most recent data from the human managing the model. Once the system finds order in past data, say from the 1980s, it is fed the record of the last several years. If it can accurately predict the 1993 outcome, based on what it found in the 1980s, then the pattern seeker has won its wings. Farmer: "The system makes twenty models. We run them each through a sieve of diagnostic statistics. Then the six of us will get together to select the one to run live." Each round of model-building may take days on the Company's computers. But once local order is detected, a prediction based on it can be spun in milliseconds.

For the final step—running it live with bundles of real money in its fists—one of the Ph.D.'s still has to hit the "enter" button. This act thrusts the algorithm into the big-league world of very fast, mind-boggling big bucks. Cut loose from theory, running on automatic, the fleshed out algorithm can only hear the murmurs of its creators: "Trade, sucker, trade!"

"If we can earn 5 percent better than what the market does, then our investors will make money," Packard says. Packard clarifies that number by explaining that they can predict 55 percent of market moves, that is, 5 percent more than by random guessing, but that when they do guess right their result can be 200 percent better. The fat-cat Wall Street financial backers who invest in the Prediction Company (currently O'Connor & Associates) get exclusive use of the algorithms in exchange for payments according to the performance of the predictions. "We have competitors," Packard states with a smile. "I know of four other companies with the same thing in mind"—capturing patterns in chaos with nonlinear dynamics and predicting from them. "Two of them are up and going. Some involve friends."

One competitor trading real money is Citibank. Since 1990, British mathematician Andrew Colin has been evolving trading algorithms. His forecasting program randomly generates several hundred hypotheses of which parameters influence currency data, and then tests the hundred against the last five years of data. The most likely influences are sent to a computer neural net which juggles the weight of each influence to better fit the data, rewarding the best combinations in order to produce better guesses. The neural net system keeps feeding the results back in so that the system can hone its guess in a type of learning. When a model fits the past data, it is sent out into the future. In 1992 the *Economist* said, "After two years of experiments, Dr. Colin reckons his computer can make returns of 25 percent a year on its notional dealing capital . . . That is several times more than most human traders hope to make." Midland Bank in London has eight rocket scientists working on prediction machinery. In their scheme, computers breed algorithms. However, just as at the Prediction Company, humans evaluate them before "hitting the return button." They were trading real money by late 1993.

A question investors like to ask Farmer is how can he prove you can make money in markets with the advantage of only a small bit of information. As an "existence proof" Farmer points to the people such as George Soros earning millions year after year trading currencies and whatnot on Wall Street. Successful traders, sniffs Farmer, "are pooh-poohed by the academics as being extremely lucky—but the evidence goes the other way." Human traders unconsciously learn how to spot patterns of local predictability streaking through the ocean of random data. The traders make millions of dollars because they detect patterns (which they cannot articulate), then make an internal model (which they are unconscious of), in order to make predictions (which they are rewarded or punished for, sharpening the feedback loop). They have no more idea of what their model or theory is than of how they catch fly balls. They just *do*. Yet both kinds of models were empirically constructed in the same inductive Ptolemaic way. And that's how the Prediction Company employs computers to build models of high-flying stocks—from the data up.

Says Farmer, "If we are successful on a broad basis in what we are doing, it will demonstrate that machines are better forecasters than people, and that algorithms are better economists than Milton Friedman. Already, traders are hesitant about this stuff. They feel threatened by it."

The hard part is keeping it simple. Says Farmer, "The more complex the problem is, the simpler the models that you end up having to use. It's easy to fit the data perfectly, but if you do that you invariably end up just fitting to the flukes. The key is to generalize."

Prediction machinery is ultimately theory-making machinery—devices for generating abstractions and generalizations. Prediction machinery chews on the mess of seemingly random chicken-scratched data produced by complex and living things. If there is a sufficiently large stream of data over time, the device can discern a small bit of pattern. Slowly the technology shapes an internal ad-hoc model of how the data might be produced. The apparatus shuns "overfitting" the pattern on specific data and leans to the fuzzy fit of a somewhat imprecise generalization. Once it has a general fit—a theory—it can make a prediction. In fact prediction is the whole point of theories. "Prediction is the most useful, the most tangible and, in many respects, the most important consequence of having a scientific theory," Farmer declares. Manufacturing a theory is a creative act that human minds excel in, although, ironically we have no theory of how we do it. Farmer calls this mysterious general-pattern-finding ability "intuition." It's the exact technology "lucky" Wall Street traders use.

Prediction machinery is found in biology, too. As David Liddle, the director of a hi-tech think tank called Interval, says, "Dogs don't do math," yet dogs can be trained to predictively calculate the path of a Frisbee and catch it precisely. Intelligence and smartness in general is fundamentally prediction machinery. In the same way, all adaptation and evolution are milder and more thinly spread apparatus for anticipation and prediction.

Farmer confessed to a private gathering of business CEOs, "Predicting markets is not my long-term goal. Frankly, I'm the kind of guy who has a hard time opening to the financial page of the *Wall Street Journal*." For an unrepentant ex-hippie, that's no surprise. Farmer sees himself working for five years on the problem of predicting the stock market, scoring big time, and then moving on to more interesting problems—such as real artificial life, artificial evolution, and artificial intelligence. Financial forecasting, like roulette, is just another hard problem. "We are interested in this because our dream is to produce prediction machinery that will allow us to predict lots of different things"—weather, global climate, epidemics—"anything generating a lot of data we don't understand well."

"Ultimately," says Farmer, "we hope to imbue computers with a crude form of intuition."

By late 1993, Farmer and Company publicly reported success in predicting markets with "computerized intuition" while trading real money. Their agreement with their investors prohibits them from talking about specific performance, as much as Farmer is dying to. He did say, though, that in a few years they should have enough data to prove "by scientific standards" that their trading success is not a statistical fluke: "We really have found statistically significant patterns in financial data. There really are pockets of predictability out there."

WHILE RESEARCHING PREDICTION and simulation machinery, I had a chance to visit the Jet Propulsion Lab in Pasadena, California, where a state-of-the-art battle simulation was under development. I came to JPL at the invitation of a computer science professor from UCLA who had been pushing the edge of computer power. Like many researchers pinched for support, this professor had to rely on military funding for his avant-garde theoretical experiments. He paid for his end of the bargain by picking a practical military problem to test his theories on.

His test-bed was to see how decentralized, massively parallel computing—what I'm calling "swarm computing"—could speed up a computer simulation of a tank battle, an application which only remotely interested him. On the other hand, I was earnestly interested to see a state-of-the-art war game.

At the busy front desk of JPL, security clearance was straightforward. Considering that I visited the national research center while American troops were on red-alert along the Iraq border, the bouncers were fairly cordial. I signed some forms swearing my allegiance and citizenship, got a substantial badge to clip on, and was escorted with the professor to his cubbyhole office on an upper floor. In a small gray conference room, I met a long-haired graduate student who used the battle simulation mathematics as an excuse to pursue some far out notions on computational theories of the universe. Then I met the JPL honcho. He was nervously uncomfortable with my presence as a journalist.

Why? my professor friend asked him. The simulation system was not classified; the results were published in the open literature. The JPL honcho replied in so many words: "Well, umm, you see, there is this war going on, and quite inadvertently the generic scenario we have been dry-running for the last year or so—a game we chose quite by accident, with no thought of prediction—is being played out now for real. When we first tested this computer algorithm we had to pick some scenario, any scenario, to try out the simulation with. So we picked a simulated desert war with . . . Iraq and Kuwait. Now we are fighting this simulation. We are a bit on the spot here. It's a little sensitive. I'm sorry."

I did not get to see that war simulation. But about a year after the Gulf War's end, I discovered that JPL was not the only place that serendipitously preenacted that war. The U.S. Military Central Command in Florida ran a second and more useful simulation of a desert battle prior to the war. Cynics interpret the fact that the U.S. government had simulated the Kuwait war twice beforehand as a mark of its imperialist and conspiratorial desire to have that war. I find the predictive scenarios spooky, strange, and instructional rather than diabolical. I use this example to portray the potential power of prediction machinery.

There are about two dozen centers around the world that are playing war games where the U.S. is Blue—the protagonist. Most of these places are small departments at military schools and training centers, such as the Wargaming Center at Maxwell Air Force Base in Alabama, the legendary Global Game room at the Naval War College in Newport, Rhode Island, or the classic "sand box" table set-ups at the Army's Combat Concepts Agency in Leavenworth, Kansas. Providing them technical support and know-how are academics and savants holed up in the numerous para-military think tanks peppering the beltway of Washington, D.C., or research alleys nested in the corridors of national laboratories like JPL and Lawrence Livermore Labs in California. The toy war simulators, of course, carry acronyms; TACWAR, JESS, RSAC, SAGA. A recent catalog of military software listed four hundred varieties of war games or other military models for sale right off the shelf.

The nerve center for any U.S. military operations is headquartered at Central Command, based in Florida. For its entire existence, Central Command, as an organ of the Pentagon, had been hawking one major scenario to Congress and the American people: Blue vs. Red—the superpower game where the only worthy opponent was the Soviet Union. When General Norman Schwarzkopf came on the scene in the 1980s, he didn't buy this story. Schwarzkopf—a thinking man's general—put out a new perspective, worded in a way that's been quoted up and down the ranks: "The Soviet dog is not going to hunt." Schwarzkopf refocused his planners' attention on alternative scenarios. High on the list was a Mid-East desert war along the border of Iraq.

In early 1989, Gary Ware, an officer at Central Command, began modeling a war based on Schwarzkopf's hunches. Ware worked with a small cell of military futurists in compiling data to create a simulated desert war. The simulation was code-named Operation Internal Look.

Any simulation is only as good as the data it is based on, and Ware wanted Operation Internal Look based on reality as much as possible. That meant collecting a hundred thousand details about current forces in the Mid-East. Most of the work was horribly dull. The war simulation needed to know the number of vehicles in the Mid-East, stockpile strengths of food and fuel, killing power of weapons, climate conditions, and so on. Most of this minutiae was not readily available, even to the military. All bits were constantly in flux.

Once Ware's team worked out a formulation of an army's organization, the war gamers compiled optical laser disc maps of the entire Gulf area. The foundation of the simulated desert war—the territory itself—was transferred from the latest satellite digitized photos. When they finished, the war gamers had the countries of Kuwait and Saudi Arabia compressed onto a CD. They were now ready to feed all this data into TACWAR, the main computerized war-gaming simulator.

In early 1990 Ware began running a desert war on the virtual battlefield of Kuwait and Saudi Arabia. In July, in a conference room in north Florida, Gary Ware summarized the results of Operation Internal Look for his superiors. They reviewed a scenario based on Iraq invading Saudi Arabia, and the U.S./ Saudi Arabia striking back. Ware's simulation forecast a fairly brief thirty-day war if anything this unlikely should occur.

Two weeks later, Saddam Hussein suddenly invaded Kuwait. At first, the upper echelons of the Pentagon had no idea they already owned a fully operational, data-saturated simulation of the war. Turn the key and it would run endless what-ifs of possible battles in that zone. When word of the prescient simulation surfaced, Ware came out smelling like roses. He admitted that "If we had to start from scratch at the time of the invasion we would have never caught up." In the future, standard army-issue preparedness may demand having a parallel universe of possible wars spinning in a box at the command center, ready to go.

Immediately after Saddam's initial invasion, the war gamers shifted Internal Look to running endless variations of the "real" scenario. They focused on a group of possibilities revolving around the variant: "What if Saddam keeps on coming right away?" It took Ware's computers about 15 minutes to run each iteration of the forecasted thirty-day war. By running those simulations in many directions the team quickly learned that airpower would be the decisive key in this war. Further refined iterations clearly showed the war gamers that if airpower was successful, the U.S. war would be successful.

Further, according to Ware's prediction machinery, if airpower could actually inflict the results assigned to it, U.S. ground forces would not sustain heavy losses. The top brass took this to mean that precise upfront airpower was the linchpin to low U.S. casualities. Gary Ware says, "Schwarzkopf was so adamant on maintaining the absolute minimum casualties of our forces that low casualities became the benchmark upon which all our analysis was done."

Predictive simulations, then, gave the command team the confidence that the U.S. could achieve success with minimum losses. This confidence led to the heavy air campaign. Says Ware, "The simulations definitely had an impact on our thinking [at Central Command]. Not that Schwarzkopf didn't have prior strong feelings, but the model gave us confidence that we could carry through the concepts."

As a prediction, Operation Internal Look got good marks. Despite some shifts in the initial balance of forces, the thirty-day simulated air and ground campaign was pretty close to the real sequence, although the percentage of air and ground action was slightly different. The ground battle pretty much unfolded as forecasted. Like everyone outside the field, the simulators were surprised by how fast Schwarzkopf's end run around the front lines went. Says Ware, "I have to tell you, though, that we did not expect to get so far [on the battlefield] as we did in a hundred hours. As I recall, we forecasted a six-day land battle instead of a hundred-hour [four-day] battle. The ground commanders had told us that they envisioned moving faster than the simulation indicated they would. So they moved exactly as fast as they predicted."

The war game prediction machinery figured greater resistance from the Iraqis than the Iraqis actually gave. That's because every combat simulation assumes that the enemy will employ all of its available systems. But Iraq never pushed hard at all. The war gamers cheekily joked that *no* model reflects the white flag as a weapons system.

The war moved so fast the simulationists never got around to the obvious next step in simulations: daily modeled forecasts of the battle in progress. Although the planners recorded every day's events as best as they could, and they *could* project out into the future from any moment, they felt "it didn't take a genius to figure what was going on after about the first 12 hours."

IF SILICON CHIPS ARE ENOUGH of a crystal ball to help steer a superarmy war, and algorithms coursing through small computers are enough predictive technology to outguess the stock market, then why not reconfigure a supercomputer to predict the rest of the world? If human society is just a large distributed system of agents and machines, why not construct an apparatus to forecast its future?

Even a cursory study of past predictions shows why not. On the whole, cultural predictions historically have been worse than random guesses. Old books are a graveyard of prophesied futures that never came to pass. A few prophecies hit the bullseye, but there is no way to discern beforehand the rare right one from the plentiful wrong ones. Since predictions are so often wrong, and since believing erroneous predictions is so tempting and so misleading, some professional futurists avoid predictions altogether on principle. To emphasize the corrupting unreliability of trying to prophesy, these futurists prefer to state their prejudice in deliberate exaggeration: "All predictions are wrong."

They have a point. So few long-term predictions prove correct that statistically they *are* all wrong. Yet, by the same statistical measure, so many *short-term* predictions are right, that all short-term predictions are right.

There is nothing more certain about a complex system than to say it will be just like it is now a moment later. This observation is nearly a truism. Systems are things that keep persisting; so it is only tautological that from one moment to the next a system—even a living thing—doesn't change much. An oak tree, the post office, and my Macintosh hardly change at all from one day to the next. I offer an easily guaranteed short-term prediction for complex things anywhere: tomorrow will be mostly like today.

Equally true is the cliché that things occasionally do change from one day to the next. But can these immediate alterations be predicted? And if they can, could you stack up a series of predictable short-term changes into a probable medium-range trend?

Yes. While long-range predictions will remain essentially unpredictable, short-range predictions for complex systems are not only possible, they are essential. Furthermore, some types of mid-range predictions are quite feasible, and becoming more so. For reasons I will explain below, the human ability to forecast aspects of our society, economy, and technology will steadily increase despite the Alice-in-Wonderland strangeness that dependable predictions will have upon present actions.

We have the technology now to forecast many social phenomena, if we can catch them at the right moment. I follow the work of Theodore Modis, whose 1992 book, *Predictions*, nicely sums up the case for utility and believability of predictions. Modis addresses three types of found order in the greater web of human interactions. Each variety forms a pocket of predictability at certain times. He applies his research to the domain of economics, social infrastructure, and technology, but I believe his findings apply to organic systems as well. The three pockets of Modis: *Invariants*, *Growth Curves*, *Cyclic Waves*.

***Invariants.*** The natural and unconscious tendency for all organisms to optimize their behavior instills in that behavior "invariants" that change very little over time. Humans in particular are certified optimizers. Twenty-four hours of time per day is an absolute invariant, so over decades people, on average, tend to spend a remarkably constant amount of time on such chores as cooking, traveling, cleaning—although the distance or what they accomplish during that time might change. If new activities (say airplane flight instead of walking) are reformulated into elemental dimensions for analysis (how much time is spent in daily moving), the new behaviors often exhibit a continuous pattern with the old that can be extrapolated (and predicted) into the future. Instead of walking a half hour to work, you now drive a half hour to work. In the future, you may fly a half hour to work. Marketplace pressures for efficiency are so relentless and unforgiving that they inevitably push human-made systems in a single (predictable) direction toward optimization. Tracing an invariant optimization point can often alert us to a clean pocket of predictability. For instance, improvement in mechanical efficiency is very slow. No system is yet over 50 percent efficient. A projected system operating on 45 percent efficiency is possible, but one that requires 55 percent is not. Therefore one can safely make a short-term prediction about fuel efficiency.

***Growth Curves.*** The larger, more layered, more decentralized a system is, the more it takes on aspects of organic growth. Growing things share several universal characteristics. Among them are a lifespan that can be plotted as an S-shaped curve: slow birth, steep growth, slow decline. The worldwide production of cars per year or the lifetime production of symphonies composed by Mozart both fit an S-curve with great precision. "The predictive power of S-curves is neither magical nor worthless," writes Modis. "What is hidden under the graceful shape of the S-curve is the fact that natural growth obeys a strict law." This law says that the shape of the ending is symmetrical to the shape of the beginning. The law is based on empirical observations of thousands of biological and institutional life histories. The law is closely related to the natural distribution of complex things as expressed in a bell curve. Growth is extremely sensitive to initial conditions; the first data points on a growth curve are almost meaningless. But once a phenomenon is on a roll, a numerical snapshot of its history can be taken and flipped over to predict the phenomenon's eventual limits and demise. One can extract from the curve a cross-over point with a competing system, or a "ceiling" and a date when the ceiling essentially flattens out. Not every system exhibits a smooth S-curve lifespan; but a remarkable variety and number do. Modis believes that more things adhere to the laws of growth than we suspect. If such growing systems are examined at the right time (midway in their history), then the presence of local order—summed up by the S-curve law—affords yet another pocket of predictability.

***Cyclic Waves.*** The apparent complex behavior of a system is partly a reflection of the complex structure of the system's environment. This was pointed out over 30 years ago by Herbert Simon, who used the journey of an ant over the ground as an illustration. The ant's jig-jagging path across the soil reflected not the ant's complex locomotion but the complex structure of its environment. According to Modis, cyclic phenomenon in nature can infuse a cyclic flavor to systems running within it. Modis is intrigued by the 56-year economic cycles discovered by economist N. D. Kondratieff. In addition to Kondratieff's economic waves, Modis adds similar 56-year cycles in scientific advances described by himself, and 56-year cycles in infrastructure replacement studied by Arnulf Grubler. The causes of these apparent waves have been hypothesized by various other authors as coming from 56-year lunar cycles, or every fifth 11-year sunspot cycle, or even from the every-other cycle of human generations—as each 28-year generational cohort swings away from the work of its parental cohort. Modis argues that primary environmental cycles trigger many secondary and tertiary internal cycles in their wake. Seekers who uncover any fragments of these cycles can use them to predict pockets of behavior.

Together, these three modes of prediction suggest that at certain moments of heightened visibility, the invisible pattern of order becomes clear to those paying attention. Like the next beat of a drum, its future can almost be heard. A moment later, the pattern is gone, muddied and overwritten by noise. Pockets of prediction won't keep away big surprises. But local predictability does point to methods that can be improved, deepened, and lengthened into bigger things.

The long odds against successful big predictions haven't discouraged hordes of amateur and full-time financial chartists attempting to extract longwave patterns from past stock market prices. Any external cyclic behavior is fair game for a chartist: the length of women's hemlines, the age of presidents, the price of eggs. Chartists are forever chasing the mythical "leading indicator" that will predict the destiny of stock prices as a number they can bet on. For many years chartists were ridiculed for their vaguely numerological approach. But in recent years academics such as Richard J. Sweeney and Blake LeBaron have shown that chartist methods often do work. A chartist's technical rule can be stunningly simple: "If the market has been going up for a while, bet that it will continue to go up. If it's on a downward trend, bet it will continue downward." Such a rule reduces the high dimensionality of a complex market to the low dimensionality of this simple two-part rule. In general, this kind of pattern-seeking works. The "up-up, down-down" pattern performs better than random chance, and thus better than the average investor. Since stasis is the most predictable thing about a system this pattern of order should not come as a surprise, even though it does.

In opposition to chartism, other financial forecasters rely on the "fundamentals" of the market in an effort to predict it. Fundamentalists, as they are called, attempt to understand the driving forces, the underlying dynamics, and the fundamental conditions of a complex phenomenon. In short they seek a theory: $f=ma$.

Chartists, on the other hand, seek a pattern from the data without concern for whether they understand why the pattern is there. If there is order in the universe, then somewhere, somehow, all complexity will disclose—at least momentarily—order that reveals its future path. One merely needs to learn what signals to disregard as noise. Chartism is organized induction in Doyne Farmer's mode. Farmer admits that he and his fellows at the Prediction Company are "statistically rigorous chartists."

In another fifty years, computerized induction, algorithmic chartism, and pocket predictionism will be respectable human endeavors. Forecasting stock markets will remain an oddball case because, more than other systems, stock markets are built out of expectations. In an expectation game, accurate predictions offer no opportunity for money-making if everyone shares the prediction. All the Prediction Company can really own is lead time. As soon as Farmer's group makes much money exploiting a pocket of predictability, others will rush in, somewhat clouding the pattern, but mostly leveling the opportunity to make any money. In a stock market, success stirs up strong self-canceling feedback currents. In other systems, such as a growing network, or an expanding corporation, anticipatory feedback is not self-canceling. Ordinarily, feedback is self-governing.

THE ORIGINAL CYBERNETICIST, Norbert Wiener, struggled to explain the immense power of feedback control. Wiener had in mind simple toilet-flusher type feedback. He noticed that delivering a constant weak trickle of information about what the system had just accomplished ("the water level is still down") into the system in some way directed the whole system. Wiener concluded that this power was a function of time-shifting. He wrote in 1954: "Feedback is a method of controlling a system by reinserting into it the results of its past performance."

There's no puzzle in a sensor sensing the present. What more does one need to know about the present other than it is here and now? It obviously pays for a system to mind the present since it has little other choice. But why expend resources on what is gone and cannot be changed? Why raid the past for present control?

A system—organism, corporate firm, computer program—spends energy feeding the past back into the present because this is an economical way for the system to deal with the future. To see into the future one must see into the past. A constant pulse of the past along feedback loops informs and controls the future.

But there is another avenue for a system to time-shift into the future. Sense organs in a body that pick up sound and light waves miles away act as meters of the present and more as gauges of the future. Events geographically distant are, for practical purposes, events that hail from the future. An image of an approaching predator becomes information about the future now. A distant roar may soon be an animal up close; a whiff of salt signals a soon-to-be change in tide. Thus an animal's eye "feed-forwards" information from a distant time/space into its here/now body.

Some philosophers say it is no coincidence that life arose on a planet bathed in two mediums—air and water—amazingly transparent in most spectrums. A cleanly transparent environment permits organs to receive data-rich signals from "distant" (future) events and process them in anticipation of a response from the organism. Eyes, ears, and noses are thus prediction machinery to peer into time.

Completely opaque water or air, according to this notion, might have squelched the development of anticipation machinery by preventing information about distant events from reaching the present. Organisms in an opaque world would be cramped in both space and time; they would lack the room to develop adaptive responses. Adaptation—at its core—requires a sense of the future. In a changing environment, either opaque or clear, systems that anticipate the future are more likely to persist. Michael Conrad writes, "At bottom adaptability is the use of information to handle environmental uncertainty." Gregory Bateson put it telegraphically when he said, "Adaptation is change in the service of nonchange." A system (nonchange by definition) adapts (changes) in order to persist (nonchange). A flamingo adapts in order to persist.

Thus, systems stuck solely in the present will more often be surprised by change, and die. Therefore, a transparent environment rewards the evolution of predictive machinery, because prediction machinery confers survivability upon complexity. Complex systems survive because they anticipate, and a transparent medium helps them anticipate. Opaqueness, on the other hand, would hinder anticipation, adaptation, and evolution of complex vivisystems altogether.

POSTMODERN HUMANS swim in a third transparent medium now materializing. Every fact that can be digitized, is. Every measurement of collective human activity that can be ported over a network, is. Every trace of an individual's life that can be transmuted into a number and sent over a wire, is. This wired planet becomes a torrent of bits circulating in a clear shell of glass fibers, databases, and input devices.

Once moving, data creates transparency. Once wired, a society can see itself. The reason the rocket scientists at the Prediction Company can fare better than the chartists of old is that they work in a more transparent medium. The billion computerized bits sloughed off by networked financial institutions clot into a transparent air through which the Company can detect unfolding patterns. The cloud of data flowing through their workstations forms a clear digital globe for them to peer into. In certain patches of the new air they can see ahead.

At the same time, industrial factories mass-produce video cameras, tape recorders, hard disks, text scanners, spreadsheets, modems, and satellite dishes. Each of these is an eye, an ear, or a neuron. Connected together they form a billion-lobed sense organ floating in the clear medium of whizzing digits. This tissue serves to feed-forward information from distant limbs into the body electric. The U.S. Command Center wargamers can use the digitized land-terrain of Kuwait, just-in-time satellite images, and the relayed reports of hand-held transmitters anchored by global positioning information (accurate to within 50 feet anywhere on Earth) to anticipate—to see in the collective mind's eye—the course of an approaching battle.

Telling the future, when it comes right down to it, is not solely a human yearning. It is the fundamental nature of any organism, and perhaps any complex system. Telling the future is what organisms are for.

My working definition of a complex system is a "thing which talks to itself." One might ask, then: What is the story that complex systems tell themselves? The answer is that they tell themselves stories of the future. Stories of what might come next—whether next is reckoned in nanoseconds or years.

IN THE 1970s, after thousands of years of telling tales about the Earth's past and creation, the inhabitants of planet Earth began to tell their first story of what might happen to the planet in the future. Rapid communications of the day gave them their first comprehensive real-time view of their home. The portrait from space was enchanting—a cloudy blue marble hanging delicately in the black deep. But down on the ground the emerging tale wasn't so pretty. Reports from every quadrant of the globe said the Earth was unraveling.

Tiny cameras in space brought back photographs of the whole Earth that were awesome in the old-fashioned sense of the word: at once inspiring and frightening. The cameras, together with reams of ground data pouring in from every country, formed a distributed mirror reflecting a picture of the whole system. The entire biosphere was becoming more transparent. The global system began to look ahead—as systems do—wanting to know what might come next, say, in the next 20 years.

The first impression arising from the data-collecting membrane around the world was that the planet was wounded. No static world map could verify (or refute) this picture. No globe could chart the ups and downs of pollution and population over time, or decipher the interconnecting influence of one factor upon another. No movie from space could play out the question, what if this continues? What was needed was a planetary prediction machine, a global what-if spreadsheet.

In the computer labs of MIT, an unpretentious engineer cobbled together the first global spreadsheet. Jay Forrester had been dabbling in feedback loops since 1939, perfecting machinery-steering servomechanisms. Together with Norbert Wiener, his colleague at MIT, Forrester followed the logical path of servomechanisms right into the birth of computers. As he helped invent digital computers, Forrester applied the first computing machines to an area outside of typical engineering concerns. He created computer models to assist the management of industrial firms and manufacturing processes. The usefulness of these company models inspired Forrester to tackle a simulation of a city, which he modeled with the help of a former mayor of Boston. He intuitively, and quite correctly, felt that cascading feedback loops—impossible to track with paper and pencil, but child's play for a computer—were the only way to approach the web of influences between wealth, population, and resources. Why couldn't the whole world be modeled?

Sitting on an airplane on the way home from a conference on "The Predicament of Mankind" held in Switzerland in 1970, Forrester began to sketch out the first equations that would form a model he called "World Dynamics."

It was rough. A thumbnail sketch. Forrester's crude model mirrored the obvious loops and forces he intuitively felt governed large economies. For data, he grabbed whatever was handy as a quick estimate. The Club of Rome, the group that had sponsored the conference, came to MIT to evaluate the prototype Forrester had tinkered up. They were encouraged by what they saw. They secured funding from the Volkswagen Foundation to hire Forrester's associate, Dennis Meadows, to develop the model to the next stage. For the rest of 1970, Forrester and Meadows improved the World Dynamics model, designing more sophisticated process loops and scouring the world for current data.

Dennis Meadows, together with his wife Dana and two other

coauthors, published the souped-up model, now filled with real data, as the "Limits to Growth." The simulation was wildly successful as the first global spreadsheet. For the first time, the planetary system of life, earthly resources, and human culture were abstracted, embodied into a simulation, and set free to roam into the future. The Limits to Growth also succeeded as a global air raid siren, alerting the world to the conclusions of the authors: that almost every extension of humankind's current path led to civilization's collapse.

The result of the Limits to Growth model ignited thousands of editorials, policy debates, and newspaper articles around the world for many years following its release. "A Computer Looks Ahead and Shudders" screamed one headline. The gist of the model's discovery was this: "If the present growth trends in world population, industrialization, pollution, food production, and resource depletion continue unchanged, the limits to growth on this planet will be reached sometime within the next 100 years." The modelers ran the simulation hundreds of times in hundreds of slightly different scenarios. But no matter how they made tradeoffs, almost all the simulations predicted population and living standards either withering away or bubbling up quickly to burst shortly thereafter.

Primarily because the policy implications were stark, clear, and unwelcome, the model was highly controversial and heavily scrutinized. But it forever raised the discussion of resources and human activity to the necessary planetary scale.

The Limits to Growth model was less successful in spawning better predictive models, which the authors had hoped to spark with their pioneer efforts. Instead, in the intervening 20 years, world models came to be mistrusted, in large part because of the controversy of Limits to Growth. Ironically, the only world model visible in the public eye now (over two decades later) is the Limits to Growth. The authors reissued it on its 20th anniversary, with only slight changes.

As currently implemented, the Limits to Growth model runs on a software program called Stella. Stella takes the dynamic systems approach worked out by Jay Forrester on mainframe computers and ports it over to the visual interface of a Macintosh.

The Limits to Growth model is woven out of an impressive web of "stocks" and "flows." Stocks (money, oil, food, capital, etc.) flow into certain nodes (representing general processes such as farming), where they trigger outflows of other stocks. For instance money, land, fertilizer, and labor flow into farms to trigger an outflow of raw food. Food, oil, and other stocks flow into factories to produce fertilizer, to complete one feedback loop. A spaghetti maze of loops, subloops, and crossloops constitute the entire world. The leverage each loop has upon the others is adjustable and determined by ratios found in real-world data: how much food is produced per hectare per kilo of fertilizer and water, generating how much pollution and waste. As is true in all complex systems, the impact of a single adjustment cannot be calculated beforehand; it must be played out in the whole system to be measured.

Vivisystems must anticipate to survive. Yet the complexity of the prediction apparatus must not overwhelm the vivisystem itself. As an example of the difficulties inherent in prediction machinery, we can examine the Limits to Growth model in detail. There are four reasons to choose this particular model. The first is that its reissue demands that it be (re)considered as a reliable anticipatory apparatus for human endeavor. Second, the model provides a handy 20-year period over which to evaluate it. Did the patterns it detected 20 years ago still prevail? Third, one of the virtues of the Limits to Growth model is that it is critiqueable. It generates quantifiable results rather than vague descriptions. It can be tested. Fourth, nothing could be more ambitious than to model the future of human life on Earth. The success or failure of this prominent attempt can teach much about using models to predict extremely complex adaptive systems. Indeed one has to ask: Can such a seemingly unpredictable process as the world be simulated or anticipated with any confidence at all? Can feedback-driven models be reliable predictors of complex phenomenon?

The Limits to Growth model has many things going for it. Among them: It is not overly complex; it is pumped by feedback loops; it runs scenarios. But among the weaknesses I see in the model are the following:

***Narrow overall scenarios.*** Rather than explore possible futures of any real diversity, Limits to Growth plays out a multitude of minor variations upon one fairly narrow set of assumptions. Mostly the "possible futures" it explores are those that seem plausible to the authors. Twenty years ago they ignored scenarios not based on what they felt were reasonable assumptions of expiring finite resources. But resources (such as rare metals, oil, and fertilizer) didn't diminish. Any genuinely predictive model must be equipped with the capability to generate "unthinkable" scenarios. It is important that a system have sufficient elbowroom in the space of possibilities to wander in places we don't expect. There is an art to this, because a model with too many degrees of freedom becomes unmanageable, while one too constrained becomes unreliable.

***Wrong assumptions.*** Even the best model can be sidetracked by false premises. The original key assumption of the model was that the world contains only a 250-year supply of nonrenewable resources, and that the demands on that supply are exponential. Twenty years later we know both those assumptions are wrong. Reserves of oil and minerals have grown; their prices have not increased; and demand for materials like copper are not exponential. In the 1992 reissue of the model, these assumptions were adjusted. Now the foundational assumption is that pollution *must* rise with growth. I can imagine that premise needing to be adjusted in the next 20 years, if the last 20 are a guide. "Adjustments" of this basic nature have to be made because the Limits to Growth model has . . .

***No room for learning.*** A group of early critics of the model once joked that they ran the Limits to Growth simulation from the year 1800 and by 1900 found a "20-foot level of horse manure on the streets." At the rate horse transportation was increasing then, this would have been a logical extrapolation. The half-jesting critics felt that the model made no provisions for learning technologies, increasing efficiencies, or the ability of people to alter their behavior or invent solutions.

There *is* a type of adaptation wired into the model. As crises arise (such as increase in pollution), capital assets are shifted to cover it (so the coefficient of pollution generated is lowered). But this learning is neither decentralized nor open-ended. In truth, there's no easy way to model either. Much of the research reported elsewhere in this book is about the pioneering attempts to achieve distributed learning and open-ended growth in manufactured settings, or to enhance the same in natural settings. Without decentralized open-ended learning, the real world will overtake the model in a matter of days.

In real life, the populations of India, Africa, China, and South America don't change their actions based upon the hypothetical projections of the Limits to Growth model. They adapt because of their own immediate learning cycle. For instance, the Limits to Growth model was caught off-guard (like most other forecasts) by global birth rates that dropped faster than anyone predicted. Was this due to the influence of doomsday projections like Limits to Growth? The more plausible mechanism is that educated women have less children and are more prosperous, and that prosperous people are imitated. They don't know about, or care about, global limits to growth. Government incentives assist local dynamics already present. People anywhere act (and learn) out of immediate self-interest. This holds true for other functions such as crop productivity, arable land, transportation, and so on. The assumptions for these fluctuating values are fixed in the Limits to Growth model, but in reality the assumptions themselves have coevolutionary mechanisms that flux over time. The point is that the learning must be modeled as an internal loop residing within the model. In addition to the values, the very structure of the assumptions in the simulation—or in any simulation that hopes to anticipate a vivisystem—must be adaptable.

***World averages.*** The Limits to Growth model treats the world as uniformly polluted, uniformly populated, and uniformly endowed with resources. This homogenization simplifies and uncomplicates the world enough to model it sanely. But in the end it undermines the purpose of the model because the locality and regionalism of the planet are some of its most striking and important features. Furthermore, the hierarchy of dynamics that arises out of differing local dynamics provides some of the key phenomena of Earth. The Limits to Growth modelers recognize the power of subloops—which is, in fact, the chief virtue of Forrester's system dynamics underpinning the software. But the model entirely ignores the paramount subloop of a world: geography. A planetary model without geography is . . . not the world. Not only must learning be distributed throughout a simulation; *all* functions must be. It is the failure to mirror the distributed nature—the swarm nature—of life on Earth that is this model's greatest failure.

***The inability to model open-ended growth of any kind.*** When I asked Dana Meadows what happened when they ran the model from 1600, or even 1800, she replied that they never tried it. I found that astonishing since backcasting is a standard reality test for forecasting models. In this case, the modelers suspected that the simulation would not cohere. That should be a warning. Since 1600 the world has experienced long-term growth. If a world model is reliable, it should be able to simulate four centuries of growth—at least as history. Ultimately, if we are to believe Limits to Growth has anything to say about future growth, the simulation must, in principle, be capable of generating long-term growth through several periods of transitions. As it is, all that Limits to Growth can prove is that it can simulate one century of collapse.

"Our model is astonishingly 'robust,'" Meadows told me. "You have to do all kinds of things to keep it from collapsing . . . Always the same behavior and basic dynamic emerges: overshoot and collapse." This is a pretty dangerous model to rely on for predictions of society's future. All the initial parameters of the system quickly converge upon termination, when history tells us human society is a system that displays marvelous continuing expansion.

Two years ago I spent an evening talking to programmer Ken Karakotsios who was building a tiny world of ecology and evolution. His world (which eventually became the game of SimLife) provides tools to god-players who can then create up to

32 virtual species of animals and 32 species of plants. The artificial animals and plants interact, compete, prey upon each other and evolve. "What's the longest you've had your world running?" I asked him. "Oh," he moans, "only a day. You know it's really hard to keep one of these complex worlds going. They do like to collapse."

The scenarios in Limits to Growth collapse because that's what the Limits to Growth simulation is good at. Nearly every initial condition in the model leads to either apocalypse or (very rarely) to stability—but never to a new structure—because the model is inherently incapable of generating open-ended growth. The Limits to Growth cannot mimic the emergence of the industrial evolution from the agrarian age. "Nor," admits Meadows, "can it take the world from the industrial revolution to whatever follows next beyond that." She explains, "What the model shows is that the logic of the industrial revolution runs into an inevitable wall of limits. The model does two things, either it begins to collapse, or we intervene as modelers and make changes to save it."

ME: "Wouldn't a better world model possess the dynamics to transform itself to the next level on its own?"

DANA MEADOWS: "It strikes me as a little bit fatalistic to think that this is designed in the system to happen and we just lean back and watch it. Instead we modeled ourselves into it. Human intelligence comes in, perceives the whole situation, and makes changes in the human societal structure. So this reflects our mental picture of how the system transcends to the next stage—with intelligence that reaches in and restructures the system."

That's Save-The-World mode, as well as inadequate modeling of how an ever complexifying world works. Meadows is right that intelligence reaches in to human culture and restructures it. But that isn't done just by modelers, and it doesn't happen only at cultural thresholds. This restructuring happens in six billion minds around the world, every day, in every era. Human culture is a decentralized evolutionary system if there ever was one. Any predictive model that fails to incorporate this distributed ongoing daily billion-headed microrevolution is doomed to collapse, as civilization itself would without it.

Twenty years later, the Limits to Growth simulation needs not a mere update, but a total redo. The best use for it is to stand as a challenge and a departure point to make a better model. A real predictive model of a planetary society would:

1) spin significantly varied scenarios,
2) start with more flexible and informed assumptions,
3) incorporate distributed learning,
4) contain local and regional variation, and
5) if possible, demonstrate increasing complexification.

I do not focus on the Limits to Growth world model because I want to pick on its potent political implications (the first version did, after all, inspire a generation of antigrowth activists). Rather, the model's inadequacies precisely parallel several core points I hope to make in this book. In bravely attempting to simulate an extremely complex adapting system (the human infrastructure of living on Earth), in order to feed-forward a scenario of this system into the future, the Forrester/Meadows model highlights not the limits to growth but the limits of certain simulations.

The dream of Meadows is the same as that of Forrester, the U.S. Command Central wargamers, Farmer and the Prediction Company, and myself, for that matter: to create a system (a machine) that sufficiently mirrors the real evolving world so that this miniature can run faster than real life and thus project its results into the future. We'd like prediction machinery not for a sense of predestiny but for guidance. And ideally it must be a Kauffman or von Neumann machine that can create things more complex than itself.

To do that, the model must possess a "requisite complexity." This is a term coined in the 1950s by the cybernetician Ross Ashby who built some of the first electronically adaptive models. Every model must distill a myriad of fine details about the real into a compressed representation; one of the most important traits it must condense is reality's complexity. Ashby concluded from his own experiments in making minimal models out of vacuum tubes that if a model simplifies the complexity too steeply, it misses the mark. A simulation's complexity has to be within the ballpark of the complexity of the modeled; otherwise the model can't keep up with the zig and zags of the thing modeled. Another cybernetician, Gerald Weinberg, supplies a fine metaphor for requisite complexity in his book *On the Design of Stable Systems*. Imagine, Weinberg suggests, a guided missile aimed at an enemy jet. The missile does not have to be a jet itself, but it must embody a requisite degree of complex flight behavior to parallel the behavior of the jet. If the missile is not at least as fast and aerodynamically nimble as the targeted jet-fighter, then it cannot hit its target.

Stella-based models such as Limits to Growth possess a remarkable surfeit of feedback circuits. As Norbert Wiener showed in 1952, feedback circuits, in all their combinatorial variety, are the fountainhead of control and self-governance. But in the forty years since that initial flush of excitement about feedback, we now know that feedback loops alone are insufficient to breed the behaviors of the vivisystems we find most interesting. There are two additional types of complexity (there may be others) the researchers in this book have found necessary in order to birth the full spectrum of vivisystem character: distributed being and open-ended evolution.

The key insight uncovered by the study of complex systems in recent years is this: the only way for a system to evolve into something new is to have a flexible structure. A tiny tadpole can change into a frog, but a 747 Jumbo Jet can't add six inches to its length without crippling itself. This is why distributed being is so important to learning and evolving systems. A decentralized, redundant organization can flex without distorting its function, and thus it can adapt. It can manage change. We call that growth.

Direct feedback models such as Limits to Growth can achieve stabilization—one attribute of living systems—but they can't learn, grow, or diversify—three essential complexities for a model of changing culture or life. Without these abilities, a world model will fall far behind the moving reality. A learning-less model can be used to anticipate the near-future where evolutionary change is minimal; but to predict an evolutionary system—if it can ever be predicted in pockets—will require the requisite complexity of a simulated, artificial evolutionary model.

But we cannot import evolution and learning without exporting control. When Dana Meadows speaks of a collective human intelligence which steps back to perceive global problems and then "reaches in and restructures the system" of human endeavor, she is pointing to the greatest fault of the Limit to Growth model: its linear, mechanical, and unworkable notion of control.

There is no control outside a self-making system. Vivisystems, such as economies, ecologies, and human culture, can hardly be controlled from any position. They can be prodded, perturbed, cajoled, herded, and at best, coordinated from within. On Earth, there is no outside platform from which to send an intelligent hand into the vivisystem, and no point inside where a control dial waits to be turned. The direction of large swarmlike systems such as human society is controlled by a messy multitude of interconnecting, self-contradictory agents who have only the dimmest awareness of where the whole is at any one moment. Furthermore, many active members of this swarmy system are not individual human intelligences; they are corporate entities, groups, institutions, technological systems, and even the nonbiological systems of the Earth itself.

The song goes: No one is in charge. We can't predict the future.

Now hear the flip side of the album: We are all steering. And we *can* learn to anticipate what is immediately ahead. To learn is to live.

23

# WHOLES, HOLES, AND SPACES

"GOOD MORNING, self-organizing systems!"

The cheerful speaker smiled with a polished ease and adjusted his tie. "I am indeed very happy to find the Office of Naval Research joining with the Armour Research Foundation in organizing this conference on what I personally consider an exceedingly important topic, and at such a well-chosen time."

It was a spring day in early May, 1959. Four hundred men from an astoundingly diverse group of scientific backgrounds had gathered in Chicago for what promised to be an electrifying meeting. Almost every major branch of science was represented: psychology, linguistics, engineering, embryology, physics, information theory, mathematics, astronomy, and social sciences. No one could remember a conference before this where so many top scientists in different fields were about to spend two days talking about one thing. Certainly there had never been a large meeting about this particular one thing.

It was a topic that only a young country flush with success and confident of its role in the world would even think about: self-organizing systems—how organization bootstraps itself to life. Bootstrapping! It was the American dream put into an equation.

"The choice of time is particularly significant in my personal life, too," the speaker continued. "For the last nine months the

Department of Defense of the United States of America has been in the throes of an organizational effort which shows reasonably clearly that we are still a long way from understanding what makes a self-organizing system."

Hearty chuckles from the early morning crowd just settling into their seats. At the podium Dr. Joachim Weyl, Research Director of the Office of Naval Research, beamed and continued. "There are three basic elements I'd like to call to your attention which can be studied best. From the area of computers we will, in the long run, draw our essential understanding of the element of memory that is absolutely and inevitably present in what you might call in the future 'self-organizing systems.' You might go so far, as I have done, as to say that a computer is nothing but a means for a memory to get from one state to another.

"The second element biologists call differentiation. In any system that will evolve it is quite clearly necessary that you have what the geneticists have called mutations, essentially random events. Some initial triggering mechanism is needed to push one group in one direction, and another in another direction. In other words, environment containing noise has to be relied on to furnish the triggering mechanism on which the long-term selection rule will operate.

"The third basic element probably presents itself most purely and most accessibly when we are dealing with large social organizations. Let me call it, for the purpose here, subordination, or if you wish, the executive function."

There they were: signal noise, mutations, executive function, self-organization. These words were spoken before the arrival of the DNA model, before digital technology, before departments of information management systems, and before complexity theory. It is difficult to imagine how alien and innovative these ideas were at the time.

And how right. In one fell swoop 35 years ago, Dr. Weyl outlined my whole 1994 book on the breaking science of adaptive, distributed systems and the emergent phenomenon they engender.

While the prescience of the 1959 meeting is remarkable, I also see something remarkable on the other side: how little our knowledge of whole systems has advanced in 35 years. Despite the great progress made recently and reported in this book, many of the basic questions about self-organization, differentiation, and subordination of whole systems still remain mysterious.

The all-star lineup who presented papers at the 1959 conference was a public rendezvous of scientists who had been convening in smaller meetings since 1942. These intimate, invitation-only gatherings were organized by the Josiah Macy, Jr. Foundation, and became known as the Macy Conferences. In the spirit of wartime urgency, the small gatherings were interdisciplinary, elite, and emphasized thinking big. Among the several dozen visionaries invited over the nine years of the conference were Gregory Bateson, Norbert Wiener, Margaret Mead, Lawrence Frank, John von Neumann, Warren McCulloch, and Arturo Rosenblueth. This stellar congregation later became known as the cybernetic group for the perspective they pioneered—cybernetics, the art and science of control.

Some beginnings are inconspicuous; this one wasn't. From the very first Macy Conference, the participants could imagine the alien vista they were opening. Despite their veteran science background and natural skepticism, they saw immediately that this new view would change their life's work. Anthropologist Margaret Mead recalled she was so excited by the ideas set loose in the first meeting that "I did not notice that I had broken one of my teeth until the Conference was over."

The core group consisted of key thinkers in biology, social science, and what we would now call computer science, although this group were only beginning to invent the concept of computers at the time. Their chief achievement was to articulate a language of control and design that worked for biology, social sciences, and computers. Much of the brilliance of these conferences came by the then unconventional approach of rigorously considering living things as machines and machines as living things. Von Neumann quantitatively compared the speed of brain neurons and the speed of vacuum tubes, boldly implying the two *could* be compared. Wiener reviewed the history of machine automata segueing into human anatomy. Rosenblueth, the doctor, saw homeostatic circuits in the body and in cells. In

Steve Heims's history of this influential circle of minds, *The Cybernetics Group*, he says of the Macy Conferences: "Even such anthropocentric social scientists as Mead and Frank became proponents for the mechanical level of understanding, wherein life is described as an entropy-reducing device and humans characterized as servomechanisms, their minds as computers, and social conflicts by mathematical game theory."

In an age when popular science fiction had just hatched, and was not the influential element it now is in modern science, the Macy Conference participants often pushed the metaphors they were playing with to extremes, much as science fiction writers do now. At one conference McCulloch said, "I don't particularly like people, never have. Man to my mind is about the nastiest, most destructive of all the animals. I don't see any reason, if he can evolve machines that can have more fun than he himself can, why they shouldn't take over, enslave us, quite happily. They might have a lot more fun, invent better games than we ever did." Humanists were horrified by such speculations, but under this nightmarish, dehumanized scenario some very important concepts were buried: that machines might evolve, that they might really be able to do practical intellectual chores better than we could, and that we share operating principles with very sophisticated machines. These are very much metaphors of the next millennium.

As Mead wrote later of the Macy Conferences, "Out of the deliberations of this (cybernetics) group came a whole series of fruitful developments of a very high order." Specifically, the ideas of feedback control, circular causality, homeostasis in machines, and political game theory were born there and gradually entered the mainstream until they became elemental, almost cliché, concepts today.

The cybernetic group did not find answers as much as they prepared an agenda for questions. Decades later scientists studying chaos, complexity, artificial life, subsumption architecture, artificial evolution, simulations, ecosystems, and bionic machines would find a framework for their questions in cybernetics. A short-hand synopsis of *Out of Control* would be to say it is an update on the current state of cybernetic research.

But therein lies a curious puzzle. If this book is really about cybernetics, why is the word "cybernetics" so absent from it? Where are the earlier practitioners of such cutting-edge science now? Why are the old gurus and their fine ideas not at the center of this natural extension of their work? What ever happened to cybernetics?

It was a mystery that perplexed me when I first started hanging out with the young generation of systems pioneers. The better-read were certainly aware of the early cybernetic work, but there was almost no one from a cybernetic background working with them. It was as if there was an entire lost generation, a hole in the transmission of knowledge.

There are three theories about why the cybernetic movement died:

- Cybernetics was starved to death by the siphoning away of its funding to the hot-shot—but stillborn—field of artificial intelligence. It was the failure of AI to produce usefulness that did cybernetics in. AI was just one facet of cybernetics, but while it got most of the government and university money, the rest of cybernetics' vast agenda withered. The grad students fled to AI, so the other fields dried up. Then, AI itself stalled.

- Cybernetics was a victim of batch-mode computing. For all its great ideas, cybernetics was mostly talk. The kind of experiments required to test its notions demanded many cycles of a computer, at its full power, in a completely exploratory mode. These were all the wrong things to ask of the priesthood guarding the mainframe. Therefore, very little cybernetic theory ever made it to experiment. When cheap personal computers hit the world, universities were notoriously slow to adopt them. So while high school kids had Apple IIs at home, the universities were still using punch cards. Chris Langton started his first a-life experiments on an Apple II. Doyne Farmer and friends discovered chaos theory by making their own computer. Real-time command of a complete universal computer was what traditional cybernetics needed but never got.

- Cybernetics was strangled by "putting the observer inside the box." In 1960, Heinz von Foerster made the brilliant suggestion that a refreshing view of social systems could be had by including the observer of the system as part of a larger metasystem. He framed his observation as Second Order Cybernetics, or the system of observing systems. The insight was useful in such fields as family therapy where the therapist had to include him- or herself in a theory of the family they were treating. But "putting the observer into the system" fell into an infinite regress when therapists video-taped patients and then sociologists taped therapists watching the tape of the patients and then taped themselves watching the therapists . . . By the 1980s the rolls of the American Society of Cybernetics were filled with therapists, sociologists, and political scientists primarily interested in the effects of observing systems.

All three reasons conspired so that by the late 1970s cybernetics had died of dry rot. Most of the work in cybernetics was at the level of the book you are now reading: armchair attempts to weave a coherent big picture together. Real researchers were bumping their heads in frustration in AI labs, or working in obscure institutes in Russia, where cybernetics did continue as a branch of mathematics. I don't believe a single formal textbook on cybernetics was ever written in English.

In the fabric of knowledge we call science, there was a rent here, a hole. It was filled by young enthusiasts not burdened by wise old men. This gap made me wonder about the space of science.

Scientific knowledge is a parallel distributed system. It has no center, no one in control. A million heads and dispersed books hold parts of it. It too is a web, a coevolutionary system of fact and theory interacting and influencing other facts and theories. But the study of science as a network of agents searching in parallel over a rugged landscape of mysteries is a field larger than any I've tackled here. To deal fairly with the mechanics of science alone would require a larger book than I've written so far. I can only hint at such a system in these closing pages.

Knowledge, truth, and information flow in networks and swarm systems. I have always been interested in the texture of scientific knowledge because it appears to be lumpy and uneven. Much of what we collectively know derives from a few small areas, yet between them lie vast deserts of ignorance. I can interpret that observation now as the effect of positive feedback and attractors. A little bit of knowledge illuminates much around it, and that new illumination feeds on itself, so one corner explodes. The reverse also holds true: ignorance breeds ignorance. Areas where nothing is known, everyone avoids, so nothing is discovered. The result is an uneven landscape of empty know-nothing interrupted by hills of self-organized knowledge.

Of this culturally produced space, I am most fascinated by the deserts—by the holes. What can we know about what we don't know? The greatest promise looming in evolution theory is unraveling the mystery of why organisms don't change, because stasis is more common than change yet harder to explain. What can we know about no-change in a system of change? What do the holes of change tell us about the whole of change? And so, it is the holes in the space of wholes that I'd like to explore here.

This very book is full of holes as well as wholes. What I don't know far exceeds what I know, but unfortunately, it is far easier to write about what I know than about what I don't know. By the nature of ignorance, I am, of course, not aware of all the places and gaps where my own knowledge fails. Recognizing one's own ignorance is quite a trick. That goes for science, too. Mapping the holes of ignorance is perhaps science's next advance.

Scientists today believe science is revolutionary. They explain how science works via a model of ongoing minirevolutions. According to this perspective, researchers build a theory to explain facts (for example, rainbows occur because light is a wave). The theory itself will suggest places to look for new facts (can you bend a wave?). It's the law of increasing returns again. As new facts are uncovered they are incorporated into the theory, buttressing its strength and reliability. Occasionally, scientists uncover new facts that aren't readily explained by the theory (light sometimes acts like a particle). These are called anomalies. Anomalies are set aside at first, while new facts that concur with the reigning theory continue to stream in. At some point, the accumulating anomalies prove too great, too troublesome, or too numerous to ignore. Inevitably then, some young turk proposes a revolutionary different model that explains the anomalies (such as, light is both wave and particle). The old is gone; the new quickly reigns.

In the terminology of science historian Thomas Kuhn, the reigning theory forms a self-reinforcing mindset called a paradigm that dictates what is fact and what is mere noise. From within the paradigm, anomalies are trivia, curiosities, illusions, or bad data. Research proposals endorsing the paradigm win grants, lab space, and degrees. Proposals operating outside the paradigm—those dabbling in distracting trivia—get nothing. The famous scientist who made his great revolutionary discovery while denied funds or credibility is so common it's become cliché; I've trotted out several of those cliché stories in this book. One example is the ignored work of scientists dabbling in ideas that contradict neodarwinian dogma.

Real discovery in science, according to Kuhn in his seminal *The Structure of Scientific Revolutions*, only "commences with the awareness of anomaly." Progress is an acknowledgment of the opposition. A series of established paradigms are overthrown by downtrodden and oppressed anomalies (and their finders) as they rebel and usurp the throne by their countertruth. The new ideas reign, at least for a while, until they too become ossified and insensitive to the squawks of new anomalies, and are eventually overthrown themselves.

Kuhn's model of paradigm shift in science is so convincing that it has become a paradigm itself—the paradigms of paradigms. We now see paradigms and paradigm overthrows everywhere, inside of science and out. Paradigm shifts are our paradigm. The fact that things don't really work that way is, well, an anomaly.

Alan Lightman and Owen Gingerich, writing in a 1991 *Science* article, "When Do Anomalies Begin?," claim that contrary to the reigning Kuhnian model of science, "certain scientific anomalies are recognized only *after* they are given compelling explanations within a new conceptual framework. Before this recognition, the peculiar facts are taken as givens or are ignored in the old framework." In other words, the real anomalies that eventually overthrow a reigning paradigm are at first *not even perceived as anomalies*. They are invisible.

A few brief examples of "retrorecognition," based on Lightman's and Gingerich's article:

- The fact that the shape of South America and Africa fit together like a lock and key did not bother any pre-1960s geologists. There was nothing troubling to them or their theories of continent formation in this observation, or in the observed ridges down the center of the oceans. Although the remarkable fit had been noticed since the Atlantic Ocean was first mapped, it was a fact that did not even need an explanation. Only later was the fit retrorecognized as something to explain.

- Newton precisely measured the inertial mass of a great many objects (what it took to get them moving, as in getting a pendulum started) and their gravitational mass (how fast they fell to the Earth), to determine that the two forces were equal, if not equivalent, and could be canceled out when doing physics. For hundreds of years this relationship was not questioned. Einstein, however, was struck that "the law has not found any place in the foundations of our edifice of the physical universe." Unlike others, he was perplexed by this observation which he successfully explained in his revolutionary general theory of relativity.

- For decades, the almost exact balance between the universe's kinetic and gravitational energies—a pair of forces that kept the expanding universe balanced between blowing up or collapsing—was noted in passing by astronomers. But it was never a "problem" until the revolutionary "inflationary universe" model came along in 1981 and made this fact a troubling paradox. The observation of the balance did not begin to be an anomaly until after the paradigm shift, when in retrospect, it was seen as a troublemaker.

The common theme in each example is that anomalies begin as observed facts that don't require any explanation at all. They are not troublesome facts; they just are. Rather than the cause of a paradigm shift, anomalies are the result of the shift.

In a letter to *Science*, David P. Barash tells of his own experience with nonanomalies. He wrote a textbook of sociobiology in 1982, where he stated that "evolutionary biologists, beginning with Darwin, have been troubled by the fact that animals often do things that appear to benefit others, often at great cost to themselves." Sociobiology was launched by the 1964 publication of William Hamilton's inclusive fitness theory, which provided a workable, though controversial, way to interpret animal altruism. Barash writes, "However, stimulated by the Lightman-Gingerich thesis, I have reviewed numerous pre-1964 textbooks of animal behavior and evolutionary biology and have discovered that, in fact—and contrary to my own above-cited assertion—before Hamilton's insight, evolutionary biologists were *not* very much troubled by the occurrence of apparently altruistic behavior among animals (at least they did not devote much theoretical or empirical attention to the phenomenon)." He ends his letter by suggesting, half in jest, that biologists "teach a course in what we *don't* know about, say, animal behavior."

The final section in my book is a short course in what we, or at least I, don't know about complex adaptive systems and the nature of control. It's a list of questions, a catalog of holes. A lot of the questions may seem silly, obvious, trivial, or hardly worth worrying about, even for nonscientists. Scientists in the pertinent fields may say the same: these questions are distractions, the ravings of an amateur science-groupie, the ill-informed musing of a techno-transcendentalist. No matter. I am inspired to follow this unorthodox short course by a wonderful paragraph written by Douglas Hofstadter in a foreword to Pentti Kanerva's obscure technical monograph on sparse distributed computer memory. Hofstadter writes:

> I begin with the nearly trivial observation that members of a familiar perceptual category automatically evoke the name of the category. Thus, when we see a staircase (say), no matter how big or small it is, no matter how twisted or straight, no matter how ornamented or plain, modern or old, dirty or clean, the label "staircase" spontaneously jumps to center stage without any

> conscious effort at all. Obviously, the same goes for telephones, mailboxes, milkshakes, butterflies, model airplanes, stretch pants, gossip magazines, women's shoes, musical instruments, beachballs, station wagons, grocery stores, and so on. This phenomenon, whereby an external physical stimulus indirectly activates the proper part of our memory, permeates human life and language so thoroughly that most people have a hard time working up any interest in it, let alone astonishment, yet it is probably the most key of all mental mechanisms.

To be astonished by a question no one else can get worked up about, or to be astonished by a matter nobody considers a problem, is perhaps a better paradigm for the progress of science.

This book is based on my astonishment that nature and machines work at all. I wrote it by trying to explain my amazement to the reader. When I came to something I didn't understand, I wrestled with it, researched, or read until I did, and then started writing again until I came to the next question I couldn't readily answer. Then I'd do the cycle again, round and round. Eventually I would come to a question that stopped me from writing further. Either no one had an answer, or they provided the stock response and would not see my perplexity at all. These halting questions never seemed weighty at first encounter—just a question that seems to lead to nowhere for now. But in fact they are protoanomalies. Like Hofstadter's unappreciated astonishment at our mind's ability to categorize objects before we recognize them, out of these quiet riddles will come future insight, and perhaps revolutionary understanding, and eventually recognition that we *must* explain them.

Readers may be perplexed themselves when they see that most of these questions appear to be the very ones I seemed to have answered in the preceding chapters! But really all I did was drive around these questions, surveying their girth, hill-climbing up them until I was stuck on a false summit. In my experience most good questions come while stuck on a partial answer somewhere else. This book has been an endeavor to find interesting questions. But on the way, some of the rather ordinary questions stopped me. They follow below.

• I often use the word "emergent" in this book. As used by the practitioners of complexity, it means something like: "that organization which is generated out of parts acting in concert." But the meaning of emergent begins to disappear when scrutinized, leaving behind a vague impression that the word is, at bottom, meaningless. I tried substituting the word "happened" in every instance I used "emerged" and it seemed to work. Try it. Global order happens from local rules. What do we mean by emergent?

• And what is "complexity" anyway? I looked forward to the two 1992 science books identically titled *Complexity*, one by Mitch Waldrop and one by Roger Lewin, because I was hoping one or the other would provide me with a practical measurement of complexity. But both authors wrote books on the subject without hazarding a guess at a usable definition. How do we know one thing or process is more complex than another? Is a cucumber more complex than a Cadillac? Is a meadow more complex than a mammal brain? Is a zebra more complex than a national economy? I am aware of three or four mathematical definitions for complexity, none of them broadly useful in answering the type of questions I just asked. We are so ignorant of complexity that we haven't yet asked the right question about what it is.

• If evolution tends to grow more complex, why? And if it really does not, then why does it appear to? Is complexity in fact more efficient than simplicity?

• There seems to be a "requisite variety"—a minimum complexity or diversity of parts—for such processes as self-organization, evolution, learning, and life. How do we know for sure when enough variety is enough? We don't even have a good measure for diversity. We have intuitive feelings but we can't translate that into anything very precise. What is variety?

• The "edge of chaos" often sounds like "moderation in all things." Is it merely playing Goldilocks to define the values at which systems are maximally adaptable, as "just right for adaptation?" Is this yet another necessary tautology?

• In computer science there is a famous conjecture called the Church/Turing hypothesis which undergirds much of the reasoning in artificial intelligence and artificial life. The hypothesis says: a universal computing machine can compute anything that another universal computing machine can compute, given unlimited time and an infinite tape. But my goodness! Unlimited time and space is the precise difference between the living and the dead. The dead have infinite time and space. The living live in finitude. So while, within a certain range, computational processes are independent of the hardware they run on (one machine can emulate anything another can), there are real limits to the fungibility of processes. Artificial life is based on the premise that life can be extracted from its carbon-based hardware and set to run on a different matrix somewhere else. The experiments so far have shown that to be true more than was expected. But where are the limits in real time and real space?

• What, if anything, cannot be simulated?

• The quest for artificial intelligence and artificial life is wrapped up (some say bogged down) in the important riddle of whether a simulation of an extremely complex system is a fake or something real in its own right. Maybe it is hyperreal, or maybe the term "hyperreality" just ducks the question. No one doubts the ability of a model to imitate an original thing. The questions are: What sort of reality do we assign a simulation of a thing? What, if any, are the distinctions between a simulation and a reality?

• How far can you compress a meadow into seeds? This was the question the prairie restorers inadvertently asked. Can you reduce the treasure of information contained in an entire ecosystem into several bushels of seeds, which, when watered, would reconstitute the awesome complexity of prairie life? Are there important natural systems which simply cannot be reduced and modeled accurately? Such a system would be its own smallest expression, its own model. Are there any artificial large systems that cannot be compressed or abstracted?

• I'd like to know more about stability. If we build a "stable" system, is there some way we can define that? What are the boundary conditions, the requirements, for stable complexity? When does change cease to be change?

• Why do species ever go extinct? If all of nature is hourly working to adapt, never resting in its effort to outwit competitors and exploit its environment, why do certain classes of species fail? Perhaps some certain organisms are better adapted than others. But why would the universal mechanism of nature sometimes work and sometimes not for entire *types* of organisms, allowing particular groups to lag and others to advance? More precisely, why would the dynamics of adaptation work for some organisms but not others? Why does nature allow some biological forms to be pushed into forms that are inherently inefficient? There is a case of an oysterlike bivalve that evolved a more and more spiraled shell until, just before extinction, the valves could barely open. Why doesn't the organism return to the range of the workable? And why does extinction run in families and groups, as if bad genes may be responsible? How could nature produce a group of bad genes? Perhaps, extinctions are caused by something outside, like comets and asteroids. Paleontologist Dave Raup postulates that 75 percent of all extinction events were caused by asteroid impacts. If there were no asteroids would there be no extinctions? If there were no extinctions of species on Earth, what would life look like now? Why, for that matter, do complex systems of any sort fail or die?

• On the other hand, why, in this coevolutionary world, is anything at all stable?

• Every figure I've heard for both natural and artificial self-sustaining systems puts the self-stabilizing mutation rate between 1 percent and 0.01 percent. Are mutation rates universal?

• What are the down sides of connecting everything to everything?

• In the space of all possible lifes, life on Earth is but a tiny sliver—one attempt at creativity. Is there a limit to how much life a given quantity of matter can hold? Why isn't there more variety of life on Earth? How come the universe is so small?

• Are the laws of the universe evolvable? If the laws governing the universe arose from within the universe, might they be susceptible to the forces of self-adjustment? Perhaps the very foundational laws upholding all sensible laws are in flux. Are we playing in a game where *all* the rules are constantly being rewritten?

• Can evolution evolve its own teleological purpose? If organisms, which are but a federation of mindless agents, can originate goals, can evolution itself, equally blind and dumb but in a way a very slow organism, also evolve a goal?

• And what about God? God gets no honor in the academic papers of artificial lifers, evolutionary theorists, cosmologists, or simulationists. But much to my surprise, in private conversations these same researchers routinely speak of God. As used by scientists, God is a coolly nonreligious technical concept, closer to god—a local creator. When talking of worlds, both real and modeled, God is an almost algebraically precise notation standing for whatever "X" operating outside a world that has created that world. "Okay, you're God . . ." says one computer scientist during a demo when he means that I'm now setting the rules for the world. God is a shorthand for the uncreated observer making things real. God thus becomes a scientific term, and a scientific concept. It doesn't have the philosophical subtleties of prime cause, or the theological finery of Creator; it is merely a handy way to talk about the necessary initial conditions to run a world. So what are the requirements for godhood. What makes a good god?

None of these questions is new. They have been asked before in different contexts by others. If the web of knowledge were completely wired then I could tag on the appropriate historical citations at this point, and pull out the historical context for all these musings.

Researchers dream of such a heavily connected network of

data and ideas. Science today is at the other end of a connectivity limit; the nodes in the distributed network of science need to be much more connected before they reach maximum evolvability.

The first step toward a highly linked web of knowledge was made by U.S. Army medical librarians trying to unify the indexing of medical journals. In 1955, Eugene Garfield, a librarian on that project who was interested in machine indexing, developed a computer system to automatically track the bibliographic citations of every scientific paper published in medicine. Eventually he founded a commercial company in his garage in Philadelphia—the Institute of Science Information (ISI)—that would track on a computer every scientific paper published, period. Today ISI—a company with many employees and supercomputers—cross-links millions of scholarly papers with their bibliographic references.

For instance, let's take one of the papers I refer to in my bibliography: Rodney Brooks's 1990 article "Elephants Don't Play Chess." I can go to the ISI system to find "Elephants Don't Play Chess" listed under its author and read off the list of all other published scientific papers, in addition to my *Out of Control*, that have cited "Elephants" in their bibliographies or footnotes. On the premise that other researchers and authors who find "Elephants" useful may also be useful to me, I have a way to backtrack the influence of ideas. (However, books are not at the moment indexed for citations, so in reality this example would only work if *Out of Control* were an article. But the principle holds.)

This citation index allows me to track the future dissemination of my own ideas. Again, assume *Out of Control* was indexed as a paper. Every year I could consult the ISI Citation Index and get a list of all those authors who cited my work in their work. This web would bring me to many people's ideas—many of them very germane since they quote me—that I might never find otherwise.

Citation indexing is currently employed to map the breaking "hot" areas of science. Clusters of a few extremely highly cited papers can indicate a rapidly moving area of research. An unintended corollary of this system is that government fund-givers use the Citation Index to assist them in determining whose research to fund. They count the total number of citations—adjusted for the "weight" or stature of the journal publishing the paper—of an individual scientist's work in order to indicate the importance of that scientist. But like any network, citation evaluation breeds the opportunity for a positive feedback loop: the more funding, the more papers produced, the more citations garnered, the more funding secured, and so on. And it engenders the identical reverse loop of no funding, no papers, no citations, no funding.

The Citation Index can also be thought of as a footnote tracking system. If you think of each bibliographic reference as a footnote in a text, then a citation index brings you to the footnote and then permits you to chase down the footnote to the footnote. A more elegant description of that system was coined "Hypertext" by Ted Nelson in 1974. In essence, hypertext is a large distributed document. A hypertext document is a vague network of live links between its words and ideas and sources. The document has no center, no end. You read hypertext by navigating through it, taking side tours to footnotes, and to footnotes to the footnotes, following parenthetical thoughts as long and complex as the "main" text. Any other document can be linked to and become part of another text. Computerized hypertext incorporates marginalia and commentaries to the text by other writers, updates, revisions, abstracts, digests, misinterpretations, and as in citation indexing, all bibliographic references to the work.

The extent of the distributed document is thus unknowable because it is without boundaries and often multiauthored. It's a swarm text. But a single author can compile a simple hypertext document which can be read in many different directions and along many paths. Thus, *the reader of hypertext creates a different work* of the author's web depending on how she goes through the material. Therefore in hypertext, as in other distributed creations, the creator must give up some control of his creation.

Hypertext documents of various depths have existed for ten years. In 1988, I was involved in developing one of the first commercial hypertext works—an electronic version of the *Whole Earth Catalog*, rendered in HyperCard on the Macintosh computer.

Even in this relatively small network of texts (there were 10,000 microdocuments; and millions of ways to travel through them), I got a sense of this new space of interlinked ideas.

For one thing, it was easy to get lost. Without the centering hold of a narrative, everything in a hypertext network seems to have equal weight and appears to be the same wherever you go, as if the space were a suburban sprawl. The problem of locating items in a network is substantial. It harks back to the days of early writing when texts in a 14th-century scriptorium were difficult to locate since they lacked cataloging, indexes, or tables of contents. The advantages which the hypertext model offers over the web of oral tradition is that the former can be indexed and cataloged. An index is an alternative way to read a printed text, but it is only one of many ways to read a hypertext. In a sufficiently large library of information without physical form—as future electronic libraries promise to be—the lack of simple but psychologically vital clues, such as knowing how much of the total you've read or roughly how many ways it can be read, is debilitating.

Hypertext creates its own possibility space. As Jay David Bolter writes in his outstanding, but little known book, *Writing Spaces*:

> In this late age of print, writers and readers still conceive of all texts, of text itself, as located in the space of a printed book. The conceptual space of a printed book is one in which writing is stable, monumental, and controlled exclusively by the author. It is the space defined by perfect printed volumes that exist in thousands of identical copies. The conceptual space of electronic writing, on the other hand, is characterized by fluidity and an interactive relationship between writer and reader.

Technology, particularly the technology of knowledge, shapes our thought. The possibility space created by each technology permits certain kinds of thinking and discourages others. A blackboard encourages repeated modification, erasure, casual thinking, spontaneity. A quill pen on writing paper demands care, attention to grammar, tidiness, controlled thinking. A printed page solicits rewritten drafts, proofing, introspection, editing. Hypertext, on the other hand, stimulates yet another way of thinking: telegraphic, modular, nonlinear, malleable, cooperative. As Brian Eno, the musician, wrote of Bolter's work, "[Bolter's thesis] is that the way we organize our writing space is the way we come to organize our thoughts, and in time becomes the way which we think the world itself must be organized."

The space of knowledge in ancient times was a dynamic oral tradition. By the grammar of rhetoric, knowledge was structured as poetry and dialogue—subject to interruption, questioning, and parenthetical diversions. The space of early writing was likewise flexible. Texts were ongoing affairs, amended by readers, revised by disciples; a forum for discussions. When scripts moved to the printed page, the ideas they represented became monumental and fixed. Gone was the role of the reader in forming the text. The unalterable progression of ideas across pages in a book gave the work an impressive authority—"authority" and "author" deriving from a common root. As Bolter notes, "When ancient, medieval, or even Renaissance texts are prepared for modern readers, it is not only the words that are translated: the text itself is translated into the space of the modern printed book."

A few authors in the printed past tried to explore expanded writing and thinking spaces, attempting to move away from the closed linearity of print and into the nonsequential experience of hypertext. James Joyce wrote *Ulysses* and *Finnegans Wake* as a network of ideas colliding, cross-referencing, and shifting upon each reading. Borges wrote in a traditional linear fashion, but he wrote *of* writing spaces: books about books, texts with endlessly branching plots, strangely looping self-referential books, texts of infinite permutations, and the libraries of possibilities. Bolter writes: "Borges can imagine such a fiction, but he cannot produce it . . . Borges himself never had available to him an electronic space, in which the text can comprise a network of diverging, converging, and parallel times."

I live on computer networks. The network of networks—the Internet—links several millions of personal computers around the world. No one knows exactly how many millions are connected, or even how many intermediate nodes there are. The Internet Society made an educated guess in August 1993 that the Net was made up of 1.7 million host computers and 17 million users. No one controls the Net, no one is in charge. The U.S. government, which indirectly subsidizes the Net, woke up one day to find that a Net had spun itself, without much administration or oversight, among the terminals of the techno-elite. The Internet is, as its users are proud to boast, the largest functioning anarchy in the world. Every day hundreds of millions of messages are passed between its members, without the benefit of a central authority. I personally receive or send about 50 messages per day. In addition to the vast flow in individual letters, there exist between its wires that disembodied cyberspace where messages interact, a shared space of written public conversations. Every day authors all over the word add millions of words to an uncountable number of overlapping conversations. They daily build an immense distributed document, one that is under eternal construction, constant flux, and fleeting permanence. "Elements in the electronic writing space are not simply chaotic," Bolter wrote, "they are instead in a perpetual state of reorganization."

The result is far different from a printed book, or even a chat around a table. The text is a sane conversation with millions of participants. The type of thought encouraged by the Internet hyperspace tends toward nurturing the nondogmatic, the experimental idea, the quip, the global perspective, the interdisciplinary synthesis, and the uninhibited, often emotional, response. Many participants prefer the quality of writing on the Net to book writing because Net-writing is of a conversational peer-to-peer style, frank and communicative, rather than precise and overwritten.

A distributed dynamic text, such as the Net and a number of new books in hypertext, is an entirely new space of ideas, thought, and knowledge. Knowledge shaped by the age of print birthed the very idea of a canon, which in turn implied a core set of fundamental truths—fixed in ink and perfectly duplicated—from which knowledge progressed but never retreated. The job of every generation of readers was to find the canonical truth in texts.

Distributed text, or hypertext, on the other hand supplies a new role for readers—every reader codetermines the meaning of a text. This relationship is the fundamental idea of postmodern literary criticism. For the postmodernists, there is no canon. They say hypertext allows "the reader to engage the author for control of the writing space." The truth of a work changes with each reading, no one of which is exhaustive or more valid than another. Meaning is multiple, a swarm of interpretations. In order to decipher a text it must be viewed as a network of idea-threads, some threads of which are owned by the author, some belonging to the reader and her historical context and others belonging to the greater context of the author's time. "The reader calls forth his or her own text out of the network, and each such text belongs to one reader and one particular act of reading," says Bolter.

This fragmentation of a work is called "deconstruction." Jacques Derrida, the father of deconstructionism, calls a text (and a text could be any complex thing) "a differential network, a fabric of traces referring endlessly to something other than itself, to other differential traces," or in Bolter's words "a texture of signs that point to other signs." This image of symbols referring to other symbols is, of course, the archetypal image of the infinite regress and the tangled recursive logic of a distributed swarm; the banner of the Net and the emblem of everything connected to everything.

The total summation we call knowledge or science is a web of ideas pointing to, and reciprocally educating each other. Hypertext and electronic writing accelerate that reciprocity. Networks rearrange the writing space of the printed book into a writing space many orders larger and many ways more complex than of ink on paper. The entire instrumentation of our lives can be seen as part of that "writing space." As data from weather sensors, demographic surveys, traffic recorders, cash registers, and all the millions of electronic information generators pour their "words" or representation into the Net, they enlarge the writing space. Their information becomes part of what we know, part of what we talk about, part of our meaning.

At the same time the very shape of this network space shapes us. It is no coincidence that the postmodernists arose in tandem as the space of networks formed. In the last half-century a uniform mass market—the result of the industrial thrust—has collapsed into a network of small niches—the result of the information tide. An aggregation of fragments is the only kind of whole we now have. The fragmentation of business markets, of social mores, of spiritual beliefs, of ethnicity, and of truth itself into tinier and tinier shards is the hallmark of this era. Our society is a working pandemonium of fragments. That's almost the definition of a distributed network. Bolter again: "Our culture is itself a vast writing space, a complex of symbolic structures . . . Just as our culture is moving from the printed book to the computer, it is also in the final stages of the transition from a hierarchical social order to what we might call a 'network culture.'"

There is no central keeper of knowledge in a network, only curators of particular views. People in a highly connected yet deeply fragmented society can no longer rely on a central canon for guidance. They are forced into the modern existential blackness of creating their own culture, beliefs, markets, and identity from a sticky mess of interdependent pieces. The industrial icon of a grand central or a hidden "I am" becomes hollow. Distributed, headless, emergent wholeness becomes the social ideal.

The ever insightful Bolter writes, "Critics accuse the computer of promoting homogeneity in our society, of producing uniformity through automation, but electronic reading and writing have just the opposite effect." Computers promote heterogeneity, individualization, and autonomy.

No one has been more wrong about computerization than George Orwell in *1984*. So far, nearly everything about the actual possibility-space which computers have created indicates they are the end of authority and not its beginning.

Swarm-works have opened up not only a new writing space for us, but a new thinking space. If parallel supercomputers and online computer networks can do this, what kind of new thinking spaces will future technologies—such as bioengineering—offer us? One thing bioengineering could do for the space of our thinking is shift our time scale. We moderns think in a bubble of about ten years. Our history extends into the past five years and our future runs ahead five years, but no further. We don't have a structured way, a cultural tool, for thinking in terms of decades or centuries. Tools for thinking about genes and evolution might change this. Pharmaceuticals that increase access to our own minds would, of course, also remake our thinking space.

One last question that stumped me, and halted my writing: How large is the space of possible ways of thinking? How many, or how few, of all types of logic have we found so far in the Library of thinking and knowledge?

Thinking space may be vast. The number of ways to overcome a problem, or to explore a notion, or to prove a statement, or to create a new idea, may be as large as the number of ideas itself. Contrarily, thinking space may be as small and narrow as the Greek philosophers thought it was. My bet is that artificial intelligence, when it comes, will be intelligent but not very humanlike. It will be one of many nonhuman methods of thought that will probably fill the library of thinking space. This space will also hold types of thinking that we simply cannot understand at all. But still we will use them. Nonhuman cognitive methods will provide us wonderful results beyond and out of our control.

Or we may surprise ourselves. We may have a brain that, like a Kauffman machine, is able to generate all types of thinking and never-seen-before complexity from a small finite set of instructions. Perhaps the space of possible cognition is *our* space. We could then climb into whatever kind of logic we can make, evolve, or find. If we can travel anywhere in cognitive space, we would be capable of an open-ended universe of thoughts.

I think we'll surprise ourselves.

24

# THE NINE LAWS OF GOD

OUT OF NOTHING, nature makes something.

First there is hard rock planet; then there is life, lots of it. First barren hills; then brooks with fish and cattails and redwinged blackbirds. First an acorn; then an oak tree forest.

I'd like to be able to do that. First a hunk of metal; then a robot. First some wires; then a mind. First some old genes; then a dinosaur.

How do you make something from nothing? Although nature knows this trick, we haven't learned much just by watching her. We have learned more by our failures in creating complexity and by combining these lessons with small successes in imitating and understanding natural systems. So from the frontiers of computer science, and the edges of biological research, and the odd corners of interdisciplinary experimentation, I have compiled *The Nine Laws of God* governing the incubation of somethings from nothing:

- Distribute being
- Control from the bottom up
- Cultivate increasing returns
- Grow by chunking
- Maximize the fringes
- Honor your errors
- Pursue no optima; have multiple goals

- Seek persistent disequilibrium
- Change changes itself.

These nine laws are the organizing principles that can be found operating in systems as diverse as biological evolution and SimCity. Of course I am not suggesting that they are the only laws needed to make something from nothing; but out of the many observations accumulating in the science of complexity, these principles are the broadest, crispest, and most representative generalities. I believe that one can go pretty far as a god while sticking to these nine rules.

***Distribute being.*** The spirit of a beehive, the behavior of an economy, the thinking of a supercomputer, and the life in me are distributed over a multitude of smaller units (which themselves may be distributed). When the sum of the parts can add up to more than the parts, then that extra being (that something from nothing) is distributed among the parts. Whenever we find something from nothing, we find it arising from a field of many interacting smaller pieces. All the mysteries we find most interesting—life, intelligence, evolution—are found in the soil of large distributed systems.

***Control from the bottom up.*** When everything is connected to everything in a distributed network, everything happens at once. When everything happens at once, wide and fast moving problems simply route around any central authority. Therefore overall governance must arise from the most humble interdependent acts done locally in parallel, and not from a central command. A mob *can* steer itself, and in the territory of rapid, massive, and heterogeneous change, only a mob can steer. To get something from nothing, control must rest at the bottom within simplicity.

***Cultivate increasing returns.*** Each time you use an idea, a language, or a skill you strengthen it, reinforce it, and make it more likely to be used again. That's known as positive feedback or snowballing. Success breeds success. In the Gospels, this principle of social dynamics is known as "To those who have, more will be given." Anything which alters its environment to increase production of itself is playing the game of increasing returns. And all large, sustaining systems play the game. The law operates in economics, biology, computer science, and human psychology. Life on Earth alters Earth to beget more life. Confidence builds confidence. Order generates more order. Them that has, gets.

***Grow by chunking.*** The only way to make a complex system that works is to begin with a simple system that works. Attempts to instantly install highly complex organization—such as intelligence or a market economy—without growing it, inevitably lead to failure. To assemble a prairie takes time—even if you have all the pieces. Time is needed to let each part test itself against all the others. Complexity is created, then, by assembling it incrementally from simple modules that can operate independently.

***Maximize the fringes.*** In heterogeneity is creation of the world. A uniform entity must adapt to the world by occasional earth-shattering revolutions, one of which is sure to kill it. A diverse heterogeneous entity, on the other hand, can adapt to the world in a thousand daily minirevolutions, staying in a state of permanent, but never fatal, churning. Diversity favors remote borders, the outskirts, hidden corners, moments of chaos, and isolated clusters. In economic, ecological, evolutionary, and institutional models, a healthy fringe speeds adaptation, increases resilience, and is almost always the source of innovations.

***Honor your errors.*** A trick will only work for a while, until everyone else is doing it. To advance from the ordinary requires a new game, or a new territory. But the process of going outside the conventional method, game, or territory is indistinguishable from error. Even the most brilliant act of human genius, in the final analysis, is an act of trial and error. "To be an Error and to be Cast out is a part of God's Design," wrote the visionary poet William Blake. Error, whether random or deliberate, must become an integral part of any process of creation. Evolution can be thought of as systematic error management.

***Pursue no optima; have multiple goals.*** Simple machines can be efficient, but complex adaptive machinery cannot be. A complicated structure has many masters and none of them can be served exclusively. Rather than strive for optimization of any function, a large system can only survive by "satisficing" (making "good enough") a multitude of functions. For instance, an adaptive system must trade off between exploiting a known path of success (optimizing a current strategy), or diverting resources to exploring new paths (thereby wasting energy trying less efficient methods). So vast are the mingled drives in any complex entity that it is impossible to unravel the actual causes of its survival. Survival is a many-pointed goal. Most living organisms are so many-pointed they are blunt variations that happen to work, rather than precise renditions of proteins, genes, and organs. In creating something from nothing, forget elegance; if it works, it's beautiful.

***Seek persistent disequilibrium.*** Neither constancy nor relentless change will support a creation. A good creation, like good jazz, must balance the stable formula with frequent out-of-kilter notes. Equilibrium is death. Yet unless a system stabilizes to an equilibrium point, it is no better than an explosion and just as soon dead. A Nothing, then, is both equilibrium and disequilibrium. A Something is persistent disequilibrium—a continuous state of surfing forever on the edge between never stopping but never falling. Homing in on that liquid threshold is the still mysterious holy grail of creation and the quest of all amateur gods.

***Change changes itself.*** Change can be structured. This is what large complex systems do: they coordinate change. When extremely large systems are built up out of complicated systems, then each system begins to influence and ultimately change the organizations of other systems. That is, if the rules of the game are composed from the bottom up, then it is likely that interacting forces at the bottom level will alter the rules of the game as it progresses. Over time, the rules for change get changed themselves. Evolution—as used in everyday speech—is about how an entity is changed over time. Deeper evolution—as it might be formally defined—is about how the rules for changing entities over time change over time. To get the most out of nothing, you need to have self-changing rules.

These nine principles underpin the awesome workings of prairies, flamingoes, cedar forests, eyeballs, natural selection in geological time, and the unfolding of a baby elephant from a tiny seed of elephant sperm and egg.

These same principles of bio-logic are now being implanted in computer chips, electronic communication networks, robot modules, pharmaceutical searches, software design, and corporate management, in order that these artificial systems may overcome their own complexity.

When the Technos is enlivened by Bios we get artifacts that can adapt, learn, and evolve. When our technology adapts, learns, and evolves then we will have a neo-biological civilization.

All complex things taken together form an unbroken continuum between the extremes of stark clockwork gears and ornate natural wilderness. The hallmark of the industrial age has been its exaltation of mechanical design. The hallmark of a neo-biological civilization is that it returns the designs of its creations toward the organic, again. But unlike earlier human societies that relied on found biological solutions—herbal medicines, animal proteins, natural dyes, and the like—neo-biological culture welds engineered technology and unrestrained nature until the two become indistinguishable, as unimaginable as that may first seem.

The intensely biological nature of the coming culture derives from five influences:

- Despite the increasing technization of our world, organic life—both wild and domesticated—will continue to be the prime infrastructure of human experience on the global scale.
- Machines will become more biological in character.
- Technological networks will make human culture even more ecological and evolutionary.

- Engineered biology and biotechnology will eclipse the importance of mechanical technology.
- Biological ways will be revered as ideal ways.

In the coming neo-biological era, all that we both rely on and fear will be more born than made. We now have computer viruses, neural networks, Biosphere 2, gene therapy, and smart cards—all humanly constructed artifacts that bind mechanical and biological processes. Future bionic hybrids will be more confusing, more pervasive, and more powerful. I imagine there might be a world of mutating buildings, living silicon polymers, software programs evolving offline, adaptable cars, rooms stuffed with coevolutionary furniture, gnatbots for cleaning, manufactured biological viruses that cure your illnesses, neural jacks, cyborgian body parts, designer food crops, simulated personalities, and a vast ecology of computing devices in constant flux.

The river of life—at least its liquid logic—flows through it all.

We should not be surprised that life, having subjugated the bulk of inert matter on Earth, would go on to subjugate technology, and bring it also under its reign of constant evolution, perpetual novelty, and an agenda out of our control. Even without the control we must surrender, a neo-biological technology is far more rewarding than a world of clocks, gears, and predictable simplicity.

As complex as things are today, everything will be more complex tomorrow. The scientists and projects reported here have been concerned with harnessing the laws of design so that order can emerge from chaos, so that organized complexity can be kept from unraveling into unorganized complications, and so that something can be made from nothing.

# ACKNOWLEDGMENTS

HARDLY AN IDEA in this volume is mine alone. In addition to the books and papers annotated in my bibliography, the concepts I present here have largely been condensed, paraphrased, or quoted from conversations, correspondence and lengthy interviews with the following people. Each, without exception, was extremely generous with his time and patient with my endless questions. They are, of course, not responsible for my idiosyncratic interpretation of their ideas. Some of the interviewees offered valuable corrections and comments to the work in progress. In addition, those indicated by asterisk were kind enough to review portions of the final manuscript. *Thank you.*

Ralph Abraham
David Ackley
Ormond Aebi
John Allen
Norberto Alvarez
Robert Axelrod
Howard Baetjer
Will Baker
John Perry Barlow
Joseph Bates
Mark Bedau
Russell Brand
Stewart Brand
Jim Brooks
Rod Brooks
Amy Bruckman
Tony Burgess*
Arthur Burks
L.G. Callahan
William Calvin
David Campbell
Peter Cariani
Mike Cass
David Chaum
Steve Cisler
Michael Cohen
Robert Collins
Michael Conrad
Neale Cosby
George Cowan
Brad Cox
Jim Crutchfield
Paul Davies
Richard Dawkins
Brad de Graf
Bill Dempster
Daniel Dennett*
Jamie Dinkelacker
Jim Drake

Gary Drescher
K. Eric Drexler
Kathy Dyer
Esther Dyson
Doyne Farmer*
David Fine
Paul Fishwick
Anita Flynn
Heinz von Foerster
Walter Fontana
Stephanie Forrest
Jay Forrester
John Gall
Eugene Garfield
John Geanokoplos
Murray Gell-Mann
George Gilder*
John Gilmore
Narenda Goel
Lloyd Gomez
Stephen Jay Gould
Ralph Guggenheim
Jeff Haas
Stuart Hameroff
Dan Harmony
Phil Hawes
Neal Hicks
Danny Hillis*
Carl Hodges
Malone Hodges
Douglas Hofstadter
John Holland*
John Hopfield
Eric Hughes
David Jefferson
Bill Jordan
Gerald Joyce
Ted Kaehler
James Kalin
Mitch Kapor
Ken Karakotsios
Stuart Kauffman*
Alan Kaufman
Ed Knapp
Barry Kort
John Koza
Bob Lambert
David Lane
Chris Langton*
Jaron Lanier
William Latham
Don Lavoie
Mike Leibhold
Linda Leigh
Steven Levy
Kristian Lindgren
Seth Lloyd
James Lovelock
Pattie Maes
Tom Malone
Lynn Margulis
Maja Matrick
Tim May
David McFarland
John McLeod
H. R. McMasters
Dana Meadows
Dennis Meadows
Ralph Merkle
Gavin Miller
John Miller
Mark Miller
Scott Miller
Marvin Minsky
Melanie Mitchell
Max More
Mark Nelson
Ted Nelson
Tim Oren
Norm Packard
Steve Packard
John Patton
Mark Pauline
Jim Pelkey
Stuart Pimm*
Charlie Plott
Przemyslaw Prusinkiewicz
Steen Rasmussen
Tom Ray
Mitchel Resnick
Craig Reynolds
Howard Rheingold*
David Rogers
Rudy Rucker
Jonathan Schull
Ted Schultz
Barry Silverman
Herbert Simon
Karl Sims*
Peter Sprague
Bruce Sterling*
Steve Strassman
Chuck Taylor
Mark Thompson
Hardin Tibbs
Mark Tilden
Ralph Toms
Joe Traub
Michael Travers
Dean Tribble
Roy Valdes
Francisco Varela
Michael Wahrman
Roy Walford
Gary Ware
Peter Warshall
Mark Weiser
Jordan Weisman
Jim Wells
Lawrence Wilkinson
Greg Williams
Christopher Wills
Johnny Wilson
Stewart Wilson
David Wingate
Ben Wintraub
Ben Wise
Steven Wolfram
Will Wright
Larry Yaeger
David Zeltzer

# Annotated Bibliography

Ahmadjian, Vernon. *Symbiosis: An Introduction to Biological Associations*. Hanover, 1986.

A comprehensive text on symbiosis which is clear and crammed with insights.

Alberch, Pere. "Orderly Monsters: Evidence for internal constraint in development and evolution." In *The construction of organisms: Opportunity and constraint in the evolution of organic form.*, Thomas, R. D. K., and W. E. Reif, eds. In press.

One of the most amazing papers I have ever read. Explains why monstrosities in living creatures are so similar and "orderly" given all possibilities.

Aldersy-Williams, Hugh. "A solid future for smart fluids." *New Scientist*, 17 March 1990.

Fluids and gel that change their state when signaled. Engineers can use this response to make them "smart."

Allen, John. *Biosphere 2: The Human Experience*. Penguin, 1991.

Coffee table book on the making of Biosphere 2 by its original visionary. Good history on how the idea arose and was tried out. It covers the experiment until shortly before it "closed."

Allen, Thomas B. *War Games: The Secret World of the Creators, Players, and Policy Makers Rehearsing World War III Today*. McGraw-Hill, 1981.

Fascinating history and insider's view of the large-scale simulations which the U.S. military agencies run to decipher the life-and-death complexity of war.

Allen, T. F. H., and Thomas B. Starr. *Hierarchy: Perspectives for Ecological Complexity*. University of Chicago Press, 1982.

Very ambitious book, but a bit soft in its arguments and clarity. The main point: patterns in ecological systems can only be perceived if viewed or measured at the appropriate scale.

Allman, William F. *Apprentices of Wonder: Inside the Neural Network Revolution*. Bantam Books, 1989.

Neural networks are the paramount example of connectionism and bottom-up control. A light journalistic treatment of the major players in the field; a good intro.

Amato, Ivan. "Animating the Material World." *Science*, 255; 17 January 1992.

Brief report on various experiments to put smartness into inanimate materials.

———. "Capturing Chemical Evolution in a Jar." *Science*, 255; 14 February 1992.

Self-replicating RNA which can generate mutant forms.

Anderson, Philip W., Kenneth J. Arrow, and David Pines. *The Economy as an Evolving Complex System*. Addison-Wesley, 1988.

A landmark series of papers in the esoteric realm of physics, math, computer science, and economics. Does a great job in reinventing how we think of the economy. The central shift is away from classical equilibrium. For lay readers the summary is in English and newsworthy.

Aspray, William, and Arthur Burks, eds. *Papers of John von Neumann on Computers and Computer Theory*. MIT Press, 1967.

If you are math-challenged (as I am), you need only read the fine introduction and summary by Burks.

Axelrod, Robert. *The Evolution of Cooperation*. Basic Books, 1984.

Lucid account of how Prisoner's Dilemma and other open-ended games can illuminate political and social thought.

Badler, Norman I., Brian A. Barsky, and David Zeltzer, eds. *Making Them Move: Mechanics, Control, and Animation of Articulated Figures*. Morgan Kaufmann Publishers, 1991.

To best exploit the technical details outlined in this book, you'll need the accompanying video of experimental figures trying to move. Some move quite well.

Bajema, Carl Jay, ed. *Artificial Selection and the Development of Evolutionary Theory*. Hutchinson Ross Publishing, 1982.

A banquet of benchmark papers on what artificial selection (breeding) has to say about natural selection. What is most noticeable is how paltry the feast is.

Basalla, George. *The Evolution of Technology*. Cambridge University Press, 1988.

Makes the case (with fascinating examples) that all innovation is incremental and not abrupt. Emphasizes the importance of novelty in technological change.

Bass, Thomas A. *The Eudaemonic Pie*. Houghton Mifflin, 1985.

The bizarre true story of how a California hippie commune of physicists and computer nerds beat Las Vegas using chaos theory. Addresses the problem of time-series predictions. An overlooked great read.

———. "Road to Ruin." *Discover*, May 1992.

Story about Joel Cohen's expansion of the Braess paradox—that adding more roads to a network may slow it down.

Bateson, Gregory. *Steps to an Ecology of Mind*. Ballentine, 1972.

A great book about the parallels between evolution and the mind. Of particular interest is the chapter on "The Role of Somatic Change in Evolution."

———. *Mind and Nature*. Dutton, 1979.

Bateson stresses and stretches the similarities between mind and evolution in nature.

Bateson, Gregory, and Mary Catherine Bateson. *Angels Fear: Toward an Epistemology of the Sacred*. Macmillan, 1987.

Interwoven between the final writings of Gregory Bateson—completed posthumously by his daughter Mary Catherine—are dialogues between father and daughter that convey Gregory's deep ideas of sacrament, communication, intelligence, and being.

Bateson, Mary Catherine. *Our Own Metaphor*. Smithsonian, 1972.

Mary Catherine Bateson's personal account of an informal conference on evolution, progress, and learning in human adaptation organized by her father, Gregory Bateson. The meeting was held to deal with the role of conscious purpose in such complex systems. Every conference should have such a document.

———. *With a Daughter's Eye*. William Morrow, 1984.

A memoir of Margaret Mead and Gregory Bateson that is more than a memoir. Written by daughter Mary Catherine, who is an intellectual of equal caliber to her parents, this is a book of cybernetic family stories.

Baudrillard, Jean. *Simulations*. Semiotext(e), Inc., 1983.

Short, very French, very dense, very poetic, very impenetrable, and somewhat useful in that he attempts to wring some meaning out of simulations.

Beaudry, Amber A., and Gerald F. Joyce. "Directed Evolution of an RNA Enzyme." *Science*, 257; 31July 1992.
Elegant experimental results of directed breeding of RNA molecules.

Bedau, Mark A. "Measurement of Evolutionary Activity, Teleology, and Life." In *Artificial Life II*, Langton, Christopher G., ed. Addison-Wesley, 1990.
A most intriguing attempt to quantify direction in evolutionary activity.

———. "Naturalism and Teleology." In *Naturalism: A Critical Appraisal*, Wagner, Steven, and Richard Warner, eds. University of Notre Dame Press, 1993.
Can natural systems have purpose? Yes.

Bedau, Mark A., Alan Bahm, and Martin Zwick. "The Evolution of Diversity." 1992.
Offers a metric for measuring diversity in an evolutionary system.

Bell, Gordon. "Ultracomputers: A Teraflop Before Its Time." *Science*, 256; 3 April 1992.
A bet whether parallel computers will beat serial computers in the race for power.

Bergson, Henri. *Creative Evolution*. Henry Holt, 1911.
A classic of philosophy about the idea that evolution proceeds by some vital force.

Berry, F. Clifton. "Re-creating History: The Battle of 73 Easting." *National Defense*, November 1991.
Blow-by-blow account of the pivotal Gulf War battle that has been recreated as a Pentagon simulation.

Bertalanffy, Ludwig von. *General System Theory*. George Braziller, 1968.
For many years this was the cybernetic bible. It's still one of the few books on whole systems or "systems in general." But it seems to me to be vague even in the places I agree with. And Bertalanffy's signature idea—equifinality—I think is wrong, or at least incomplete.

Biosphere 2 Scientific Advisory Committee. "Report to the Chairman, Space Biosphere Ventures." *Space Biosphere Ventures*, 1992.
Evaluates the validity and quality of Biosphere 2's first nine months from a scientific viewpoint.

Bolter, Jay David. *Writing Space: The Computer, Hypertext, and the History of Writing*. Lawrence Erlbaum Associates, 1991.

A marvelous, overlooked little treasure that outlines the semiotic meaning of hypertext. The book is accompanied by an expanded version in hypertext for the Macintosh. I consider it a seminal work in "network culture."

Bonner, John Tyler. *The Evolution of Complexity, by Means of Natural Selection*. Princeton University Press, 1988.

A pretty good argument that evolution evolves toward complexity.

Botkin, Daniel B. *Discordant Harmonies: A New Ecology for the Twenty-First Century*. Oxford University Press, 1990.

Essays in natural history by an ecologist who has a fresh view of nature as a disequilibrial system.

Bourbon, W. Thomas, and William T. Powers. "Purposive Behavior: A tutorial with data." Unpublished, 1988.

An intriguing claim that much behavior is not "caused" but emanates from emergent internal purposes. Illustrated with a simple experiment.

Bowler, Peter J. *The Eclipse of Darwinism: Anti-Darwinian Evolution Theories in the Decades around 1900*. The Johns Hopkins University Press, 1983.

This history serves as an excellent primer on alternative scientific theories to strict neodarwinism.

———. *The Invention of Progress*. Basil Blackwell, 1989.

A fascinating scholarly examination of how during the Victorian era evolutionary theory initially created a notion of progress, a legacy only now eroding.

Braitenberg, Valentino. *Vehicles: Experiments in Synthetic Psychology*. The MIT Press, 1984.

Shows how very simple circuits can produce the appearance of complicated behaviors and movement. The experiments were eventually implemented in tiny model cars.

Brand, Stewart. *II Cybernetic Frontiers*. Random House, 1974.

A curious, small book that is pleasantly two-faced. One-half is the first published report on computer hackers playing computer games, and the other is Gregory Bateson talking about evolution and cybernetics.

———. *The Media Lab: Inventing the Future at MIT*. Viking, 1987.

Although about media future, there are enough gems of insight about the future of interconnectivity to keep this rich book ahead of the curve.

Bratley, Paul, Bennet L. Fox, and Linus E. Schrage. *A Guide to Simulation*. Springer-Verlag, 1987.

The best overview of the role and dynamics of simulations in theory and practice.

Briggs, John. *Turbulent Mirror*. Harper & Row, 1989.

Goes from the theory of chaos to the "science of wholeness." Pretty good introduction to the strange behavior of complex systems, with many wonderful pictures and diagrams. Emphasizes the turbulent chaotic side, rather than the self-organizing side of wholeness.

Brooks, Daniel, and R. E. O. Wiley. *Evolution as Entropy*. The University of Chicago Press, 1986.

An important book although I have read only a little of it. I wish I had a more technical and mathematical background to plunge deeper into it and to appreciate its attempt to be a "unified theory of biology."

Brooks, Rodney A. "Elephants Don't Play Chess." *Robotics and Autonomous Systems*, 6; 1990.

Instead elephants wander around doing things in the real world. This paper summarizes Brooks's lab's attempts (about eight robots so far) to make intelligence situated in the real physical environment.

———. "Intelligence without representation." *Artificial Intelligence*, 47; 1991.

Treats the evolutionary aspects of bottom-up control in robots.

———. "New Approaches to Robotics." *Science*, 253; 1991.

Summary of Brooks's subsumption architecture for robots.

Brooks, Rodney A., and Anita Flynn. "Fast, Cheap and Out of Control: A Robot Invasion of the Solar System." *Journal of The British Interplanetary Society*, 42; 1989.

About "invading a planet with millions of tiny robots." This is the source of my book title.

Brooks, Rodney A., Pattie Maes, Maja J. Mataric, and Grinell More. "Lunar Base Construction Robots." *IROS*, IEEE International Workshop on Intelligence Robots & Systems, 1990.

Proposal for a swarm of minibulldozers, saturated with "collective intelligence."

Bruckman, Amy. "Identity Workshop: Emergent Social and Psychological Phenomena in Text-Based Virtual Reality." Unpublished, 1992.

Excellent study of the new sociology of teenage obsessives building and playing online MUDs.

Buss, Leo W. *The Evolution of Individuality*. Princeton University Press, 1987.

Difficult book to grasp. The introductory and summary chapters are clear and fascinating, and probably important in understanding hierarchical evolution. Buss is onto something vital: that the individual is not the only unit of selection in evolution.

Butler, Samuel. "Darwin Among the Machines." In *Canterbury Settlement*. AMS Press, 1923.

An essay written in 1863, by the author of *Erewhon*, suggesting the biological nature of machines.

———. *Evolution, Old and New*. AMS Press, 1968.

An early (1879), but still persuasive, philosophical rant against Darwinism penned by an early supporter of Darwin who renegaded into a fierce anti-Darwinian stance.

Cairns-Smith, A.G. *Seven Clues to the Origin of Life*. Cambridge University Press, 1985.

The freshest book to date on the puzzle of the origin of life. Written as a scientific detective story. Digests in lay terms his more technical treatment in *Genetic Takeover*.

Card, Orson Scott. *Ender's Game*. Tom Doherty Associates, 1985.

A science fiction novel about kids trained to fight real wars while playing simulated war games.

Casdagli, Martin. "Nonlinear Forecasting, Chaos and Statistics." In *Nonlinear Modeling and Forecasting*, Casdagli, M., and S. Eubank, eds. Addison-Wesley, 1992.

Some heavy-duty algorithms for extracting order from irregularity.

Cellier, François E. *Progress in Modelling and Simulation*. Academic Press, 1982.

Deals with the practical problems of computers modeling ill-defined systems.

Chapuis, Alfred. *Automata: A Historical and Technological Study*. B. T. Batsford, 1958.

Amazing details of amazing clockwork automatons in history, both European and Asian. Can be thought of as a catalog of early attempts at artificial life.

Chaum, David. "Security Without Identification: Transaction Systems to Make Big Brother Obsolete." *Communications of the ACM*, 28; 10 October 1985.

Highly detailed explanation of how an ID-less electronic money system works. Very readable and visionary. A revised version is even clearer. Worth seeking out.

Cherfas, Jeremy. "The ocean in a box." *New Scientist*, 3 March 1988.
Journalistic report on Walter Adey's synthetic coral reefs.

Cipra, Barry. "In Math, Less Is More—Up to a Point." *Science*, 250; 23 November 1990.
Report on Hwang and Du's proof of shortening a network by adding more nodes.

Clearwater, Scott H., Bernardo A. Huberman, and Tad Hogg. "Cooperative Solution of Constraint Satisfaction Problems." *Science*, 254; 22 November 1991.
Pioneer work on cooperative problem solving. Tells how managing "hints" for a swarm of cooperating agents trying to solve a problem is vital to the agents' success.

Cohen, Frederick B. *A Short Course on Computer Viruses*. ASP Press, 1990.
The scoop from the guy who coined the term "computer virus."

Cole, H. S. D., et al. *Models of Doom*. Universe Books, 1973.
A critique of the model/book "Limits to Growth" done by an interdisciplinary team at Sussex University in England.

Colinvaux, Paul. *Why Big Fierce Animals Are Rare*. Princeton University Press, 1978.
Pure pleasure. Wonderful prose in a short book on the intricacies and complexities of ecological relationships. Based on the author's own naturalist experiences. Seeks to extract ecological principles. Best book I know of about the cybernetic connectiveness of ecological systems.

Conrad, Michael. *Adaptability: The Significance of Variability from Molecule to Ecosystem*. Plenum Press, 1983.
A good try at describing adaptation in broad terms across many systems.

———. "The brain-machine disanalogy." *Biosystems*, 22; 1989.
Argues that no machine using present day organization or materials could pass the Turing Test. In other words, human-type intelligence will only come with human-type brains.

———. "Physics and Biology: Towards a Unified Model." *Applied Mathematics and Computation*, 32; 1989.
I verge on understanding this short paper; I think there's a good idea here.

Conrad, Michael, and H. H. Pattee. "Evolution Experiments with an Artificial Ecosystem." *Journal of Theoretical Biology*, 28; 1970.
One of the earliest experiments in modeling coevolutionary behavior on a computer.

Cook, Theodore Andre. *The Curves of Life*. Dover, 1914.

The self-organizing power of living spirals, in pictures.

Crutchfield, James P. "Semantics and Thermodynamics." In *Nonlinear Modeling and Forecasting*, Casdagli, M., and S. Eubank, eds. Addison-Wesley, 1992.

Further work on an automatic method for extracting a mathematical model from a set of data over time.

Culotta, Elizabeth. "Forecasting the Global AIDS Epidemic." *Science*, 253; 23 August 1991.

Various studies take the same problem, same data, and get wildly different models. Good example of the problems inherent in simulations.

———. "Forcing the Evolution of an RNA Enzyme in the Test Tube." *Science*, 257; 31 July 1992.

Nice summary of Gerald Joyce's work.

Dadant & Sons, eds. *The Hive and the Honey Bee*. Dadant & Sons, 1946.

Bees are probably the most studied of insects. This fat book offers practical management tips for the distributed organism of bees and their hives.

Darwin, Charles. *The Origin of Species*. Collier Books, 1872.

The fountainhead of all books on evolution. Darwinism reigns in large part because this book is so full of details, supporting evidence, and persuasive arguments, all so well written, that other theories pale in comparison.

Davies, Paul. "A new science of complexity." *New Scientist*, 26 November 1988.

Nicely written overview article of the new perspective of complexity.

———. *The Mind of God*. Simon & Schuster, 1992.

I have not yet been able to say exactly why I think this book is so apt to my subject of complexity and evolution. It's about current understandings of the underlying laws of the physical universe, but Davies presents these laws in the space of all possible laws, or all possible universes, and talks about why these laws were chosen or evolved or happened. Thus one gets into the mind of God, or god. It's full of fresh perspectives and near-heretical thoughts.

Dawkins, Richard. *The Selfish Gene*. Oxford University Press, 1976.

A wholly original idea (that genes replicate for their own reasons) and brilliant exposition. Dawkins also introduces his equally original secondary idea of memes (ideas that replicate for their own reasons).

———. *The Blind Watchmaker*. W. W. Norton, 1987.

Perhaps the most neodarwinian of all books. Dawkins presents the case for a "universe without design" based entirely on natural selection. And he writes so well and clearly that his forceful ideas are hard to argue with. At the very least, this book is probably the best general introduction to orthodox evolutionary theory anywhere. Full of clever examples.

———. "The Evolution of Evolvability." In *Artificial Life*, Langton, Christopher G., ed. Addison-Wesley, 1987.

A brilliant sketch of a stunningly new idea: that evolvability can evolve.

Dempster, William F. "Biosphere II: Technical Overview of a Manned Closed Ecological System." *Society of Automotive Engineers*, 1989, SAE Technical Paper Series #891599.

Prelaunch technical details about the engineering achievements of Bio2.

Denton, Michael. *Evolution: A Theory in Crisis*. Burnett Books, 1985.

This is the best scientific critique of Darwinian evolution available. Denton does not seem to have a hidden agenda, which is refreshing in these kinds of books.

Depew, David J., and Bruce H. Weber, eds. *Evolution at a Crossroads*. The MIT Press, 1985.

A collection of scientific papers that explore fairly radical approaches to the steep conceptual problems in evolution theory.

De Robertis, Eddy M. et al. "Homeobox Genes and the Vertebrate Body Plan." *Scientific American*, July 1990.

Readable article on importance of ancient homeobox regulatory genes.

Dixon, Dougal. *After Man: A Zoology of the Future*. St. Martin's Press, 1981.

The only book I know that extrapolates evolution into the future without being capricious or superficial, that is, with some measure of scope and consistency. Although not meant to be scientific, this gorgeously illustrated book is an inspiration.

Dobzhansky, Theodosius. *Mankind Evolving*. Yale University Press, 1962.

A rather old-fashioned book in tone, geneticist Dobzhansky calmly plunges into the controversial waters of race, intelligence, personality, and evolution.

Drake, James A. "Community-assembly Mechanics and the Structure of an Experimental Species Ensemble." *The American Naturalist*, 137; January 1991.

Elegant experiments showing how the order and timing of introducing species influences the final mix of an ecological community.

Drexler, K. Eric. "Hypertext Publishing and the Evolution of Knowledge." *Social Intelligence*, 1; 2, 1991.

A thorough and enthusiastic sketch of a distributed public hypertext system and its advantages in spurring scientific knowledge.

Dupre, John, ed. *The Latest on the Best: Essays on Evolution and Optimality*. The MIT Press, 1987.

By and large these essays make a convincing case that biological systems do not optimize to the best, because the question "best for what?" can't be answered.

Dykhuizen, Daniel E. "Experimental Evolution: Replicating History." *Trends in Ecology and Evolution*, 7; August 1992.

Review and comments on laboratory studies of observed evolution within microbial populations.

Dyson, Freeman. *From Eros to Gaia*. HarperCollins, 1990.

Contains great chapter on "Carbon Dioxide in the Atmosphere and the Biosphere."

———. *Origins of Life*. Cambridge University Press, 1985.

Refreshingly lucid and orthogonal view of the origin of life problem by a noted physicist. In terms of brilliance has much in common with Schrodinger's "What is Life?"

———. *Infinite in All Directions*. Harper & Row, 1988.

An original thinker writes very lyrically on whatever interests him, which is usually what almost no one else is thinking about. Dyson can take an ordinary subject and find incredibly fresh insights in it. In this volume he considers how the universe will end.

Eco, Umberto. *Travels in Hyperreality*. Harcourt Brace Jovanovich, 1986.

The key essay in this compendium should be required reading for all Americans graduating from high school. It's about the real, the fake, and the hyperreal.

Eigen, Manfred, and Peter Schuster. *The Hypercycle: A Principle of Natural Self-Organization*. Springer-Verlag, 1979.

A powerful abstraction of cycles within cycles producing self-made stable cycles, or hypercycles.

Eldredge, Niles. *Unfinished Synthesis: Biological Hierarchies and Modern Evolutionary Thought*. Oxford University Press, 1985.

Eldredge, who coauthored punctuated equilibrium theory, here pushes evolutionary theory further in a pioneering work on hierarchies of evolutionary change. By all accounts understanding hierarchical change is the next frontier in the science of complexity.

———. *Macroevolutionary Dynamics: Species, Niches, and Adaptive Peaks.* McGraw-Hill, 1989.

A technical treatise for professionals on how emergent levels of evolution impact adaptation at the species level.

Endler, John A.*Natural Selection in the Wild.* Princeton University Press, 1986.

Endler rounds up all known studies of natural selection in nature and dissects them rigorously. In the process he arrives at refreshing insights of what natural selection is.

Flynn, Anita, Rodney A. Brooks, and Lee S. Tavrow. "Twilight Zones and Cornerstones: A gnat robot double feature." MIT Artificial Intelligence Laboratory, 1989, *A.I. Memo 1126.*

Blue-sky dreaming on why and how to build tiny gnat-sized robots—disposable, entirely self-contained autonomous critters that can do real work.

Foerster, Heinz von. "Circular Causality: Fragments." Intersystems Publications, ca. 1980.

A short chronology of the Macy Conference and the participants at each meeting, and an introduction to the seed idea of emergent "telos" or goal and purpose.

———. *Observing Systems.* Intersystems Publications, 1981.

An anthology of von Foerster's papers. These range from mathematical treatise to philosophical rants. All point to von Foerster's law that observers are part of the system.

Fogel, Lawrence J., Alvin J. Owens, and Michael J. Walsh. *Artificial Evolution Through Simulated Evolution.* Wiley & Sons, 1966.

Early connectionism that didn't produce much intelligence but did prove the worth of evolutionary programming. This is probably the first computational evolution.

Folsome, Clair E. "Closed Ecological Systems: Transplanting Earth's Biosphere to Space." *AIAA*, May 1987.

A rough sketch at what science needs to know to make a closed extraterrestrial living habitat.

Folsome, Clair E., and Joe A. Hanson. "The Emergence of Materially-closed-system Ecology." In *Ecosystem Theory and Application*, Polunin, Nicholas, ed. John Wiley & Sons, 1986.

A wonderful report on sealed jars of microbial life that keep going and going. The authors measure the energy flow and productivity of the closed system.

Forrest, Stephanie, ed. *Emergent Computation*. North-Holland, 1990.
How does collective and cooperative behavior step out of a mass of computing nodes? These proceedings from a conference on nonlinear systems round up current approaches from neural nets, cellular automata, and simulated annealing, among other computational techniques.

Frazzetta, T. H. *Complex Adaptations in Evolving Populations*. Sinauer Associates, 1975.
Realistically examines the riddle of how adaptation occurs with linked genes in real, fuzzy populations. Sort of an engineer's approach; pretty readable.

Frosch, Robert A., and Nicholas E. Gallopoulos. "Strategies for Manufacturing." *Scientific American*, September 1989.
A position paper that introduces closed loop manufacturing and the biological analog.

Gardner, M. R., and W. R. Ashby. "Connectance of Large Dynamic (Cybernetic) Systems: Critical Values for Stability." *Nature*, 228; 5273,1970.
Often cited paper on ratio between connectivity and stability.

Gelernter, David. *Mirror Worlds*. Oxford University Press, 1991.
A magically elegant vision of mirroring real systems (such as a town or hospital) with parallel real-time virtual models as a means of overseeing, managing, and exploring them.

Gell-Mann, Murray. "Simplicity and Complexity in the Description of Nature." *Engineering & Science*, 3, Spring 1988.
A not-impressive start at unraveling the difference between simplicity and complexity. But it's something.

George, F. H. *The Foundations of Cybernetics*. Gordon and Breach Science Publishers, 1977.
A lukewarm (but French!) overview of cybernetics (pretty much outdated by now) with a couple of good generalizations.

Gilder, George. *Microcosmos: The Quantum Revolution in Economics and Technology*. Simon and Schuster, 1989.
A generous and meaty book on how technology is retreating from the material realm and heading into the symbolic realm, and the economic consequences of that shift.

Gleick, James. *Chaos*. Viking Penguin, 1987.
This bestseller hardly needs an introduction. It's a model of science writing, both in form and content. Although a small industry of chaos

books has followed its worldwide success, this one is still worth rereading as a delightful way to glimpse the implications of complex systems.

Goldberg, David E. *Genetic Algorithms in Search, Optimization, and Machine Learning*. Addison-Wesley, 1989.
Best technical overview of genetic algorithms.

Goldschmidt, Richard. *The Material Basis of Evolution*. Yale University Press, 1940.
To get to the juicy parts, you have to read a lot of old-fashioned 1940s genetics. Consider this the prime source of the hopeful monster theory.

Gould, Stephen Jay. *Ever Since Darwin*. W. W. Norton, 1977.
Gould's essays never fail to inform and change my mind. In this collection, I was particularly attentive to "The Misunderstood Irish Elk."

———. *The Panda's Thumb*. W. W. Norton, 1980.
Of all Gould's anthologies of essays from his column in *Natural History*, this one has the most about macroevolutionary dynamics and new evolutionary thinking.

———. *Hen's Teeth and Horse's Toes*. W. W. Norton, 1983.
Lots of fascinating history about evolution theory in Gould's peerless style.

———. *Wonderful Life: The Burgess Shale and the Nature of History*. W. W. Norton, 1989.
A splendid masterwork. Rich, lucid, flawless, and iconoclastic. Gould's story of the painful reinterpretation of old shale fossils leading to an altered view of the history of life—that of decreasing diversity—is a mandatory read these days.

———. "Opus 200." *Natural History*, August 1991.
You'll find no better, more succinct explanation of how punctuated equilibrium works than this one from the horse's mouth. Not only the why but also a bit of history of what supporters call "punk eke" and detractors label "evolution by jerks."

Gould, Stephen Jay, and R. C. Lewontin. "The spandrels of San Marco and the Panglossian paradigm: a critique of the adaptationist programme." *Proceedings of the Royal Society of London*, B 205; 1979.
An oft-cited paper that argues against perceiving everything as the result of selective adaptation (the Panglossian paradigm). Gould makes a very readable case for a plurality of evolutionary dynamics.

Gould, Stephen Jay, and Elisabeth S. Vrba. "Exaptation—a missing term in the science of form." *Paleobiology*, 8; 1, 1982.

The term is for a feature devised as an adaptation for one reason which is then repurposed for another adaptive pressure. Using feathers devised for warmth in order to fly is the stock example.

Grasse, Pierre P. *Evolution of Living Organisms: Evidence for a New Theory of Transformation.* Academic Press, 1977.
Representative subchapters cover such juicy topics as "Limits to Adaptation," and "Forbidden Phenotypes," favorite postdarwinian challenges. Provocative book.

Hamilton, William D., Robert Axelrod, and Reiko Tanese. "Sexual reproduction as an adaptation to resist parasites (A Review)." *Proceedings of the National Academy of Science*, USA, 87; May 1990.
Not only is this a clever and convincing explanation of the origin of sex, but it is a marvelous demonstration of the power of computational biology.

Harasim, Linda M., ed. *Global Networks.* The MIT Press, 1993.
Twenty-one contributors speak on the effects seen so far of decentralized high-bandwidth communication at global scale; there is little hard data, mostly hints of opportunities and pitfalls.

Hayes-Roth, Frederick. "The machine as partner of the new professional." *IEEE Spectrum*, 1984.
Source of cute employment letter for humans.

Heeter, Carrie. "BattleTech Masters: Emergence of the First U.S. Virtual Reality SubCulture." Michigan State University, Computer Center, 1992.
Somewhere between a scholarly report and a marketing survey of the fanatical users of the first commercial networked virtual reality installation.

Heims, Steve J. *The Cybernetics Group.* The MIT Press, 1991.
An incredibly thorough history of the agenda and flavor of the Macy Conferences and vignettes of some of the illustrious participants.

Hillis, W. Daniel. *The Connection Machine.* The MIT Press, 1985.
The inventor's conceptual blueprint for the first commercial parallel processing computer and a few thoughts on what it might mean.

———. "Intelligence as an Emergent Behavior." In *Artificial Intelligence*, Graubard, Stephen, ed. The MIT Press, 1988.
In a special issue of *Daedulus* magazine which examined the state of artificial intelligence research in 1988, Hillis offers a connectionist view of possible AI, but one embedded in parallel and evolutionary processes. His are some of the most intelligent remarks I've heard on intelligence.

Hiltz, Starr Roxanne, and Murray Turoff. *The Network Nation: Human Communication via Computer* (Revised Edition). The MIT Press, 1993.

A visionary book when it was first published in 1978, it accurately forecasted many of the effects of intensely connected computer communications and distributed groups. It still has much to say about the coming network culture. A new section in the revised edition addresses the authors' current thoughts on superconnectivity.

Hinton, Geoffrey E., and Steven J. Nowlan. "How Learning Can Guide Evolution." *Complex Systems*, 1; 1987.

This very brief paper presents intriguing results of a type of Lamarckian evolution running on computers and some provocative speculations of other postdarwinian evolutions.

Ho, Mae-Wan, and Peter T. Saunders. *Beyond Neo-Darwinism*. Academic Press, 1984.

Not too many non-Darwinian books are published within science itself. This one comes from real biologists getting results that are suggestive, or merely permit a hint, of non-Darwinian evolution. This is good science at work.

Hofstadter, Douglas. *Gödel, Escher, Bach: An Eternal Golden Braid*. Basic Books, 1979.

Identical in all respects to the strangely loopy Pulitzer Prize-winning volume, *Copper, Silver, Gold: An Indestructible Metallic Alloy* by Egbert B. Gebstadter, now out of print.

Holldobler, Bert, and Edward O. Wilson. *The Ants*. Harvard University Press, 1990.

Deep, deep, rich, rich. All that is known about ants to date (including some expanded and revised sections from Wilson's earlier "Insect Societies"). A book to own and get lost in. Deserves the Pulitzer Prize it won.

Huberman, B. A. *The Ecology of Computation*. Elsevier Science Publishers, 1988.

A most interesting collection of pioneering papers on using economic and ecological dynamics within computation to manage complex computational tasks.

Johnson, Phillip E. *Darwin on Trial*. Regnery Gateway, 1991.

Johnson is a lawyer who treats neodarwinism as a defendant on trial, and subjects its evidence to the strict rules of court. He concludes that it is an unproven hypothesis that does not at this point seem to fit the evidence at hand. For the uninitiated layperson, a good first read on anti-Darwinism, but it follows lawyerly logic rather than science logic.

Kanerva, Pentti. *Sparse Distributed Memory*. MIT Press, 1988.

A dry, but daring monograph on a new architecture for computer memory, one that relies on weak associative connections. Wonderful foreword by Douglas Hofstadter, who explains the novel design's significance.

Kauffman, Stuart A. "Antichaos and Adaptation." *Scientific American*, August 1991.

A very accessible summation of Kauffman's important major ideas, with nary an equation in it. Read this one first.

———. "The Sciences of Complexity and 'Origins of Order'." Santa Fe Institute, 1991, technical report 91-04-021.

A personal and almost poetic short history of Kauffman's own idea of self-organizing order.

———. *The Origins of Order: Self Organization and Selection in Evolution*. Oxford University Press, 1993.

A sprawling, deep, massive magnum opus of a book, as dense as a dictionary. Kauffman tries to tell you everything he knows, and he's bright, so hang in there. It's about the yin and yang of natural selection and self-organization. A seminal work, not to be missed.

Kauppi, Pekka E., Karl Mielikainen, and Kullervo Kuusela. "Biomass and Carbon Budget of European Forests, 1971 to 1990." *Science*, 256; 3 April 1992.

Shows a biomass increase in Gaia which may be due to atmospheric carbon dioxide increase.

Kay, Alan C. "Computers, Networks and Education." *Scientific American*, September 1991.

A notable vision of how peer-to-peer networks might change education.

Kleiner, Art. "The Programmable World." *Popular Science*, May 1992.

About a chip that could be the basis for smart houses and distributed cooperative computing in the fabricated environment.

Kochen, Manfred. *The Small World*. Ablex Publishing Corporation, 1989.

Small world as in "there must be only 200 people in the whole world because I keep running into the same ones." If you go deeper into this incredibly rich volume of studies on social networks, you'll find it contains some of the coolest data for network culture seen yet. Here are real numbers on how many friends-of-a-friend connect us all.

Koestler, Arthur. *Janus: A Summing Up*. Random House, 1978.

No critic of Darwin in modern times has been as literate or influential as the brilliant Koestler. He spends the latter third of this book summing up his objections to Darwinism, and offering some suggestions for alternatives. His agile thinking on the subject loosened up my mind.

Korner, Christian, and John A. Arnone. "Responses to Elevated Carbon Dioxide in Artificial Tropical Ecosystems." *Science*, 257; 18 September 1992.
Where the $CO_2$ goes in closed greenhouses.

Koza, John. *Genetic Programming: On the Programming of Computers by Means of Natural Selection*. The MIT Press, 1992.
More than anyone else, Koza has tried to evolve software in systematic ways. This humongous tome is the record of his experimental details and results.

Langreth, Robert. "Engineering Dogma Gives Way to Chaos." *Science*, 252; 10 May 1991.
How engineers can outsmart chaotic vibration and injury with antichaos.

Langton, Christopher G., ed. *Artificial Life*. Addison-Wesley, 1987.
The mother of all artificial life studies. This is the proceedings of the first a-life workshop. The breadth of the articles is amazing.

Langton, Christopher, et al, eds. *Artificial Life II*. Addison-Wesley, 1992.
True news here. The most recent results of simulations of artificial evolution and protolife in computers. Original, deeply significant, and very accessible papers. Probably the most important book in this bibliography.

Lapo, Andrey. *Traces of Bygone Biospheres*. Synergetic Press, 1987.
Very Russian reclassification of life types on Earth by a sort of grand biomystic combining Chardin's "noosphere" with Lovelock's "Gaia," and Vernadsky's geochemcial vitalism. Hard to read but intriguing.

Laszlo, Ervin. *Evolution, the Grand Synthesis*. Shambhala, 1987.
New-agey speculations of the role of evolutionary change in the universe. I guess I found the freewheeling style and long view refreshing although I can't say I learned anything in particular from it.

Latil, Pierre de. *Thinking by Machine: A Study of Cybernetics*. Houghton Mifflin, 1956.
A real find. This French author had the most insightful and news-filled takes on feedback cybernetics I found anywhere. All the more amazing for having been written in 1956. I owe much to him.

Layzer, David. *Cosmogenesis: The Growth of Order in the Universe*. Oxford University Press, 1990.

Seems a bit flaky to me, but he did have an unusual idea or two that I couldn't dismiss. He came up with "reproductive instability" as a driving force in evolution.

Lenat, Douglas B. "The Heuristics of Nature: The Plausible Mutation of DNA." Stanford Heuristic Programming Project, 1980, technical report HPP-80-27.

The most heretical, yet plausible, alternative theory to Darwinian evolution I am aware of is compactly presented in this technical report from the Stanford Computer Science Department.

Leopold, Aldo. *Aldo Leopold's Wilderness: Selected Early Writings by the Author of A Sand County Almanac.* Stackpole Books, 1990.

Among many other things, this volume airs Leopold's early thoughts about the role of fire in natural systems.

Levy, Steven. *Artificial Life.* Pantheon, 1992.

An extremely enjoyable narrative of the making of the artificial life movement and a memorable overview of its central ideas and characters.

Lewin, Roger. *Complexity: Life at the Edge of Chaos.* Macmillan Publishing, 1992.

Annotated interviews with some of the central characters currently involved in making complexity itself a science. Not as deep or satisfying as Waldrop's book about the same subject, or Levy's on artificial life; this one gives a quick but superficial overview, and has a more biological, rather than mathematical, slant. Best part is the treatment of the problem of direction or trends in evolution.

Lightman, Alan, and Owen Gingerich. "When Do Anomalies Begin?" *Science,* 255; 7 February 1991.

Provocative thesis on the mechanism of progress within science.

Lima-de-Faria, A. *Evolution without Selection: Form and Function by Autoevolution.* Elsevier, 1988.

A difficult book. He seems to arrive at the same place as Kauffman but by intuitive and poetic means, rather than mathematics and science.

Lipset, David. *Gregory Bateson: The Legacy of a Scientist.* Prentice-Hall, 1980.

Bateson was interested in all things mysteriously complex. This biography of him and his interests illuminates the range of complexities that might be understood by looking at language, learning, the unconscious, and evolution.

Lloyd, Seth. "The Calculus of Intricacy." *The Sciences,* October 1990.

The best general introduction to defining complexity I have seen, and gracefully written to boot.

Lovece, Joseph A. "Commercial Applications of Unmanned Air Vehicles." *Mobile Robots and Unmanned Vehicles*, 1, July–August 1990.

Comprehensive roundup of current work-in-progress in commercial autonomous robots.

Lovelock, James. *The Ages of Gaia: A Biography of Our Living Earth*. W. W. Norton, 1988.

Lovelock rounds out his Gaia hypothesis into a theory here, and offers his best arguments and observations in support of it. He also speaks of how Gaia might have evolved.

Lovtrup, Soren. *Darwinism: The Refutation of a Myth*. Croom Helm, 1987.

This is a detailed blow-by-blow history of the ideas and personalities of anti-Darwinism. It's chock-full of delicious excerpts and quotes from past critics up until the present. It goes deep into the doubts of other experts about Darwinism.

Macbeth, Norman. *Darwin Retried*. Gambit Incorporated, 1971.

A fair "trial" of the evidence for Darwinian evolution. Short, but effective. Tends to highlight the discrepancies, but offers no alternatives.

Maes, Pattie. "How to do the Right Thing." *Connection Science*, 1; 3,1989.

Discusses an algorithm for robotic intelligence which will bias choice of action in certain directions as an ongoing "plan."

———. "Situated Agents Can Have Goals." *Robotics and Autonomous Systems*, 6; 1990.

How functional goals can emerge from a mass of simple rules in robots.

Malone, Thomas W. Joanne Yates, and Robert I. Benjamin. "Electronic Markets and Electronic Hierarchies." *Communications of the ACM*, 30; 6, 1987.

How increased use of cheap coordination technology will shift the economy away from hierarchical forms to market networks. Excellent paper.

Mann, Charles. "Lynn Margulis: Science's Unruly Earth Mother." *Science*, 252; 19 April 1991.

An entertaining account of mainstream evolutionary biologists' reaction to Margulis's ideas.

Margalef, Ramon. *Perspectives in Ecological Theory*. The University of Chicago Press, 1968.

The best treatment of ecosystems as cybernetic systems.

Margulis, Lynn, and Rene Fester, eds. *Symbiosis as a Source of Evolutionary Innovation: Speciation and Morphogenesis*. The MIT Press, 1991.

Lots of case studies on symbiotic relations. A few good chapters on reevaluating symbiosis' role in evolution.

Markoff, John. "The Creature That Lives in Pittsburgh." *The New York Times*, 21 April 1991.

About Ambler, the huge semismart walking robot built by CMU in Pittsburgh.

May, Robert M. "Will a Large Complex System be Stable?" *Nature*, 238;18 August 1972.

An early mathematical demonstration that showed that beyond a critical value, complexity unstabilizes a system.

Mayo, Oliver. *Natural Selection and its Constraints*. Academic Press, 1983.

This extremely technical book treats the genetic constraints on natural selection very seriously. Mayo asserts the constraints create narrow boundaries for evolution. He also dabbles with some alternative theories, which he woefully concludes cannot replace the current theory.

Mayr, Ernst. *Toward a New Philosophy of Biology*. The Belknap Press of Harvard University Press, 1988.

Mayr is the arch-orthodox Darwinian. Not only did he cofound the Modern Synthesis of neodarwinism, he remains its most dogmatic defender. Yet, he proposed what later became the bad-boy idea of punk-eek twenty years before Gould, and in this book he makes a strong case for radical, cohesive constraints of the gene.

Mayr, Otto. *The Origins of Feedback Control*. MIT Press, 1969.

A readable history of ancient servomechanisms and modern mechanical feedback devices, including one invented by the author's father.

———. *Authority, Liberty & Automatic Machinery in Early Modern Europe*. Johns Hopkins University Press, 1986.

How the metaphors of control shaped and were shaped by the technologies of control.

Mazlish, Bruce. *The Fourth Discontinuity: The Coevolution of Humans and Machines*. Yale University Press, 1993.

An excellent, penetrating history of the bionic convergence and its philosophical consequences. If this book had been published earlier, I would have borrowed much from it; but it came out as mine was being wrapped up.

McCulloch, Warren S. "An Account of the First Three Conferences on Teleological Mechanisms." Josiah Macy, Jr. Foundation, 1947.

A dense summary of the first three Macy Conferences, which covered an amazing range of topics, all before they hit upon the term "cybernetics."

McKenna, Michael, Steve Pieper, and David Zeltzer. "Control of a Virtual Actor: The Roach." *Computer Graphics*, 24; 2, 1990.

How to direct a virtual roach to walk where you want it to within a virtual environment.

McShea, Daniel W. "Complexity and Evolution: What Everybody Knows." *Biology and Philosophy*, 6; 1991.

A wonderful review of historical notions of increasing complexity in biological evolution ("what everybody knows"), and the author's own evidence against the idea.

Meadows, Donella H., Dennis L. Meadows, et al. *The Limits to Growth.* New American Library, 1972.

Notorious simulation from the Club of Rome which extrapolates economic and environmental trends of the whole Earth. Widely lauded and critiqued in the 1970s.

Meadows, Donella H., Dennis L. Meadows, and Jorgen Randers. *Beyond the Limits: Confronting Global Collapses, Envisioning a Sustainable Future.* Chelsea Green Publishing, 1992.

Sequel to 1972's best-selling *The Limits to Growth.*

Metropolis, N., and Gian-Carlo Rota, eds. *A New Era in Computation.* The MIT Press, 1992.

A very fine collection of essays written for the layperson which speak on the impact that parallel computing has had and will have on computer science, culture, and our own thinking.

Meyer, Jean-Arcady, and Stewart Wilson, eds. *From Animals to Animats.* The MIT Press, 1991.

The papers from a fruitful conference on the simulation of adaptive behavior, which gathered ethologists studying real animal behavior and roboticists trying to synthesize behavior in artificial "animats."

Meyer, Thomas P., and Norman Packard. "Local Forecasting of High Dimensional Chaotic Dynamics." Center for Complex Systems Research, The Beckman Institute, University of Illinois, 1991, technical report CCSR-91-1.

Theoretical underpinning for attempts to make "local" predictions in complex systems.

Midgley, Mary. *Evolution as a Religion: Strange Hopes and Stranger Fears.* Methuen & Co., Ltd., 1985.

Midgley wrestles with the philosophical consequences of "belief" in evolution, sometimes successfully and sometimes not. But she provides much to think about.

Miller, James Grier. *Living Systems*. McGraw-Hill, 1978.
A massive (we're talking about 1100 pages of minuscule type here) tome on the levels, sublevels and sub-sublevels of living systems, including organizations and such. Think of this as a printout of raw data on all living systems.

Minsky, Marvin. *The Society of Mind*. Simon & Schuster, 1985.
In 270 very readable one-page essays, Minsky presents a society of ideas about the society of mind. It is true Zen. Every page is a mob of astounding and mind-changing ideas. And at every point in thinking about complex systems I would come back to Minsky. This is the book that eventually led me to write this book.

Modis, Theodore. *Predictions*. Simon & Schuster, 1992.
In some ways a little cranky, but still useful nonetheless as a summary of technological forecasting.

Mooney, Harold A. *Convergent Evolution in Chile and California*. Dowden, Hutchinson & Ross, 1977.
Marks the parallel biological forms in two continents. Primarily ascribes this similarity to the orthodox explanation of similar climate. Does not address the alternative theory of internalist reasons for convergent evolution.

Morgan, C. Lloyd. *Emergent Evolution*. Henry Holt and Company, 1923.
A very early and not very successful stab at trying to articulate emergent control in evolution.

Moss, J. Eliot B. *Nested Transactions: An Approach to Reliable Distributed Computing*. The MIT Press, 1985.
Practical use of hierarchy.

Motamedi, Beatrice. "Retailing Goes High-Tech." *San Francisco Chronicle*, 8 April 1991.
Story on real-time trend-spotting, inventory stocking and manufacturing in the top retailers using intensive networked communications.

Needham, Joseph. *Science and Civilisation in China*. Cambridge at the University Press, 1965.
The ancient Chinese invented remarkably sophisticated mechanical devices, and this series of awesome books tracks each invention in mind-boggling detail. It's like having a patent registry for the Han people.